Editors

W. Bandlow, F. Kaudewitz, R.J. Schweyen, K. Wolf
Genetisches Institut der Universität München,
Maria-Ward-Str. 1a,
D-8000 München 19, Germany

CIP-Kurztitelaufnahme der Deutschen Bibliothek

Genetics and biogenesis of mitochondria:
proceedings of a colloquium held at Schliersee,
Germany, August 1977 / Mitochondria 1977. Ed. W. Bandlow...
- 1 Aufl. - Berlin, New York: de Gruyter, 1977.
ISBN 3-11-007321-8
NE: Bandlow, Wolfhard [Hrsg.]; Mitochondria
<1977, München>

Library of Congress Cataloging in Publication Data

Colloquium on Genetics and Biogenesis of Mitochondria,
Schliersee, Ger., 1977.
Mitochondria 1977.

"Sponsored by the 'Deutsche Forschungsgemeinschaft'
and the 'Verein zur Förderung der Pilzgenetik."
Bibliography: p.
Includes index.
1. Mitochondria–Congresses. 2. Cytoplasmic inheritance–
Congresses. 3. Microbial genetics–Congresses. I. Bandlow, W., 1938.
II. Deutsche Forschungsgemeinschaft (Founded 1949). III. Verein
zur Förderung der Pilzgenetik. IV. Title.
QH603.M5C64 1977 574.8'734 77-25262

 Printing: Mercedes-Druck, Berlin. - Binding: Lüderitz & Bauer, Buchgewerbe GmbH, Berlin. Printed in Germany.

Preface

Considerable interest in our international course on Methods in Yeast Genetics and Molecular Biology, held in 1975, prompted EURATOM to lend financial support to a similar course given by members of the Genetisches Institut of the University of Munich in July and August 1977. Also this time it was followed by a "Colloquium on Genetics and Biogenesis of Mitochondria" sponsored by the "Deutsche Forschungsgemeinschaft" and the "Verein zur Förderung der Pilzgenetik". Its twofold task was to acquaint the students who had attended the course with recent developments in mitochondrial genetics and – beyond this – to serve as a forum for the presentation and discussion of most of the outstanding work in today's mitochondrial research. The location of the conference, Schliersee/Oberbayern, guaranteed a maximum of informality permitting intensive discussions amongst the 70 participants.

The genetics and biogenesis of mitochondria in this colloquium were looked upon from three different points of view: The first day was dedicated to contributions mainly dealing with experimental data more or less connected with the mapping of mitDNA. The construction of a general map of mitDNA of *Saccharomyces cerevisiae* is almost completed. The climax of this type of genetic work is reached. The identification and structural analysis of mitochondrial gene products, the main topics of the following two sessions, now are in rapid progress, broadening our knowledge of the coding capacity of mitDNA on mitochondrial transcripts and translation products. The fourth session finally was dedicated to a group of independent studies approaching a rather new field of research in mitochondrial genetics, the regulation of mitochondrial gene activities. For the moment still in the very beginning, it may become a considerable part of future scientific efforts.

We would like to thank the contributors for the prompt submission of their papers, which enables the editors and the publisher to guarantee rapid publication of the proceedings. We also would like to thank the participants of the course and the colloquium for their active and lively participation, and last not least, all our colleagues of the Genetisches Institut for their enthusiastic and devoted help in the organization of the course and colloquium.

F. Kaudewitz

Contents

Preface V

Contributors XII

Session 1. Mitochondrial DNA, Mutations and Genetic Mapping

Chairman: P. P. Slonimski

Mitochondrial DNA and the Heritable Unit of the Yeast Mitochondrial Genome: A Review
D. H. Williamson, L. H. Johnston, K. M. V. Richmond, J. C. Game 1

Two Aspects of Mitochondrial DNA Structure: The Occurrence of Two Types of Mitochondrial DNA in Rat Liver and the Isolation From Rat Liver of DNA Complexes of High Buoyant Density
J. F. Francisco, F. F. Vissering, M. V. Simpson 25

Transposable Segments of Mitochondrial DNA: A Unitary Hypothesis for the Mechanism of Mutation, Recombination, Sequence Reiteration and Suppressiveness of Yeast "Petite Colony" Mutants
P. P. Slonimski, J. Lazowska 39

Control of the Genetic Recombination of Mitochondrial Genes: Preliminary Approaches
B. Dujon, H. Baranowska, G. Dujardin 53

Extrachromosomal Inheritance in *Schizosaccharomyces pombe.* V. On a Possible Correlation Between Bias in Transmission, Spontaneous mit^--Production, and Formation of Uniparental Zygotic Clones
K. Wolf, F. Kaudewitz 65

Mitochondrial Mutants Isolated by a New Screening Method Based Upon the Use of the Nuclear Mutation *op1*
Z. Kotylak, P. P. Slonimski 83

Control of Mitochondrial Inactivation of Temperature-sensitive *Saccharomyces cerevisiae* Nuclear Petite Mutants
E. Schweizer, W. Demmer, U. Holzner, H. w. Tahedl 91

Biochemical Characterization of a Temperature Conditional Mutant Having a Mutation in the ω-Controlled Region of the Yeast Mitochondrial Genome and Effects of the Mutation on Mitochondrial Recombination
H. B. Lukins, J. T. English, T. W. Spithill, R. J. Devenish, R. M. Hall, P. Nagley, A. W. Linnane 107

Preferential Loss of a Specific Region of Mitochondrial DNA by *rho*$^-$ Mutation
H. Fukuhara, M. Wesolowski 123

Preferential Loss or Retention of Mitochondrial Genes in *rho*$^-$ Clones
S. Mathews, R. J. Schweyen, F. Kaudewitz 133

The Genetic Map of the Mitochondrial Genome, Including the Fine Structure of cob and oxi Clusters
R. J. Schweyen, B. Weiss-Brummer, B. Backhaus, F. Kaudewitz 139

Extrachromosomal Inheritance in *Schizosaccharomyces pombe.* VI. Preliminary Genetic and Biochemical Characterization of Mitochondrially Inherited Respiratory-deficient Mutants
G. Seitz, G. Lückemann, K. Wolf, F. Kaudewitz, M. Boutry, A. Goffeau 149

Fine Structure Genetic Map of the Mitochondrial DNA Region Controlling Coenzyme QH_2: Cytochrome *c* Reductase
Z. Kotylak, P. P. Slonimski 161

Cytochrome *c* Reductase and Cytochrome Oxidase Formation in Mutants and Revertants in the "box" Region of the Mitochondrial DNA
P. Pajot, M. L. Wambier-Kluppel, P. P. Slonimski 173

Mapping of Drug-resistant Loci in the Coenzyme QH_2: Cytochrome *c* Reductase Region of the Mitochondrial DNA Map in *Saccharomyces cerevisiae*
A. M. Colson, P. P. Slonimski 185

Resistance to High Concentrations of Antimycin A in *Saccharomyces cerevisiae:* Synergistic Interaction of Nuclear and Mitochondrial Mutations
E. Pratje, G. Michaelis 199

Mutations to Drug Resistance in the cob Region of the Mitochondrial Genome
G. Burger, B. Lang, B. Backhaus, K. Wolf, W. Bandlow, F. Kaudewitz 205

Session 2. Organization of the Mitochondrial Genome, Transcription and Translation Products
Chairman: D. Wilkie

The Physical Map of Yeast Mitochondrial DNA Anno 1977 (A Review)
P. Borst, J. L. Bos, L. A. Grivell, G. S. P. Groot, C. Heyting, A. F. M. Moorman, J. P. M. Sanders, J. L. Talen, C. F. van Kreijl and G. J. B. van Ommen 213

Physical Characterization of the Difference Between Yeast mitDNA Alleles ω^+ and ω^-
C. Jacq, C. Kujawa, C. Grandchamp and P. Netter 255

Physical Mapping of Mitochondrial Genes and Transcripts in *Saccharomyces cerevisiae*
A. Lewin, R. Morimoto, S. Merten, N. Martin, P. Berg, T. Christianson, D. Levens, M. Rabinowitz 271

Properties and Genetic Localization of Mitochondrial Transfer RNAs of *Neurospora crassa*
P. Terpstra, H. de Vries, A. M. Kroon 291

The Genetic Organization of Rat Liver Mitochondrial DNA
C. Saccone, G. Pepe, H. Bakker, A. M. Kroon 303

Genetic, Physical and Biochemical Analysis of a Mitochondrial Gene
R. A. Butow, R. D. Vincent, R. L. Strausberg, E. Zanders, P. S. Perlman 317

Mutations at Anyone of Three Unlinked Mitochondrial Genetic Loci, BOX1, BOX4 and BOX6, Modify the Structure of Cytochrome *b* Polypeptide(s)
M. L. Claisse, A. Spyridakis, P. P. Slonimski 337

Structural and Regulatory Mutations Affecting Mitochondrial Gene Products
H. R. Mahler, D. Hanson, D. Miller, T. Bilinski, D. M. Ellis, N. J. Alexander, P. S. Perlman 345

A Structural Analysis of the oxi3 Region on Yeast mitDNA
L. A. Grivell, A. F. M. Moorman 371

Yeast Mitochondrial Messenger RNA is not Correctly Translated in Heterologous Cell-free Systems
A. F. M. Moorman, F. N. Verkley, F. A. M. Asselbergs, L. A. Grivell 385

Effect of Mitochondrial Mutations on Cytochrome *c* Oxidase in Yeast
F. Cabral, M. Solioz, D. Deters, Y. Rudin, G. Schatz, L. Clavilier, O. Groudinski, P. P. Slonimski 401

Transcription Map of Yeast Mitochondrial DNA
G. J. B. van Ommen, G. S. P. Groot 415

A One-step Method for *rho*$^+$cRNA/*rho*$^-$DNA Hybridization in Mapping of the Mitochondrial Genome of *Saccharomyces cerevisiae*
G. Grosch, I. Doxiadis, B. Lang, R. J. Schweyen, F. Kaudewitz 425

Session 3. Mitochondrial Components and Assembly
Chairman: C. Saccone

Altered Amino Acid Sequence of the DCCD-binding Protein of the Nucelar, Oligomycin-resistant Mutant AP-2 From *Neurospora crassa*
W. Sebald, M. Sebald-Althaus, E. Wachter 433

Altered Amino Acid Sequence of the DCCD-binding Protein in the oli 1-resistant Mutant D 273-10 b/A 21 of *Saccharomyces cerevisiae*
E. Wachter, W. Sebald, A. Tzagoloff 441

Structure of the Mitochondrial and Plasma-membrane ATPases of the Yeast *Schizosaccharomyces pombe*
A. Goffeau, M. Boutry, J. P. Dufour 451

Partial Identification, Stoicheiometry and Site of Translation of the Sub-units of Ubiquinone : Cytochrome *c* Oxidoreductase
H. Weiss, B. Ziganke 463

Isolation and Hydrodynamic Properties of the bc_1 Complex from Beef Heart Prepared by Hydroxyapatite Chromatography in Triton X-100
G. von Jagow, H. J. Kolb 473

A Cell-free System to Study Synthesis and Transport of Cytoplasmically Translated Mitochondrial Proteins
R. Zimmermann, H. Korb, W. Neupert 489

DNA-binding Properties of Mitochondrially Synthesized Proteins
G. Rödel, W. Bandlow 503

Session 4. Mitochondrial-Nucleocytoplasmic Interactions
Chairman: C. Saccone

Some Preliminary Observations on the Effect of Cerulenin on the Formation of Mitochondrial and Cytoplasmic Precursors of Electron Transfer Complexes III and IV in Adapting, Lipid-limited *Saccharomyces cerevisiae*
W. Rouslin 511

An Inductor-Repressor Model Co-ordinating the Expression of Mitochondrial and Nuclear Genes that Specify Mitochondrial Proteins in Yeast (*Candida utilis*)
E. Keyhani, J. Keyhani 523

Derepression of Mitochondrial Enzymes in Yeast Wild Type and a Mitochondrial Conditional Mutant, tsm8, Retarded in Oxidative Adaptation
W. Bandlow 531

Isolation and Preliminary Characterization of Yeast Mutants Affected in Carbon Catabolite Repression
M. von Ciriacy 543

The Role of Mitochondria in the Mechanism of Initiation of Differentiation in *Saccharomyces cerevisiae*
H. J. Rhaese, R. Groscurth, R. Scheckel 551

Mitochondria-Cell Relationships in *Saccharomycec cerevisiae:* Aspects of Sporulation
E. S. Diala, D. Wilkie 563

Differences in Mitochondrial Isoaccepting tRNAs from *Saccharomyces cerevisiae* as a Function of Growth Conditions
G. Baldacci, C. Falcone, L. Frontali, G. Macino, C. Palleschi 571

Appendix

The Mitochondrial Genetic Map of *Saccharomyces cerevisiae:* A Literature Compilation Towards a Unique Map
B. Dujon, A. M. Colson, P. P. Slonimski 579

Contributors

Speakers are accentuated with italics

Alexander, N. J., Department of Genetics, The Ohio State University, Columbus, Ohio 43210, USA

Asselbergs, F. A. M., Department of Biochemistry, University of Nijmwegen, Geert Grooteplein Noord 21, Nijmwegen, The Netherlands

Backhaus, B., Genetisches Institut der Universität München, Maria-Ward-Str. 1 a, D-8000 München 19, Germany

Bakker, H., Laboratory of Physiological Chemistry, State University, Bloemsingel 10, Groningen, The Netherlands

Baldacci, G., Centro Acidi Nucleici, Istituto di Fisiologia Generale, University of Rome, I-00100 Rome, Italy

Bandlow, W., Genetisches Institut der Universität München, Maria-Ward-Str. 1 a, D-8000 München 19, Germany

Baranowska, H., Institute of Biochemistry and Biophysics, Polish Academy of Sciences, Rakowiecka 36, 02-532 Warszawa, Poland

Berg, P., Department of Biology, Medicine and Biochemistry, and the Franklin McLean Memorial Research Institute, University of Chicago, Chicago, Illinois 60637, USA

Bilinski, T., Institute of Biochemistry and Biophysics, Polish Academy of Sciences, Rakowiecka 36, 02-532 Warszawa, Poland

Borst, P., Section for Medical Enzymology and Molecular Biology, Laboratory of Biochemistry, University of Amsterdam, Eerste Constantijn Huygensstraat 20, Amsterdam, The Netherlands

Bos, J. L., Section for Medical Enzymology and Molecular Biology, Laboratory of Biochemistry, University of Amsterdam, Eerste Constantijn Huygensstraat 20, Amsterdam, The Netherlands

Boutry, M., Laboratoire d'Enzymologie, Université de Louvain, Place Croix du Sud 1, B-1348 Louvain-la-Neuve, Belgium

Burger, G., Genetisches Institut der Universität München, Maria-Ward-Str. 1 a, D-8000 München 19, Germany

Butow, R. A., Department of Biochemistry, University of Texas, Health Science Center at Dallas, Dallas, Texas 75235, USA

Cabral, F., Department of Biochemistry, Biozentrum der Universität Basel, Klingelbergstr. 20, CH-4056 Basel, Switzerland

Christianson, T., Departments of Biology, Medicine and Biochemistry and The Franklin McLean Memorial Research Institute, University of Chicago, Chicago, Illinois 60637, USA

von Ciriacy, M., Institut für Mikrobiologie, Schnittspahnstr. 10, D-6100 Darmstadt, Germany

Claisse, M. L., Centre Génétique Moléculaire du C.N.R.S., F-91190 Gif-sur-Yvette, France

Clavilier, L., Centre Génétique Moléculaire du C.N.R.S., F-91190 Gif-sur-Yvette, France

Colson, A. M., Laboratoire d'Enzymologie, Université de Louvain, Place Croix du Sud 1, B-1348 Louvain-la-Neuve, Belgium

Demmer, W., Institut für Mikrobiologie und Biochemie, Lehrstuhl für Biochemie der Universität Erlangen, Egerlandstr. 7, D-8520 Erlangen, Germany

Deters, D., Department of Biochemistry, Biozentrum der Universität Basel, Klingelbergstr. 20, CH-4056 Basel, Switzerland

Devenisch, R. J., Department of Biochemistry, Monash University, Clayton, Victoria 3168, Australia

Diala, E. S., Department of Botany and Microbiology, University College London, Gower Street, London WC 1 6BT, England

Doxiadis, I., Genetisches Institut der Universität München, Maria-Ward-Str. 1 a, D-8000 München 19, Germany

Dufour, J. P., Laboratoire d'Enzymologie, Université de Louvain, Place Croix du Sud 1, B-1348 Louvain-la-Neuve, Belgium

Dujardin, G., Centre Génétique Moléculaire du C.N.R.S., F-91190 Gif-sur-Yvette, France

Dujon, B., Centre Génétique Moléculaire du C.N.R.S., F-91190 Gif-sur-Yvette, France

Ellis, D. M., Department of Genetics, The Ohio State University, Columbus, Ohio 43210, USA

English, J. T., Department of Biochemistry, Monash University, Clayton, Victoria 3168, Australia

Falcone, C., Centro Acidi Nucleici, Istituto di Fisiologia Generale, University of Rome, I-00100 Rome, Italy

Faye, G., Fondation Curie, Istitut du Radium, Section de Biologie, F-91405 Orsay, France

Francisco, J. F., Department of Biochemistry, State University of New York, Stone Brook, N.Y. 11794, USA

Frontali, L., Centro Acidi Nucleici, Istituto di Fisiologia Generale, University of Rome, I-00100 Rome, Italy

Fukuhara, H., Fondation Curie, Institut du Radium, Section de Biologie, F-91405 Orsay, France

Game, J. C., National Institute for Medical Research, The Ridgeway, Mill Hill, London SW7 1AA, England

Goffeau, A., Laboratoire d'Enzymologie, Université de Louvain, Place Croix du Sud 1, B-1348 Louvain-la-Neuve, Belgium

Grandchamp, C., Centre Génétique Moléculaire du C.N.R.S., F-91190 Gif-sur-Yvette, France

Grivell, L. A., Section for Molecular Biology, Institute of Animal Physiology, Kruislaan 320, Amsterdam, The Netherlands

Groot, G. S. P., Section for Molecular Biology, Institute of Animal Physiology, Kruislaan 320, Amsterdam, The Netherlands

Grosch, G., Genetisches Institut der Universität München, Maria-Ward-Str. 1 a, D-8000 München 19, Germany

Groscurth, R., Institut für Mikrobiologie, Arbeitsgruppe Molekulare Genetik, Universität Frankfurt, Siesmayerstr. 70, F-6000 Frankfurt/M., Germany

Groudinsky, O., Centre Génétique Moléculaire du C.N.R.S., F-91190 Gif-sur-Yvette, France

Hall, R. M., Department of Biochemistry, Monash University, Clayton, Victoria 3168, Australia

Hanson, D., Chemistry Department, Indiana University, Bloomington, Indiana 47401, USA

Heyting, C., Section for Medical Enzymology and Molecular Biology, Laboratory of Biochemistry, University of Amsterdam, Eerste Constantijn Huygensstraat 20, Amsterdam, The Netherlands

Holzner, U., Institut für Mikrobiologie und Biochemie, Lehrstuhl für Biochemie der Universität Erlangen, Egerlandstr. 7, D-8520 Erlangen, Germany

Jacq, C., Centre Génétique Moléculaire du C.N.R.S., F-91190 Gif-sur-Yvette, France

von Jagow, G., Institut für Physiologische Chemie, Physikalische Biochemie und Zellbiologie, Universität München, Goethestr. 33, D-8000 München 2, Germany

Johnston, L. H., National Institute for Medical Research, The Ridgeway, Mill Hill, London SW7 1AA, England

Kaudewitz, F., Genetisches Institut der Universität München, Maria-Ward-Str. 1 a, D-8000 München 19, Germany

Keyhani, E., Department of Biochemistry, St. Jude's Children's Research Hospital, Memphis, Tennessee 38101, USA

Keyhani, J., Department of Biochemistry, St. Jude's Children's Research Hospital, Memphis, Tennessee 38101, USA

Kolb, H. J., Institut für Diabetesforschung, Kölner Platz 1, D-8000 München 40, Germany

Korb, H., Institut für Physiologische Chemie, Physikalische Biochemie und Zellbiologie, Universität München, Goethestr. 33, D-8000 München 2, Germany

Kotylak, Z., Institute of Microbiology, Department of Microbiology, Kanonia 6/8, 50-328 Wrocław, Poland

van Kreijl, C. F., Section for Medical Enzymology and Molecular Biology, Laboratory of Biochemistry, University of Amsterdam, Eerste Constantijn Huygensstraat 20, Amsterdam, The Netherlands

Kroon, A.M., Laboratory of Physiological Chemistry, State University, Bloemsingel 10, Groningen, The Netherlands

Kujawa, C., Centre Génétique Moléculaire du C.N.R.S., F-91190 Gif-sur-Yvette, France

Lang, B., Genetisches Institut der Universität München, Maria-Ward-Str. 1a, D-8000 München 19, Germany

Lazowska, J., Centre Génétique Moléculaire du C.N.R.S., F-91190 Gif-sur-Yvette, France

Levens, D., Departments of Biology, Medicine and Biochemistry and The Franklin McLean Memorial Research Institute, University of Chicago, Chicago, Illinois 60637, USA

Lewin, A., Department of Medicine, University of Chicago, P.O. Box 4007, 950 E. 59th St., Chicago, Illinois 60637, USA

Linnane, A. W., Department of Biochemistry, Monash University, Clayton, Victoria 3168, Australia

Lückemann, G., Genetisches Institut der Universität München, Maria-Ward-Str. 1 a, D-8000 München 19, Germany

Lukins, H. B., Department of Biochemistry, Monash University, Clayton, Victoria 3168, Australia

Macino, G., Centro Acidi Nucleici, Istituto di Fisiologia Generale, University of Rome, I-00100 Rome, Italy

Mahler, H. R., Chemistry Department, Indiana University, Bloomington, Indiana 47401, USA

Martin, N., Departments of Biology, Medicine and Biochemistry and The Franklin McLean Memorial Research Institute, University of Chicago, Chicago, Illinois 60637, USA

Mathews, S., Genetisches Institut der Universität München, Maria-Ward-Str. 1a, D-8000 München 19, Germany

Merten, S., Departments of Biology, Medicine and Biochemistry and The Franklin McLean Memorial Research Institute, University of Chicago, Chicago, Illinois 60637, USA

Michaelis, G., Fakultät für Biologie, Universität Bielefeld, Universitätsstr., D-4800 Bielefeld 1, Germany

Miller, D., Chemistry Department, Indiana University, Bloomington, Indiana 47401, USA

Moorman, A. F. M., Section for Molecular Biology, Institute of Animal Physiology, Kruislaan 320, Amsterdam, The Netherlands

Morimoto, R., Departments of Biology, Medicine and Biochemistry and The Franklin McLean Research Institute, University of Chicago, Chicago, Illinois 60637, USA

Nagley, P., Department of Biochemistry, Monash University, Clayton, Victoria 3168, Australia

Netter, P., Centre Génétique Moléculaire du C.N.R.S., F-91190 Gif-sur-Yvette, France

Neupert, W., Institut für Physiologische Chemie, Physikalische Biochemie und Zellbiologie, Universität München, Goethestr. 33, D-8000 München 2, Germany

van Ommen, G. J. B., Section for Molecular Biology, Institute of Animal Physiology, Kruislaan 320, Amsterdam, The Netherlands

Pajot, P., Centre Génétique Moléculaire du C.N.R.S., F-91190 Gif-sur-Yvette, France

Palleschi, C., Centro Acidi Nucleici, Istituto di Fisiologia Generale, University of Rome, I-00100 Rome, Italy

Pepe, G., Istituto Chimica Biologica dell'Università, Via Amendola 165/A, I-70126 Bari, Italy

Perlmann, P. S., Department of Genetics, Ohio State University, 1735 Neil Avenue, Columbus, Ohio 43210, USA

Pratje, E., Fakultät für Biologie, Universität Bielefeld, Universitätsstr., D-4800 Bielefeld 1, Germany

Rabinowitz, M.; Departments of Biology, Medicine and Biochemistry and the Franklin McLean Memorial Research Institute, University of Chicago, Chicago, Illinois 60637, USA

Rhaese, H. J., Institut für Biologie, Arbeitsgruppe Molekulare Genetik, Universität Frankfurt, Siesmayerstr. 70, D-6000 Frankfurt/M., Germany

Richmond, K. M. V., National Institute for Medical Research, The Ridgeway, Mill Hill, London SW7 1AA, England

Rödel, G., Genetisches Institut der Universität München, Maria-Ward-Str. 1a, D-8000 München 19, Germany

Rouslin, W., Department of Pharmacology and Cell Biophysics, The University of Cincinnati, College of Medicine, Cincinnati, Ohio 45267, USA

Rudin, Y., Biochemistry Department, Biozentrum der Universität Basel, Klingelbergstr. 20, CH-4056 Basel, Switzerland

Saccone, C., Istituto Chimica Biologica dell'Università, Via Amendola 165/A, I-70126 Bari, Italy

Sanders, J. P. M., Gist Brocades N. V., Research Secretariat, P.O. Box 1, Delft, The Netherlands

Schatz, G., Biochemistry Department, Biozentrum der Universität Basel, Klingelbergstr. 20, CH-4056 Basel, Switzerland

Scheckel, R., Institut für Mikrobiologie, Arbeitsgruppe Molekulare Genetik, Universität Frankfurt, Siesmayerstr. 70, D-6000 Frankfurt/M., Germany

Schweizer, E., Institut für Mikrobiologie und Biochemie, Lehrstuhl für Biochemie der Universität Erlangen, Egerlandstr. 7, D-5820 Erlangen, Germany

Schweyen, R. J., Genetisches Institut der Universität München, Maria-Ward-Str. 1a, D-8000 München 19, Germany

Sebald, W., Institut für Physilogische Chemie, Physikalische Biochemie und Zellbiologie, Universität München, Goethestr. 33, D-8000 München 2, Germany

Sebald-Althaus, M., Institut für Physiologische Chemie, Physikalische Biochemie und Zellbiologie, Universität München, Goethestr. 33, D-8000 München 2, Germany

Seitz, G., Genetisches Institut der Universität München, Maria-Ward-Str. 1a, D-8000 München 19, Germany

Simpson, M. V., Department of Biochemistry, State University of New York, Stony Brook, N.Y. 11794, USA

Slonimski, P. P., Centre Génétique Moléculaire du C.N.R.S., F-91190 Gif-sur-Yvette, France

Solioz, M., Biochemistry Department, Biozentrum der Universität Basel, Klingelbergstr. 20, CH-4056 Basel, Switzerland

Spithill, T. W., Department of Biochemistry, Monash University, Clayton, Victoria 3168, Australia

Spyridakis, A., Centre Génétique Moléculaire du C.N.R.S., F-91190 Gif-sur-Yvette, France

Strausberg, R. L., Department of Biochemistry, University of Texas, Health Science Center at Dallas, Dallas, Texas 75235, USA

Tahedl, H. W., Institut für Mikrobiologie und Biochemie, Lehrstuhl für Biochemie der Universität Erlangen, Egerlandstr. 7, D-8500 Erlangen, Germany

Talen, J. L., Section for Medical Enzymology and Molecular Biology, Laboratory of Biochemistry, University of Amsterdam, Eerste Constantijn Huygensstraat 20, Amsterdam, The Netherlands

Terpstra, P., Laboratory of Physiological Chemistry, State University, Bloemsingel 10, Groningen, The Netherlands

Tzagoloff, A., Department of Biochemistry, The Public Health Research Institute of the City of New York, Inc., New York, N.Y. 10016, USA

Verkley, F. N., Section for Molecular Biology, Institute of Animal Physiology, Kruislaan 320, Amsterdam, The Netherlands

Vincent, R. D., Department of Biochemistry, University of Texas Health Science Center at Dallas, Dallas, Texas 75235, USA

Vissering, F. F., Department of Biochemistry, State University of New York, Stony Brook, N.Y. 11794, USA

de Vries, H., Laboratory of Physiological Chemistry, State University, Bloemsingel 10, Groningen, The Netherlands

Wachter, E., Institut für Physiologische Chemie, Physikalische Biochemie und Zellbiologie, Universität München, Goethestr. 33, D-8000 München 2, Germany

Wambier-Kluppel, M. L., Centre Génétique Moléculaire du C.N.R.S., F-91190 Gir-sur-Yvette, France

Weiss, H., European Molecular Biology Laboratory, P.O. Box 102209, D-6900 Heidelberg, Germany

Weiss-Brummer, B., Genetisches Institut der Universität München, Maria-Ward-Str. 1a, D-8000 München 19, Germany

Wesolowski, M., Fondation Curie, Institut du Radium, Section de Biologie, F-91405 Orsay, France

Wilkie, D., Department of Botany and Microbiology, University College, London, Gower Street, London WC1 6BT, England

Williamson, D. H., National Institute for Medical Research, The Ridgeway, Mill Hill, London SW7 1AA, England

Wolf, K., Genetisches Institut der Universität München, Maria-Ward-Str. 1a, D-8000 München 19, Germany

Zanders, E., Department of Biochemistry, University of Texas Health Science Center at Dallas, Dallas, Texas 75235, USA

Ziganke, B., European Molecular Biology Laboratory, P.O.Box 102209, D-6900 Heidelberg, Germany

Zimmermann, R., Institut für Physiologische Chemie, Physikalische Biochemie und Zellbiologie, Universität München, Goethestr. 33, D-8000 München 2, Germany

MITOCHONDRIAL DNA AND THE HERITABLE UNIT OF THE YEAST MITOCHONDRIAL GENOME: A REVIEW

D. H. Williamson, L. H. Johnston, K. M. V. Richmond, J. C. Game
National Institute for Medical Research, Mill Hill, London

Current rapid developments in the field of mitochondrial genetics owe much to a trivial property of the yeast *Saccharomyces cerevisiae*. This is the fact, a matter of common experience in many laboratories, that it is relatively easy to isolate mutants coded in mitochondrial DNA (mitDNA). If this were not the case, progress in this area would have been greatly retarded, and meetings such as the one that generated this volume relatively rare.

The basis of this flexibility of the yeast mitochondrial system is complex and only imperfectly understood. It is however compatible with the idea that the system has a relatively low "ploidy" or, in other words, that it carries a relatively low number of genetically active copies of the mitochondrial genome. By itself, this is not a specially remarkable property. It gains interest, however, when contrasted with our knowledge that the yeast cell carries a relatively large number of physical copies of the genome, in the form of DNA molecules. Since these are packaged in an organellar system which is equipped for their expression, there is an obvious gap in our understanding of how the system operates, particularly in regard to 'transmission genetics' (1). This is the problem we intend to explore here.

THE NUMBER OF HERITABLE UNITS

Information about the number of heritable units in the yeast mitochondrial system is hard to come by. Observations on the cytoplasmic petite mutation are best ignored, since this occurs with such an abnormally high frequency and is so unusual in character that it probably constitutes a special case; evidence based on its spontaneous origin or induction by mutagenic agents could therefore be misleading.

With regard to vegetative cells in culture, two groups have shown by plating on media containing erythromycin that spontaneous mutants resistant to this drug occur in normal wild-types with frequencies of around 10^{-7} (2, 3), but there is no firm information as to rates of mutation.

Genetic data from analysis of crosses is understandably much more abundant, but, as it turns out, only a little more helpful. Many groups have carried out pedigree analyses of zygotes in multifactorial crosses (for recent references, see (3)), but despite the resulting wealth of knowledge there is little hard evidence as to the ploidy of the mitochondrial system. One general conclusion is that mitochondrial genotypes appear to segregate out very rapidly in zygotic clones. Usually all the progeny cells comprising such a clone on a plate are homoplasmic, and the zygote itself often becomes homoplasmic after only a few buds have been produced. It also looks as though the number of genotypes detectable (by subsequent segregation out) in any individual bud issuing from a zygote may be rather limited, which suggests that these buds only inherit a small number of determinants. However this is weak evidence, since few individual zygotes in multifactorial crosses generate the whole range of possible recombinants, and the possible occurence of recombination in the progeny after separation from the zygote is a complicating factor.

More pertinent to the question of mitochondrial ploidy is the observation by Dujon et al. (4) of the frequency with which homoplasmic first buds arose in a non-polar cross. On the basis of straightforward assumptions about the mechanisms operating in such crosses, and armed with an understanding of the input of alleles into the cross, these authors calculated the average number of heritable units entering the bud to be around 3. Recently, more detailed calculations have been made, leading to essentially the same result ((5) and Dujon, personal communication). Bearing in mind that the amount of data analysed was relatively small, and allowing for the undoubted variability that occurs between crosses, it is probably reasonable to assume that the average number of heritable units entering the bud is less than 5.

THE NUMBER OF mitDNA MOLECULES IN THE YEAST CELL

The only species of DNA molecule involved in specifying the yeast mitochondrial genome and known to be carried in mitochondria is the so-called "25 μm circle" first recognised by Hollenberg et al. (6), and whose detailed physical and genetic structure is the subject of many of the articles in this and other recent symposia (7, 8, 9). There is abundant evidence in these sources that this molecule carries a complete set of the genes comprising the mitochondrial genome* in an order that is fixed, even though the physical distances between markers may vary slightly amongst different wild-types (14, 15, 16). Strain-dependent variation of this type also extends to the fine structure of the molecule and to its size (17, 18, 19), a fact that may account for the range of sizes (ca. 21 - 27 μm) noted by different workers (6, 20, 21). A recent estimate (14) based on analysis of restriction enzyme fragments arrived at a figure nicely in the middle of this range, namely 22 μm, corresponding roughly to a molecular weight in the order of 45×10^6 daltons.

Given this size, determination of the number of mitDNA molecules per cell requires an estimate of the absolute content per cell of mitDNA. Accurate estimation of any sort of DNA is rarely easy, and for technical reasons the usual expedient in measuring yeast mitochondrial DNA is to extract total DNA and then assess the proportion which is mitochondrial by isopycnic centrifugation (22, 23, 24, 25). The absolute amount of mitDNA may then be calculated from knowledge of the total amount of DNA per cell, this figure being arrived at by colorimetric analysis. In practice, this last datum is often not available and calculations are made using an assumed content of nuclear DNA based on current dogma as to the size of the nuclear genome. This in itself is uncertain (26, 27, 28) and in the case of growing cells requires secondary assumptions to account for cell cycle effects. Currently the size of the nuclear

---

* This ignores the separate genetic determinant recognised by Griffiths and co-workers (10, 11, 12), which is genetically and physically distinct from the 25 μm circle but which carries certain drug resistance markers relevant to mitochondria. The physical nature and intracellular location of this determinant is not known but its existence in no way conflicts with the arguments we discuss here. Equally, we need not concern ourselves with other non-mendelian determinants (13) which are thought to have an extra-mitochondrial location.

genome is conveniently accepted as being around 10^{10} daltons, a figure based largely on renaturation kinetics (28). We shall use this figure in this article, but it should be borne in mind that it may be subject to future modification.

It will be realised from the foregoing remarks that accurate counting of mitDNA molecules is full of pitfalls. However, these difficulties lose much of their importance because it transpires that unlike nuclear DNA, the amount of mitDNA per cell is highly variable. The root cause of this variability is the fact that the synthesis of mitDNA is controlled and regulated separately from that of nuclear DNA. This is clearly shown by the effects of various conditional mutants which selectively block the synthesis of nuclear DNA, leaving that of the mitochondrial molecules unaffected, at least in the short term (29, 30, 31). An example is shown in Fig. 1. Inhibitors such as cycloheximide and alpha-factor (32, 33) similarly affect nuclear DNA synthesis in a selective way, while a claim has also been made for a selective inhibition of mitochondrial DNA synthesis by a mixture of ethidium bromide and nalidixic acid (34).

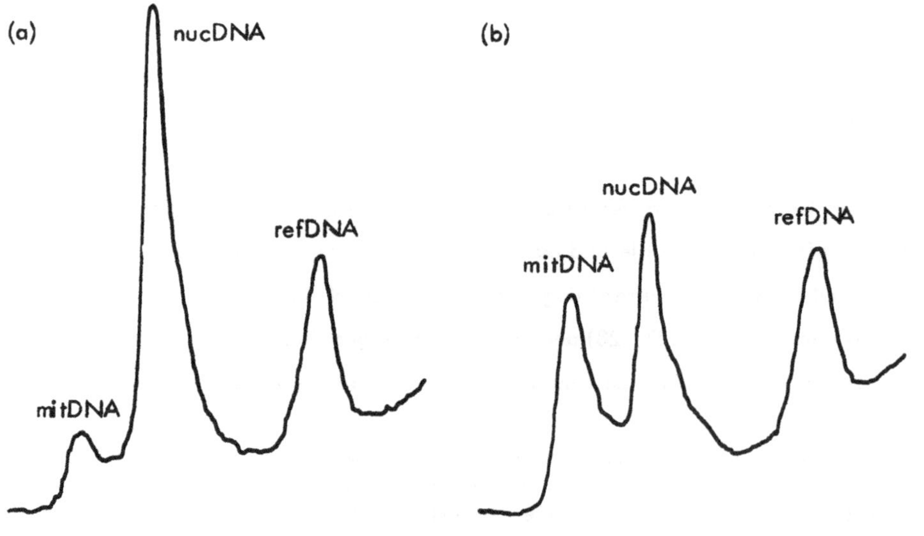

Fig. 1. Analytical CsCl gradients of yeast DNA: (a) wild-type strain; (b) conditional mutant ts47 incubated for 6 hr at non-permissive temperature (unpublished data of K. M. V. Richmond).

This dissociability of the synthetic mechanisms of nuclear and mitochondrial DNA molecules is most strikingly underlined by direct analytical evidence that in meiosis (35), in the mitotic cell cycle (36, 37, 38, 39) and in zygotes (40) the replication of mitDNA is not mandatorily coupled with that of nuclear DNA. The case of the mitotic cycle is particularly important because it means that cell cycle effects alone could cause the fluctuations in the relative proportions of nuclear and mitochondrial DNA sometimes seen in batch cultures (41).

However this whole area is further complicated for the yeast species we are considering by the operation of carbon catabolite repression, more usually known as glucose repression. The glucose repressibility of yeast mitochondrial enzymes and bioenergetic activities is well documented and is discussed, for instance, in some of the contributions to this volume. However, the glucose repressibility or otherwise of mitDNA levels has been the subject of some controversy. One report (23) claimed no effect of glucose on mitDNA, whilst others have reported widely varying degrees of repression (24, 25, 36, 42, 43, 44, 45). Some of this uncertainty may have resulted from differences in experimental technique between operators. However, in view of the relatively large number of competent analyses now to hand, it seems more likely that the reported variations in glucose repressibility of mitDNA levels largely reflect inter-strain variability. In any case there is now no doubt that some strains do show a marked effect of glucose on mitDNA levels. Good examples have been provided by Bleeg et al. (24) and Goldthwaite et al. (25), both these groups using more than one method of analysis. This last group reported what seems to be the most repressible strain observed to date, strain 20BB, which showed an overall range in mitDNA proportion between 2.2% in highly glucose repressed log phase, to 36% in derepressed (glycerol grown) stationary phase.

Glucose repression should obviously be taken into account in considering models for regulation of mitDNA. The fact that some sort of regulatory mechanism exists and is influenced by genetic background is attested by Hall et al. (46) and by the experience of our own and other laboratories that, despite the variability we have been discussing, particular strains tend to fluctuate round characteristic modal values. The nature of

Table 1. The Average Proportion of mitDNA and the Number of mitDNA Molecules per Haploid Cell

		Fully Repressed (glucose)	De-Repressed (glycerol)
Stationary phase (G1)*	proportion:	13.3%	35.6%
	mols./cell:	34	123
Log phase*	proportion:	2.2%	16.7%
	mols./cell:	8	77
Zygotes+	molecules:	112	
First zygotic bud+	molecules:	44-67	

* The proportions of mitDNA (i.e. mitDNA/total DNA) are the experimental data of Goldthwaite et al. (25) for strain 20BB. The numbers of molecules are our calculations, using the conventions described in the text.
\+ The calculations are from Sena et al. (40), based on a mitDNA molecular weight of 50×10^6 daltons. The figures quoted are just after fusion (for the zygote) and just after cell separation in the case of the zygotic bud.

the underlying regulatory influences is not known. A theoretical model has been proposed (41) but little direct experimental work has been carried out in this area and as yet no mutants specifically affected in mitDNA levels have been isolated. The genetic analysis of Hall et al. (46) using strains differing moderately in average mitDNA levels led only to the conclusion that regulation was under the control of more than one gene. Whatever the nature of the process, it is likely, in view of the observations of Lee et al. (47), to embody some element involving cell size. This is also suggested by observations that the proportion of mitDNA is independent of nuclear ploidy (at least for haploids and diploids) so that the absolute amount of mitDNA per cell is roughly proportional to ploidy (41, 48, 49), itself correlated with cell size (26).

In Table 1 we present our calculations for the average cell contents of mitDNA molecules under different conditions. We based these largely on the experimental analyses of Goldthwaite et al. (25) for their highly repressible strain 20BB, since this strain effectively defines the approximate global limits of mitDNA levels to be found amongst supposedly 'normal' wild-types. As already noted, we used a molecular weight of 45×10^6 daltons for the mitDNA molecule and assumed the size of the haploid nuclear genome to be 10^{10} daltons. In the case of the stationary phase cells, the calculations were straightforward, though unlike Goldthwaite and her colleagues we assumed the amount of nuclear DNA in G1 to be a constant 10^{10} daltons, only the amount of mitDNA being variable. Partly as a consequence of this, our estimates for numbers of molecules per cell are appreciably higher than theirs, especially in the derepressed state. Growing cells presented more of a problem, and any average figure must of course hide a wide range of variation between individual cells. Moreover the calculation is complicated by the fact that the average nuclear DNA content per cell varies with cell cycle changes in different media in a way that we can only approximate. To do this, we have taken the data of Barford and Hall (50) for the relative lengths of cell cycle phases for cultures growing on glucose. They gave no figures for a glycerol culture, and in this case we assumed 15% of the population would be in G1. To simplify the calculation we have ignored the age distribution of log phase cultures (51) and have assumed that all the cells in S have a nuclear DNA content of 1.5×10^{10} daltons.

The results of these calculations (Table 1) probably give a fair indication of the extremes of variability the system can display, and they extend somewhat the range published previously (41). On the other hand, by stressing the extremes they may be somewhat misleading, and in any case, for ease of discussion, it is convenient to have some sort of modal figure to refer to - what one might call a "conversational average". In our experience few strains are as repressible as strain 20BB, and mostly we find cells in conventional glucose media to fall in the range of 10 - 15% mitDNA in log phase and perhaps 15 - 20% in stationary phase. On this basis, using the conventions adopted above, we arrive at figures close to 50 molecules per cell for both growing and stationary haploids, and about 100 molecules per cell in diploids.

In Table 1 we have also listed the special case of the zygote, using for this purpose the elegant analyses of synchronized zygote cultures reported by Sena et al. (40), to which the reader is referred for full details. In this case we report their calculated numbers of molecules, based on a size of 50×10^6 daltons for the mitochondrial molecule.

THE NATURE OF THE PROBLEM

The significance of the numbers presented above is not so much their absolute value, but rather the fact that in contrast to the observed genetic behaviour of the system, as described above, they seem rather high. There is roughly an order of magnitude difference between the number of mitDNA molecules per haploid cell (about 50) and the number of heritable units (<5).

The problem this poses rests on the fact that mitochondria are believed to be able to transcribe and translate the DNA molecules they contain, and there is no obvious reason why these 50 molecules should not be potentially capable of expression. Therefore, provided they can all be replicated and transmitted to offspring, they should all constitute independent genetic determinants for the information they carry.

The problem is illustrated schematically in Fig. 2, which we show to emphasise the

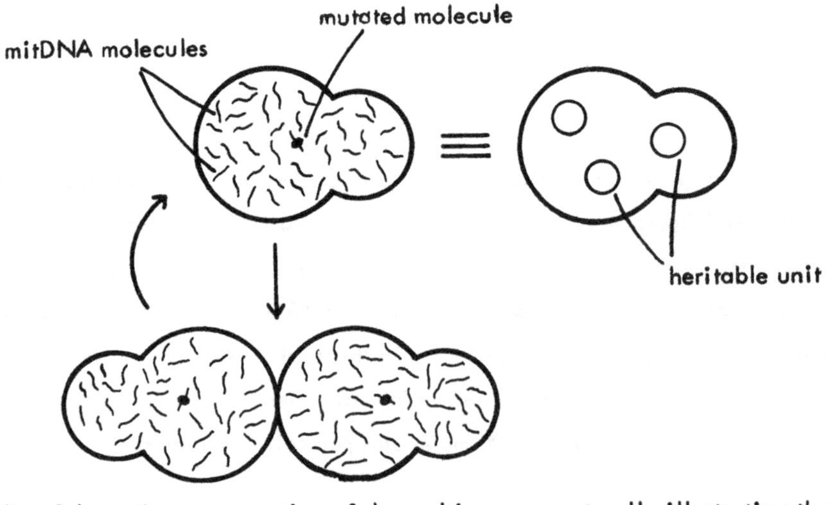

Fig. 2. Schematic representation of the multi-copy yeast cell, illustrating the central problems discussed here. These are: (a) how can we relate the large number of physical copies of the genome to the apparently low number of heritable units?, and (b) how does a mutant genome find expression?

major difficulty facing this type of system. If a mutation were introduced into one of the molecules without altering its replicative capacity, the mutated molecule would be propagated indefinitely. Moreover, as long as there were no direct physical interaction with other molecules, it would seldom if ever segregate out to give a homoplasmic clone, and since this would apply to any similar mutation, the system would be slow to evolve. In fact, as we have seen, segregation occurs rapidly, thus providing ample opportunity for evolution to occur.

Various solutions to this problem may be considered. In the first place, short of a special segregation mechanism, any multi-organellar system is likely to show some randomness of inheritance, sufficient to lead to a slow 'sorting out' process and eventual segregation of homoplasmic cells. However, as discussed by Michaelis (52) and Birky (1), genetic drift of this nature could not possibly account for the rapidity of segregation observed in practice. A frequently discussed alternative involves the idea of grossly unequal segregation of mitDNA molecules at cell division. If, for instance, daughter cells regularly inherited only one out of fifty of the parents' complement of mitDNA molecules, the probability of generating a homoplasmic mutant daughter cell would be 1 in 50 per generation. This would help to speed up segregation somewhat, but any such extreme asymmetry of inheritance would necessitate that the inherited determinants went through a rapid burst of replication in order to maintain their numbers. Thus a population of growing cells would contain a sub-set of rapidly replicating molecules, and density labelling experiments ((53) and see below) have failed to reveal this. On the other hand, some asymmetry probably does occur in growing cells, since daughters are usually a little smaller than their parents (54), but the magnitude of this phenomenon is insufficient to resolve the problem.

Reference should be made to the special case of the zygote. Sena et al. (40) recently published a detailed study of the replication of both nuclear and mitochondrial DNA during the formation and first budding of synchronized zygotes. As regards the mitochondrial system, the information is complex, but it appears that there is considerable asymmetry of inheritance. At the time when fusion of parents was just complete, the zygote contained a little over 100 mitDNA molecules. By the time the first zygotic bud was released by cytokinesis, this bud was carrying round 50 mitDNA molecules, but since replication was continuing throughout the growth of the bud, we can only guess at the number of molecules actually inherited. However there was no evidence that any mitDNA molecules were replicating unduly rapidly, so one might surmise that the bud inherited around 25 molecules from the zygote. This would only be about a quarter of the number originally present at the time of fusion, but it is nearly ten times the number of heritable units postulated to enter the bud (4) on the basis of genetic analysis. Thus in the zygote as in the vegetative cell, unequal segregation of mitDNA molecules does not answer the central problem.

Another attempt to resolve this problem involves the so-called master copy hypothesis. A number of such hypotheses are extant, all revolving around the central idea that one of the mitDNA copies in the cell is for some reason 'special' and it is this copy that determines the genetic behaviour of the system (34, 55, 56). There is at present no definitive evidence for the existence of such a copy or for a mechanism whereby it could be specially conserved, replicated or inherited. Moreover, from a philosophical standpoint, the master copy idea,which seeks to explain the mitochondrial system in terms of single copy genetics,as developed for nuclear inheritance on mendelian principles,seems to be an evasion of reality. The likelihood that the multiple mitochondrially packaged mitDNA copies of the genome are not genetically competent is remote. It seems far more likely that the system is, both genetically and physically, a multi-copy one, and any attempt to explain its behaviour in transmission should be based on that central principle.

With that in mind, we are finally led to consider the possibility, previously discussed in outline form (57, 58) that the problem we are discussing could be resolved if the cell had a mechanism designed to maintain the genetic homogeneity of its mitochondrial system. This should allow rapid acceptance or rejection of a new mutational event on any particular mitDNA copy. Acceptance or rejection should be on a random basis, since otherwise the cell would have to have the capacity for value judgements. If accepted, a mutation would be rapidly transmitted to other copies so that whole cell would soon become homoplasmic and the mutation guaranteed expression;

if rejected, the cell would remain homoplasmic for wild-type. Clearly a mechanism with these properties would be advantageous, by allowing the mitochondrial system to evolve. It would also of course lead to relatively rapid segregation of mitochondrial mutants.

The most likely candidate for such a mechanism would be the process of recombination and gene conversion. It is now generally accepted that recombination between DNA duplexes involves the formation of a heteroduplex molecule, followed by correction of any mismatches resulting from heterozygosity of the recombining molecules (59, 60). The mismatch correction process can lead to gene conversion in which, more or less randomly, alleles of only one of the two types entering into the cross are recovered. If all mitDNA molecules in the cell frequently entered into recombination, with the possibility of gene conversion at each recombinational event, the population would tend to drift towards homogeneity at any particular locus at a rate dependent on the frequency of recombination and the size of the population.

This principle is discussed in some detail, in a general context, by Birky and Skavaril (61), to whom reference will be made below. It is also implicit in the model for mitochondrial recombination in zygotes put forward by Dujon et al. (4). Though focussing largely on the nature of polar ($\omega^+ \times \omega^-$) crosses, this model centres round the idea that the mitDNA molecules in the zygote form a panmictic pool undergoing multiple random rounds of recombination, any molecule having, in principle, the chance of recombining with any other. In non-polar regions of the genome, or in non-polar crosses, random gene conversion would help in homogenising the gene pool. With some reservations (1) this model fits the genetic data closely and there is no doubt it explains the zygote situation very well. It implies a high degree of mixing of mitDNA molecules in the zygote, and high freedom of access between molecules in the pool. It is not clear at present how this is brought about. Degradation of mitochondrial structures in zygotes has been reported (62), but at present our information about mitochondrial morphology in the zygote is fragmentary. In any case the zygote might be a special case in which internuclear and intermitochondrial fusion is specially favoured in order to facilitate recombination between the appropriate DNA molecules.

We wish to address ourselves particularly to the situation in vegetatively growing cells, several generations removed from the zygote. Here we do not have a body of genetic information comparable with that for the zygote. However, there is strong physical evidence for the occurrence of extensive mitochondrial recombination in vegetatively growing cells, and recent advances in our knowledge of the overall structure of the mitochondrial system raise the possibility that mitDNA molecules may be remarkably free to interact with each other, regardless of spatial distribution within the cell.

RECOMBINATION AND THE ORGANISATION OF THE MITOCHONDRIAL SYSTEM

Seeking to explore the overall mode of generation of mitDNA molecules in growing cells, two groups (53, 63) have carried out density-shift experiments of the Meselson and Stahl (64) type. Both groups noted that the replication of mitDNA showed a 'dispersive' pattern in distinct contrast to the normal semi-conservative pattern shown by the nuclear molecules. Williamson and Fennell (53) concluded that the observed pattern resulted from the occurrence of multiple recombinational events between mitDNA molecules; in order to account for their data, each molecule would have to undergo several recombinational exchanges in each cell cycle. These experiments incidentally suggest with near certainty that all the mitDNA molecules in the cell normally undergo replication, and by ruling out the occurrence of any sub-set of rapidly replicating molecules, they effectively dispose of models of inheritance involving grossly unequal segregation, as discussed above. However they do not throw any light on the degree of freedom with which any given molecule can recombine with any other in the cell. Such recombination might be geographically restricted and occur only within small pools of molecules or even, in the extreme, only between the true sister molecules resulting from replication of an individual duplex.

Happily for our present thesis however, we can now at least envisage the possibility that the recombinational interactions seen in the density-shift experiments need not be restricted in any way; as in the zygote, the vegetative cells' mitochondrial system may even constitute a panmictic pool. This conclusion stems from current knowledge concerning the overall structure of the system.

In 1973, Hoffman and Avers, using electron microscopy of serial sections (65), reported that in contrast to an earlier report based on light microscopy (66) the yeast mitochondrion consisted of a single 'giant' and much branched organelle. Soon after, Grimes et al. (48), also using electron microscopy, reported the presence in lactate-grown cells of about 10 mitochondria in a haploid and around 20 in a diploid. These apparently conflicting reports have now been largely resolved by Stevens (67, 68) in what is the most comprehensive examination of this problem so far to appear. It turns out that the number of mitochondria is not fixed, but depending on strain and physiological conditions can vary between 1 and about 50 per cell. The reader is referred to (68) for details, but in essence the situation is that the number of organelles per cell tends to be markedly higher in stationary phase than in log phase, particularly in glucose media. Stationary phase glucose grown cells may have nearly 50 small organelles, while log phase cells in glucose usually have less than 5, and sometimes only one. In these growing cells there is , characteristically, one large branched mitochondrion, the others (if any) being smaller but showing a wide range of sizes. This is a marked departure from conventional ideas about the mitochondrial system which tend to depict a number of discrete uniform entities reproducing by a process resembling binary fission. The data of Stevens suggest rather strongly that mitochondria arise by irregular fragmentation much more than by equal binary fission. Moreover the possibility is raised, though it cannot be established by electron microscopy alone, that the organelles in the system undergo repeated fusions and dis-aggregations during normal vegetative growth.

Whether or not this is so, this new knowledge of the mitochondrial system at least implies that there is much less compartmentalisation of mitDNA molecules than one might have supposed. There is no obvious barrier to recombinational interactions between the molecules within a single organelle, and we now know that at least in some individual cells, those with a unitary mitochondrion, this can encompass the cells' entire mitDNA complement. In other cells (or situations) in which several organelles are seen, repeated fusions and dis-aggregations could allow for a great deal of mixing of mitDNA molecules from all parts of the cell. The extent of such mixing in vegetative cells is not known, though there is some indirect support for its occurrence (58, 68). In any event, however, this new 'fluid' concept of the mitochondrial system gives us reason to consider that widespread recombinational interactions and gene conversion events could take place in growing cells. The question we shall now consider is whether or not such interactions could be frequent enough to account for the rapidity of segregation that we observe.

THE EFFICIENCY OF THE GENE CONVERSION MECHANISM

In 1976, Birky and Skavaril (61) discussed in a general way the possibility that multicopy genetic systems like mitochondria are maintained in a genetically homogeneous state by the operation of just the sort of mechanism we have been discussing - i.e. multiple rounds of mating with gene conversion. They coined the acronym SMAC (Stochastic Matching and Conversion) and set up computer simulations of populations of molecules in an attempt to see how rapid the genetic drift resulting from such a process would be. To simplify the models, replication and segregation of molecules were not considered, so that the simulations were admittedly fairly crude approximations to reality. Heteroduplex formation was assumed to be symmetric, and most of the simulations were based on conversion occurring with equal probability in either direction. In one such simulation, involving an input of one mutant allele to 49 wild-types, populations of molecules homogeneous for the majority allele were generated after only a few rounds of mating, but a population homogeneous for the minority allele only appeared after 312 mating rounds. Thus in most cases, the mutation was rapidly lost and pure mutant populations were slow to appear. However, the authors demonstrated that the process would work much faster with smaller populations. They considered, for instance, a mixture consisting of one mutant and three wild-type molecules. In 26% of runs, this population became homogeneous for wild-type after only a single mating round, and pure mutant populations were also generated correspondingly rapidly.

These observations led the authors to the view that SMAC would work only rather slowly with the mitDNA population of a whole cell, but that it could be an effective agent in maintaining the genetic homogeneity of the molecules within a mitochondrion, thus making the organelle the effective unit of segregation, rather than the mitDNA molecule.

In reaching this conclusion, Birky and Skavaril made use of the data of Grimes et al. (48), who reported that the mitochondria of lactate grown haploid cells contained about 4 mitDNA molecules each, a number obviously low enough to favour the operation of SMAC. However it should be recognised that in these de-repressed cells, the number of organelles was several fold higher than the number of heritable units that might be inferred from the data of Dujon et al. (ref. 4) for zygotes. The situation is no better for repressed cells. As shown by Barbara Stevens (68), the number of mitochondria in log phase cells on glucose (1 - 5) does fit closely with the Dujon estimate of heritable units. However, except in highly repressible strains, the organelles in these cells would contain around 16 mitDNA molecules each, a number far too high, it would seem, for the effective operation of SMAC. Of course these numerical arguments should not be stressed too much, since we are rather loosely comparing genetic and physical numbers derived from zygotes on the one hand, and vegetative cells of unrelated strains in different media on the other. But in any case, there is one more fact about the mitochondrion that we have so far ignored, and which may have an important bearing on the whole situation. This is the way the mitDNA molecules are distributed within the mitochondria.

THE INTRAMITOCHONDRIAL DISTRIBUTION OF mitDNA MOLECULES: THE CHONDRIOLITE

Using a fluorescent dye specific for DNA, Williamson and Fennell(69) showed that mitDNA in yeast rarely occurs as isolated molecules. More usually it is distributed in aggregates termed "chondriolites" (58). Detailed results of a study of the numbers and behaviour of chondriolites will be published elsewhere (70), but a preliminary report has been given in (58). It has been found that the number of chondriolites per cell varies in much the same way as the number of mitochondria (68), and the size of the chondriolites ranges between 1 and 8 mitDNA molecules depending on the environment and culture phase. In general terms, an average log phase cell in glucose might have around 2 - 5 mitochondria (68), carrying a total of 10 chondriolites, each consisting of 4 - 5 molecules. In de-repressed log phase, a normal strain might have about 10 organelles, each of which would contain 1 - 2 chondriolites, again consisting of about 4 molecules.

Could the chondriolite be the heritable unit? Obviously it is small enough for the effective operation of SMAC, and it is hard to imagine there would be any hindrance to recombination between its constituent molecules. Moreover in the zygote, the numbers involved are acceptable. The data of Sena et al. (40) suggested that a zygotic bud could inherit around 25 mitDNA molecules, and it is not wildly unrealistic to imagine these packaged in 4 or 5 average sized chondriolites. Unfortunately, however, this straightforward proposal fails badly when we consider logarithmically growing cells, for in these the number of chondriolites (15 - 20) is far too high to be equated easily with 3 heritable units. Nevertheless, as pointed out above, we have no good genetic basis on which to quantify the number of heritable units in growing cells, and until we have more information about the stability of chondriolites, their reproduction, and their freedom to interact with each other under different circumstances, the 'numbers game' we have indulged in here should not be taken too seriously. More important perhaps is the recognition that the chondriolite is one more factor to be taken into account in trying to understand the genetic behaviour of the mitochondrial system. It is part of a growing body of evidence which is eroding the concept of the mitochondrion as a neat, regular and uniform entity which never interacts with its fellows. Rather we should accept the mitochondrial system as a 'fluid' whole, continually undergoing fragmentation and fusion, in which local irregularities and heterogeneities are commonplace. To illustrate this concept we now present, in summary form, a model of the yeast mitochondrion that incorporates, as far as possible, all the relevant information we have discussed in this article.

A MODEL YEAST MITOCHONDRION

The main features of the model are shown in Fig. 3. Most of them follow directly from the information discussed above. One additional concept, at the moment entirely theoretical, is that for at least part of its life-span, the chondriolite is in some

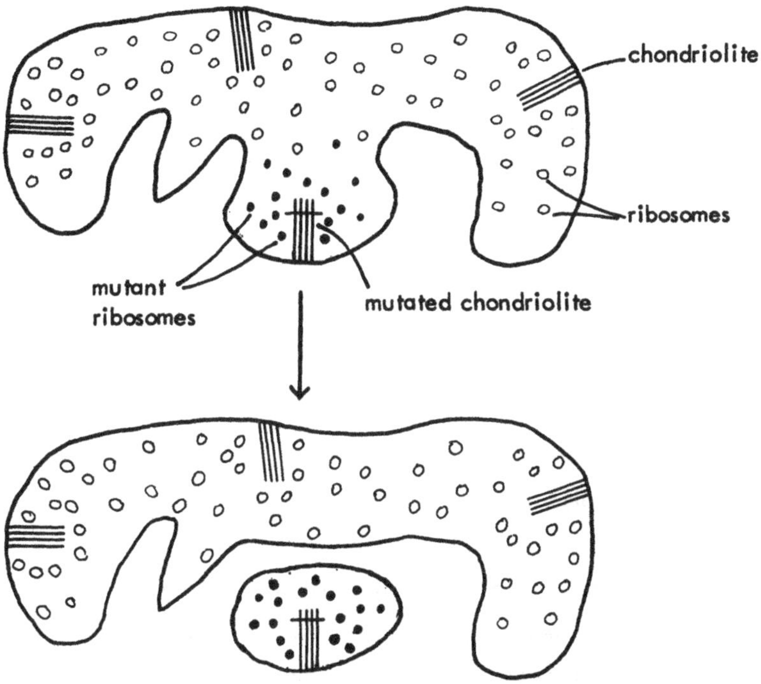

Fig. 3. A diagrammatic model of the yeast mitochondrion based on the information and ideas discussed in the text. The irregularly shaped giant organelle ie shown reproducing by fragmentation of a region containing a chondriolite. For illustrative purposes this is shown carrying a mutation affecting mitoribosomes. The chondriolites are kept genetically homogeneous by repeated rounds of recombination and gene conversion ("SMAC" (61)). Transcription and translation products derived from a particular chondriolite tend to be utilised in the region immediately surrounding that chondriolite so that (excluding components coded in the nucleus) the chondriolite genetically determines its own local environment within the organelle.

way localised within the organelle, possibly by attachment to the membrane. This attachment cannot be permanent under all circumstances, since it has to allow for reproduction of the chondriolite and, possibly, for interactions between them in both zygotes and vegetative cells. We assume that the chondriolite is maintained in a state of genetic homogeneity by the operation of SMAC, and thus constitutes, at least within the confines of the organelle, a genetically competent unit. We further suppose that transcripts from a given chondriolite are all used within the region of the chondriolite, as are polypeptides resulting from translation. This would result in each chondriolite having its own "zone of influence" within the organelle, i.e. a region dominated by self-coded products. Reproduction of the organelle by the fragmentation process illustrated in Fig. 3 would tend to generate mini-organelles, each of rather homogeneous genetic origin. This would go some way to explaining problems that we have not discussed here concerning the selection of certain types of mutant. This and other aspects of the model will be discussed in more detail elsewhere (71).

THE NATURE OF THE HERITABLE UNIT

It will be seen from the above that we have considered, as possible candidates for the heritable unit, each known member of the hierarchy comprising the mitochondrial system - the mitDNA molecule, the chondriolite, and the organelle. Each, in turn, has been found wanting. This leads us to make a somewhat revolutionary suggestion. This is that, like Thurber's unicorn (72), the heritable unit we have been seeking is a mythical beast. More prosaically, we suggest that there is no single stable physical entity which we can identify as being the heritable unit, and in seeking such an entity we are betraying our prejudiced upbringing, which would have us explain genetic systems in terms of single copy mendelian principles. In mitochondria, and no doubt in other organellar systems, we have a multi-copy system whose genetic behaviour can be understood only as the net outcome of various interlocking factors such as recombination, gene conversion and the 'fluid' nature of the organelles. We are right to look for the heritable unit, because doing so will lead us to a deeper understanding of the system. But we should not be disappointed if it eludes our grasp.

REFERENCES

1. Birky, C. W. Jr.: Mitochondrial genetics in fungi and ciliates. In: Genetics and biogenesis of mitochondria and chloroplasts (Birky, C. W. Jr., Perlman, P. S., Byers, T. J., Eds.), Ohio State University Press, Columbus, 1975, pp. 182-224.
2. Birky, C. W. Jr.: On the origin of mitochondrial mutants: evidence for intracellular selection of mitochondria in the origin of antibiotic-resistant cells in

yeast. Genetics 74, 421-432 (1973).

3. Dujon, B., Bolotin-Fukuhara, M., Coen, D., Deutsch, J., Netter, P., Slonimski, P. P., Weill, L.: Mitochondrial genetics XI. Mutations at the mitochondrial locus ω affecting the recombination of mitochondrial genes in Saccharomyces cerevisiae. Molec. gen. Genet. 143, 131-165 (1976).
4. Dujon, B., Slonimski, P. P., Weill, L.: Mitochondrial genetics IX: A model for recombination and segregation of mitochondrial genomes in Saccharomyces cerevisiae. Genetics 78, 415-437 (1974).
5. Dujon, B., Slonimski, P. P.: Mechanisms and rules for transmission, recombination and segregation of mitochondrial genes in Saccharomyces cerevisiae. In: Genetics and biogenesis of chloroplasts and mitochondria (Bücher, Th., Neupert, W., Sebald, W., Werner, S., Eds.) North-Holland Publishing Co., Amsterdam, 1976, pp. 393-403.
6. Hollenberg, C. P., Borst, P., Van Bruggen, E. F. J.: Mitochondrial DNA V. A 25-μ closed circular duplex DNA molecule in wild-type yeast mitochondria. Structure and genetic complexity. Biochim. Biophys. Acta 209, 1-15 (1970).
7. Genetics, biogenesis and bioenergetics of mitochondria (Bandlow, W., Schweyen, R. J., Thomas, D. Y., Wolf, K., Kaudewitz, F., Eds.) Walter de Gruyter, Berlin, 1976.
8. Genetics and biogenesis of chloroplasts and mitochondria (Bücher, Th., Neupert, W., Sebald, W., Werner, S., Eds.) North-Holland Publishing Co., Amsterdam, 1976.
9. The genetic function of mitochondrial DNA (Saccone, C., Kroon, A. M., Eds.) North-Holland Publishing Co., Amsterdam, 1976.
10. Griffiths, D. E., Lancashire, W. E., Zanders, E. D.: Evidence for an extrachromosomal element involved in mitochondrial function: A mitochondrial episome? FEBS Lett. 53, 126-130 (1975).
11. Lancashire, W. E., Griffiths, D. E.: Studies on energy-linked reactions: Isolation, characterisation and genetic analysis of tri-alkyltin resistant mutants of Saccharomyces cerevisiae. Eur. J. Biochem. 51, 377-392 (1975).
12. Lancashire, W. E., Griffiths, D. E.: Studies on energy-linked reactions: Genetic analysis of venturicidin-resistant mutants. Eur. J. Biochem. 51, 403-

413 (1975).

13. Guerineau, M., Slonimski, P. P., Avner, P. R.: Yeast episome: Oligomycin resistance associated with a small covalently closed non-mitochondrial circular DNA. Biochem. Biophys. Res. Commun. 61, 462-469 (1974).
14. Sanders, J. P. M., Borst, P., Weijers, P. J.: The organisation of genes in yeast mitochondrial DNA II. The physical map of Eco RI and Hind II + III fragments. Molec. gen. Genet. 143, 53-64 (1975).
15. Sanders, J. P. M., Heyting, C., Borst, P.: The variability of the mitochondrial genome of Saccharomyces strains. In: Genetics and biogenesis of chloroplasts and mitochondria (Bücher, Th., Neupert, W., Sebald, W., Werner, S., Eds.) North-Holland Publishing Co., Amsterdam, 1976, pp. 511-518.
16. Sanders, J. P. M.: This volume.
17. Bernardi, G., Prunell, A., Kopecka, H.: An analysis of the mitochondrial genome of yeast with restriction enzymes. In: Molecular biology of nucleo-cytoplasmic relationships (Puiseux-Dao, S., Ed.) Elsevier Scientific Publishing Co., Amsterdam, 1975, pp. 85-90.
18. Prunell, A., Kopecka, H., Strauss, F., Bernardi, G.: The mitochondrial genome of wild-type yeast cells V. Genome evolution. J. Mol. Biol. 110, 17-52 (1977).
19. Prunell, A., Bernardi, G.: The mitochondrial genome of wild-type yeast cells VI. Genome organization. J. Mol. Biol. 110, 53-74 (1977).
20. Petes, T. D., Byers, B., Fangman, W. L.: Size and structure of yeast chromosomal DNA. Proc. Nat. Acad. Sci. USA 70, 3072-3076 (1973).
21. Blamire, J., Cryer, D. R., Finkelstein, D. B., Marmur, J.: Sedimentation properties of yeast nuclear and mitochondrial DNA. J. Mol. Biol. 67, 11-24 (1972).
22. Williamson, D. H., Moustacchi, E., Fennell, D.: A procedure for rapidly extracting and estimating the nuclear and cytoplasmic DNA components of yeast cells. Biochim. Biophys. Acta 238, 369-374 (1971).
23. Fukuhara, H.: Relative proportions of mitochondrial and nuclear DNA in yeast under various conditions of growth. Eur. J. Biochem. 11, 135-139 (1969).
24. Bleeg, H. S., Leth Bak, A., Christiansen, C., Smith, K. E., Stenderup, A.:

Mitochondrial DNA and glucose repression in yeast. Biochem. Biophys. Res. Commun. 47, 524-530 (1972).

25. Goldthwaite, C. D., Cryer, D. R., Marmur, J.: Effect of carbon source on the replication and transmission of yeast mitochondrial genomes. Molec. gen. Genet. 133, 87-104 (1974).

26. Ogur, M., Minckler, S., Lindegren, G., Lindegren, C. C.: The nucleic acids in a polyploid series of Saccharomyces. Archs. Biochem. Biophys. 40, 175 (1952).

27. Parry, E. M., Cox, B. S.: The general properties of a strain of yeast with less DNA than normal. J. gen. Microbiol. 70, 129-149 (1972).

28. Bicknell, J. N., Douglas, H. C.: Nucleic acid homologies among species of Saccharomyces. J. Bacteriol. 101, 505-512 (1970).

29. Wintersberger, U., Hirsch, J., Fink, A. M.: Studies on nuclear and mitochondrial DNA-replication in a temperature-sensitive mutant of Saccharomyces cerevisiae. Molec. gen. Genet. 131, 291-299 (1974).

30. Newlon, C. S., Fangman, W. L.: Mitochondrial DNA synthesis in cell cycle mutants of Saccharomyces cerevisiae. Cell 5, 423-428 (1975).

31. Richmond, K. M. V.: Uncoupling of nuclear and mitochondrial DNA synthesis in a temperature sensitive mutant of Saccharomyces cerevisiae. Heredity 38, 277 (1977).

32. Grossman, I., Goldring, E. S., Marmur, J.: Preferential synthesis of yeast mitochondrial DNA in the absence of protein synthesis. J. Mol. Biol. 46, 367-376 (1969).

33. Petes, T. D., Fangman, W. L.: Preferential synthesis of yeast mitochondrial DNA in α-factor-arrested cells. Biochem. Biophys. Res. Commun. 55, 603-609 (1973).

34. Whittaker, P. A., Hammond, R. C., Luha, A. A.: Mechanism of mitochondrial mutation in yeast. Nature New Biology 238, 266-268 (1972).

35. Pinon, R., Salts, Y., Simchen, G.: Nuclear and mitochondrial DNA synthesis during yeast sporulation. Exptl. Cell Res. 83, 231-238 (1974).

36. Cottrell, S. F., Avers, C. J.: Evidence of mitochondrial synchrony in synchronous cell cultures of yeast. Biochem. Biophys. Res. Commun. 38, 973-980

(1970).

37. Williamson, D. H., Moustacchi, E.: The synthesis of mitochondrial DNA during the cell cycle in the yeast Saccharomyces cerevisiae. Biochem. Biophys. Res. Comm. 42, 195-201 (1971).

38. Wells, J. R.: Mitochondrial DNA synthesis during the cell cycle of Saccharomyces cerevisiae. Exptl. Cell Res. 85, 278-286 (1974).

39. Sena, E. P., Welch, J. W., Halvorson, H. O., Fogel, S.: Nuclear and mitochondrial deoxyribonucleic acid replication during mitosis in Saccharomyces cerevisiae. J. Bacteriol. 123, 497-504 (1975).

40. Sena, E., Welch, J., Fogel, S.: Nuclear and mitochondrial DNA replication during zygote formation and maturation in yeast. Science 194, 433-435 (1976).

41. Williamson, D. H.: The effect of environmental and genetic factors on the replication of mitochondrial DNA in yeast. In: Control of organelle development, Symp. Soc. Exp. Biol. XXIV (Miller, P. L., Ed.) University Press, Cambridge, 1970, pp. 247-276.

42. Moustacchi, E., Williamson, D. H.: Physiological variations in satellite components of yeast DNA detected by density gradient centrifugation. Biochem. Biophys. Res. Commun. 23, 56-61 (1966).

43. Nagley, P., Linnane, A. W.: Cellular regulation of mitochondrial DNA synthesis in Saccharomyces cerevisiae. Cell Differentiation 1, 143-148 (1972).

44. Mian, F. A., Kuenzi, M. T., Halvorson, H. O.: Studies on mitochondrial membrane proteins in Saccharomyces cerevisiae under different degrees of glucose repression. J. Bacteriol. 115, 876-881 (1973).

45. Böker, E., Kaudewitz, F., Richmond, V., Schweyen, R., Thomas, D. Y.: Experimental variation of the number of mitochondrial genomes participating in mitochondrial genetics. In: Genetics, biogenesis and bioenergetics of mitochondria (Bandow, W., Schweyen, R. J., Thomas, D. Y., Wolf, K., Kaudewitz, F., Eds.) Walter de Gruyter, Berlin, 1976, pp. 99-115.

46. Hall, R. M., Nagley, P., Linnane, A. W.: Biogenesis of mitochondria XLII. Genetic analysis of the control of cellular mitochondrial DNA levels in Saccharomyces cerevisiae. Molec. gen. Genet. 145, 169-175 (1976).

47. Lee, E.-H., Johnson, B. F.: Volume-related mitochondrial deoxyribonucleic

acid synthesis in zygotes and vegetative cells of Saccharomyces cerevisiae. J. Bacteriol. 129, 1066-1071 (1977).

48. Grimes, G. W., Mahler, H. R., Perlman, P. S.: Nuclear gene dosage effects on mitochondrial mass and DNA. J. Cell Biol. 61, 565-574 (1974).

49. Gunge, N.: Effects of elevation of strain-ploidy on transmission and recombination of mitochondrial drug resistance genes in Saccharomyces cerevisiae. Molec. gen. Genet. 146, 5-16 (1976).

50. Barford, J. P., Hall, R. J.: Estimation of the length of cell cycle phases from asynchronous cultures of Saccharomyces cerevisiae. Exptl. Cell Res. 102, 276-284 (1976).

51. Mitchison, J. M.: The biology of the cell cycle, The University Press, Cambridge, 1971.

52. Michaelis, P.: Über Gesetzmässigkeiten der Plasmon - Umkombination und über eine Methode zur Trennung einer Plastiden-, Chondriosomen-, resp. Sphaerosomen-, (Mikrosomen)- und einer Zytoplasmavererbung. Cytologia 20, 315-338 (1955).

53. Williamson, D. H., Fennell, D. J.: Apparent dispersive replication of yeast mitochondrial DNA as revealed by density labelling experiments. Molec. gen. Genet. 131, 193-207 (1974).

54. Johnson, B. F.: Morphometric analysis of yeast cells. Adult volume of Saccharomyces cerevisiae. Exptl. Cell Res. 39, 577-585 (1965).

55. Wilkie, D.: The induction by monochromatic U.V. light of respiratory-deficient mutants in aerobic and anaerobic cultures of yeast. J. Mol. Biol. 7, 527-533 (1963).

56. Oliver, S. G.: On the mutability of the yeast mitochondrial genome. J. theor. Biol. 67, 195-201 (1977).

57. Williamson, D. H.: In: Molecular biology of nucleocytoplasmic relationships (Puiseux-Dao, S., Ed.) Elsevier Scientific Publishing Co., Amsterdam, 1975, pp. 95-100.

58. Williamson, D. H.: Packaging and recombination of mitochondrial DNA in vegetatively growing yeast cells. In: Genetics, biogenesis, and bioenergetics of mitochondria (Bandlow, W., Schweyen, R. J., Thomas, D. Y., Wolf, K.,

Eds.) Walter de Gruyter, Berlin, 1976, pp. 117-123.

59. Holliday, R.: A mechanism for gene conversion in fungi. Genet. Res. Camb. 5, 282-304 (1964).

60. Meselson, M. S., Radding, C. M.: A general model for genetic recombination. Proc. Nat. Acad. Sci. USA 72, 358-361 (1975).

61. Birky, C. W. Jr., Skavaril, R. V.: Maintenance of genetic homogeneity in systems with multiple genomes. Genet. Res. Camb. 27, 249-265 (1976).

62. Smith, D. G., Wilkie, D., Srivastava, K. C.: Ultrastructural changes in mitochondria of zygotes in Saccharomyces cerevisiae. Microbios 6, 231-238 (1972).

63. Sena, E.: Thesis, University of Wisconsin, Madison (1972).

64. Meselson, M., Stahl, F. W.: The replication of DNA in Escherichia coli. Proc. Nat. Acad. Sci. USA 44, 671-682 (1958).

65. Hoffman, H. P., Avers, C. J.: Mitochondrion of yeast: Ultrastructural evidence for one giant, branched organelle per cell. Science 181, 749-751 (1973).

66. Avers, C. J., Lin, F. H., Pfeffer, C. R.: Histochemical studies of mitochondrial variation during anaerobic growth of respiration-normal Bakers' Yeast. J. Histochem. Cytochem. 13, 344-349 (1965).

67. Stevens, B. J.: Variation in mitochondrial numbers and volume in yeast according to growth conditions. J. Cell Biol. 63, 336A (1974).

68. Stevens, B. J.: Variation in number and volume of the mitochondria in yeast according to growth conditions. A study based on serial sectioning and computer graphics reconstitution. Biologie Cellulaire 28, 37-56 (1977).

69. Williamson, D. H., Fennell, D. J.: The use of fluorescent DNA binding agent for detecting and separating yeast mitochondrial DNA. In: Methods in cell biology (Prescott, D. M., Ed.) Academic Press, New York, 1975, Vol. 12, pp. 335-351.

70. Williamson, D. H., Fennell, D. J.: In preparation.

71. Williamson, D. H., Johnston, L. H., Richmond, K. M. V., Game, J. C.: In preparation.

72. Thurber, J.: The unicorn in the garden. In: The Thurber Carnival, Harper and Row Publishers Inc., New York, 1945.

TWO ASPECTS OF MITOCHONDRIAL DNA STRUCTURE: THE OCCURRENCE OF TWO TYPES OF MITOCHONDRIAL DNA IN RAT LIVER AND THE ISOLATION FROM RAT LIVER OF DNA COMPLEXES OF HIGH BUOYANT DENSITY

J.F. Francisco, F.F. Vissering and M.V. Simpson
Department of Biochemistry, State University of New York,
Stony Brook, N.Y. 11794, U.S.A.

The studies to be reported here concern two aspects of mitochondrial DNA (mtDNA). The first relates to the occurrence of two types of mtDNA, as distinguished by Eco RI analysis, distributed throughout the population of Rattus norwegicus, the common laboratory rat. The second concerns preliminary work on a mtDNA-containing complex of very high buoyant density isolated from rat liver mitochondria. The complex consists of a hazy core containing areas of high density and surrounded by many DNA loops partially buried in the core.

OCCURRENCE OF TWO TYPES OF RAT LIVER mtDNA

Recently, Potter et al. (1) have shown that on the basis of Hae III endonuclease analysis, multiple mtDNA species exist in human and horse populations (individuals, however, possessing only a single type). The scope of this phenomenon appeared to be limited, however, since twelve other quite diverse mammalian species studied including the laboratory rat did not show such mtDNA multiplicity. Moreover, in several more recent studies on the rat (2-4), in which the goal was to obtain a physical map of mtDNA through the use of endonuclease Eco RI, no such DNA diversity was reported despite the fact that the fragments generated were examined both by gel electrophoresis (3,4) and by electron microscopy (2).

Our involvement in this problem stemmed from our studies

on the origin and role of "8S" mtDNA which we first observed some years ago (5,6). This material is made up of a class of small DNA molecules, somewhat heterogeneous in size, which become rapidly and highly labeled on incubation of mitochondria with labeled DNA precursors. We had already established that this material hybridizes with high molecular weight mtDNA and wished to inquire into its possible homology with a particular domain of this DNA. For this purpose, we required a fair amount of material and so made use of a pool of about 20 livers from which mtDNA was isolated (6) and purified (7). Some of the results presented below on this and other mtDNA preparations analyzed by restriction enzymes have been reported in a preliminary communication (7).

The Eco RI pattern obtained from the hydrolysis of the pooled DNA (Fig. 1, Channel II) shows eight clearly visible bands. When the bands were examined with respect to molar ratios and molecular sizes, the total size value obtained was 31 kilobases (kb). This figure corresponds to two complete mtDNA molecules of 15.5 kb each. Inasmuch as a number of controls were run (Eco RI treatment of phage lambda, addition of more enzyme and reincubation, use of specially purified mtDNA) which indicated that we were not dealing with partials, with nuclease contamination, or with DNA other than closed circular mtDNA, we could infer that two different kinds of DNA were present in the pooled sample. On the basis of the results of Potter et al. (1) and those on goats and sheep by Upholt and Dawid (8), our data seemed to indicate that two types of mtDNA were distributed through the rat population rather than being present in a single individual.

The use of individual rather than pooled livers confirmed this view (Fig. 1, Channels III and IV), and we term the two types A and B. We define type A DNA as those circles which yield seven fragments upon Eco RI treatment and type B DNA as those circles which yield five fragments. Although only six and

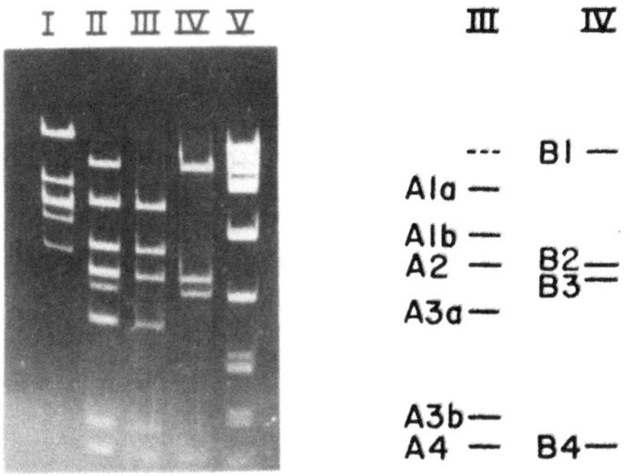

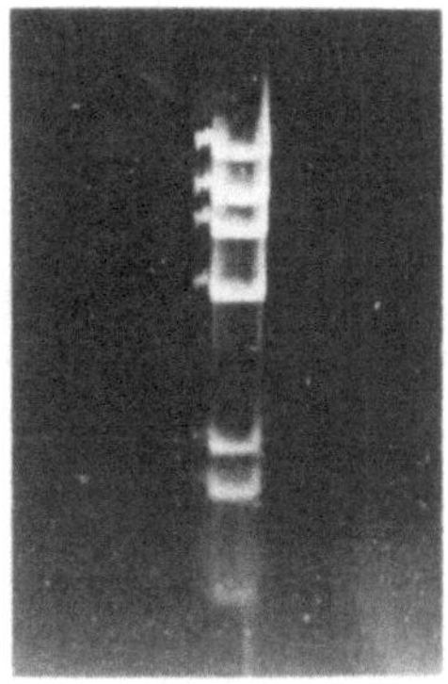

Fig. 1. Agarose gel electrophoresis of a complete Eco RI endonuclease digest of mtDNA from the livers of Sprague Dawley rats. For the pooled liver experiments (channel II), the extracted mtDNA from 20 livers was purified by sucrose density gradient centrifugation followed by CsCl isopycnic centrifugation. In the individual animal experiments, the latter step was omitted. A diagrammatic representation of the gel patterns is shown to the right of the gel photograph since the fainter bands may not reproduce clearly in half tone. The dashed line in channel III represents a "partial" digest fragment. Channel I, marker fragments from λDNA (wild type); channel II, mtDNA from pooled livers; channel III, type A mtDNA from an individual liver; Channel IV, type B mtDNA from an individual liver; Channel V, marker fragments from F factor DNA.

Fig. 2. Visualization of fast moving band on agarose gel electrophoresis. Conditions were as described in Fig. 1 except that the gel was loaded more heavily with sample and the time of electrophoresis was decreased by 25%.

four fragments, respectively, are depicted in Fig. 1, we have reported (7) the presence of a fast moving fragment which is partly obscured because of its entry into the more optically dense trapping gel and which is derived from both types A and B DNA. The fragment becomes clearly visible when the gel is slightly overloaded and is run for a shorter period of time (Fig. 2). It is present in equimolar amount and its size is about 160 base pairs. All fragment sizes appear in Table 1. We have now analyzed the mtDNA from the livers of 97 individual rats using Eco RI. In no individual liver did we find more than one type of DNA.

Table I

FRAGMENT SIZES OF PRODUCT OF COMPLETE Eco RI DIGESTION OF TYPES A AND B RAT LIVER mtDNA

Type A DNA		Type B DNA	
Fragment	Size	Fragment	Size
A1a (2)*	6.03	B1 (1)	9.79
A1b (3)	3.76	B2 (4)	2.89
A2 (4)	2.88	B3 (5)	2.45
A3a (6)	1.77	B4 (8)	0.41
A3b (7)	0.67	B5	0.16
A4 (8)	0.42		
A5	0.16		

*Numbers in parentheses refer to the nomenclature used for the fragments in the pooled liver experiment (Fig. 1, channel II).

Table II

THE OCCURRENCE OF TYPES A AND B DNA IN THE LIVER MITOCHONDRIA OF DIFFERENT RAT POPULATIONS

Rat Strain	Source	Number of rats possessing DNA of	
		Type A	Type B
Sprague	CR*	50	5
Dawley	TF	6	2
Wistar	DR	5	16
	MA	3	3
Long Evans	CR	0	1
Hooded	MBF	4	2

*CR, Charles River Laboratories; TF, Taconic Farms; M.A., Microbiological Associates; MBF, Marland Breeding Farms.

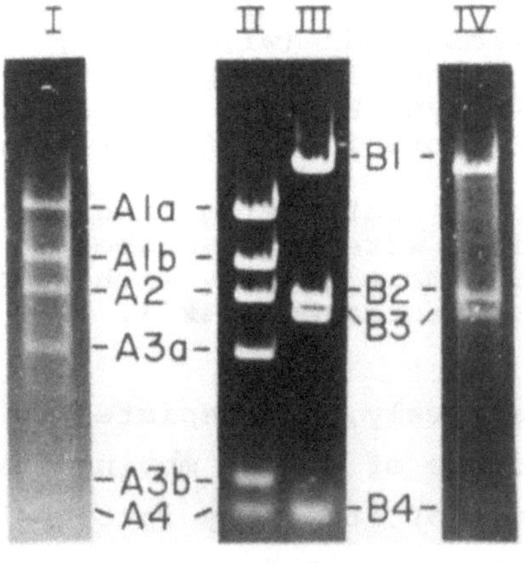

Fig. 3. Agarose gel electrophoresis of a complete Eco RI endonuclease digest of mtDNA from the livers of Long-Evans Hooded rats. Conditions of the experiment are described in the legend to Fig. 1 except that, for direct comparison, the marker fragments used were an Eco RI digest of either type A or type B liver mtDNA obtained from Sprague Dawley rats. Faint bands may not appear on half tone reproduction. Channel I, Sprague Dawley type A marker fragments; Channel II, Long Evans Hooded type A mtDNA; Channel III, Long Evans Hooded type B mtDNA; Channel IV, Sprague Dawley type B marker fragments.

Up to this point, we had been using only Sprague-Dawley rats, and it was of interest to discover whether the type of DNA might be a function of the strain of rat from which it originated. To this end, the livers of individual animals of two additional strains were examined. The Eco RI patterns showed that types A and B DNA occur in both the Long Evans (Fig. 3) and the Wistar populations (not shown) as well as the Sprague Dawley. Moreover, all three strains of rats were obtained from four different suppliers and the mtDNA types which occur appear to be independent of the commercial source insofar as we can judge from the relatively small number of rats examined of a particular strain from a particular supplier. This conclusion is substantiated by our recent finding that, of a second batch of Wistar animals obtained from the Charles River Laboratories, five were type A. In the first batch of nine animals reported on previously (7), all possessed type B DNA. All of the strain and supplier results are summarized in Table II.

While our initial report (7) of this work was in press, a paper by Parker and Watson appeared (9) reporting a physical map of mitochondrial DNA. The authors discuss the lack of accord between their map and that constructed by Saccone et al. (3) (both of which, in turn, differ from that of Koike et al. (2)). A perusal of the Eco RI gel patterns of these groups makes it evident to us that the preparations of Kroon et al. and Koike et al. were primarily composed of type A DNA while Parker and Watson were using type B. However, the preparations of both Kroon et al. and Parker and Watson contain small amounts of types B and A respectively. The latter authors entertain the possibility that the small 160 base pair fragment (Table I) which they have also detected but which was not noted by Saccone et al. might resolve the differences between the two maps. However, it may well be that in the light of the disagreement between all three maps, the presence of the two types of DNA in the mtDNA preparations could lead to error in the interpretation of results. Alternatively, the two DNA's might possess certain differences in structure, e.g., certain regions might be transposed. We are engaged in studying the structural differences between these two molecules.

We have also begun the physical mapping of the two types of DNA. We can report at this point that hybridization studies with isolated 7S D-loop DNA and Eco RI fragments, using the technique of Southern (10), confirm the results of Koike et al. (11). The 7S fragment lies somewhere within the 6030 base pair fragment of type A DNA. Results of experiments with "8S" DNA will be described elsewhere.

THE HIGH BUOYANT DENSITY mtDNA COMPLEX

When rat liver mitochondria are incubated with a labeled deoxynucleotide triphosphate precursor and the isolated mtDNA subjected to CsCl isopycnic centrifugation, a peak of radioactivity is observed at a buoyant density of 1.760 g/ml in addition to the well known peak at 1.700 (Fig. 4). The size of the 1.760 peak can range from a frequently observed small fraction of the radioactivity in the 1.700 peak to occasionally an almost equal amount of radioactivity.

A density of 1.760 g/ml clearly is too high to reflect DNA alone (or even poly GC) but could point to the presence of RNA. Moreover, the s value of the material is considerably greater than 39S, the value for closed circular mtDNA, suggesting the occurrence of some type of complex. Some electron micrographs of this material can be seen in Fig. 5. These structures were selected to indicate variants in a spectrum of structures. All structures observed are characterized by a kind of cluster of loops surrounding a hazy undefined core containing varying amounts of electron opaque material. A part of each loop is buried in the core.

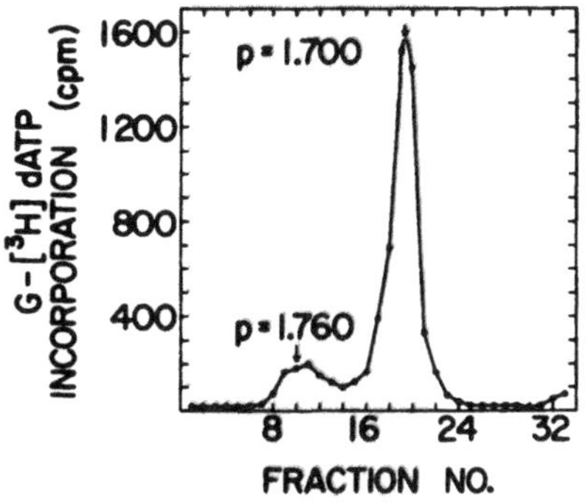

Fig. 4. CsCl isopycnic centrifugation of mtDNA labeled *in vitro*. mtDNA was isolated, centrifuged in a sucrose gradient, and all fractions greater than and inclusive of 27 S were pooled, the DNA precipitated in ethanol, and subjected to CsCl isopycnic centrifugation, all as described previously (6).

In Fig. 5A, the core is unusually small with the loops numbering somewhat less than average. The total contour length of the visible loops in this structure is equivalent to 3.5 complete molecules of mtDNA, making the assumption that the loops are completely DNA and completely double stranded. The great majority of structures observed have larger cores than this one with the contour lengths varying between the equivalent of 1 and 9 mtDNA molecules. An occasional molecule (Fig. 5C) possesses as many as 13 mtDNA equivalents and a huge core containing a very large amount of electron opaque material. A distribution of the most frequently seen types is shown in Fig. 5B; the large molecule close to the center of the field possesses about 7 mtDNA equivalents.

We have just begun to study some of the biochemical properties of the 1.760 material. When it was treated with Eco RI, it gave a pattern identical to that of mtDNA (both types, showing) that its DNA component is in fact mtDNA and not a nuclear or viral contaminant. Subjecting the material to conditions known to thermally denature DNA leaves the core essentially intact but seems to partly cleave off most or all of the loops, leaving a short legged centipede structure (Fig. 6). The for-

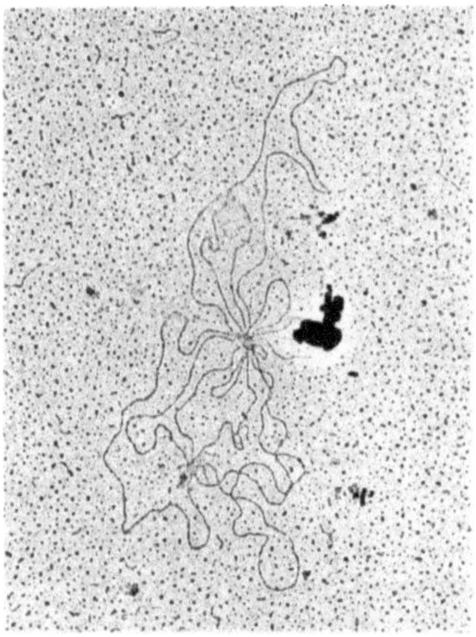

5A

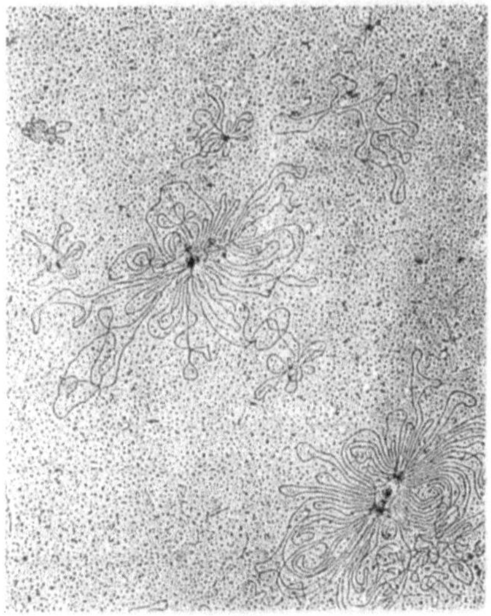

5B

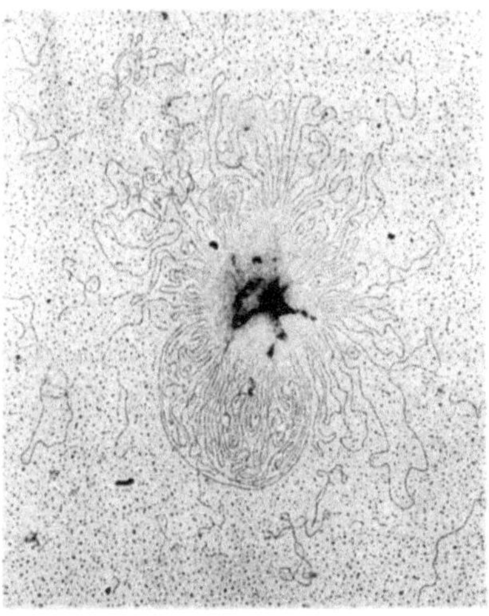

5C

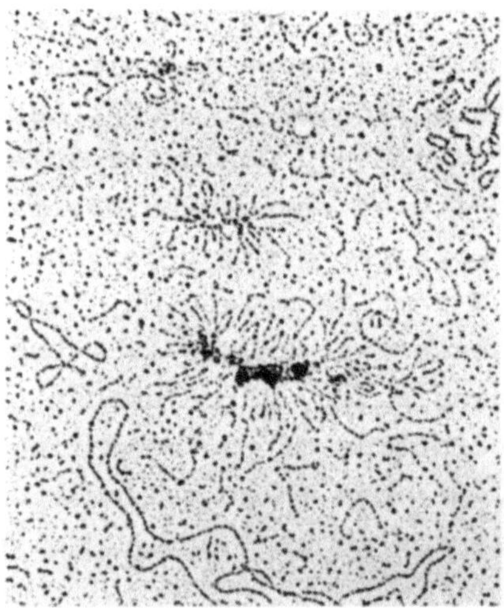

6

Fig. 5. Electron microscopy of high buoyant density DNA. Fractions of the 1.760 DNA from the CsCl run were pooled, dialyzed in STE buffer (6), precipitated in ethanol (6), and taken up in 0.1 concentration of STE buffer. The DNA was spread by the formamide technique (12). The magnification was 12,000X.

Fig. 6. Electron microscopy of high buoyant density DNA following short thermal denaturation. Conditions were as described in Fig. 5 except that the DNA was subjected to a temp. of 100° for 60 sec prior to the CsCl centrifugation.

mation of this structure would be explicable if the DNA loops contained multiple nicks so that upon denaturation, stretches of DNA would be lost from both strands. The presence of nicks is in accord with the relaxed (i.e., not supercoiled) nature of most of the loops. This explanation would be equally applicable if some or all of the loops were DNA-RNA hybrids. Indeed, preliminary experiments in which the 1.760 material was treated with RNase A also yielded a centipede structure. This could indicate the presence of RNA (which would need to be single stranded at the points of attack of RNase A). However, pure mtDNA is subject to RNase action (13,14) albeit slowly, because of its content of ribonucleotides. Experiments with RNase H are in progress. No effect was seen with proteinase K.

Thus far, we have no clue as to the composition of the core. It is tempting to speculate that it consists of mitochondrial membrane to which replicating or transcribing DNA is attached, and, in fact, D loops are often seen in the structure. The electron opaque material in the core is more difficult to explain. One of a number of possibilities is that it is mtDNA itself which, for some reason, has undergone a phase change and has essentially precipitated out as a solid. This, rather than RNA might account for the high buoyant density of this material.

The question naturally arises as to whether the 1.760 material could be an artifact, particularly since Ott et al. (15) have recently observed the formation of structures not too dissimilar to these by treatment of linear lambda DNA with magnesium perchlorate (a denaturing agent) or with high concentrations of ammonium acetate. In our case, the dense material can be obtained without exposure to denaturants or even to the high concentration of CsCl used in the isopycnic centrifugation. It can, for example, be isolated from a crude mitochondrial extract upon gradient centrifugation in sucrose containing only 10 mM Tris. The material appears in fractions estimated to be about 45-50 S. Moreover, the cluster is stable to dialysis against dilute salt whereas the complex of Ott et al. reversibly dissociates to linear DNA when the magnesium perchlorate is removed. It is conceivable, however, that the large structures such as the one illustrated in Fig. 5B might be artifactual conglomerates. Such a naturally occurring structure would demand that some mitochondria possess as many as 13 DNA molecules per mitochondrion. This is unlikely but not ruled out.

DNA-containing structures possessing various kinds of cores have previously been found in biological systems. The membrane-associated folded E. coli chromosome is one such system (16-20). Here, one sees a definitive electron opaque core holding together many loops, the entire structure being stabilized by RNA. The large hazy core is not evident, however. In the case of adenovirus DNA, a protein covalently attached to the 5' end of the molecule serves to condense or fold the molecule by interaction at many points in the DNA chain (21). There is no core apparent in such structures, however, other than the single protein molecule. Finally, a membrane-like structure is associated with the D-loop site of mtDNA (22). This structure bears no obvious resemblance to ours but further work may indicate whether or not the two structures are functionally related.

At this point, it is idle to speculate on the function of this complex. However, the specific radioactivity of the material is not less than that of closed circular DNA and we are therefore encouraged to continue these studies.

ACKNOWLEDGEMENTS

We owe a debt of gratitude to Dr. Martha Simon of Brookhaven Laboratories for her interest and her invaluable advice on the electron microscopy of the dense complex. This work was supported by research grants from the NIGMS, National Institutes of Health (GM2233311) and the American Cancer Society (NP87H). One of us (J.F.F.) was supported by a National Institutes of Health Training Grant (GM07212).

REFERENCES

1. Potter, S.S., Newbold, J.E., Hutchison III, C.A., and Edgell, M.H. Specific cleavage analysis of mammalian mitochondrial DNA. Proc. Nat. Acad. Sci. USA 72, 4496-4500 (1975).
2. Koike, K., Kobayashi, M., Tanaka, S., and Mizusawa, H. Physical map and replication of rat mitochondrial DNA. In "Genetics and Biogenesis of Chloroplasts and Mitochondria" (Bucher, T., Neupert, W., Sebald, W., and Werner, S., eds.) North Holland Publishing Co., Amsterdam. pp. 593-596.
3. Saccone, C., Pepe, G., Cantatore, P., Terpstra, P., and Kroon, A.M. Mapping of the transcription products of rat liver mitochondria by hybridization. In "The Genetic Function of Mitochondrial DNA" (Saccone, C. and Kroon, A.M., eds.) North Holland Publishing Co., Amsterdam, pp. 27-36.
4. Kroon, A.M., Bakker, H., Holtrop, M. and Terpstra, P. The restriction endonuclease cleavage map of rat liver mitochondrial DNA. Biochim. Biophys. Acta 474, 61-68 (1977).
5. Hamilton, F.D., Robins, D.R., and Simpson, M.V. The synthesis of 8S DNA fragments in isolated mitochondria. Fed. Proc. 29, 726 (1970).
6. Van Tuyle, G.C., Hamilton, F.D., Vissering, F.F., and Simpson, M.V. Biosynthesis of mitochondrial DNA. Is 8S DNA an artifact? J. Biol. Chem., 252, 2984-2991 (1977).

7. Francisco, J.R. and Simpson, M.V. The occurrence of two types of mitochondrial DNA in rat populations as detected by Eco RI endonuclease analysis. FEBS Letters, in press.
8. Upholt, W.B. and Dawid, I.B. Comparative restriction endonuclease analysis and mapping of sheep and goat mitochondrial DNA. Cell, in press.
9. Parker, R.C. and Watson, R.M. Restriction endonuclease cleavage maps of rat and mouse mitochondrial DNAs. Nucleic Acids Res. 4, 1291-1299 (1977).
10. Southern, E.M. Detection of specific sequences among DNA fragments separated by gel electrophoresis. J. Mol. Biol. 98, 503-517 (1975).
11. Koike, K., Kobayashi, M., and Tanaka, S. Replication of mitochondrial DNA and its initiation site, in "DNA Synthesis and its Regulation", ICN-UCLA Symposium on Molecular and Cellular Biology, 3 (Goulian, M., Hanawalt, P., and Fox, C.F., eds.) W.A. Benjamin, Inc., Menlo Park, California, 1975, pp. 578-580.
12. Davis, R.W., Simon, M., and Davidson, N. Electron microscope heteroduplex methods for mapping regions of base sequence homology in nucleic acids, in Methods in Enzymology (Grossman, L. and Moldave, K., eds.) Academic Press, New York 21, pp. 413-428.
13. Koch, Jurgen. Introduction of "nicks" and "chops" into human mitochondrial DNA in vivo and in vitro. Eur. J. Biochem. 33, 98-103 (1973).
14. Miyaki , M., Koide, K., and Ono, T. RNase and alkali sensitivity of closed circular mitochondrial DNA of rat ascites hepatoma cells. Biochem. Biophys. Res. Commun. 50, 252-258 (1973).
15. Ott, G.S., Bastia, D., and Bauer, W. Private communication from Dr. Bauer.
16. Delius, H. and Worcel, A. Electron microscope studies on the folded chromosome of E. coli. Cold Spring Harbor Symp. Quant. Biol. 38, 53-58 (1973).
17. Pettijohn, D.E. and Hecht, R. RNA molecules bound to the

folded bacterial genome stabilize DNA folds and segregate domains of supercoiling. Ibid., pp. 31-41.

18. Hecht, R.M. and Pettijohn, D.E. RNA molecules attached to DNA in isolated bacterial nucleoids: their possible role in stabilizing the condensed chromosome, in "DNA Synthesis and its Regulation," ICN-UCLA Symposium on Molecular and Cellular Biology, 3, (Goulian, M., Hanawalt, P., and Fox, C.F., eds.). W.A. Benjamin, Inc., Menlo Park, California, 1975, pp. 122-137.
19. Drlica, K. and Worcel, A. Viscometric analysis of conformational transitions in the Escherichia coli chromosome. Ibid., pp. 138-158.
20. Ryder, O.A., Kavenoft, R., and Smith, D.W., Ibid, pp. 159-186.
21. Rekosh, D.M.K., Russel, W.C., Bellet, A.J.D., and Robinson, A.J. Identification of a protein linked to the ends of adenovirus DNA. Cell, 11, 283-295 (1977).
22. Albring, M., Griffith, J., and Attardi, G. Association of a protein structure of probable membrane derivation with HeLa cell mitochondrial DNA near its origin of replication. Proc. Natl. Acad. Sci. USA, 74, 1348-1352 (1977).

TRANSPOSABLE SEGMENTS OF MITOCHONDRIAL DNA: A UNITARY HYPOTHESIS FOR THE MECHANISM OF MUTATION, RECOMBINATION, SEQUENCE REITERATION AND SUPPRESSIVENESS OF YEAST "PETITE COLONY" MUTANTS

P.P. SLONIMSKI and J. LAZOWSKA
Centre de Génétique Moléculaire du C.N.R.S. Gif sur Yvette 91190 - France

Since their discovery by EPHRUSSI and his collaborators some thirty years ago (10), the cytoplasmic petite colony mutants of yeast continue to attract attention and interest. This is attested by the fact that in numerous publications dealing with cytoplasmic inheritance and organelle biogenesis the petite mutants constitute either a direct target or a preferred tool of investigations. In the two most recent symposia on "The Genetic Function of Mitochondrial DNA" (1) and "The Genetic and Biogenesis of Chloroplast and Mitochondria" (2), some 40 publications, out of a hundred devoted to mitochondria in general, have dealt with or made use of petites.

Some ten years ago it was shown that mtDNA of cytoplasmic petites (i.e. ρ^-) was so drastically changed that its buoyant density was grossly modified although its total amount per cell did not vary (28). This in turn was explained by demonstrating that ρ^- mutants result from very large deletions of the wild type mtDNA compensated by numerous periodic repetitions of the non-deleted sequence (14). The deletion-repetition explanation of the petite mtDNA is now widely accepted and numerous investigations are concerned with the organization of this DNA as an aim in itself or as a tool to explore the structure and function of the wild type mtDNA in general (cf.1, 2, 15, 23). There remains however, a large set of phenomena for which there is no satisfactory explanation. Some were observed in the very early days of organelle genetics like the high spontaneous mutation ρ^+ to ρ^-, the massive induction by various agents or conditions, and suppressivness (cf.23, 33). Others were described more recently, like the marker transmission in ρ^- by ρ^+ crosses, the genetic and molecular recombination in ρ^- by ρ^- crosses and the formation of multimeric and/or additive circular mtDNA molecules (1, 2, 6, 8, 14, 18, 19, 20, 21, 22, 25, 26, 27, 34, 35). No unifying mechanism integrating these different features has as yet emerged.

Such a unification is the purpose of this article. We shall present a model aimed at explaining most of the facets of the petite phenomenon. The main idea of the model is that the repetitive segments of petites are equivalent to the transposable or the translocatable elements in bacteria (see 5, 17, 32, 38 for reviews) and behave as discrete genetic and physical entities which can move from one position to another and from one molecule to another in the mitochondrial genome. These translocation events would be promoted by a recombination machinery other than that involved in the general homologous recombination of the wild type mtDNA. The rules that govern the latter have been described by DUJON et al (9) and need not be rewieved here.

I - PREMISSES OF THE MODEL -

In this section premisses of the model will be presented without discussion. The comparison of the model with the experimental data and predictions which can be derived from it will be discussed later on.

1) We postulate first that yeast cells use two different mechanisms to recombine their mtDNA. The first mechanism (referred to henceforth as A)cf.7) necessitates long stretches of homologous sequences while the second mechanism (referred to as B), involves highly specific sites but does not require an extensive sequence homology. By comparison with what is known in prokaryotes (cf. 5, 17, 32, 38) the mitochondrial mechanism A is thought to be analogous to the bacterial rec A dependent "general" or "ordinary", homologous recombination while the mitochondrial mechanism B is analogous to the rec A independent, "illegitimate" recombination.
2) We postulate further that mechanism B is entirely coded by the nuclear genome and the enzymes involved are synthesized on cytoplasmic ribosomes. On the contrary, at least one element of mechanism A, an enzyme or a protein subunit of an enzymic complex, is coded by the mitochondrial genome and/or synthesized by mitochondrial ribosomes.
3) The last postulate concerns the nature of the specific sites which are involved in the mitochondrial recombination occuring through mechanism B. We think that these specific sites are precisely the sequences present at the covalent junctions of repetitions in ρ^- molecules. The fundamental repeat of a ρ^- molecule can be thus described as P_1' a b c P_1 where P_1 and P_1' are the sequences at the termini of the repeating unit or,

in other terms, the sequences surrounding the junctions. A circular monomeric molecule of a ρ^- mtDNA will be $\overline{P_1' \text{ a b c } P_1}$; a circular dimer, in tandem direct arrangement, will be $\overline{P_1' \text{ a b c } P_1 \; P_1' \text{ a b c } P_1}$ and a circle of two inverted repeats will be $\overline{P_1' \text{ a b c } P_1 \; P_1 \text{ c b a } P_1'}$ etc ... The exact nature of the sequences P_1 P_1' will be discussed later. It suffices to say that we postulate that these sequences are relatively short, from a few base pairs to a few dozen base pairs as an order of magnitude. Furthermore, there is more than one kind of these specific sites and the fundamental repeats of various petites may be described as P_1' a b c P_1 or P_2' g h i P_2 or P_3' x y z P_3 where $P_1 \neq P_2 \neq P_3$. Because it is well established that mtDNA sequences of petites are mainly, if not exclusively, segments of the mtDNA present in the wild type (1, 14), it follows logically that the wild type molecule must contain numerous specific recombination sites and can be described as a sequence ... a b c P_1 P_1' ... g h i P_2 P_2' ... x y z P_3 P_3' ..., some 75000 base pairs long.

II - THE MODEL EXPLAINS MOST OF THE OBSERVATIONS ALREADY MADE

In order to make explicit the main ideas of our model we shall describe first the situation which occurs when a grande ρ^+ cell is crossed with a petite ρ^- cell i.e. "suppressivness" and "markers transmission". Afterwards, we shall envisage in the light of the model the various facets of the petite phenomena such as the spontaneous production of petites, induction of ρ^- mutants, formation of hybrid molecules by recombinaison between different ρ^- and the repetitive nature of their mtDNA. Finally, more general features of mtDNA in yeast, such as its mode of replication and evolution of its structure will be discussed.

1. Suppressivness and Marker Transmission of Petites.

When a population of ρ^- cells (issued from a single haploid clone) is crossed to a population of ρ^+ cells (issued from a single haploid clone) two things happen. A certain proportion of the diploid progeny is composed of ρ^- cells. This is called "suppressivness" (12). A certain proportion of the remaining ρ^+ cells carries a mitochondrial allele brought into the cross by the petite parent, when the cross was heterozygous. This is referred to as "marker transmission" or "marker rescue" (4, 6, 8, 14, 26, 27, 34, 35, 37). A petite clone is said to be "suppressive" when most of the total progeny is ρ^- (more than 99% in some instances) or "neutral" when most of the progeny is ρ^+. A petite clone is said to transmit highly when most of the ρ^+ progeny carries the allele incoming from the petite parent (more than 80% in some cases).

Suppressivness was first described and studied extensively by EPHRUSSI et al. (11, 12, 13). They have shown that a) all descendants of a clone issued from an individual zygote can be ρ^-. The normal ρ^+ factor can be thus eliminated or destroyed completely by the suppressive ρ^- factor; b) the degree of suppressivness is a hereditary property of each ρ^- clone, for although subject to drift and/or mutation, this degree is nevertheless relatively stable in vegetative multiplication; c) the suppressive ρ^- factor does not destroy the respiratory enzymes of the zygote but provokes an unstable, long-lasting cellular state during which the zygote and its vegetative progeny can switch into the more stable ρ^+ or to the entirely stable ρ^- condition. During this unstable state the outcome of zygotic clones can be manipulated by external conditions such as temperature, sugar concentration or the presence of sodium acetate.

Transmission by petites of their mitochondrial alleles to the ρ^+ progeny of a cross has been studied first with antibiotic resistance mutations used as genetic markers (4, 6, 8, 14, 18, 26, 27, 34, 35) and more recently with wild type alleles of various mit^- mutations located in cytochrome oxidase, coenzyme QH_2 - cytochrome c reductase, ATPase genes(1, 19, 37). Depending on the petite clone studied any segment of mtDNA can be transmitted and each clone has its own hereditary characteristics stable in vegetative multiplication. These characteristics are subject to genetic drift and mutation (8, 14, 35).

The premisses of our model explain simultaneously "marker transmission" and "suppressivness". Consider a zygote formed by fusion of a ρ^+ cell with a ρ^- one (Fig.1). In the zygote both recombination mechanisms are active, the mechanism A enzymes being provided by the ρ^+ parent. MtDNA molecules of the petite can recombine with the wild type molecules by either one of the two mechanisms. In pathway A, pairing takes place through homology and "ordinary" recombination (possibly via non-reciprocal events, cf.9) occurs.

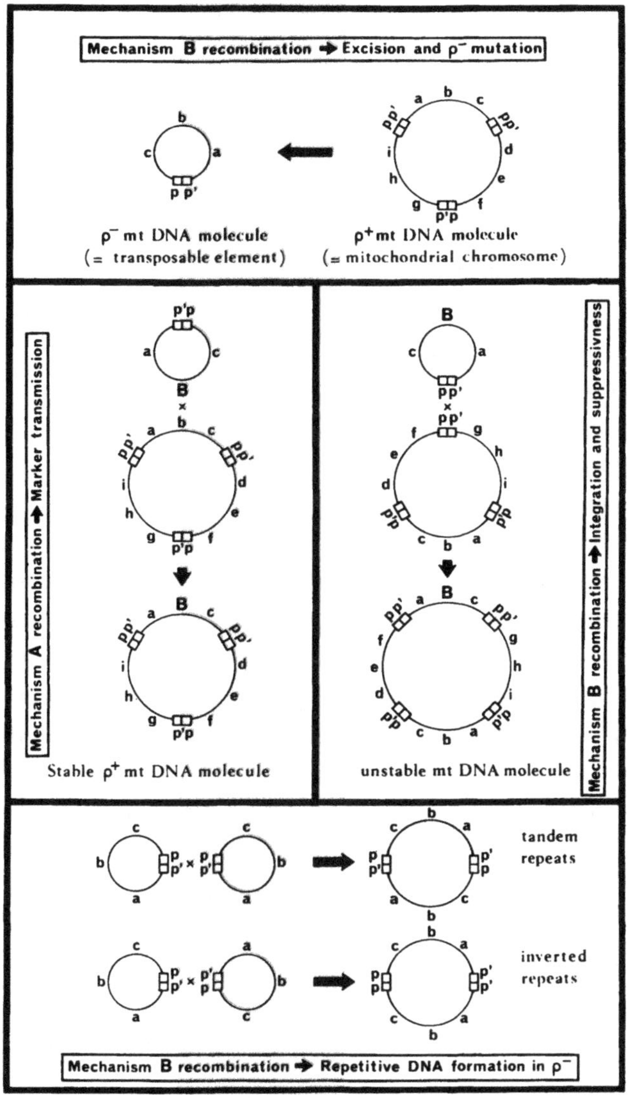

Fig. 1. General scheme for production of ρ^- molecules, (upper panel) for marker transmission and suppressivness in ρ^+ by ρ^- crosses (middle panels) and for repetitive mtDNA formation (lower panel)

An allele from the ρ^- molecule can be integrated into the ρ^+ one in a non destructive manner. The outcoming ρ^+ molecule remains as intact and stable after the recombinational event as it was before it. This molecule, when segretated out into a pure cell line after successive cell divisions, will give rise to a stable ρ^+clone with the mitochondrial marker transmitted from the petite. However, a ρ^- molecule can also recombine with a ρ^+ one by pathway B. Integration occurs (via probably a reciprocal recombination event) between the specific sites PP' on the ρ^- molecule (i.e. the ends of the repeating unit) and the corresponding specific sites on the ρ^+ molecule. We do not postulate a very precise mechanism for this integration but the most plausible one appears to be analogous to the CAMPBELL (1962) model for λ integration (cf. 38 for a recent review). By virtue of this integration the ρ^+ molecule becomes unstable and has a high probability of undergoing eventually a transformation into a ρ^- molecule. The instability may be due to several reasons : a)imprecise excision of the integrated ρ^- segment with concomittant deletions and/or inversions of adjacent ρ^+ sequences (via mechanism B), the excised segment being able, in a new round of recombination, to integrate another ρ^+ molecule ; b) deletion formation in the vicinity, generated in one direction or another, possibly from the specific sites themselves (via mechanism B); c) sequence rearrangements like deletions and inversions, occuring via mechanisms A, since homologous recombination can now occur between a preexisting segment of the wild type sequence (eg. a b c) and the integrated segment a b c from the petite which is misplaced by site-specific B recombination; d) translocation of the integrated ρ^- segment into other sites on the mitochondrial chromosome. The occurence of all these phenomena is well documented in the case of transposable elements in bacteria, like IS sequences, drug resistance elements and bacteriophages (cf. 5, 17, 32, 38).

In yeast mitochondria, we are dealing with populations of hundreds of molecules per cell, multiple rounds of pairing and recombination events and continuous sorting out (cf.9). Depending on the respective intensity of the recombination pathways A and B the final outcome of a cross will balance between neutrality with a high "marker transmission" and "suppressivness" with little or no transmission at all. The unstable cellular state of EPHRUSSI et al. (13) will correspond to heteroplasmic cells with a pool of unstable molecules in a dynamic equilibrium between antinomic A and B recombination events. The environmental effects will favor one pathway or another (eg. sodium acetate favors the A pathway). Finally, different hereditary degrees of suppressivness are easily explained by the very nature of terminal sequences of ρ^- molecules. Two parameters should be considered. Firstly, a ρ^- molecule with repetition termini P_1 P_1', should be suppressive if the ρ^+ molecule has numerous specific sites corresponding to the P_1 P_1' ends, since the probability of integration leading to the unstable molecule would be high. Conversly, a ρ^- molecule with repetition ends P_2 P_2' which are rare in the ρ^+ molecule should be less suppressive. Secondly, the influence of the location of integration sites on the expression of mitochondrial genes necessary for the activity of recombination A mechanism should be considered. Since this mechanism counterbalances the suppressive pathway B an integration with neighbouring formation of deletions which impair mechanism A (eg. by destroying genes involved in the mitochondrial protein synthesis) may be more suppressive prone than integration located far from these genes.

2. Spontaneous and Induced Formation of Petites.

Both mechanisms A and B are present in the wild type cell and mtDNA molecules undergo recombination events by either one of these two pathways. Our model explains the ρ^+ to ρ^- mutation in the following way (Fig. 1). At any moment a segment of the wild type sequence can be excised with the formation of two molecules, one with a deletion and another which behaves now like a transposable element. The excision occurs through mechanism B by a reversal of the process envisaged for the integration of a ρ^- molecule into the ρ^+ one. The excised segment acting as a transposable element can in turn infect another ρ^+ molecule in a manner identical to the one discussed in the preceding section. Thus, as pointed out previously (6), suppressivness in $\rho^+ \times \rho^-$ crosses and ρ^+ to ρ^- mutations are two facets of the same basic process. In the first facet the transposable element is added externally, while in the second one it is created internally.

Several observations can be readily explained :

a) the fact that all ρ^+ molecules can be destroyed by hitting a single target in a cell which contains 50 to 100 molecules (6); this is to be expected, if the first transposable element created is highly suppressive for the reasons discussed previously, the B mechanism is very active, the A mechanism is inactive and many cycles of excision-integration take place.

b) the fact that a population of many different ρ^- molecules is generated from a single ρ^+ cell (6, 14, 35); this is to be expected, if the cycle excision-integration occurs at different sites in different molecules and creates various deletions in the ρ^+ molecules.

c) the observation that the average size of the extracted mtDNA diminishes temporarily as a consequence of the induction (16, 23);

d) the constant experience of every laboratory that various, non-isogenic ρ^+ strains greatly differ in the spontaneous rate of the ρ^- production and that numerous nuclear genes controlling radiation sensitivity (cf. 23, 29) influence it; this is to be expected if the recombination pathways A and B counteract each other, so that conditions and/or mutations which favor one in respect to another modify eventually the overall ρ^+ vs ρ^- balance.

e) among the dozens of diverse agents which induce ρ^- mutants (see 23 for review) two classes are of particular interest, those which inhibit the expression of mitochondrial genes and therefore diminish the pathway A and those which are known to provoke excision of prophages and would favor the pathway B. Erythromycin (39) is an example of the first class while U.V. is an example of the second class (31). It is possible that some agents like ethidium bromide owe their particulary high ρ^+ to ρ^- mutagenecity (36) to the fact that they simultaneously stimulate the excision process by covalently binding to mtDNA (24) and activating an enzyme of the B pathway (30)as well as by inhibiting the mitochondrial transcription and translation necessary for the pathway A (23).

3. Formation of Hybrid Molecules by Recombination between Different Petites and Repetitive Structure of their MtDNA.

It was shown previously that genetic and molecular recombination occurs between different petites. The recombinant petite clones did not contain a mixture of the two parental DNA's but a new molecular species of mtDNA (26). Organization of such hybrid molecules constitutes the first direct evidence in favor of our model (20, 25). We shall therefore describe these molecules in some details. The full account of the analysis will be

published elsewhere (20).
MtDNA's of parental and recombinant clones were studied first by DNA-DNA hybridization, high resolution melting and reassociation kinetics (25). The organization of molecules was further investigated by electron microscopy analysis of circle lengths, partial denaturation maps and self-annealing experiments (20). Recombinant mtDNA's display a highly ordered arrangement of repeats in which two inverted repeats of one parent follow two inverted repeats of the other parent. The overall arrangement is a palindrome. Of particular importance is the fact that parental mtDNA sequences practically do not overlap in the recombined molecule. The sequence common to the two parents is at most a length of 0.05 µm and possibly much less. This is in excellent agreement with DNA-DNA filter hybridization results which have shown that <u>ca</u> 1% (some 20 to 30 base pairs only) of one parental sequence are homologous to the other parental sequence. Thus recombination between different mtDNA sequences consists of an end to end covalent joining of parental segments and is not due to an extensive sequence homology. If, as postulated in our model, one parental repeat is $\overline{P_1 a\ b\ c\ P_1}$ and another is $\overline{P_1\ d\ e\ f\ P_1}$ a recombination event, by mechanism B (the only one occuring in petites), would lead precisely to an additive molecule $\overline{P_1\ a\ b\ c\ P_1\ P_1\ d\ e\ f\ P_1}$.
Site specific recombination explains the formation of multimeric series of circular molecules frequently observed in petites (14, 18, 19, 20, 21). Consider two monomeric circles with identical sequences (Fig. 1). An intermolecular recombination event between the junctions PP' would form a circular dimer, a dimer with a monomer would form a trimer etc ... An intramolecular recombination event occuring in a dimer could reform two monomeric circles etc ... Thus, a population of ρ^- molecules would be in a dynamic equilibrium through continuous site specific inter- and intra-molecular recombinations occuring at the junctions. Frequency distribution of the lengths of circular molecules is consistent with this mechanism (19). Furthermore, some ρ^- mutants display multiple multimeric series of different circle circumference and the longest fundamental circle length appears as to be equal to the sum of the two shorter circle lengths (19). In conclusion, repetitive DNA formation i.e. reiteration of the same sequence and

hybrid DNA formation i.e. recombination of different sequences would result from the same process based upon exchanges at the junctions.
There is very little information on the nature of the sites involved in mechanism B recombination. In recombinant petites studied until now the sites have a two-fold axis of symmetry (20). This may be of importance because inverted repeats are frequently found in transposable elements in bacteria (cf. 5, 17). However there exist petites in which the repeats display a direct tandem arrangement rather than an inverted repeat arrangement (1, 14, 18, 19, 21, 22). It is possible, nevertheless, that even in those petites a local, short palindrome is present at the repeat junctions.
Evidence exists that not all junction sequences are identical : a) were they identical, a mtDNA arranged only in palindromic repetition should not exist because recombinations between "head to head" and "tail to tail" junctions would have lead to both straight and inverted repeat organization; this is not the case (18, 19, 20, 22). b) there exist combinations of petites where no genetically detectable recombinants are found (27).
Conversly, evidence exists that recombination between petites carrying different segments of wild type mtDNA does occur, which suggest that such petites may have the same junction sequences. A case in point is ρ^- clone KL14-4A/A311 which carries the P^R marker and has mtDNA composed of very short repeats 0.3 μm long (1, 19). This clone gives genetic recombinants with 8 different petites carrying the C^R marker (27) which are of different repeat length. This means that a true wild type sequence could not have been reconstitued in all the eight combinations studied, which in turn implies that numerous identical,or similar recombination sites are present in the wild type sequence. In our opinion, analysis of the recombination matrix between petites will provide information on the similarity of specific sites rather than on the overall sequence homology.

III - PREDICTIONS OF THE MODEL -

The model presented here makes a number of predictions. We shall cite but a few.

1. Mutations located in the nuclear DNA should exist which by impairing mechanism B would lead to the absence of repetitive sequences in ρ^- mtDNA and to the absence of recombination in ρ^- by ρ^- crosses, but would

not impair marker transmission in ρ^- by ρ^+ crosses. Conversly, one would expect mutations to be located in mitochondrial DNA which by impairing mechanism A would prevent the marker transmission, but would not impair the suppressivness in ρ^+ by ρ^- crosses. For approaches to this problem cf(7).

2. Appropriate treatment with inhibitors of mitochondrial protein synthesis (eg. chloramphenicol, erythromycin) should be able to prevent the homologous recombination of mitochondrial genomes.
3. Petite clones which retain long sequences of homologous mtDNA but possess different junction termini should be unable to recombine among themselves because of the absence of mechanism A and the non fulfillment of conditions for mechanism B. Such petites should nevertheless be able to recombine their mtDNA in ρ^- by ρ^+ crosses. Conversly, all petites which recombine among themselves should present identical, or similar sequence at the ends of their repetitive segments.
4. Finally, it will be of interest to see whether mitochondrial DNA segments of petites could behave as translocable genetic elements in bacteria. Genetic engineering technology renders this question testable. A positive answer would be of importance for the understanding of the evolution of recombination mechanism in prokaryotes and eukaryotes as well as for the evolution of organelles in general.

The only strong experimental evidence in favor of the model presented here comes from the direct study of recombined molecules (20, 25). Other evidence should be regarder rather as circumstantial or even conjectural. Furthermore, we have presented our model in a purposely schematic manner in order to sift out the main ideas from a considerable body of experimental findings. We are aware that amendements and readjustements may be necessary. It is our feeling, however, that the analogy between transposable elements in bacteria and mitochondrial DNA in yeast petite mutants is not a trivial one.

Acknowledgments.

The authors take pleasure in thanking Dr. B. DUJON for stimulating discussions. This work was been supported by a D.G.R.S.T. Grant N° 75.70748.

References.

1. "The Genetic Function of Mitochondrial DNA". Eds. C. SACCONE and A.M. KROON. (1976). Elsevier/North Holland Biomedical Press, Amsterdam, pp. 1-354.
2. "Genetic and Biogenesis of Chloroplasts and Mitochondria". Eds. Th. BUCHER et al.(1976). Elsevier/North Holland Biomedical Press, Amsterdam, pp. 1-895.
3. CAMPBELL, A.M. (1962) Adv. Genet. 11, 101-145.
4. COEN, D., DEUTSCH, J., NETTER, P., PETROCHILO, E., SLONIMSKI, P.P. (1970) Symposia of the Society for Experimental Biology, N° 24, Cambridge University Press, pp. 449-496.
5. COHEN , S.N. (1976) Nature, 263, 731-738.
6. DEUTSCH, J., DUJON, B., NETTER, P., PETROCHILO, E., SLONIMSKI, P.P., BOLOTIN-FUKUHARA, M., COEN, D. (1974) Genetics 76, 195-219.
7. DUJON,B., BARANOWSKA, H., DUJARDIN, G. (1977). This symposium.
8. DUJON, B., MICHEL, F. (1976) in "The Genetic Function of Mitochondrial DNA". Eds. C. SACCONE and A.M. KROON.Elsevier/North Holland Biomedical Press, Amsterdam. pp. 175-184.
9. DUJON, B., SLONIMSKI, P.P., WEILL, L. (1974) Genetics, 78, 415-437.
10. EPHRUSSI, B. (1949) in "Unités Biologiques douées de Continuité Génétique". C.N.R.S. Paris, pp. 165-180.
11. EPHRUSSI, B., GRANDCHAMP, S. (1965) Heredity, 20, 1-7.
12. EPHRUSSI, B., HOTTINGUER, H., ROMAN, H. (1955) Proc. Nat. Acad. Sci. USA, 41, 1065-1071.
13. EPHRUSSI, B., JAKOB, H., GRANDCHAMP, S. (1966) Genetics, 54, 1-29.
14. FAYE, G., FUKUHARA, H., GRANDCHAMP, C., LAZOWSKA, J., MICHEL, F., CASEY, J., GETZ, G.S., LOCKER, J., RABINOWITZ, M., BOLOTIN-FUKUHARA, M., COEN, D., DEUTSCH, J., DUJON, B., NETTER, P., SLONIMSKI, P.P. (1973) Biochimie, 55, 779-792.
15. GILLHAM, N. (1974). Ann. Rev. Genetics, 8, 347-393.
16. GOLDRING, E.S., GROSSMAN, L.I., KRUPNICK D., CRYER, D.R., MARMUR, J. (1970). J. Mol. Biol. 52, 323-335.

17. KLECKNER N., (1977) Cell. 11, 11-23.
18. LAZOWSKA, J., JACQ, C., CEBRAT, S., SLONIMSKI, P.P. (1976) in "The Genetic Function of Mitochondria DNA, eds. C. SACCONE and A.M. KROON. Elsevier/North Holland Biomedical Press, Amsterdam, pp. 325-336.
19. LAZOWSKA, J., SLONIMSKI, P.P. (1976) Mol. Gen. Genet. 146, 61-78.
20. LAZOWSKA, J., SLONIMSKI, P.P. Mole. Gen. Genet. submitted.
21. LOCKER, J., RABINOWITZ, M., GETZ, G.S. (1974). J. Mol. Biol. 88, 489-507
22. LOCKER, J., RABINOWITZ, M., GETZ, G.S. (1974) Proc. Nat. Acad. Sci. USA, 71, 1366-1370.
23. MAHLER, H.R. (1973) Critic. Rev. Biochem., 1, 381-400.
24. MAHLER, H.R., BASTOS, R.N. (1974) F.E.B.S. Letters, 39, 27-34.
25. MICHAELIS, G., MICHEL, F., LAZOWSKA, J., SLONIMSKI, P.P. (1976). Mol. Gen. Genet. 149, 125-130.
26. MICHAELIS, G., PETROCHILO, E., SLONIMSKI, P.P. (1973)Mol. Gen. Genet. 123, 51-65.
27. MICHAELIS, G., PRATJE, E., DUJON, B., WEILL, L. (1976) in "Genetics, Biogenesis and Bioenergetics of Mitochondria". Eds W. BANDLOW et al. Walter de Gruyter, Berlin, New-York, pp. 49-56.
28. MOUNOLOU, J.C., JAKOB, H., SLONIMSKI, P.P. (1966) Biochem. Biophys. Res. Commun., 24, 218-224.
29. MOUSTACCHI, E., WATERS, R. , HENDE, M., CHANET, R. (1974). Proc. of Vth International Congress on Radiation Research (Seattle) Radiation Res.
30. PAOLETTI, C., COUDERC, H., GUERINEAU, M. (1972) Biochim. Biophys. Res. Com. 48, 950-958.
31. RAUT, C., SIMPSON, W.L. (1955) Arch. Biochem. Biophys. 57, 218-228.
32. REANNEY, D. (1976) Bacterial. Rev. 40, 552-590.
33. SAGER, R. (1972). Cytoplasmic Genes and Organelles. Academic Press Inc. New-York and London.
34. SAUNDERS, G.W., GINGOLD, F.B., TREMBATH, M.K., LUKINS, H.B., LINNANE, A.W. (1971) in "Autonomy and Biogenesis of Mitochondria and Chloroplasts" North Holland Publish. Co pp. 185-193.
35. SCHWEYEN, R.J., STEYRER, U., KAUDEWITZ, F., DUJON, B., SLONIMSKI, P.P. (1976) Molec. Gen. Genet. 146, 117-132.
36. SLONIMSKI, P.P., PERRODIN, G., CROFT, J.H.,(1968). Biochem. Biophys. Res. Commun. 30, 232-239.

37. SLONIMSKI, P.P., TZAGOLOFF, A. (1976). Eur. J. Biochem. 61, 27-41.
38. WEISBERG, R.A., GOTTESMAN, S., GOTTESMAN, M.E. (1977) Comprehensive Virology, in Press.
39. WILLIAMSON, D.H., MAROUDAS, N.G., WILKIE, D. (1971) Mol. Gen. Genet. 111, 209-223.

CONTROL OF THE GENETIC RECOMBINATION OF MITOCHONDRIAL GENES : PRELIMINARY APPROACHES.

B. DUJON[(+)], H. BARANOWSKA[(*)], G. DUJARDIN[(+)]
(+) Centre de Génétique Moléculaire du C.N.R.S., 91190 - Gif-sur-Yvette - FRANCE and
(*) Institute of Biochemistry and Biophysics Polish Academy of Sciences, 02532 Warszawa - POLAND

INTRODUCTION

Genetic recombination between mitochondrial genetic markers in crosses of S. cerevisiae is now well demonstrated and has been documented by many studies (1 to 18). In particular it has been shown that genetic recombination occurs very actively in every $\rho^+ \times \rho^+$ cross studied so far and that multiple events of recombination take place between the different mit-DNA molecules of the same yeast cell (10, 11, 14, 18). Similarly genetic recombination between wild type mt-DNA molecules and molecules composed of regular repeats of a small part of the wild type sequence (in ρ^- cells) is regarded as responsible for the efficient integration of the markers of the ρ^- genome into the ρ^+ genome (19, 20, 21). Finally, in crosses between two given ρ^- strains, very efficient genetic recombination between different repeated molecules has been clearly demonstrated as leading to the physical integration of the two different repeating units (7, 23, 24, 25). However, in crosses between other ρ^- mutants such a process, tested by the appearance in the progeny of recombinant ρ^- clones carrying the genetic markers of both parents, could not be detected, suggesting thereby a certain specificity of the phenomenon (7, 25).

Although genetic recombination of mitochondrial genes has been extensively studied, little is known until now about the mechanisms and the enzymes involved in, or the control of, these processes. In $\rho^+ x \rho^+$ crosses, however, the phenomenon of polarity of recombination determined by the ω locus (3,16) suggests the existence of non reciprocal events occurring by correction or conversion of mismatched regions in relatively long heteroduplexes (10, 11). On the contrary, in $\rho^- x \rho^-$ crosses, the absence of long stretches of sequence homology between the mtDNA of the two parents rather suggests a mechanism of integration without long heteroduplex formation (23, 24, 26).

It is obvious that considerable progress in the understanding of the mechanisms involved should result from the finding of mutations of S. cerevisiae (either nuclear or mitochondrial ones) affecting genetic recombination of mitochondrial genes. However, the direct screening of such types of mutants involves crosses with tester strains and, thus, only dominant mutations would be detectable although the majority of such mutations are expected to be recessive ones. We have therefore recently investigated this problem by the following approaches :

1. A systematic analysis of the effects of nuclear mutations conferring UV and/or X ray sensitivity (rad mutations) on the recombination of mitochondrial genes.
2. The selection of mutants affected in the process of $\rho^+ \rightarrow \rho^-$ mutation induced by ethidium bromide and the analysis of recombination frequencies in crosses with such mutants.

The aim of this article is to summarize briefly the present status of these two approaches and to discuss their significance for the mechanisms of genetic recombination of mitochondrial genes.

II. EFFECTS OF RAD MUTATIONS ON THE RECOMBINATION OF MITOCHONDRIAL GENETIC MARKERS.

A series of mutations conferring UV and/or X ray sensitivity has been described (27 to 31). They have been shown to belong to several non allelic nuclear genes (32) which are involved in different pathways of repair or genetic recombination of nuclear DNA (see for review 33). Effects on mitochondrial recombination of X ray sensitive mutations in the loci rad50, rad51, rad52, rad53, rad54 and rad56 have been systematically studied in $\rho^+ x \rho^+$ crosses. For the purpose of this study a series of isomitochondrial strains, all carrying the same mitochondrial markers (C^R_{321}, E^R_{221} and O^R_1), and each a different rad^- mutation has been constructed by crosses and meiosis as will be explained elsewhere (Baranowska et al. in prep.). For all rad mutations homosexual ($\omega^+ x \omega^+$) and heterosexual ($\omega^+ x \omega^-$) three point crosses have been performed. In each case the results of two isomitochondrial crosses (at least) were compared : crosses homozygous for the rad mutation on the one hand and crosses heterozygous for the rad mutation on the other hand. In the first case (homozygous crosses) the two parents carry the same rad^- mutation and all diploid cells formed in the progeny are homozygous (rad^-/rad^-). In the second case (heterozygous crosses) only one of the parents has the rad^- mutation, the other one being a rad^+ tester strain, and all diploid cells formed in the progeny are heterozygous (rad^-/rad^+).

The figure 1 which, for each cross, represents the frequency of recombinants as a function of input, summarizes the main features observed in this series of crosses (the complete results will be presented elsewhere : Baranowska et al., in prep.). In the case of homosexual crosses (fig. 1a,1c), within the scattering of experimental data (due mainly to the sampling error on the colony-score), the frequencies of recombinants are not distinguishable between homozygous and heterozygous crosses for all the rad mutants studied. Therefore these rad mutations do not affect the recombination of mitochondrial genes in non polar crosses. In the case of heterosexual crosses (fig. 1b, 1d) it can be seen that homozygous and heterozygous crosses involving the mutations in loci rad50, rad51, rad52, rad53 and rad54 form a unique population of crosses in which the frequencies of recombinants vary as a function of input as predicted from the model

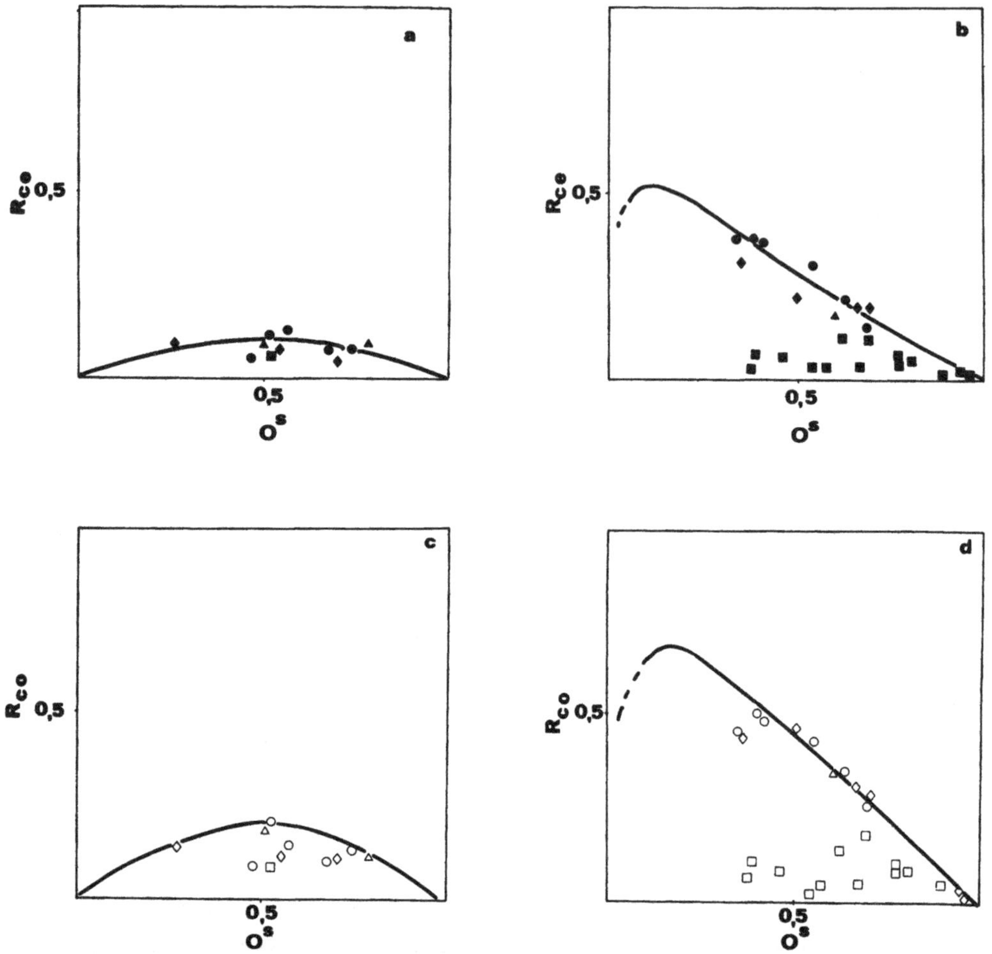

Fig. 1 : This figure represents for each cross the frequencies of recombinants between pairs of the markers $C^R{}_{321}$, $E^R{}_{221}$, $O^R{}_1$ (ordinates) as a function of the input (determined by the transmission of the O^s allele) (abscissae). Each point represents the results of a single cross (data taken from Baranowska et al., in prep.) : ◇ ; ○ . Respectively homozygous and heterozygous crosses for the mutations in rad50, rad51, rad52, rad53, and rad54 loci ; □ ; △ respectively homozygous and heterozygous crosses for the rad56 mutation .Full and open symbols refer respectively to recombinant between C and E or C and O. Fig. 1a and 1c homosexual crosses. Fig. 1b and 1d heterosexual crosses. The lines are drawn according to the predictions of the model for normal crosses (10, 14).

for normal crosses (10).Therefore, mutations at these 5 rad loci do not affect the recombination of mitochondrial genes in polar crosses. On the contrary, if one considers the rad56 mutation it can be seen on fig. 1b and 1d that for a similar input the frequency of recombinants is significantly reduced (about 2 to 3 times) in the homozygous crosses as compared to that of the heterozygous one where the recessive rad mutation was not expressed (the latter is similar to the values expected from the model). Several homozygous crosses between different strains, all carrying the rad56 mutation but different nuclear genetic background, have been performed. Furthermore, homozygous crosses have been made in which one of the parents was treated by hydroxyurea prior to mating in order to change the input (method developped by Dujardin et al., in prep.). In all cases, frequencies of recombinants are significantly lower than expected. Nevertheless, the degree of polarity of recombination does not seem greatly modified (as far as can be judged from such a low frequency of recombinants). In conclusion, among the 6 rad mutations studied, a single one (rad56) had a lowering effect on the frequency of recombinants. As expected for a recessive mutation only the homozygous combination was efficient. In all heterosexual crosses performed so far with the rad56 mutation, frequencies of recombinants are always significantly diminished but recombination is not completely abolished. In the homosexual crosses the frequencies of recombinants do not seem greatly affected but, due to the relatively important sampling error, an effect of a factor of about 2 cannot be easily excluded.

III. SELECTION OF MUTATIONS AFFECTING THE PROCESS OF ρ^- FORMATION AND THEIR EFFECTS OF THE RECOMBINATION OF MITOCHONDRIAL GENETIC MARKERS.

We have undertaken a systematic research for mutants affected in the process of $\rho^+ \rightarrow \rho^-$ mutation, taking advantage of the fact that some drugs (e.g. ethidium bromide see 34) induce this mutation with very high efficiencies (close to 100% of ρ^- mutant per generation). Ethidium bromide therefore offers a screening procedure for mutants which remain ρ^+ under conditions where normal strains are entirely converted to ρ^- cells. After EMS[(+)] mutagenesis a series (more than 30) of independent mutations (called ebi^-) conferring a high degree of resistance to the induction of the $\rho^+ \rightarrow \rho^-$ mutation by EB[(+)] has been selected (Dujardin et al., in prep.).

The production of ρ^- cells (either spontaneous or EB induced) has been determined for given number of generation in glucose medium respectively with or without ethidium bromide. Results are given by Table 1.

It can be seen that without EB (column 1) for some of the ebi^- mutants the frequency of ρ^- cells is similar to that of the wild type. In the same medium for other ebi^- mutants a considerable accumulation of ρ^- cells is found suggesting that the spontaneous $\rho^+ \rightarrow \rho^-$ mutation rate is affected in these mutants. It can also be seen that with EB (column 2) all ebi^- mutants are highly insensitive to the EB induced $\rho^+ \rightarrow \rho^-$ mutagenesis since a relatively high frequency of ρ^+ cells remains after several cellular generations.

For all ebi^- mutations studied so far, tetrad analysis or random spore analysis indicate a nuclear genetic determinism (data not shown). In order to examine the effects of the ebi^- on mitochondrial recombination in $\rho^+ x \rho^+$ crosses a series of isomitochondrial strains all carrying the same mitochondrial markers (C^R_{321}, O^R_1, P^R_{454}) and each a different ebi^- mutation, has been constructed by crosses and meiosis, as will be explained elsewhere (Dujardin et al., in prep.). Preliminary results (data not shown) indicate that ebi^- mutations both in homozygous or heterozygous crosses do not reduce the frequency of recombinants in homosexual $\rho^+ x \rho^+$ crosses. Their effect on heterosexual crosses have not yet been examined. Some peculiarities of recombination (bias in the frequency of

---

(+) EMS : ethyl methane sulfonate ; EB : ethidium bromide.

recombinants) have been found in some homosexual crosses but preliminary evidence suggests that they may not necessarily be associated with the ebi⁻ mutations.

TABLE 1 : SPONTANEOUS AND ETHIDIUM BROMIDE-INDUCED PRODUCTION OF ρ^- CELLS IN CULTURES OF EBI⁻ MUTANTS.

		COLUMN 1		COLUMN 2	
		Nb of generations in NO medium	% of ρ^+ cells	Nb of generations in NOEB medium	% of ρ^+ cells
Wild type	IL216-1B	7.4	98.1	8.0	<0.01
ebi⁻ mutants	---/08	7.6	96.4	8.1	26.8
	---/011	7.3	20.0	8.0	14.0
	---/N20	7.1	56.3	7.9	9.2

Cells of the wild type strain (IL216-1B α his_1 ura_1 $\omega^+ C^S E^R_{221} O^S$) and of the ebi⁻ mutants derived from it have been cultured in N3 medium (Yeast Extract 1 %, Bactopeptone 1 % ; glycerol 2 %, Sörensen buffer pH = 6.2 ; 0.05 M. Aliquots of these cultures have been inoculated in 10 ml of NO medium (same as N3 except 2 % of glucose instead of glycerol) and 10 ml of NOEB medium (same as NO plus 20 µg/ml of EB) and grown at 28° C with aeration. Growth of the cultures were determined by Klett Summerson spectrophotometer. After several cell generations (as indicated) aliquots were withdrawn, diluted and plated out onto N2 medium (same as N3 plus 0.1 % glucose) and incubated at 28° C for 5 to 6 days. Colonies are then scored as ρ^+ or ρ^-.

DISCUSSION

The systematic analysis of mitochondrial crosses performed with strains carrying X ray sensitive mutations clearly demonstrates that for five of the rad loci considered no effect on mitochondrial recombination is observed, indicating that the genes concerned are not involved in the pathways of mitochondrial recombination. However for one of these

(rad56), a systematic correlation between the presence of the mutant allele in both parents and a significant decrease of the frequencies of recombinants is established in heterosexual crosses. Nevertheless, even in this case recombination is not completely abolished.
It has been shown previously (10, 14, 18) that the frequency of recombinants in a mitochondrial cross is the final results of reiteration of multiple recombination events between the mtDNA molecules and thus a function of several parameters (input, average number of mating rounds and probabilities of genetic exchanges). Therefore many hypotheses can account for a reduction of the frequency of recombinants in mitochondrial crosses. Among these, two extreme classes can be distinguished : 1. The efficiency of the mechanism of recombination itself is reduced whereas all other biological parameters remain unchanged ; 2. On the contrary, other processes than the recombination itself are modified, the mechanism of recombination being unaffected. Experiments are in progress to determine whether the reduced frequencies of recombinants observed with rad56 mutation actually reflect an effect (direct or indirect) of the rad56 locus on the mechanism of mitochondrial recombination or on other biological processes (such as reduction of the average number of mating rounds; incomplete panmixie leading preferentially to the segregation of pure parental types during budding ; production of ρ^- mutants, etc ...). If the first type of hypothesis is correct this finding may offer the first opportunity to analyze the pathways and the control of mitochondrial recombination.
The systematic research of mutants affected in the process of $\rho^+ \rightarrow \rho^-$ mutation induced by EB has led to stable mutants (ebi^-) almost entirely insensitive to the EB induction of ρ^-, even under conditions of active multiplication. Several hypotheses as to the site of action of these mutations can obviously be proposed : modification of permeability to the drug, modification of the fixation of EB on mit DNA,

diminution of the transcription blocking effect, etc ... and finally alteration of the process of formation of ρ^- mitDNA molecules itself. Now the ebi^- mutants (in contrast to the B^R mutants of Gouhier et al. 35, 36) are unable to grow on an EB medium containing glycerol as carbon source. This argues against the first 3 above mentioned hypotheses and rather suggests an alteration in the sequence of events leading from wild type mtDNA molecules to the creation of massive deletions and to the formation of regularly repeated sequences. Although these mechanisms are still not fully understood two main steps are generally considered : a first one reversible and a second one irreversible (37). The fact that ebi^- mutations are insensitive to EB under growing conditions rather suggests that they are affected in the non reversible step. Preliminary analysis of ebi^- mutants shows that it is possible to affect the processes of $\rho^+ \rightarrow \rho^-$ mutation without affecting the genetic recombination in $\rho^+ x \rho^+$ crosses. This suggests that either the two processes are entirely independent or at least that the mutations concern a part of the pathways not common to both mechanisms.

As already mentioned in the introduction the comparison of the genetic evidences obtained mainly from $\rho^+ x \rho^+$ crosses (10) with the molecular evidences obtained in analyzing the mtDNA of ρ^- recombinants in $\rho^- x \rho^-$ crosses (7, 24) suggests that two different mechanisms of recombination between mitDNA molecules might exist. The first one, which would operate predominantly in $\rho^+ x \rho^+$ crosses, leads to a low maximum of recombinants (18). It is supposed to be involved in the polarity of recombination determined by the ω locus, suggesting therefore the formation and correction of long heteroduplexes (10). No mutants of this mechanisms are known as yet although the rad56 mutation might interfere with it. The second one, which is known in $\rho^- x \rho^-$ crosses, creates recombinants by integration of two different sequences with no or very little sequence homology such that no long heteroduplex can be formed. By opposition to the first one this mechanism is apparently able to create recombinants with extremely high frequency (7, 25). It is tempting to hypothesize as it has been developped in (26) that the integration of two different sequences in $\rho^- x \rho^-$ crosses and the formation of ρ^- deletions and repetitions during $\rho^+ \rightarrow \rho^-$ mutagenesis occurs by the same basic mechanism operating in opposite directions. If so it is conceivable that some of the ebi$^-$ mutants might be involved in this second mechanism .

Acknowledgments

The authors take pleasure in thanking Pr. P. Slonimski and Dr. A. Putrament for fruitful discussions, Dr. R. Mortimer for the generous gift of rad$^-$ mutants, Miss. S. Robineau for her skillful technical assistance and Dr. M. Somlo for her critical reading of the manuscript. H.B. gratefully acknowledges the receipt of a fellowship from the Convention d'Echange C.N.R.S. Polish Academy of Sciences ; G.D. is a Boursière de la Ligue Nationale contre le Cancer. This work has been supported by a Grant n° 75.70748 from the D.G.R.S.T.

References

1. THOMAS, D.Y., WILKIE, D. (1968). Biochem.Biophys. Res. commun. 30, 368-372.

2. COEN, D., DEUTSCH, J., NETTER, P., PETROCHILO, E., SLONIMSKI, P.P. (1970). Symposia of the Society for Experimental Biology, n° 24, Cambridge University Press, pp. 449-496.

3. BOLOTIN, M., COEN, D., DEUTSCH, J., DUJON, B., NETTER, P., PETROCHILO, E., SLONIMSKI, P.P. (1971). Bull. Inst. Pasteur, 69, 215-239.

4. KLEESE, R.A., GROTBECK, R.C., SNYDER, J.R. (1972). J. of Bacter. 112, 1023-1025.

5. AVNER, P.R., COEN, D., DUJON, B., SLONIMSKI, P.P. (1973). Mol. Gen. Genet. 125, 9-52.

6. HOWELL, N., TREMBATH, M.K., LINNANE, A.W., LUKINS, H.B. (1973) Mol. Gen. Genet. 122, 37-51.

7. MICHAELIS, G., PETROCHILO, E., SLONIMSKI, P.P. (1973) Mol. Gen. Genet. 123, 51-65.

8. RANK, G.H. (1973) Heredity, 30, 265-271.

9. WOLF, K., DUJON, B., SLONIMSKI, P.P. (1973). Mol. Gen. Genet. 125, 53-90.

10. DUJON, B., SLONIMSKI, P.P., WEILL, L. (1974). Genetics, 78, 415-437.

11. DUJON, B. (1974) "Mechanisms in Recombination" R.P. Grell Ed.; Plenum Publ. Co., pp. 307-316.

12. NETTER, P., PETROCHILO, E., SLONIMSKI, P.P., BOLOTIN-FUKUHARA, M., COEN, D., DEUTSCH, J., DUJON, B. (1974). Genetics, 78, 1063-1100.

13. SUDA, K., UCHIDA, A. (1974). Mol. Gen. Genet. 128, 331-339.

14. DUJON, B., KRUSZEWSKA, A., SLONIMSKI, P.P., BOLOTIN-FUKUHARA, M., COEN, D., DEUTSCH, J., NETTER, P., WEILL, L. (1975). Mol. Gen. Genet. 137, 29-72.

15. GUNGE, N. (1975). Mol. Gen. Genet. 139, 189-202.

16. DUJON, B., BOLOTIN-FUKUHARA, M., COEN, D., DEUTSCH, J., NETTER, P., SLONIMSKI, P.P., WEILL, L. (1976). Mol. Gen. Genet. 143, 131-165.

17. DUJON, B., MICHEL, F. (1976) "The genetic function of mitochondrial DNA". eds. C. SACCONE and A.M. KROON.; Elsevier/North Holland Biomedical Press, Amsterdam, pp. 175-184.

18. DUJON, B., SLONIMSKI, P.P. (1976) "Genetics and Biogenesis of chloroplasts and mitochondria" Eds Th. BUCHER et al.; Elsevier/North Holland Biomedical Press, Amsterdam, pp. 393-403.

19. FAYE, G., FUKUHARA, H., GRANDCHAMP, C., LAZOWSKA, J., MICHEL, F., CASEY, J., GETZ, G., LOCKER, J., RABINOWITZ, M., BOLOTIN-FUKUHARA, M., COEN, D., DEUTSCH, J., DUJON, B. NETTER, P., SLONIMSKI, P.P. (1973). Biochimie, 55, 779-792.

20. DEUTSCH, J., DUJON, B., NETTER, P., PETROCHILO, E., SLONIMSKI, P.P., BOLOTIN-FUKUHARA, M., COEN, D., (1974). Genetics 76, 195-219.

21. BOLOTIN-FUKUHARA, M., FUKUHARA, H. (1976). Proc. Natl. Acad. Sci. U.S.A., 73, 4608-4612.

22. SCHWEYEN, R.J., STEYRER, U., KAUDEWITZ, F., DUJON, B., SLONIMSKI, P.P. (1976). Mol. Gen. Genet. 146, 117-132.

23. LAZOWSKA, J., SLONIMSKI, P.P. (1976). Mol. Gen. Genet. 146, 61-78 and (1977) Mol.Gen.Genet. (submitted)

24. MICHAELIS, G., MICHEL, F., LAZOWSKA, J., SLONIMSKI, P.P. (1976). Mol. Gen. Genet. 149, 125-130.

25. MICHAELIS, G., PRATJE, E., DUJON, B., WEILL, L. (1976). "Genetics and Biogenesis and Bioenergetics of Mitochondria" Eds W. BANDLOW et al. Walter de Gruyter, Berlin, N.Y.,

pp. 49-56.

26. SLONIMSKI, P.P., LAZOWSKA, J. (1977) This volume.

27. HUNNABLE, E.G., COX, B.S. (1971). Mutat. Res. 13, 297-309.

28. GAME, J.C., COX, B.S. (1973). Mutat. Res., 16, 353-362.

29. GAME, J.C., COX, B.S. (1973). Mutat. Res., 20, 35-44.

30. COX, B.S., GAME, J.C. (1974). Mutat. Res., 26, 257-264.

31. HAYNES, R.H. (1975) "Molecular Mechanisms for Repair of DNA" Part B. Ed. P.C. HANAWALT, R.B. SETLOW, New York, Plenum, pp. 529-540.

32. GAME, J.C., MORTIMER, R.K. (1974). Mutat. Res. 24, 281-292.

33. BAKER, B.S., CARPENTER, A.T.C., ESPOSITO, M.S., ESPOSITO, R.E., SANDLER, L. (1976). Ann. Rev. of Genetics, 10, 53-134.

34. SLONIMSKI, P.P., PERRODIN, G., CROFT, J. (1968). Biochem. Res. Commun., 30, 232.

35. GOUHIER, M., MOUNOLOU, J.C. (1973). Mol. Gen. Genet. 122, 149-164.

36. GOUHIER-MONNEROT, M. (1974). Mol. Gen. Genet. 130, 65-79.

37. MAHLER, H.R. (1973) Critic. Rev. Biochem., 1, 381-400.

EXTRACHROMOSOMAL INHERITANCE IN SCHIZOSACCHAROMYCES POMBE V. ON A POSSIBLE CORRELATION BETWEEN BIAS IN TRANSMISSION, SPONTANEOUS MIT$^-$- PRODUCTION, AND FORMATION OF UNIPARENTAL ZYGOTIC CLONES

K. Wolf and F. Kaudewitz

INTRODUCTION

Mitochondrial and chloroplast genes are transmitted to progeny predominantly or entirely by only one parent in a number of different organisms. This uniparental inheritance is generally through the maternal parent in organisms with oogamy, and in such cases may be attributable to the male gamete containing many fewer organelles than the egg, or to failure of the organelles from the male gamete to enter the egg (for review see (1)). In Pelargonium, however, many crosses involving chloroplast mutants produce three classes of zygotes: those transmitting only maternal markers, those transmitting only paternal markers, and those transmitting both to various extents (2). In the alga Chlamydomonas reinhardtii (3,4) chloroplast markers are transmitted principally from one parent, but in this isogamous species the zygote contains the entire chloroplast of both parent gametes. In this case, it is clear that the failure of uniparental zygotes to transmit organelle markers from one parent must be due to the failure of those markers to replicate in the zygote and their loss by dilution or enzymatic degradation, or to the conversion of the markers to the corresponding alleles from the other parent. In Saccharomyces cerevisiae, studies on the inheritance of mitochondrial genes controlling antibiotic resistance have shown that most, and perhaps all, crosses can produce uniparental zygotes, provided that one parent contributes substantially more mitochondrial DNA molecules to the zygote than does the other parent (5,6). In such a biassed cross, some zygotes are uniparental for mitochondrial alleles from the majority parent, and the frequency of such zygotes increases with increasing bias.

In this paper we describe a possible correlation between transmissional bias and uniparental zygote formation, involving three different mitochondrial mutants of the petite-negative fission yeast Schizosaccharomyces pombe: ana^r- 8, ana^r- 14 (resistant to antimycin), and diu^r- 301 (resistant to diuron). In addition we propose a hypothesis which correlates the rate of spontaneous mit^-- formation with transmissional bias and formation of uniparental zygotes.

RESULTS

1) Isolation and genetic characterization of the mitochondrially inherited antibiotic resistant (ant^r) mutants ana^r- 8, ana^r- 14, and diu^r- 301

Mutants ana^r- 8 and ana^r- 14 were derived from strain ade 7 - 50 h^- by nitrosoguanidine mutagenesis (7,8). Diu^r- 301 is a spontaneous mutant of ana^r- 14 which has lost the original antimycin resistance after subcloning on diuron - containing medium. Whereas these mutants give rise to spontaneous respiratory - deficient mutants of the mit^-- type very frequently, their antibiotic - resistance conferring mutational sites seem to be stable (Table 1). By random spore -, tetrad -, mitotic segregation -, and mitotic haploidization analysis the three different mutational sites were found to be located on mitochondrial DNA (8,9)

2) Random spore analysis of crosses between the three antibiotic resistant mutants

To determine, whether these mutants carry different mutational sites, crosses were set up in different pairwise combinations. Diploids of two different antibiotic resistant mutants carrying opposite mating types and complementing auxotrophic markers were grown in minimal medium for 6 generations and then allowed to sporulate. Vegetative cells were killed with 30 % ethanol and spores were either plated directly or allowed to grow for another 10 generations in glucose - complete - medium or glycerol - complete - medium enriched with 0.1 % glucose (0.1 e GLY). In the crosses ana^r- 14 x diu^r- 301 and ana^r- 8 x diu^r- 301, A^{R}D^R clones do not grow on double medium. By resuspension and replating cells of such colonies, clones could be detected which still grew on both antibiotic media, but failed to grow on double medium. This implies that a mixture of two mitochondrial DNA populations is present in the cells. It cannot, however, be decided, if this is due to a medium effect or to very low rates of the double resistant recombinant type. The percentages of double sensitive recombinants were 2.4 % for the cross ana^r- 8 x diu^r- 301, 0.8 % for ana^r- 14 x diu^r- 301, and 0.06 % for the two antimycin - resistant mutants (Table 2)

TABLE 1
Spontaneous revertibility from ant^r to ant^s in three different antibiotic resistant mutants

mutant	ant^r	ant^s	RD	total	
diu^r- 301	13380	0	40	13420	col.
	99.7	0.0	0.3	100.0	%
ana^r- 14	7554	0	36	7590	col.
	99.5	0.0	0.5	100.0	%
ana^r- 8	14575	0	1310	15885	col.
	91.8	0.0	8.2	100.0	%

3) Transmission of the ant^r- phenotypes in standard mitochondrial crosses

A standard mitochondrial cross in Schizosaccharomyces pombe is equivalent to that developed for Saccharomyces cerevisiae except that diploids have to be stabilized by the mutation mei 1, which prevents sporulation. A suspension of zygotes from a cross ant^r x mei 1 is plated onto minimal medium to form a dense lawn of prototrophic diploid clones. This lawn is washed off, diluted and cells plated on minimal medium containing phloxin, which differentially stains diploid colonies. Colonies are then replicated onto the respective anti-

TABLE 2: Random spore analysis of crosses between antibiotic resistant mutants

cross	growth of spores in	A^RD^S	A^RD^R	A^SD^S	A^SD^R	RD	total RC	
ana^r- 8 x	-	282	301+	8	291	14	882	colonies
diu^r- 301		32.0	34.1	0.9	33.0	-	100.0	%
	10 generations in	1887	2294	138	1511	42	5830	colonies
	0.1 e GLY	32.4	39.3	2.4	25.9	-	100.0	%
ana^r- 14 x	-	117	115+	0	163	1	395	colonies
diu^r- 301		29.6	29.1	0.0	41.3	-	100.0	%
	10 generations in	726	764	13	128	1	1631	colonies
	0.1 e GLY	44.5	46.8	0.8	7.9	-	100.0	%
ana^r- 14 x	-	3420	-	1	-	14	3421	colonies
ana^r- 8		99.97	-	0.03	-	-	100.0	%
	10 generations in	9072	-	5	-	14	9077	colonies
	0.1 e GLY	99.94	-	0.06	-	-	100.0	%

+ not growing on medium containing both antimycin and diuron

biotic media. Transmission rates were determined using glycerol- or glucose - grown precultures. Data presented in Table 3 show that transmission rates of ana^r- 14 and ana^r- 8 are significantly higher for glucose - grown precultures, whereas transmission rates of diu^r- 301 are approximately 50 % for either preculture. Growing strain diu^r- 301 under glucose repression and mei 1 under derepressed conditions in glycerol medium does not change transmission rate. This is in marked contrast to Saccharomyces, where transmission of markers can be influenced by glucose repression. Comparing transmission rates for the three mutants precultured in glucose medium, there is an increase in transmission from ana^r- 8 to ana^r- 14 to diu^r- 301. To test if the diploid clones of a standard cross are homoplasmic, resistant and sensitive diploid clones of a standard cross were analyzed by resuspension and replating. While sensitive clones were already pure, about 1 - 3 among 20 resistant diploid clones were still heteroplasmic (Table 4).

4) Mit$^-$- production in the three ant^r- mutants

The 3 mutants were grown for 10 - 12 generations in static cultures containing 10 % glucose and 2 % yeast extract. In late logarithmic phase cultures were plated onto glucose - complete - medium (GLU) and colonies replicated onto 0.1 e GLY. The percentage of respiratory deficient (RD) mutants varied from only 0.2 % in diu^r- 301 to ana^r- 14 with 2.9 % and ana^r- 8 with 20.1 % (Table 5).

5) Identification of uniparental zygotes

Uniparental zygotes can be identified either by analysis of zygotic clones or by aliquot plating. In a zygotic clone analysis diploid colonies issued from single zygotes were picked, resuspended, and several hundred cells plated onto GLU - medium. Clones pure for either the antibiotic - resistant or the antibiotic - sensitive allele could be identified in this way, i. e. clones uniparental for alleles from either parent. However, the sample size is very small and it is not possible to detect alleles transmitted by a zygote to less

TABLE 3: Transmission of ant^r in standard crosses with strain mei 1 - 102 ura 1 - 171 arg 1 - 1

cross	precultures	ant^r	ant^s	total	
x diu^r- 301	3% GLY/3% GLY	3971	4193	8164	col.
		48.6	51.4	100.0	%
	3% GLU/3% GLU	2359	2545	4904	col.
		48.1	51.9	100.0	%
	10% GLU/3% GLY	6061	6189	12250	col.
		49.5	50.5	100.0	%
x ana^r- 14	3% GLY/3% GLY	2565	4775	7340	col.
		34.9	65.1	100.0	%
	3% GLU/3% GLU	2071	2274	4345	col.
		47.7	52.3	100.0	%
x ana^r- 8	3% GLY/3% GLY	221	5569	5790	col.
		3.8	96.2	100.0	%
	3% GLU/3% GLU	795	4889	5684	col.
		14.0	86.0	100.0	%

than 0.5 - 1 % of its progeny. More precisely, the 95 % confidence interval for identification of a uniparental zygote may be estimated by 3.4/N where N is the sample size; the range of N is about 100 - 500, so this method may overlook 0.7 - 3.4 % marker transmission. Consequently, a certain fraction of zygotes scored as uniparental may actually be biparental.

Using the aliquot plating technique, samples of approximately 10^5 cells from each zygotic clone were plated directly onto antibiotic plates. No growth on the antibiotic plate indicates that the zygotic clone was uniparental for the antibiotic - sensitive allele, having transmitted the resistant allele to less than 0.01 to 0.001 % of its progeny. Growth on the antibiotic plate means that the zygotic clone was either biparental or was uniparental but transmitted only the

TABLE 4: Purity of diploid clones issued from standard crosses with strain mei 1 - 102 ura 1 - 171 arg 1 - 1

cross	phenotype of diploid clone	number of diploid secondary clones	% ant^r	% ant^s
x diu^r- 301	resistant	18	100.0	0.0
		1	99.7	0.3
		1	82.6	17.4
	sensitive	18	0.0	100.0
x ana^r- 14	resistant	17	100.0	0.0
		1	99.8	0.2
	sensitive	21	0.0	100.0
x ana^r- 8	resistant	17	100.0	0.0
		1	98.2	1.8
		1	97.3	2.7
		1	88.4	11.6
	sensitive	19	0.0	100.0

Sample size: 400 - 1200 colonies per clone

TABLE 5: Mit^-- production in ant^r- mutants

mutant	RC	RD	total	
diu^r- 301 ade 7 - 50 h^-	2381	5	2386	col.
	99.8	0.2	100.0	%
ana^r- 14 ade 7 - 50 h^-	2698	82	2780	col.
	97.1	2.9	100.0	%
ana^r- 8 ade 7 - 50 h^-	2417	609	3026	col.
	79.9	20.1	100.0	%

antibiotic - resistance allele to its progeny. This technique can thus identify only one of the two possible classes of uniparental zygotic clones; i. e. those uniparental for the sensitive allele. For this class, however, the method provides a much more rigorous criterion of uniparental inheritance than zygote clone analysis. The experimental procedure for both methods is illustrated in Fig. 1.

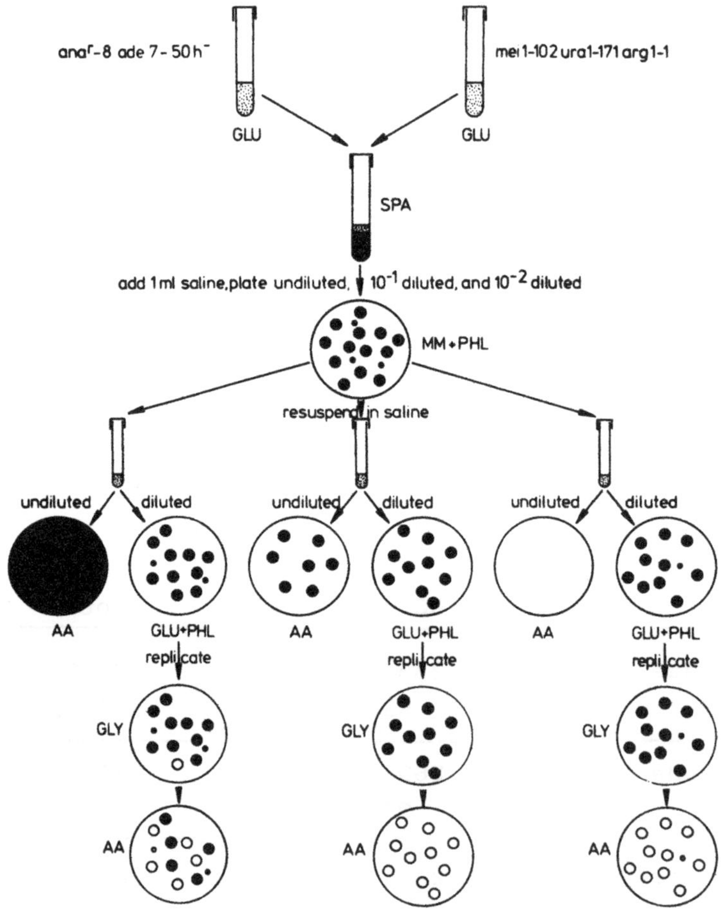

Figure 1: Diagramatic presentation of methods to detect uniparental zygotic clones.
Lower part: aliquot plating (left),
replica plating (right).

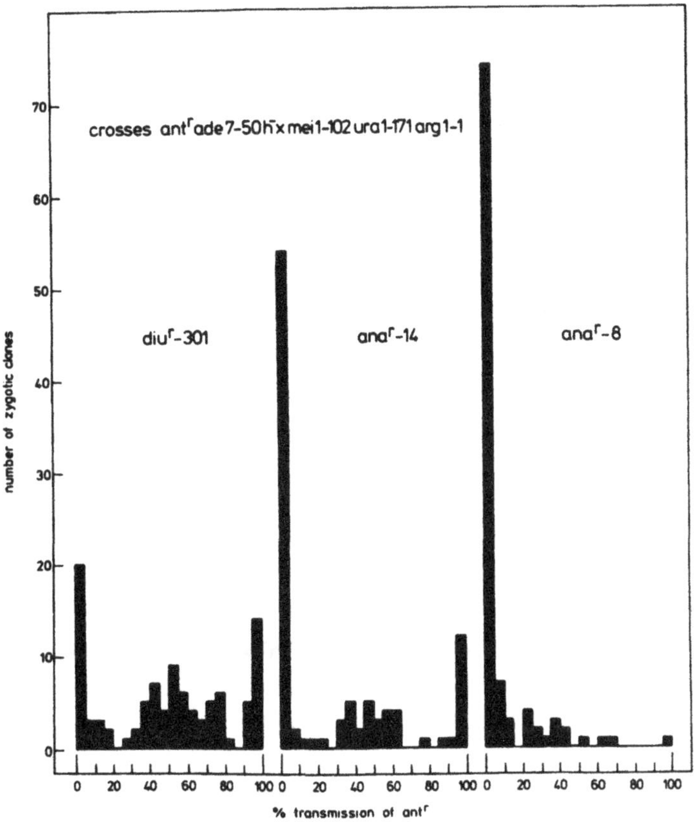

Figure 2: Transmissional patterns of ant^r in crosses ant^r x mei 1, analyzed by the replica - plating technique

Fig. 2 shows the results of an analysis of 100 individual zygotic clones from crosses ant^r x mei 1. The transmission rates are plotted in intervals of 5 %. This representation gives a rough estimate of the distribution of transmissional classes among zygotic clones. A more refined method is provided by the gaussian scale. In Figures 3a, b, and c the percentage of transmission is plotted versus the cumulative percentage of zygotic clones. (This coincides with the actual number, since in all three crosses 100 zygotic clones were analyzed).

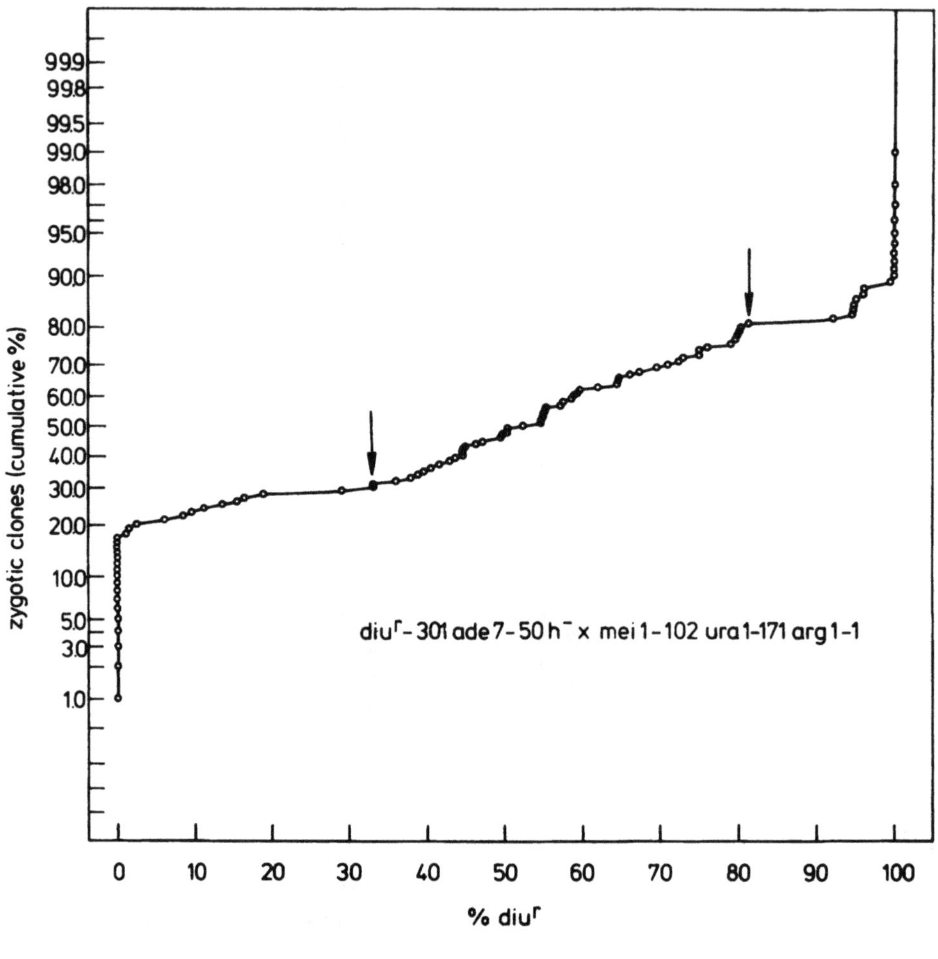

Figures 3a, b, and c: Analysis of transmission of ant^{r} in individual zygotic clones issued from crosses mei 1 x diu^{r}- 301 (a), mei 1 x ana^{r}- 14 (b), and mei 1 x ana^{r}- 8 (c) by use of the cumulative gaussian scale. The range of continuous gaussian distributions is indicated by arrows.

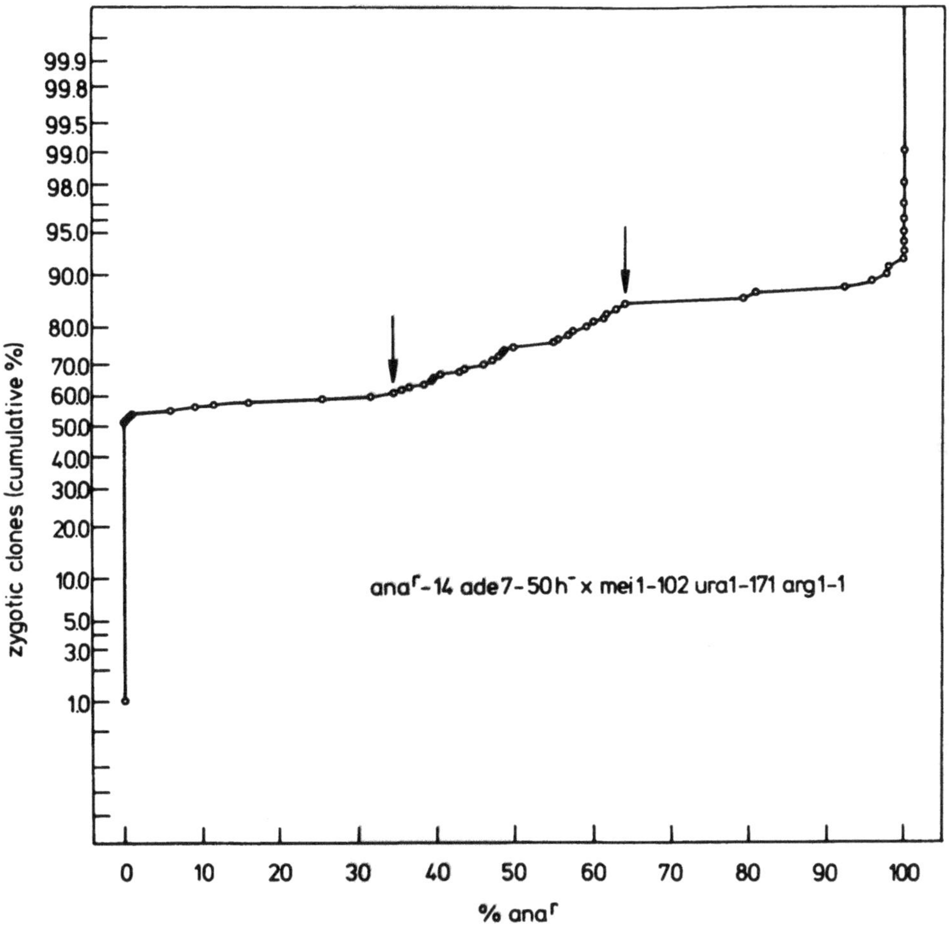

In the case of zygotic clones issued from cross diu^{r}- 301 x mei 1 there is a continuous broad gaussian distribution between 35 % and 80 % transmission. Zygotic clones from crosses ana^{r}- 14 x mei 1 exhibit a shorter stretch of gaussian distribution (from 35 % to 65 %), accompanied by an increase in uniparental, predominantly uniparental sensitive clones. In clones issued from cross ana^{r}- 8 x mei 1 no single continuous gaussian distribution could be detected.

The question arises, if 100 zygotic clones are a representative sample of the total population of zygotic clones found in a standard cross. The agreement of both sets of

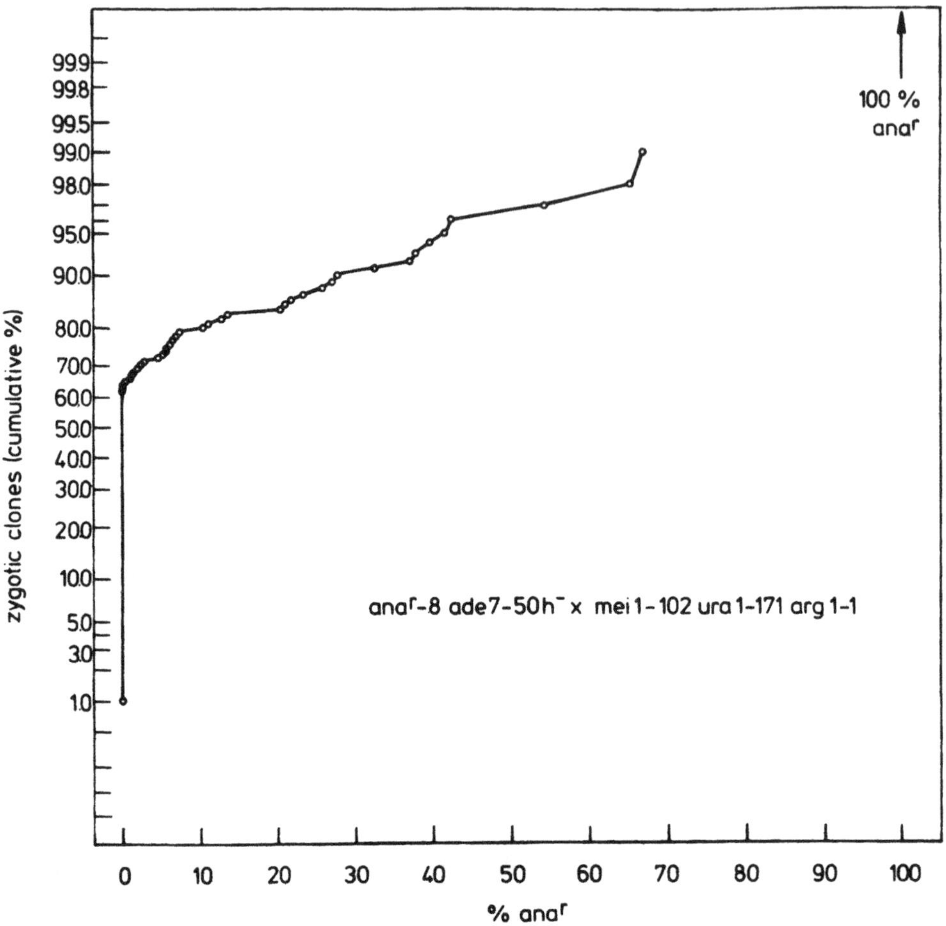

data given in Table 6 confirms that both methods lead to comparable results. The fractions of uniparental sensitive, uniparental resistant, and biparental clones, analyzed by replica - plating are given in Table 7. There is an increase in uniparental sensitive clones from diu^{r}- 301 to ana^{r}- 14 and finally ana^{r}- 8, accompanied by a decrease in uniparental resistant clones.

Since the method of replica - plating does not allow to detect transmission rates from zygotes to diploid progeny of less than 0.5 %, apparent uniparental sensitive zygotic clones were analyzed by the aliquot plating technique.

TABLE 6: Comparison of ant^r- transmission in standard crosses and analysis of single zygotic clones from crosses x mei 1

cross	type of analysis	% ant^r	% ant^s
x diu^r- 301	standard cross+	48.6	51.4
	individual zygotes++	49.6	50.4
x ana^r- 14	standard cross	34.9	65.1
	individual zygotes	27.8	72.2
x ana^r- 8	standard cross	3.8	96.2
	individual zygotes	6.8	93.2

+ Values from Table 3

++ Sum of 100 individual zygotic clones; 300 - 500 colonies scored per zygotic clone.

TABLE 7: Frequency of uniparental zygotes, analyzed by replica - plating

cross	uniparental sensitive	uniparental resistant	biparental
diu^r- 301 x mei 1	17	11	72
ana^r- 14 x mei 1	51	9	40
ana^r- 8 x mei 1	62	1	37

Sample size: 300 - 500 colonies

Clones appearing uniparental by this analysis contain less than 1 resistant per 10^5 sensitive zygotes. Results presented in Table 8 again show an increase in uniparental sensitive zygotic clones from diu^r- 301 to ana^r- 14 to ana^r- 8.

TABLE 8: Analysis of transmission of ant^r- phenotype by aliquot - plating of zygotes from crosses ant^r x mei 1

cross	% ant^r- diploids	number of zygotic clones
x diu^r- 301	0	2
	0.001 - 0.01	6
	0.01 - 0.1	10
	0.1 - 0.5	4
x ana^r- 14	0	4
	0.001 - 0.01	18
	0.01 - 0.1	24
	0.1 - 0.5	6
x ana^r- 8	0	38
	0.001 - 0.01	18
	0.01 - 0.1	8
	0.1 - 0.5	7

6) Rescue of genomes by meiosis and spore formation

It is now of interest, if markers are already lost in the zygote of in the successive diploid generations. We have tried to answer the first part of the question, if both markers involved are still present in the zygotes. Diploids from h^+ x h^- - crosses usually undergo sporulation immediately after zygote formation. Thus the diploid state is confined to the zygote. We have crossed the three different antibiotic resistant mutants with a sensitive h^+- strain and analyzed random spore clones. Results summarized in Table 9 show an approximate 50 : 50 distribution of resistant versus sensitive spore clones. This demonstrates that loss

TABLE 9: Random spore analysis of crosses ant^r x ura 1 - 171 h^+

cross	ant^r	ant^s	URA	ura	ADE	ade	total	
x diu^r- 301	1230	1290	1380	1240	1300	1320	2620	colonies
	50.8	49.2	52.7	47.3	49.6	50.4	100.0	%
x ana^r- 14	1110	1212	1232	1090	1214	1108	2322	colonies
	47.8	52.2	53.1	46.9	52.3	47.7	100.0	%
x ana^r- 8	644	752	682	714	720	676	1396	colonies
	46.1	53.9	48.9	51.1	51.6	48.4	100.0	%

of genomes has not occurred in the zygotes. This, however, does not exclude the possibility that genomes are determined for loss or retention in the zygote by some mechanism discussed below.

DISCUSSION

In this paper we have provided good evidence that uniparental inheritance also occurs in Schizosaccharomyces pombe. The occurrence of uniparental zygotic clones in a fission yeast confirms the results obtained by Birky et al. (10) that the formation of uniparental zygotes is independent from the bud site and therefore cannot be attributed to the budding process itself.

On the basis of the results presented in this paper, we want to put forward a hypothesis, where transmissional bias, mit^-- formation, and uniparental inheritance have a common basis. It is reasonable to assume that the mutator, acting in all three ant^r- mutants, produces defective genomes, some of them are "mit^-- genomes", some of them may have suffered other, perhaps lethal, mutational changes. An increase in the amount of mit^-- genomes in a cell is therefore due to an increase in "defective" genomes. Due to the mutator effect there is also an decrease in genomes that are capable to transmit antibiotic - resistance to the progeny. Therefore, mit^-- production and reduced transmission of ant^r- markers may have a common origin.

In analogy to Saccharomyces cerevisiae any bias in transmission of one parent will increase the amount of zygotes uniparental for markers of the other parent. Birky et al. (10) propose two general classes of models to explain the mechanism of uniparental inheritance. In the first category of models there is a random selection of a small sample of molecules for replication, as for example competition for replication sites on a membrane. The second set of models proposes differential marking of molecules from two parents, followed by a "counting" mechanism which insures that molecules of the majority type will be preferentially replicated and molecules of the minority type will be preferentially lost.

Up to now it is impossible to explain the peculiar "mutator - effect" responsible for mit^-- formation in Schizosaccharomyces pombe in molecular terms, nor is it possible to decide which of the two sets of models may be true for uniparental inheritance in both organisms.

ACKNOWLEDGEMENTS

The authors are indebted to B. Lang and Dr. G. Burger for kindly providing the mutants ana^r- 8, ana^r- 14, and diu^r- 301. They appreciate the skilful technical assistance of Miss R. Stets and Mrs. B. Sydow. This work was supported by the Deutsche Forschungsgemeinschaft.

REFERENCES

(1) Birky, C. W. jr. (1976) BioScience 26, 26

(2) Tilney - Basset, R. A. E. (1970) Genet. Res. 16, 49

(3) Sager, R. (1975) In: Genetics and Biogenesis of Mitochondria and Chloroplasts. Edited by C. W. Birky jr., P. S. Perlman, and T. J. Byers. Ohio State University Press, Columbus

(4) Gillham, N. W. (1974) Ann. Rev. Genet. 8, 347

(5) Birky, C. W. jr. (1974) Genetics 77, S 5 (abstr.)

(6) Birky, C. W. jr. (1975) Molec. gen. Genet. 141, 41

(7) Lang, B., Burger, G., Wolf, K., Bandlow, W., Kaudewitz, F. (1975) Molec. gen. Genet. 137, 353

(8) Lang, B. et al., in preparation

(9) Wolf, K., Burger, G., Lang, B., Kaudewitz, F. (1976) Molec. gen. Genet. 144, 67

(10) Birky, C. W. jr., Demko, C. A., Perlman, P. S., Strausberg, R., (1977), in press

MITOCHONDRIAL MUTANTS ISOLATED BY A NEW SCREENING METHOD BASED UPON THE USE OF THE NUCLEAR MUTATION op1.

Z. KOTYLAK and P.P. SLONIMSKI
Institute of Microbiology, University of Wroclaw, Poland and Centre de Génétique Moléculaire du C.N.R.S., Gif-sur-Yvette 91190, France

We have developped a new screening method which greatly facilitates isolation, handling and analysis of mitochondrial mutants in Saccharomyces cerevisiae. It is based upon the use of the nuclear mutation op1. This method was presented at the Pre-Federation of European Biochemical Societies Symposium "Structure, Synthesis and Function of Nucleic Acids in Organelles",Paris, July 19, 1975 and at the Symposium "Genetics, Biogenesis and Bioenergetics of Mitochondria", Munich, September 13, 1975. It was briefly described in (1) and some mutants isolated by this method have been analyzed in (1, 2, 3, 10). The purpose of the present article is threefold : (i) to give a full account of the method, (ii) to describe 121 mitochondrial mutants isolated with its help, and (iii) to present a method of localization of the mutants obtained by a set of 4 judiciously chosen ρ^- deletion tester strains. This permits to allocate rapidly most of the mutants isolated into the main genetic regions of the mitDNA molecule.

RATIONALE OF THE METHOD

The nuclear mutation op1 (4) makes ρ^- cells lethals (5). In its presence a culture of S. cerevisiae is composed of 100 % ρ^+ cells. The op1 cells do not grow on non-fermentable substrates (more precisely, they grow at an extremely slow rate, cf 4). However, when crossed to the wild type OP1 partner, the resulting diploid cells, heterozygous op1/OP1, recover the normal capacity to grow on glycerol or lactate media (4). These observations constitute the rationale of our method : if an op1 viable haploid clone is hybridized with an OP1 ρ^0 haploid clone and the resulting op1/OP1 cells do not grow on glycerol, then the op1 clone must carry a mutation located in the mitDNA. This mutation cannot be a mitochondrial ρ^- deletion and cannot be a recessive nuclear mutation. Theoretically, it could be a dominant nuclear mutation. Crosses with appropriate OP1 ρ^- testers permit rapidly to eliminate this latter possibility. No such case was found in the present work.

TECHNIQUE USED

The haploid strain 777-3A, α ade1 op1 ρ^+ was used all throughout this work to obtain mitochondrial mutants. Overnight culture of this strain (density 5 X 10^7 to 10^8 cells/ml) in the YPGA medium (Yeast extract 1 %, bactopeptone 1 %, glucose 2 %, adenine 20 mg/ml) was inoculated (1 to 10 or 1 to 50, vol/vol, depending on the experiment) into the same medium containing in addition 7 mM $MnCl_2$ or Mn SO_4. The culture was shaken at 28°C. Various incubation times were used (6 to 8 hrs, 24 and 48 hrs) with no major differences. The mutagenized culture, (cf 6), was diluted and spread on solid YP10 medium (Yeast extract 1 %, bactopeptone 1 %, glucose 10 %) to give approximately 100 colonies per plate. After 4 days of incubation at 28°C, colonies were picked and transferred in gridded patterns on YPGA plates. The parental strain gives large red colonies on a medium with high concentration of glucose. We picked three classes of colonies : those which were of smaller size, those which were less coloured and those which resembled the untreated culture. Mitochondrial mutants were found in all three classes, although they may be more frequent among the first two classes. Master YPGA plates with gridded colonies were replica-crossed (cf 7) on a lawn of OP1 ρ^0 tester strains of opposite mating type and auxotrophic requirements. Two testers were used : 55R5-3C/51 a ura1 and KL14-4A/60 a his1 trp2, with identical results. The resulting diploid prototrophic colonies were selected on the minimal, WO, medium (incubation for 2-3 days) and replicated on the Yeast extract, bactopeptone, glycerol medium, N3 (cf 7). Replicas were incubated at 28°C for 5 days. Diploids that failed to grow on glycerol were traced back to haploid clones which were isolated and further subcloned on a glucose medium.

CHARACTERIZATION AND GENETIC LOCALIZATION OF MUTANTS

The list of isolated mutants is given in the accompanying table. They were characterized by their cytochrome absorption spectra recorded at liquid nitrogen temperature (cf 8). In order to minimize the glucose repression effects,the cells were harvested from galactose grown cultures like in (2). Most of the mutants analyzed were deficient in cytochrome aa_3 only (76 mutants), while mutants deficient in cytochrome b only were less frequent (10 mutants). Twenty mutants displayed a deficiency in both cytochromes and 5 mutants showed a spectrum with both cytochromes present.

Mutants were allocated to various regions of the mitDNA molecule by recombination with discriminating ρ^- clones according to the method developped in (9). Four ρ^- testers were sufficient to localize 115 mutants out of 121 isolated. This represents a considerable economy of time and labour in comparaison with previous work (9) and analogous procedures adopted later in other laboratories. All ρ^- tester strains derive from the same parental ρ^+ strain KL14-4A. The tester RP131 (cf 9) restores all the previously studied mutations localized in the OXI3 region and none in the OXI1, OXI2, COB and BOX regions. The tester RP617/A1B1/PM1 (cf 10) was selected in order to restore all the mutations localized in the COB and BOX regions and none in the others. The tester DT4A1 was selected in order to restore mutations in the OXI1 locus only and the tester A211 restores mutations previously (9) localized in OXI1 and OXI2 loci. The difference between DT4A1 and A211 permits to allocate mutations in the OXI2 locus. Among the 121 mutants studied 110 displayed a clear-cut growth response (as <u>OP1/op1</u> diploids) on glycerol. Among the remaining 11 mutants, partial deficiencies, residual or "leaky" growth as well as temperature dependance of the deficient phenotype were observed.

Allocation of mutants into different loci is summarized below :

LOCUS	MUTANTS	
	CLEAR	LEAKY and TD
OXI3	73	6
BOX	23	3
OXI1	4	0
OXI2	3	2
NEW	6	0
OXI3, BOX	1	0
	110	11

SYMBOL OF STRAIN	LOCUS ALLOCATION	RHO – TESTERS				CYTOCHROME	
		RP 131	RP 617/A1B1	DT4 A1	A 211	CYTOCHROME aa_3	CYTOCHROME b
G 20	OXI 3	+++	−	−	−	−	+
G 55	BOX	−	+++	−	−	−	−
G 74	OXI 3	+++	−	−	−	−	+
G 107	NEW	−	−	−	−	−	−
G 108	OXI 3	+++	−	−	−	−	+
G 110	OXI 3	+++	−	−	−	−	+
G 112	OXI 3 LEAKY	+++	Σ	+	+		
G 171	BOX	−	+++	−	−	−	−
G 179	NEW	−	−	−	−		
G 183	NEW	−	−	−	−		
G 192	OXI 3	+++	−	−	−	−	+
G 199	OXI 2	−	−	−	+	−	+
G 205	OXI 3 LEAKY	++	PAPILLES	PAPILLES	PAPILLES	−	+
G 225	OXI 3	+	−	−	−	−	+
G 247	OXI 3 LEAKY	+++	Σ	Σ	Σ	−	+
G 291	OXI 3	+++	−	−	−	−	+
G 341	OXI 3	+++	−	−	−	−	−
G 389	OXI 2 LEAKY	Σ	Σ	+	++		
G 390	OXI 3	Σ	−	−	−		
G 392	BOX	−	+++	−	−	−	−
G 410	BOX LEAKY	Σ	+++	−	−	+	−
G 421	OXI 3	+++	−	−	−	−	+
G 433	OXI 3	+++	−	−	−	−	+
G 442	BOX LEAKY	−	+++	−	−	−	+
G 450	OXI 3	+++	−	−	−	−	+
G 476	BOX	−	+++	−	−	−	−
G 481	OXI 3	+++	−	−	−	−	+
G 565	OXI 1	−	−	+	++	−	+
G 588	OXI 3	+++	−	−	−	−	+
G 625	BOX	−	+++	−	−	−	−
G 656	OXI 3	+++	−	−	−	−	+
G 689	OXI 3	+++	−	−	−	−	+
G 696	OXI 3	+++	−	−	−	−	+
G 706	BOX	−	+++	−	−	−	−
G 728	OXI 3	+++	−	−	−	−	+
G 789	OXI 3	+++	−	−	−	−	+
G 839	BOX	−	+++	−	−	−	−
G 922	OXI 3	+++	−	−	−	−	±
G 925	BOX	−	+++	−	−	+	+
G 950	OXI 2 LEAKY	Σ	Σ	+	+++		
G 1003	OXI 3	+++	−	−	−	−	+
G 1055	OXI 3	+++	−	−	−	−	+
G 1089	OXI 3	+++	−	−	−	−	+
G 1099	OXI 3	+++	−	−	−	−	+
G 1160	OXI 3	+++	−	−	−	−	+
G 1181	OXI 3	+++	−	−	−	−	+
G 1186	OXI 3	+++	−	−	−	−	+
G 1194	OXI 3	+++	−	−	−	−	+
G 1264	OXI 3	+++				−	+
G 1273	OXI 3	+++	−	−	−	−	−
G 1280	OXI 3 LEAKY	+++	Σ/+	+	+(WEAK)	−	+
G 1287	OXI 3	+++	−	−	−	−	+
G 1305	OXI 3 LEAKY	+++	−	−	Σ	−	+
G 1334	BOX	−	+++	−	−	−	−
G 1342	OXI 1	−	−	+++	++	−	+
G 1370	BOX	−	+++	−	−	−	−
G 1374	OXI 1	−	−	+++	++	−	+
G 1375	OXI 3	+++	−	−	−	−	±
G 1547	OXI 3	+++	−	−	−	−	+
G 1659	BOX	−	+++	−	−	−	−
G 1683	OXI 3	+++	−	−	−	−	+

SYMBOL OF STRAIN	LOCUS ALLOCATION	RHO – TESTERS				CYTOCHROME	
		RP 131	RP 617/A1B1	DT4 A1	A 211	CYTOCHROME aa_3	CYTOCHROME b
G 1734	OXI 3	+++	–	–	–	–	+
G 1765	OXI 3	+++	–	–	–	–	+
G 1810	OXI 2	–	–	–	+++	±	+
G 1859	OXI 3	+++	–	–	–	–	+
G 1893	OXI 1	–	–	+++	Σ	–	+
G 1909	BOX	–	+++	–	–	–	–
G 1921	OXI 3	+++				–	+
G 1979	OXI 3	+++					
G 1988	BOX	–	+++	–	–	+	–
G 2026	BOX LEAKY	–	+++	–	–	–	–
G 2027	OXI 3	+++	–	–	–	–	+
G 2029	OXI 3	+++	–	–	–	–	+
G 2084	OXI 3	+++	–	–	–	–	+
G 2118	NEW	–	–	–	–	+	+
G 2124	OXI 3	+++	–	–	–	–	+
G 2133	OXI 3	+++	–	–	–	–	+
G 2166	OXI 3	+++	–	–	–	–	+
G 2190	OXI 3 LEAKY	+++	Σ	Σ	–	–	+
G 2220	OXI 3	+++	–	–	–	–	+
G 2276	OXI 3	+++	–	–	–	–	+
G 2283	NEW	–	–	–	–	–	+
G 2286	BOX	–	+++	–	–	+	–
G 2291	OXI 3	+++	–	–	–	–	+
G 2312	OXI 3	+++	–	–	–	–	+
G 2342	OXI 3	+++	–	–	–	–	+
G 2343	OXI 3	+++	–	–	–	–	+
G 2354	OXI 3	+++	–	–	–	–	+
G 2394	OXI 3	+++				–	+
G 2399	OXI 3	+++	–			–	+
G 2443	OXI 3	+++				–	+
G 2457	BOX	–	+++	–	–	+	–
G 2461	OXI 3	+++	–	–	–	–	+
G 2469	OXI 3	+++	–	–	–	–	+
G 2494	BOX	–	+++	–	–	+	–
G 2496	OXI 3	+++				–	+
G 2508	OXI 3	+++	–	–	–	–	+
G 2567	OXI 3 LEAKY	+++	Σ	Σ/+	1c	–	+
G 2590	BOX	–	+++	–	–	+	–
G 2592	OXI 3	+++	–	–	–	–	–
W 7	BOX	–	+++	–	–	+	+
W 19	BOX	–	+++	–	–	+	–
W 22	NEW	–	–	–	–	–	–
W 48	OXI 3	+++	–	–	–	–	+
W 69	OXI 3	+++	–	–	–	–	+
W 91	BOX	–	+++	–	–	–	–
W 109	OXI 3	+++	–	–	–		
W 113	OXI 3	+++	–	–	–		
W 164	OXI 3	+++	–	–	–		
W 165	OXI 3	+++	–	–	–	–	+
W 201	OXI 3	+++	–	–	–	–	+
W 231	OXI 3 ; BOX	+++	+++	–	–	–	–
W 235	OXI 2	–	–	–	+++	±	+
W 246	OXI 3	+++	–	–	–	–	+
W 248	OXI 3	+++	–	–	–	–	+
W 264	OXI 3	+++	–	–	–	–	–
W 267	BOX	–	+++	–	–	+	–
W 300	BOX	–	+++	–	–	+	–
W 364	OXI 3	+++	–	–	–	–	+
W 373	OXI 3	+++	–	–	–	–	+
W 401	BOX	–	+++	–	–	–	–

DISCUSSION

Mutants isolated by the <u>op1</u> screen clearly belong to the class of mitochondrially deficient and mitochondrially coded mutations designated by the generic term <u>mit</u>$^-$ (9, 11). In the original procedure of TZAGOLOFF et al.(11) the distinction between the ρ^- and the <u>mit</u>$^-$ mutants was achieved by measurements of mitochondrial protein synthesis. In the present method the ρ^- mutants are automatically eliminated. Thus the proportion of finally obtained <u>mit</u>$^-$ mutants is enriched and time consuming biochemical operations are eliminated. This is attested by the fact that ca 1.5 % of isolates are <u>mit</u>$^-$ (12) when starting with <u>OP1</u> strains as compared to ca 5 % isolates (121 out of 2400 in the present work) when starting with <u>op1</u> strains. No major bias is apparent in any of the methods employed. Mutants belonging to the OXI3 locus are, by far, the most numerous ones in both cases (cf 9, 12 and the present work) and mutants allocated to the COB, BOX regions appear at the second place. It will be of interest to study whether mutants of the <u>syn</u>$^-$ class or ATPase deficient mutants can be isolated from the <u>op1</u> strains. They are apparently very rare (cf 12) even in the <u>OP1</u> context.

There is an additional advantage in the isolation of <u>mit</u>$^-$ mutations in the <u>op1</u> nuclear genetic background. In the wild type nuclear context the <u>mit</u>$^+$ revertant cells have a very strong selective advantage over the <u>mit</u>$^-$ cells, even on glucose media. Thus, a culture of a <u>mit</u>$^-$ clone can be rapidly overgrown by revertants. This is not the case in the <u>op1</u> <u>mit</u>$^-$ cells, the <u>op1</u> <u>mit</u>$^+$ revertants being devoid of a selective advantage (<u>op1</u> is epistatic over <u>mit</u>$^+$) and the revertants do not accumulate readily. Furthermore, in the wild type nuclear context the ρ^- cells accumulate frequently and, depending on the <u>mit</u>$^-$ mutant studied, may represent an important fraction of the total cell population. This can make biochemical characterizations ambiguous. No ρ^- mutants accumulate in <u>op1</u> strains and an <u>op1</u> <u>mit</u>$^-$ <u>culture is composed of a pure population of mit</u>$^-$ <u>cells</u>.

<u>Acknowledgments</u>

Z. KOTYLAK greatfully acknowledges the receipt of an EMBO long term fellowship held at Gif in 1975-76. This work has been supported by a D.G.R.S.T. Grant N° 75.70748. Thanks are due to Patrick PAJOT for his help in determination of absorption spectra and to G. PERRODIN and J. LUTY for technical assistance.

References.

1. KOTYLAK, Z. and SLONIMSKI P.P. (1976) in "The Genetic Function of Mitochondrial DNA" Eds. C. SACCONE and A.M. KROON. Elsevier/North Holland Biomedical Press, Amsterdam, pp. 143-154.
2. PAJOT, P., WAMBIER-KLUPPEL, M.L., KOTYLAK, Z., and SLONIMSKI, P.P. (1976) in "Genetic and Biogenesis of Chloroplasts and Mitochondria" Eds. Th. BUCHER et al., Elsevier/North Holland Biomedical Press, Amsterdam, pp. 443-451.
3. SCHWEYEN, R.J., WEISS-BRUMMER, B., BACKHAUS, B. and KAUDEWITZ, F. (1976) in "The Genetic Function of Mitochondrial DNA" Eds. C. SACCONE and A.M. KROON. Elsevier/North Holland Biomedical Press, Amsterdam,pp.251-258.
4. KOVAC, L., LACHOWICZ, T.M., SLONIMSKI, P.P. (1967) Science 158, 1564-1567.
5. KOVACOVA, V., IRMLEROVA, J., KOVAC, L. (1968) Biochem. Biophys. Acta 162, 157-163.
6. PUTRAMENT, A., BARANOWSKA, H. and PRAZMO, W. (1973) Mol. Gen. Genetics 126, 357-366.
7. DEUTSCH, J., DUJON, B., NETTER, P., PETROCHILO, E., SLONIMSKI, P.P. and BOLOTIN-FUKUHARA, M. (1974) Genetics, 76, 195-219.
8. CLAISSE, M.L., PERE-AUBERT, G.A., CLAVILIER, L. and SLONIMSKI, P.P. (1970) Eur. J. Biochem. 16, 430-438.
9. SLONIMSKI, P.P. and TZAGOLOFF, A. (1976) Eur. J. Biochem. 61, 27-41.
10. KOTYLAK, Z. and SLONIMSKI, P.P. (1977), This volume.
11. TZAGOLOFF, A., AKAI, A., NEEDELMAN, R.B., ZULCH, G. (1975). J. Biol. Chem. 250, 8236-8242.
12. TZAGOLOFF, A., FOURY, F. and AKAI, A. (1976) in "The Genetic Function of Mitochondrial DNA" Eds. C. SACCONE and A.M. KROON. Elsevier/North Holland Biomedical Press, Amsterdam, pp. 155-161.

CONTROLLED MITOCHONDRIAL INACTIVATION OF TEMPERATURE-SENSITIVE *SACCHAROMYCES CEREVISIAE* NUCLEAR PETITE MUTANTS

E. Schweizer, W. Demmer, U. Holzner and H.W. Tahedl
Lehrstuhl für Biochemie der Universität Erlangen-Nürnberg
8520 Erlangen, W.-Germany

Abstract Temperature-sensitive nuclear *petite* mutants genetically attributable to 106 different complementation groups have been studied biochemically by using either whole cells or isolated mitochondria obtained in temperature shift-up experiments (22°C ⟶ 36°C) with continuous chemostat cultures. The following results were obtained:

1. In 20 of the 106 complementation groups studied respiratory competence is retained at 36°C, although cells were unable to grow on lactate as a carbon source. Depending on the complementation group studied, P/O-ratios were either normal or reduced by a factor of 40 - 70 percent. ADP-dependent respiratory control was normal in some groups and absent in others with either both or only one of the two substrates used, succinate and α-ketoglutarate. Only a single strain (ts-1416) exhibited the characteristics expected of a partially uncoupled mutant.
2. Of 12 complementation groups studied 2 are deficient in mitochondrial protein synthesis. Probably as a consequence of this defect, the ρ^+-character is lost in these mutants after 2-3 generations growth at 36°C.
3. Two groups specifically deficient in cytochrome oxidase and one other with a specific defect in the NADH-cytochrome c oxidoreductase were characterized. Upon growth at 36°C all other respiratory-deficient nuclear *petites* lose cytochrome oxidase activity, but at the same time - or after a slight delay - they also lose NADH/cytochrome c oxidoreductase and ATPase or NADH/cytochrome c oxidoreductase, ATPase and ρ^+-character, or NADH/cytochrome c oxidoreductase alone. One strain, possibly a cytochrome c mutant, retains all three activities although respiration is blocked at 36°C.
4. A sizeable portion of the *petite*-ts mutants studied undergoes a $\rho^+ \longrightarrow \rho^-$ transition after 2-4 generations growth at 36°C. Careful kinetic studies indicate that in most cases this transition is the consequence of, and not the reason for, the loss of respiratory competence in these strains. In 2 mutants studied, the loss of ρ^+-character could be prevented by performing the shift-up experiment in lactate instead of glucose-containing medium.

INTRODUCTION

Until recently, the preferred object of studies on mitochondrial biogenesis has been the DNA contained in the mitochondria itself. Especially by the use of appropriate respiration-deficient rho⁻ and mit⁻-yeast mutants considerable progress has been made towards the genetic and functional characterization of this DNA. Thus, cistrons for the two large mitochondrial rRNA species as well as for tRNA's for most of the 20 different amino acids have been located on yeast mitochondrial DNA (1,2). Similarly, distinct subunits of the cytochrome oxidase, the ATPase and the NADH-Cytochrome c reductase enzyme complexes apparently are encoded on the mitochondrial genome, too (3). On the other hand, the interest in nuclear Saccharomyces cerevisiae petite-mutants affected in mitochondrial oxidative phosphorylation has been considerably increased during the past several years. Work in various different laboratories has led to the isolation and characterization of nuclear ATPase, NADH-cytochrome c reductase, cytochrome c and cytochrome oxidase mutants (3-7). However, it has been known for a long time, that petite-mutants with specific, monofunctional defects are rare, the majority of them being pleiotropically affected in several respiratory functions at once and sometimes even devoid of functional mitochondrial DNA (8). Obviously, the characterization of these mutants by classical gene-enzyme relation studies is difficult. Nevertheless, the study of these functionally pleiotropic mutants is expected to be especially rewarding with respect to possible interrelationships between different components of the inner mitochondrial membrane. These interrelationships may exist for the biosynthesis, the integration or the biological activity of the membrane-bound multienzyme system of oxidative phosphorylation.

As an approach to the study of both specific and pleiotropic nuclear petite-mutants, we recently initiated the isolation and characterization of temperature-sensitive nuclear petite-mutants in yeast. Under appropriate experimental conditions, these mutants may provide information on the sequence of events leading from the intact inner mitochondrial membrane present at permissive temperature to the pleiotropically deficient membrane at non-permissive conditions. In an earlier publication, the petite ts-mutants available in this laboratory have been attributed to 106 different complementation groups (9). In the present paper, these complementation groups have been characterized with respect to various different biochemical parameters, like the ATPase, cytochrome oxidase and NADH-cytochrome c reductase activities, the capacity for mitochondrial protein synthesis and the tendency of losing ρ^+-character.

MATERIALS AND METHODS

The isolation of the Saccharomyces cerevisiae segregational petite ts-mutants as well as their attribution to distinct genetic complementation groups has been reported previously (9). Similarly, respiration rates and ρ^+-character were determined as described earlier (9).

Isolation of mitochondria. Cells from a freshly isolated single clone were inoculated into 30 ml of yeast extract-peptone-dextrose (YEP) medium containing 0.5 % saccharose, 0.4 % yeast extract, 0.6 % peptone. The culture was grown on a rotary shaker for about 48 hr up to early stationary phase (cell density 0.5-1.0 x 10^8 cells/ml). Depending on the growth temperature, either 3 ml (at 22°C) or all 30 ml (at 36°C) of this culture were used to inoculate another 300 ml of YEP medium equilibrated at the desired temperature. Growth on a rotary shaker was continued for another 24 hr. Subsequently, cells (16-13 g wet weight) were harvested by centrifugation, washed once and used for the isolation of mitochondria according to Kovac et al. (10).

Chemostat cultivations and enzyme assays. ATPase, cytochrome oxidase and NADH-cytochrome c reductase activities were determined in submitochondrial particles essentially as described by Needleman and Tzagoloff (11). For this, cells were grown from a pure single clone in 50 ml minimal (MIN) medium containing 0.5 % saccharose, 0.7 % yeast nitrogen base w/o amino acids (Difco) on a rotary shaker at 22°C up to a density of 0.5 x 10^8 cells/ml. This culture was then inoculated into a chemostat with another 100 ml of MIN-medium. Incubation was continued at 22°C until a cell density of 0.5 x 10^8 cells/ml was obtained. Then, operation of the chemostat was started adjusting the addition of fresh MIN-medium to a rate which maintained the cell density at about 0.5 x 10^8 cells/ml. After 1-2 generations at 22°C, the temperature was shifted to 36°C within 15-20 min. 10-20 ml-samples were withdrawn from the culture at least once per generation and used for the preparation of submitochondrial particles according to Needleman and Tzagoloff (11). At the same time, cellular respiration rates and the proportion of ρ^+/ρ^--cells were determined separately in each sample. If not used immediately for the preparation of submitochondrial particles, cells were collected by centrifugation and kept frozen at -16°C.

Cytochrome spectra were recorded in an Aminco-DW 2 UV/VIS-spectrophotometer with dithionite-reduced versus Fe^{3+}-oxidized yeast cells after freezing in liquid nitrogen. Alternatively, cytochrome spectra of intact yeast cells were examined using a Zeiss low-dispersion spectroscope which was fitted on a microscope (8).

Mitochondrial protein synthesis was determined for each mutant studied in two 8 ml parallel cultures grown aerobically in 1 % glucose-containing YEP medium at 22°C and 36°C, respectively. As soon as respiration in the 22°C culture had become fully derepressed, 8 x 10^7 cell-samples were taken from both cultures, the cells collected by centrifugation were washed twice with 0.7 % yeast nitrogen base (Difco) and subsequently resuspended in 1 ml MIN-medium (0.1 % glucose). Both samples were divided into 2 equal portions and to one portion of each sample 200 µg/ml of cycloheximide were added. After 45 min. incubation at 22°C and 36°C, respectively, 1 µCi/ml 1-^{14}C-leucine (NEN-Chemicals, 50 mCi/mmole) was added to all four assay mixtures. After 40 min. incubation at 22°C and 36°C, respectively, the reaction was stopped by adding 5 ml 5 % trichloroacetic acid, containing 3 µg/ml unlabeled leucine, to each assay mixture. After remaining overnight at 22°C the precipitate formed was collected by centrifugation and washed twice with 5 ml 5 % trichloroacetic acid (containing 3 µg/ml unlabeled leucine), each time. Subsequently, the precipitate was dissolved in 0.2 ml 2 % sodium dodecyl sulfate/60 mM Tris and, after addition of 5 ml Bray-solution (12), the radioactivity was determined by liquid scintillation counting.

P/O ratios and respiratory control. In a volume of 2.2 ml containing 0.25 M saccharose the isolated mitochondria (1-2 mg protein) were incubated for 7 min at either 22°C or 36°C together with 4 µmoles $MgCl_2$, 0.96 µmoles ATP, 20 µmoles succinate, 60 units of dialyzed hexokinase (Boehringer), 4 mg bovine serum albumin (Serva) and 32 µmoles $^{32}PO_4$ (spec. act. 0.8-2.0 x 10^5 cpm/µmole). The reaction was stopped by the addition of 0.5 ml of 15 % trichloroacetic acid. Excess inorganic phosphate was removed as ammonium phosphomolybdate (13) by two successive extractions with water-saturated iso-butanole/benzene (1:2). $^{32}PO_4$ incorporated into organic compounds was subsequently determined in the aqueous phase by scintillation counting. In parallel assays, respiration rates of the mitochondrial preparations used were determined with an oxygen electrode.

Respiratory control measurements were performed essentially as indicated by Kovac & Hrusovska (14). The isolated mitochondria (200-400 µg protein) were suspended in 1 ml of 0.6 M mannitol containing 20 mM KCl, 1 mM EDTA, 10 mM tris-maleate buffer, pH 6.4, 10 mM K-phosphate pH 6.4. ADP was added as a 0.02 M solution and succinate and α-ketoglutarate as 1 M solutions in the above-mentioned buffer. In the presence of either substrate the response of respiration rates upon addition of ADP was followed on an oxygen electrode.

Membrane potential determinations. In separate experiments, mitochondria were isolated from cells grown at both, permissive and non-permissive temperatures as described above. The intramitochondrial accumulation of ^{14}C-labeled triphenyl-methyl-phosphonium ($TPMP^+$) was determined essentially as described by Klingenberg and Phaff (15). ^{14}C-$TPMP^+Br^-$ was prepared from $^{14}CH_3Br$ (NEN-Chemicals, spec. act. 1 Ci/mole, cf. 16). The isolated mitochondria were suspended in 0.25 M saccharose containing 10^{-3} M Tris · H_2SO_4 pH 7.5, 5×10^{-3} M $MgCl_2$, 5×10^{-3} succinate. In three parallel experiments 0.5 ml portions of this suspension, containing 200-800 µg protein, were added with 14×10^{-6} M ^{14}C-$TPMP^+$ in the first, 100.000 cpm ^{14}C-dextrane (NEN-Chemicals, 2×10^{-6} Ci/mg) in the second and 130.000 cpm tritium water in the third experiment. The samples were transferred to polypropylene tubes containing 1 ml of silicone oil (CR 50, Wacker-Chemie) overlaid to 0.5 ml of a 10 % CsCl (w/v) solution. After 10 min incubation at room temperature the mitochondria were spun through the silicone oil layer by a 5 min centrifugation at 3.000 x g. The two aqueous phases were collected separately and, after addition of 15 ml Bray-solution (12), the radioactivity contained in each phase was determined by liquid scintillation counting. The values obtained were corrected for a quenching factor determined in a separate experiment. The amount of intramitochondrially accumulated $TPMP^+$ was calculated according to Riebeling et al. (17). From this, the membrane potential was determined by the equation $\Delta\Psi = RT/\ln F \times \ln \text{conc } TPMP^+_{in} / \text{conc } TPMP^+_{out}$. Whenever the uncoupler carbonyl cyanide m-chlorophenyl hydrazon (CCCP) was applied its concentration was 5×10^{-5} M for respiration and $^{32}PO_4$-incorporation studies, and 1×10^{-4} M for the $\Delta\Psi$ determination studies.

RESULTS AND DISCUSSION

1. Cytochrome spectra and enzymes of oxidative phosphorylation. In a first series of experiments all petite ts-mutants available to us were spectroscopically screened for their cytochrome spectra after the cells had been grown on YEP-agar at both, permissive and non-permissive temperatures. The data listed in Table 1 indicate, that in most mutants only at 22°C a complete set of cytochromes is present, while in cells grown at 36°C usually at least one of the cytochrome absorption bands is missing. In spite of experimental errors and uncertainties probably being rather high in these first experiments, several conclusions may already be drawn: (a) the spectral characteristics observed are complementation group-specific, (b) at 36°C only respiratory competent petites exhibit normal cytochrome spectra, (c) cytochrome a deficiencies are observed with almost all respiratory

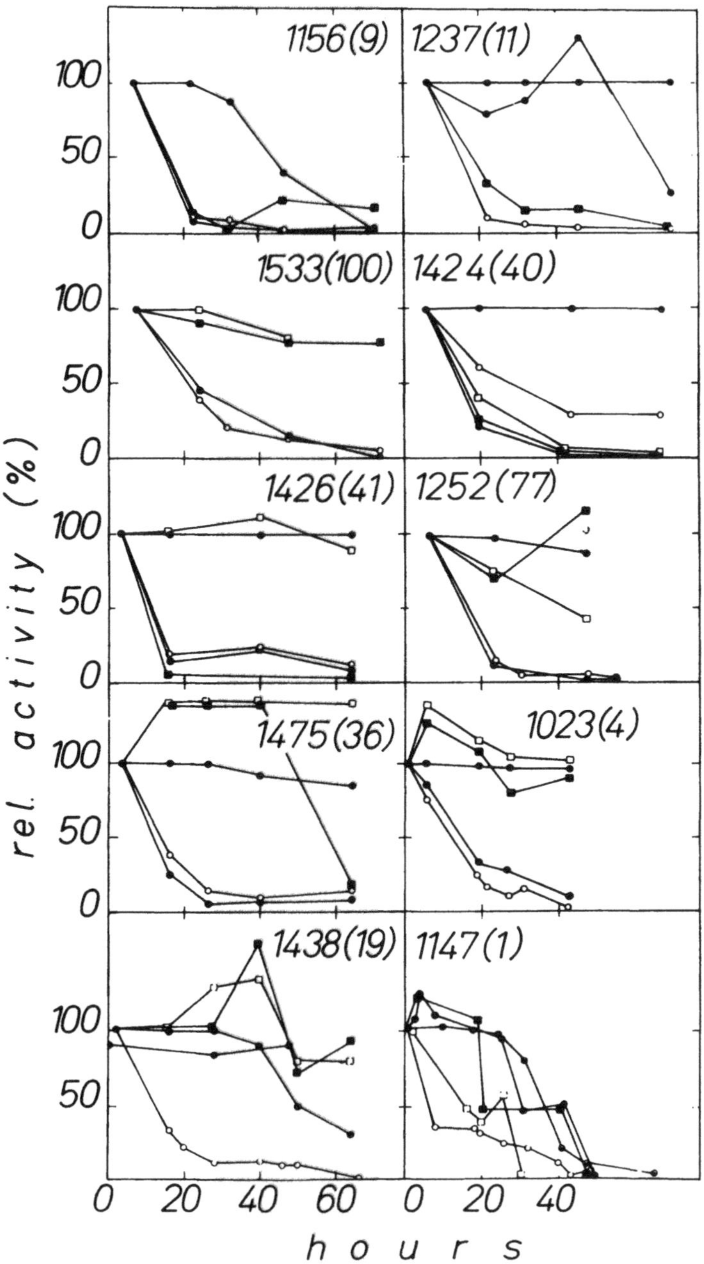

Fig. 1 Inactivation Kinetics of respiration rates (-o-), rho$^+$-character (-⊗-), Cytochrome oxidase (-●-), ATPase (-□-) and NADH-cytochrome c reductase (-■-) activities at 36°C. Cells were grown in 150 ml chemostat cultures. Exp.details as described in Materials and Methods. Temperature shift-up at time = o. Compl.group No.in brackets.

TABLE 1. Biochemical characterization of petite ts-mutants. Experimental details are given in Materials and Methods. Spectral characteristics were visually recorded with a spectroscope and qualitatively compared with wild type data.

n.d. = not determined. 0 = weakly active. α-KG = α-Ketoglutaric acid.

ts-mutant No.	Complementation group	Cytochrome content at 36°				Cytochrome content at 23°				O_2-Consumption at 36°C (per 1 hour)		Mitochondrial $P^{32}O_4$-incorporation at 36°C		Respiratory control		Rel. increase in respiration upon addition of ADP		Mitochondrial protein-synthesis (% wild type activ.)		ATPase (particle-bound)	Cytochrome c-oxidase	Cytochrome c-reductase	rho⁻-cells
		a	b	c_1	c	a	b	c_1	c	µl per 10^8 cells	µmoles per mg mitoch.	nmoles $mg^{-1} hr^{-1}$	P/O	Succ.	α-KG	Succ.	α-KG	23°	36°				%
wild type	-	+++	+++	+++	+++	+++	+++	+++	+++	123	4.3	13.3	1.5	1.9	2.8	1.7	2.1	100	100	+	+	+	
86	1				++	+++	+++	+++	+++	0										0	-	+	100
321					++	++	+++	+++	+++	0										+	-	+	100
1016					+++	+++	+++	+++	+++	0										+	-	+	1
1147					++	++	+++	+++	+++	0										-	-	-	1
1201					+++	++	+++	+++	+++	0										0	-	-	10
1266					+++	+++	+++	+++	+++	0										-	-	+	1
1286					+++	++	+++	+++	+++	0										+	-	-	80
1296					+++	+++	+++	+++	+++	0										+	-	0	50
1316					+++	+++	+++	+++	+++	0										+	-	-	
1335					+++	++	+++	+++	+++	0								45	7	+	-	-	
1363					++	+++	+++	+++	+++	0													
1418					++	+	+++	+++	+++	0										+	-	0	100
1420					+++	++	+++	+++	+++	0										-	-	+	10
1425		n.d.				n.d.																	
1463		++	+++	+++	+++	++	+++	+++	+++	0								82	6	+	-	-	
1513					++	+++	+++	+++	+++	0										-	-	-	90
1547					+++	+	+++	+++	+++	0													
1508	2				+++	+++	+++	+++	+++	0										+	-	-	92
1549					++	+	+++	+++	+++	0										+	-	-	
1583					++	+	+++	+++	+++	0										+	-	-	10
413	3				++	+	+++	+++	+++	0										+	0	0	1
1094					+	+	+++	+++	+++	0													
1098					++	+	+++	+++	+++	0								10	7	+	-	0	2
1350			++	+++	+++	++	+++	+++	+++	0											-	-	
1596			++	+++	+++	++	+++	+++	+++	0								12	8	+	-	0	100
1414					+	+	+++	+++	+++	0										0	-	0	1
17	4		+++	+++	+++	+++	+++	+++	+++	5								41	25	+	-	+	1
106			++	+++	+++	+	+++	+++	+++	0										+	-	+	1
108			+++	+++	+++	+	+++	+++	+++	0										+	-	+	1
203			+++	+++	+++	+	+++	+++	+++	2								46	110	+	-	+	1
634			++	+++	+++	+	+++	+++	+++	0								18	37	+	-	+	1
1202		+	+++	+++	+++	+	+++	+++	+++	22								25	61	+	-	+	1
1336			++	+++	+++	+	+++	+++	+++	0								120	54	+	-	+	1
1541			+++	+++	+++	+++	+++	+++	+++	0								21	45	+	-	+	1
1575			+	+++	+++	+	+++	+++	+++	0										+	-	+	1
545	5	++	+++	+++	+++	+++	+++	+++	+++	4											+	+	
1100		n.d.				n.d.				77	0.5	0.6	0.7	0.8	3.9	-	1.3	84	65		+	+	
1119		+++	+++	+++	+++	+++	+++	+++	+++	74	1.4	4.7	1.9	0.8	1.8	-	-	88	54		+	+	
1244		+++	+++	+++	+++	+++	+++	+++	+++	42	0.8	3.1	2.0	0.8	6.6	-	1.3				+	+	
453	7	+++	+++	+++	+++	+++	+++	+++	+++	30	2.4	2.4	0.5	1.0	3.2	-	0.9				+	+	
1023	8	+	++	+++	+++	++	+++	+++	+++	0										+	-	+	20
1325				+	+++	+++	+++	+++	+++	0								27	20	+	-	0	1
1371		+	++	+++	+++	+++	+++	+++	+++	0										+	-	+	1
1396					++	+	+++	+++	+++	0								23	20	0	-	0	20
320	9			+	+++	++	++	+++	+++	0								97	24	+	-	-	
1156					+++	+++	+++	+++	+++	0								23	19		-	-	100
1260					++	+++	+++	+++	+++	0								52	21		-	-	1
1479			+	+++	+++	++	+++	+++	+++	0											-	-	
724	10	n.d.				n.d.				19	0.9	2.5	1.5	1.0	1.0	-	-			+	+	+	
1227		++	+++	+++	+++	+++	+++	+++	+++	68	0.5	1.7	1.9	1.0	1.0	-	-	70	46	+	+	+	
1537		n.d.				n.d.				21	3.5	4.7	0.7	1.0	1.0	-	-	72	43	+	+	+	
1552		+	+++	+++	+++	++	+++	+++	+++	30	1.2	3.6	1.5	n.d.	1.9	n.d.	1.9			+	+	+	
267	11	++	+++	+++	+++	n.d.				0								31	16	+	+	-	1
1087		+++	+++	+++	+++	+++	+++	+++	+++	0										+	+	-	100
1237		++	++	+++	+++	++	+++	+++	+++	0								48	28	-	-	-	1
1524		+++	+++	+++	+++	+++	+++	+++	+++	0								35	5	+	-	-	
578	12	++	+++	+++	+	+++	+++	+++	+++	29								32	22		+	+	
1360		+	+++	+++	+	++	+++	+++	+++	24	0.8	2.0	1.3	1.0	1.92	-	2.0	24	20		+	+	
1597		+	+++	+++	+	++	+++	+++	+++	25	1.5	2.4	0.8	1.0	1.5	-	-				+	+	
1033	13									31	1.7	3.3	1.0	1.3	1.2	-	-				+	+	
1580			++	+++	+++	++	+++	+++	+++	0											-	+	
383	14	+++	+++	+++	+++	++	+++	+++	+++	65	1.5	4.3	1.4	1.5	1.5	1.2	-				+	+	
591		n.d.				n.d.				40	2.23	2.79	1.25	-	2.2	-	-				+	+	
1221		+	++	+++	+++	+++	+++	+++	+++	42	2.3	6.4	1.4	1.4	1.8	0.9	1.7						
1276		n.d.				n.d.				72	1.3	3.5	1.4	1.5	1.7	0.9	1.9				+	+	

1042	17	+ +++ +++	+ +++ +++ +++	0											-	+	
1594		+ + ++	+ +++ +++ +++	0											-	-	
1194	18	+ +++ +++ +++	++ +++ +++ +++	0								-	-				
1503		+ +++ +++ +++	++ +++ +++ +++	31	2.12	3.3	1.56	1.11	1.66	-	3.2	93	60	+	-	+	
1604		++ +++ +++ +++	++ +++ +++ +++	0								95	72	+	-	-	1
1438	19	n.d.	n.d.	0										+	+	+	1
74	20		+++ +++ +++ +++	11											+	+	
1592		+ +++ +++ +++	+++ +++ +++ +++	13											+	+	
1416	21	+++ +++ +++ +++	+++ +++ +++ +++	67	1.2	1.0	0.4	1.0	1.0	-	-				+	+	
1274	22	+ +++ +++ +++	+++ +++ +++ +++	0													
1017	23	+ +++ +++	+++ +++ +++ +++	0													
1187		+ ++	++ +++ +++ +++	2										-	-	-	1
1563	24	+ +++ +++ +++	+ +++ +++ +++	0													
1571		n.d.	n.d.	0											+	+	
9	25	+ + +++ +++	++ +++ +++ +++	0								86	63	+	-	+	80
1440		++ +++ +++	++ +++ +++ +++	0								58	45	0	+	+	1
1226	27	n.d.	n.d.	37	2.2	7.5	1.7	1.5	1.2	0.9	-						
1014	29	n.d.	n.d.	71	4.0	11.5	1.3	1.3	2.4	-	2.2				+	+	
1022	30	n.d.	n.d.	0											+	+	
1527				14											+	+	
1048	31	+++ +++ +++ +++	+++ +++ +++ +++	105	8.5	28.0	1.6	2.1	1.0	0.9	-	90	82		+	+	
1074	32	++ +++ +++ +++	+++ +++ +++ +++	47	2.27	3.4	1.49	1.0	2.76	-	1.4				+	+	
197	33	++	+ +++ +++ +++	0													
1107	34	+ ++ +++ +++	+ +++ +++ +++	0											-	-	
1578		++ +++ +++	++ +++ +++ +++	0										0	-	0	1
1475	36	+++ +++ +++	+ +++ +++ +++	0										+	-	+	15
1331	39	++ +++ +++ +++	+++ +++ +++ +++	25	1.3	3.3	0.8	2.2	4.0	1.98	1.7				-	-	
1424	40	+ +++ +++ +++	+++ +++ +++ +++	0											-	-	
1426	41	+++	+++ +++ +++ +++	0										+	-	-	1
1476	43	+ + + +++	++ +++ +++ +++	0										+	+	-	20
1010	47	+ +++ +++	++ +++ +++ +++	2										0	0	0	1
1012	48	n.d.	n.d.	0										+	-	-	1
1035	49	+ ++ +++	+ ++ +++ +++	0										+	-	-	1
1050	50	+++ +++ +++ +++	+++ +++ +++ +++	50	6.5	28.0	2.2	2.4	1.4	0.9	-				+	+	
1080	56	+ + +++	n.d.	0											-	-	
1099	57	n.d.	n.d.	0											+	+	
1106	58	+++ +++ +++ +++	+++ +++ +++ +++	56	2.5	5.3	1.0	1.2	0.7	-	-				+	+	
1124	61	+++ +++ +++ +++	+++ +++ +++ +++	n.d.													
1144	63	+++ +++ +++ +++	+++ +++ +++ +++	0										-	+	-	20
1160	64	+++	+ +++ +++ +++	0										+	+	0	100
1186	67	++ +++ +++	+ +++ +++ +++	0										-	-	-	10
1192	68	+++	+ ++ +++	0											-	+	100
1200	70	n.d.	n.d.	0											-	-	
1204	71	n.d.	n.d.	78	8.9	7.9	1.12	1.17	2.85	-	-				+	+	
1213	73	n.d.	n.d.	46	3.5	12.3	1.7	3.7	1.9		1.98				+	+	
1231	75	+ + +++	+++ +++ +++ +++	0											-	-	
1236	76	n.d.	n.d.	144	1.4	3.0	1.0	1.4	1.4	-	-				+	+	
1252	77	++	+++ +++ +++ +++	0											-	-	
1282	79	n.d.	n.d.	69	2.7	5.9	1.1	1.0	3.3	-	0.7				+	+	
1284	80	++ ++ +++ +++	+++ +++ +++ +++	33	1.4	7.2	1.2	0.9	8.4	-	-						
1294	82	+ + +++	+++ +++ +++ +++	1											+	+	
1347	86	+++ +++ +++ +++	+++ +++ +++ +++	4											-	+	
1359	87	+ +++ +++ +++	++ +++ +++ +++	n.d.													
1365	88	++ ++ +++ +++	+++ +++ +++ +++	0										+	-	-	1
1384	89	+ +++ +++	+++ +++ +++ +++	0											-	-	
1391	90	+ ++ +++ +++	+++ +++ +++ +++	21								25	20		-	+	
1412	91	+++ +++ +++ +++	+++ +++ +++ +++	14											+	+	
1074	93	++ +++ +++ +++	+++ +++ +++ +++	47	1.1	3.4	1.5	1.0	2.8	-	1.4				+	+	
1462	95	n.d.	n.d.	0													
1475	96	n.d.	n.d.	0													
1476	97	n.d.	n.d.	2													
1496	98	n.d.	n.d.	0													
1533	100	n.d.	n.d.	4										+	-	+	100
1570	103	+++ +++ +++ +++	+++ +++ +++ +++	19											+	+	
1590	105	n.d.	n.d.	46	1.0	3.0	1.4	1.0	1.1	-	-				-	-	
1599	106	n.d.	n.d.	0											-	-	

deficient mutants, (d) only few instances of specific cytochrome a, b or c-deficiencies are found, most mutants having lost, though at varying relative extent, - all three cytochromes at once. These results essentially agree with those reported earlier by Sherman on a series of non-conditial nuclear *petite* strains (8).

ATPase, cytochrome oxidase and NADH-cytochrome c reductase activities have been determined in submitochondrial particles isolated from a large number of *petite* ts-mutants representative for most of the complementation groups reported earlier (9). In this series of experiments, chemostat cultures were first grown at 22°C and then shifted to 36°C. The mutant character becoming expressed while growth at 36°C was continued was determined in samples taken at regular time intervals. Examples of some typical mutant enzyme patterns are depicted in Fig. 1. The results obtained with all other mutants are listed in Table 1. Obviously, there are at least three different complementation groups specifically deficient in cytochrome oxidase (groups 4, 36 and 100) and one in cytochrome b (group 11). There is one group (No. 19) which becomes respiratory deficient without loosing any of the three activities studied. This mutant may still be deficient in cytochrome c since cytochrome c was added to all assay mixtures employed and, thus, its disappearance would remain undetected in this study. All other mutants are pleiotropic, that is they lose several functions at once, be it either cytochrome oxidase and ATPase (group 9 and possibly 77), or cytochrome oxidase and NADH-cytochrome c reductase (group 41), or cytochrome oxidase NADH-cytochrome c reductase and ATPase (group 40) or, finally, cytochrome oxidase, NADH-cytochrome reductase, ATPase and ρ^+-character altogether (group 1). Other groups (not shown) either exhibited inconsistent patterns within the same group, or the mutant character finally acquired at 36°C was similar to one of the patterns shown in Fig. 1 but the various characters disappeared sequentially. To define the primary lesion in these and in the above-mentioned pleiotropic mutants, a further refinement of the inactivation kinetics should be achieved. Furthermore, it remains to be studied whether together with the overall function of a specific enzyme all its component protein subunits disappear from the mutant cell, too.

2. Respiratory competent *petite* ts-mutants. As it is indicated in Table 1 some of the mutants isolated, though unable to grow on non-fermentable carbon sources like lactate or ethanol, still exhibit appreciable CN^--sensetive respiration rates. Studies to characterize the mutational defect in these strains, didn't yield conclusive results, so far. As listed in Table 1, the incorporation of $^{32}PO_4$ into organic compounds, though definitely lower than in wild type mitochondria, was not completely blocked in any of the respective complementation groups. In all cases, the P/O ratios calculated were between 30 and 100 percent normal (Table 1). On the other hand, only in some of these mutants respiration rates of isolated mitochondria could still be stimulated by the addition of ADP (Table 1). In other cases, no respiratory control was observed with neither succinate nor α-ketoglutarate (Fig. 2) or it could be achieved only with one of the two substrates but not with the other.

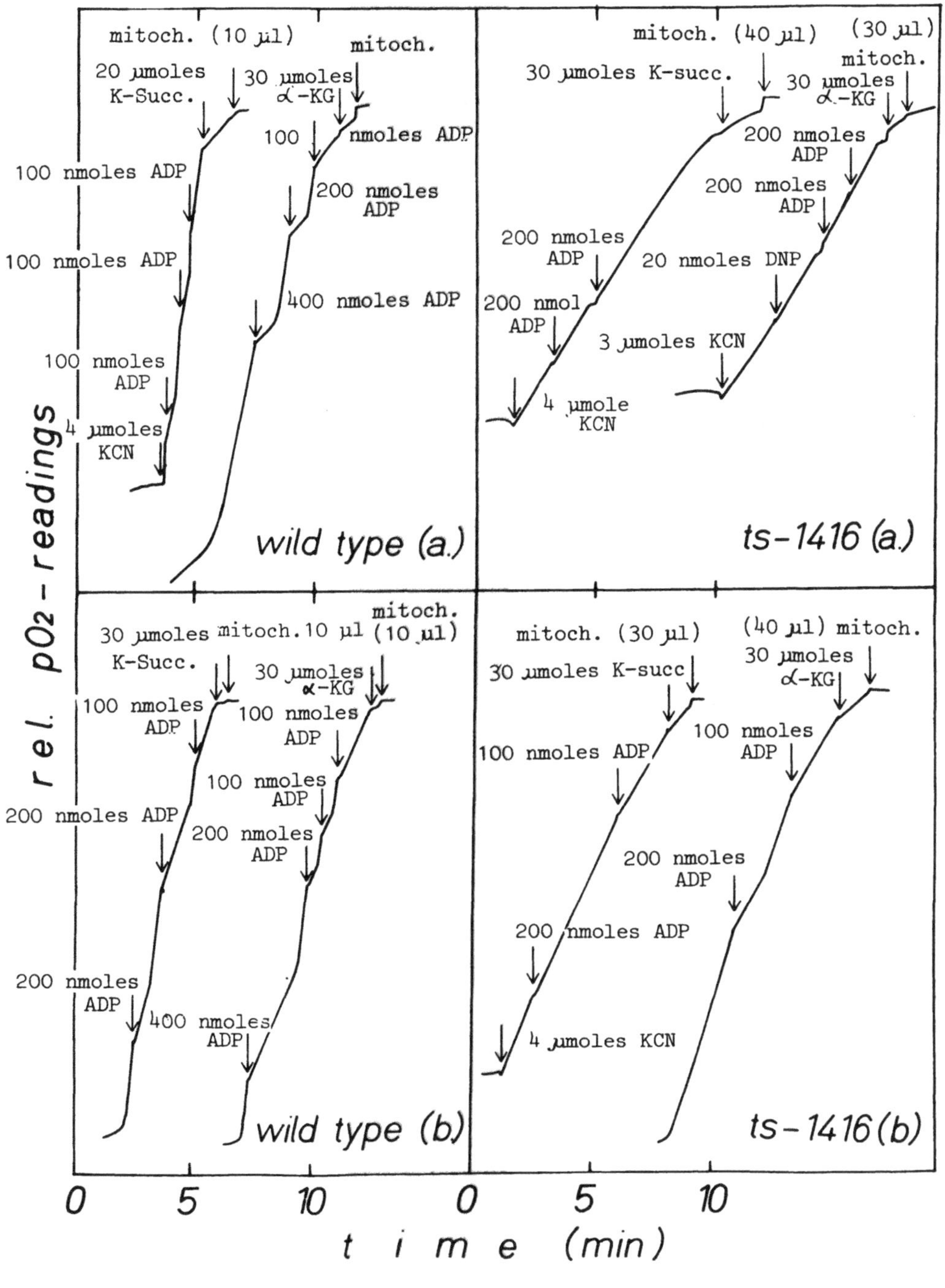

Fig. 2 Respiratory control determinations at 36°C (a) and 23°C (b).

TABLE 2. Oxidative phosphorylation in isolated ts-1416 mitochondria. Experimental details are described in Materials and Methods.

growth temperature	23°C				36°C			
substrate for mitochondria	α-KG		Succ		α-KG		Succ	
CCCP added	-	+	-	+	-	+	-	+
	% of wild type							
Respiration	32	43	20	22	20	22	25	28
$^{32}P_i$-incorporation	22	7	9	4	6	0	7	0.3
$TPMP^+_{in}/TPMP^+_{out}$	20	7	-	-	13	5	-	-
membrane potential	65	35	-	-	52	32	-	-

Mutant ts-1416 which exhibited especially low $^{32}PO_4$-incorporation rates and showed no respiratory control with either substrate has been studied further as a possible candidate for an uncoupled mutant. With this mutant, also the membrane potential of respiring mitochondria has been determined as an additional criterium for a mutational defect in the oxidative phosphorylation system. However, as it is shown in Tables 1 and 2, all parameters studied were comparatively low in the mutant even at permissive temperature and, on the other hand, no complete deficiency could be demonstrated at 36°C with any of the functions examined. Thus, it remainsto be studied further, whether the primary mutational lesion in this and in other respiratory competent mutants resides within or outside the oxidative phosphorylation pathway.

3. Transition of rho^+ into rho^--cells. It is a well known characteristic of many *petite* mutants to aquire, as the result of a nuclear mutation, cytoplasmically inherited rho^--characteristics (8). As it has been reported earlier, this behaviour was found to be a complementation group-specific characteristic of some of the ts-mutants isolated in this laboratory, too (9). In subsequent studies, the transition into respiratory deficient and rho -cells at 36°C were followed kinetically with a representative collection of *petite* ts-mutants of this type. It may be seen from the results shown in Fig. 3 and 4, that, at least in most of the mutants studied, the respiratory deficiency is the primary mutational lesion and that the loss of rho^+-character is the consequence of, and not the reason for the loss of respiratory competence. Usually, respiration decreases below 10 percent of the initial activity within 2-4 generations. This is consistent with a replacement of preformed active by new inactive components of the respiratory chain. Surprisingly, however, the content of rho^+-cells in the culture

TABLE 3. Mitochondrial protein synthesis in ts-mutants of different complementation groups.

Strain	Complementation-group	C^{14}-leucine-incorporation				ρ^{-}-cells present after growth at 36°C
		at 23°C		at 36°C		
		total	mitochondria	total	mitochondria	
		cpm/8x10^{7}-cells		cpm/8x10^{7}-cells		%
wild type	-	350.000	8.500	360.000	9.000	100
ρ^{o}-mutant	-	13.000	1.400	14.000	1.500	0
ts-1335	1	218.386	4.520	13.139	1.864	97
ts-1463	1	275.322	8.282	18.910	1.819	92
ts-1519	1	264.348	5.116	5.487	1.342	40
ts-1016	1	296.678	6.778	3.296	1.641	39
ts-1363	1	256.045	4.363	7.796	1.965	15
ts-1336	4	344.000	9.900	67.500	4.500	96
ts-9	25	233.200	6.700	19.400	3.950	93

Fig. 3 Inactivation Kinetics of respiration and rho^+-character at non-permissive temperature. Cells were grown in 500 ml chemostat cultures on YEP-0.1 % glucose media. Complementation groups are indicated in brackets. Temperature shift-up to 36°C at generation time = o.

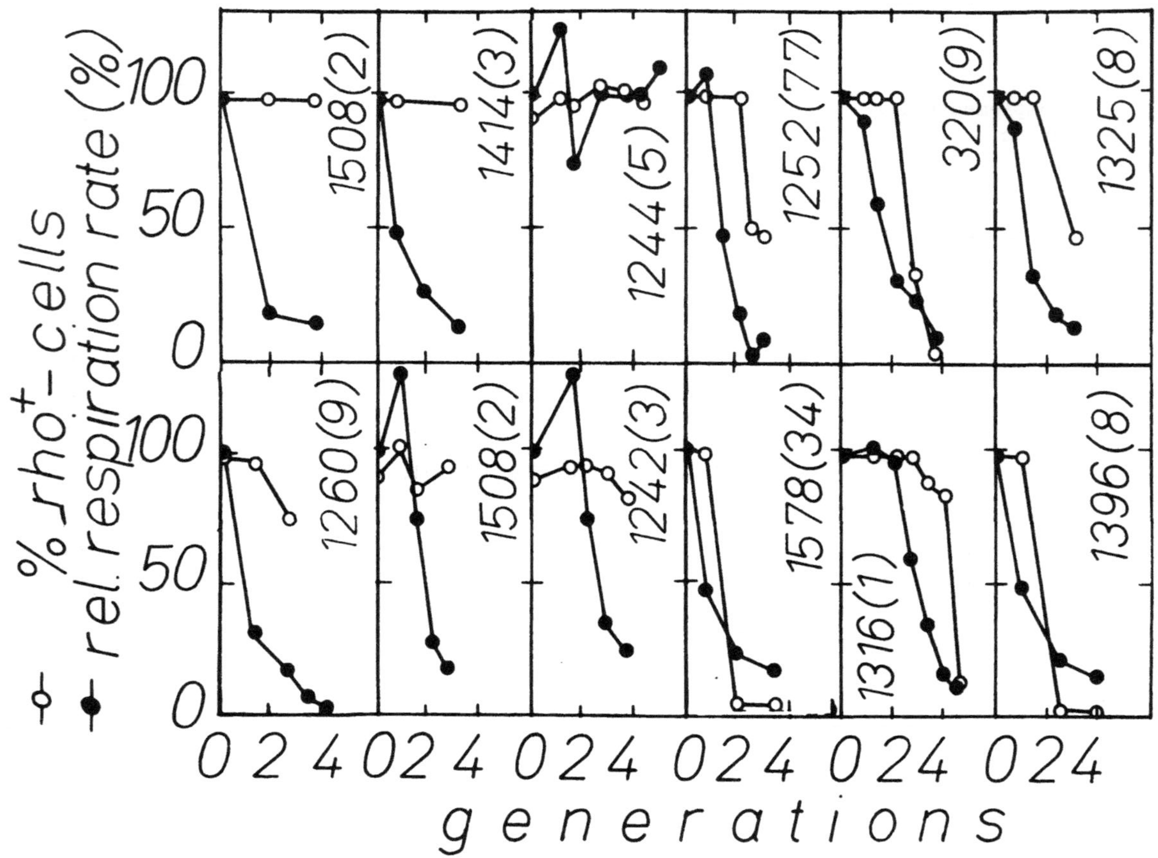

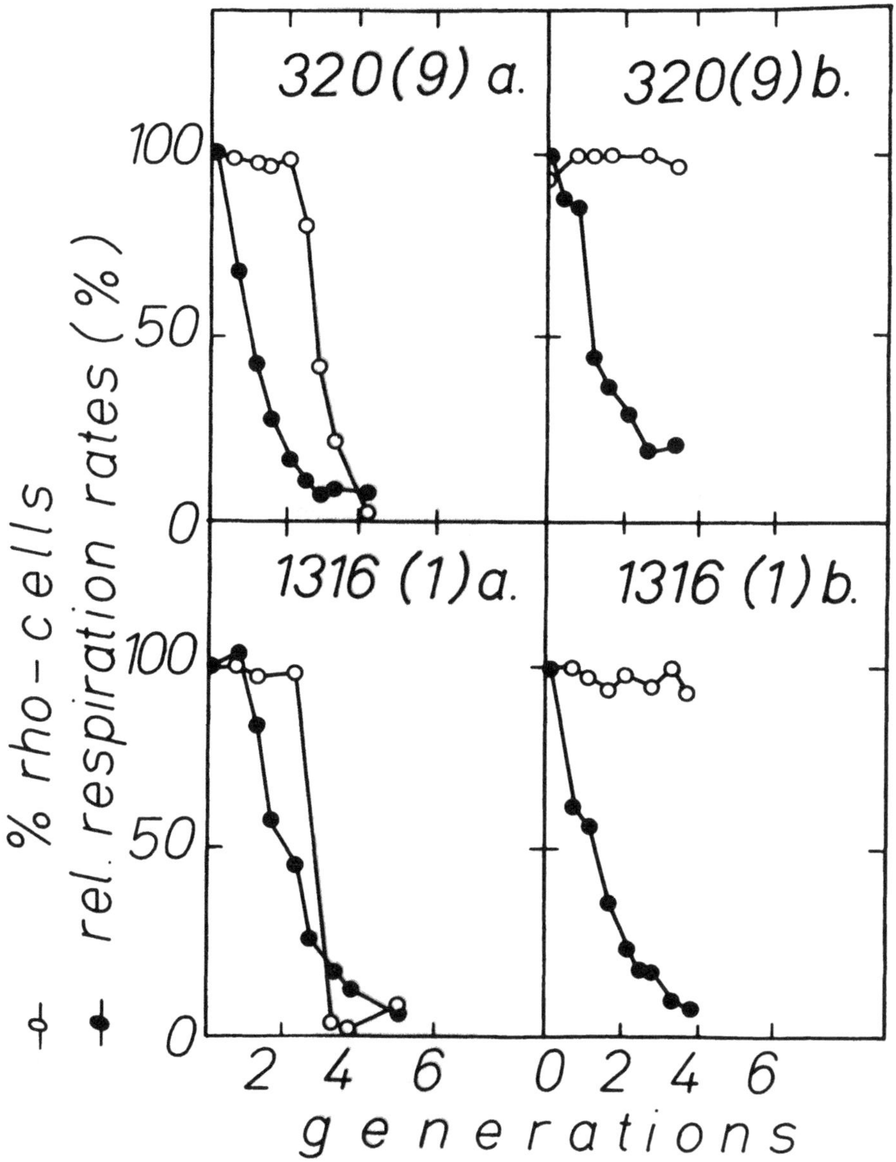

Fig. 4 Inactivation Kinetics of respiration and rho^{+}-character at non-permissive temperature in YEP-medium containing 0.1 % glucose (a) or 3 % glycerol (b) as a carbon source. For experimental conditions see Fig. 3.

decreases from 100 to below 10 percent sometimes within less than 1 generation (Fig. 3). This rather indicates an inactivation of functional mitochondrial DNA and not a passive dilution by cell division. It remains to be studied whether this is due to specific DNA degradation or to recombination with new and possibly nonfunctional DNA.

In two mutants studied, the rho^+-character could be preserved even after several generations at 36°C, if the cells were grown on lactate instead of glucose (Fig. 4). It remains to be studied further whether this effect is a general characteristic of all mutants of this phenotype and what may be the reason for this. If the inactivation of mitDNA is an ATP-dependent process like DNA-synthesis it should be specifically inhibited by the much more drastic decrease of ATP levels in _petite_ cells grown on lactate compared to those which grow on glucose.

4. Mitochondrial protein synthesis. So far, ts-mutants of 12 different _petite_-complementation groups have been studied for a possible defect in mitochondrial protein synthesis. Among these groups, apparently only two (group 1 and 2) were deficient in this process (Table 3). It may be worth noting that both groups affected but none of the unaffected ones tend to loose rho^+-character at 36°C (cf. Table 1). It remains to be studied with a larger sample of mutants whether this is a general phenomenon and what may the reason for this. So far, it may only be concluded, that under appropriate experimental conditions, mitochondrial protein synthesis activity is lost _before_ the cells become rho^-. (Table 3). Finally, it should also be kept in mind, that a defect in mitochondrial RNA synthesis would lead to a similar phenotype. This possibility should be studied in future experiments, too.

ACKNOWLEDGEMENT

We thank Dr. Fred Sherman, University of Rochester, for the possibility to use his Hartree low-dispersion spectroscope (Beck Ltd., London) for our first spectral screening experiments. This work was supported by the Deutsche Forschungsgemeinschaft and by the Stiftung Volkswagenwerk.

REFERENCES

1. P. Borst, J.P.M. Sanders, and C. Heyting, in "Genetics, Biogenesis and Bioenergetics of Mitochondria", (W. Bandlow, R.J. Schweyen, D.Y. Thomas, K. Wolf and F. Kaudewitz, eds.) W. de Gruyter, Berlin 1976. pp. 85-97.
2. J. W. Casey, H.-J. Hsu, G.-S. Getz, M. Rabinowitz and H. Fukuhara. J. Mol. Biol. _88_, 735 (1974).
3. For review see: G. Schatz, in "Genetics, Biogenesis and Bioenergetics of Mitochondria", (W. Bandlow, R.J. Schweyen, D.Y. Thomas, K. Wolf and F. Kaudewitz, eds.) W. de Gruyter, Berlin 1976. pp. 163-177.
4. F. Sherman, J.W. Stewart, M. Jackson, R.A. Gilmore, and J.H. Parker. Genetics _77_, 255 (1974).
5. L. Kovacs. Biochem. Biophys. Acta _346_, 101 (1974).
6. R.A. Woods, H.K. Sanders, M. Briquet, F. Foury, B.-E. Drysdale, and J.R. Mattoon. J. Biol. Chem. _250_, 9090 (1975).

7. A. Tzagoloff, A. Akai, and R.B. Needleman. J. Biol. Chem. 250, 8228 (1975).
8. F. Sherman, and P.P. Slonimski. Biochim. Biophys. Acta 90, 1 (1964).
9. G. Burkl, W. Demmer, H. Holzner, and E. Schweizer, in "Genetics, Biogenesis and Bioenergetics of Mitochondria". (W. Bandlow, R.J. Schweyen, D.Y. Thomas, K. Wolf and F. Kaudewitz, eds.) W. de Gruyter, Berlin 1976. pp. 39-48.
10. L. Kovac, G.S.P. Groot, and E. Racker. Biochim. Biophys. Acta 256, 55 (1972).
11. R.B. Needleman and A. Tzagoloff. Anal. Biochem. 64, 545 (1975).
12. G. Bray. Anal. Biochem. 1, 279 (1960).
13. M. Avron. Biochim. Biophys. Acta 40, 257 (1960).
14. L. Kovac and E. Hrusovska. Biochim. Biophys. Acta 153, 43 (1968).
15. M. Klingenberg and E. Phaff. Methods Enzymol. 10, 680 (1967).
16. Houben-Weyl, Methoden der organischen Chemie, Vol. XII (4th Edn.) G. Thieme, Stuttgart 1963. p. 81.
17. V. Riebeling, R. Thauer, and K. Jungermann. Eur. J. Biochem. 55, 445 (1975).

BIOCHEMICAL CHARACTERIZATION OF A TEMPERATURE CONDITIONAL MUTANT HAVING A MUTATION IN THE ω CONTROLLED REGION OF THE YEAST MITOCHONDRIAL GENOME AND EFFECTS OF DELETION OF THE MUTATION ON MITOCHONDRIAL RECOMBINATION.

H.B. Lukins, J.T. English, T.W. Spithill, R.J. Devenish, R.M. Hall, P. Nagley and A.W. Linnane.
Department of Biochemistry, Monash University, Clayton, Vic., 3168, Australia.

In the continuing search for new mutations in yeast mitochondrial DNA (mtDNA), several laboratories including our own have been isolating mutants with a respiratory deficient phenotype conditional upon growth at elevated (heat-sensitive) or reduced (cold-sensitive) temperatures. Mutants of this type can be expected to affect a variety of mitochondrial functions and are particularly amenable to biochemical analysis. The temperature sensitive mutations so far described are widely distributed on the mitochondrial genome (1,2,3,4,5,6) and affect such properties as maintenance of the mitochondrial genome as a whole (1), cytochrome *c* oxidase synthesis (2) and assembly of oligomycin sensitive ATPase (3). In this latter case, a specific defect in the rate of synthesis of subunit 6 of the ATPase was observed (7).

One cold-sensitive mutant, out of several temperature sensitive mutants currently under investigation in our laboratory, has been shown to carry a mutation located in the ω controlled region of the mitochondrial genome. Other laboratories have also reported mutations in this region producing temperature sensitive phenotypes (4,5,6). This region of the genome is of special interest since it includes the ω locus controlling mitochondrial recombination within the region, the gene for the 21 S rRNA of the mitochondrial ribosome and several antibiotic resistance markers which affect mitochondrial protein synthesis. This polar region of the genome can be approximated as the segment encompassing the *cap 1-r* (RIB 1) and *ery 1-r* (RIB 3) loci, and recent physical measurements on the mtDNA of petite mutants has indicated this segment to be only 5% (5 map units) of the genome (8).

Both the precise location of ω within this polar region and the molecular nature of ω have yet to be determined. A size difference in the nucleotide sequences for ω^+ and ω^- is one concept regarding the nature of the ω locus and studies on the derivation of mutant ω alleles have been interpreted as indicating that ω^n and ω^+ are partial macro-lesions of ω^- (9). The location of ω was originally proposed to be outside the region bounded by the RIB1 and RIB3 loci in the order ω RIB1 RIB3 (10), but a location between *cap1* (RIB1) and *ery1* (RIB3) was considered more likely by Howell *et al.*, (11) following the analysis of the segregation of ω alleles in minority recombinants.

This paper reexamines the location of the ω locus employing strains carrying a mutation designated *tsr1*$^-$ which produces a cold sensitive, respiratory deficient phenotype. This mutation is shown to be located between *cap1-r* and *ery1-r* and the biochemical characteristics of the mutant are consistent with the mutation being in the gene for the 21 S rRNA. The deletion of the *tsr1* locus from petite mutants that still retain the *cap1-r* and *ery1-r* markers is found to be associated with marked changes in the polarity properties of the petites indicative of changes at the ω locus. The significance of this observation in regard to the location and nature of the ω locus will be discussed.

ISOLATION AND MAPPING OF THE *tsr1*$^-$ MUTATION CONFERRING A COLD-SENSITIVE RESPIRATORY DEFICIENCY.

Cold sensitive mutants were isolated from wild type strain L410 (α *ura his* [ω^+]) by mutagenesis with ethylmethane sulphonate and by enrichment at 18°C in liquid medium containing ethanol as growth substrate in the presence of nystatin. Treated cells were then screened for mutants that failed to grow on solidified medium containing ethanol at 18°, but which grew normally on glucose at 18° and on glucose and ethanol at 28°.

The cold-sensitive isolate the subject of this communication showed a complex pattern of inheritance in an initial genetic cross, but a cold-sensitive haploid derivative of this cross (strain 151K-2-2C) on further analysis showed an inheritance pattern characteristic of a single mitochondrial mutation. This mutation is designated *tsr1*$^{-}$ (wild type, cold-resistant allele *tsr1*$^{+}$); the details of the isolation and characterization of this mutation as mitochondrial is given elsewhere (12).

The cold sensitive phenotype of the mutant is shown by comparing the generation times at 28° and 18° on glucose and ethanol for strain 151K-2-2C with the corresponding values for the wild type strain L410. In media containing glucose as carbon source, the generation times for the mutant and wild type are 2.6 and 2.4 hours respectively, while at 18° the values for both strains is 5.6 hrs. With ethanol as substrate at 28°, the mutant has a slightly longer (4.3 hrs.) generation time than the wild type (3.3 hrs.), but at 18° the value for the mutant is 31 hrs., about 4-fold greater than the original parent strain (8.8 hrs.). The slow growth of the mutant on ethanol at 18° is due neither to cell death nor to a marked induction of petite cells, the petite (*rho*$^{-}$) frequency at both temperatures being less than 5%.

The mapping of the *tsr1* locus within the polar region of the genome was shown by crossing the mutant strain 151K-2-2C to a set of genetically and physically defined petite strains previously used to derive a physical map of the mitochondrial genome (8,13,14). Production of cold-resistant diploid progeny in the cross to a petite indicates retention of the *tsr1*$^{+}$ allele in the petite genome, whereas absence of the locus from the petite is shown by the production of only cold-sensitive progeny. This analysis (see Figure 1) showed that the *tsr1*

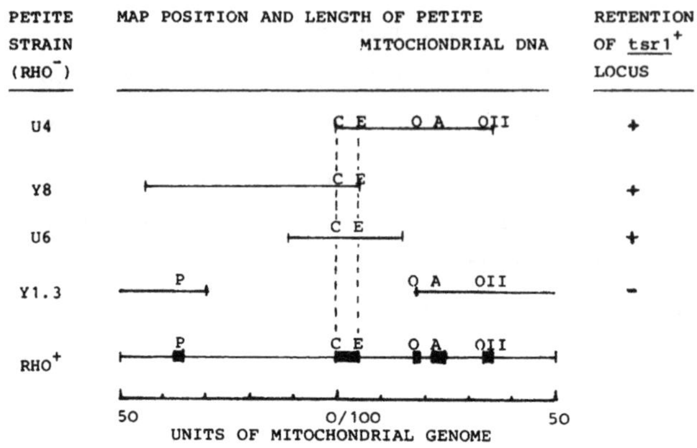

FIGURE 1. MAPPING OF *tsr1*⁻ MUTATION WITH PETITE STRAINS HAVING GENETICALLY AND PHYSICALLY DEFINED MITOCHONDRIAL GENOMES.
"+" indicates retention of the *tsr1*⁺ allele in the petite strain and "-" the absence. Retention or deletion of the *tsr1*⁺ allele in the *rho*⁻ strains determined by a cross to a tester strain carrying the *tsr1*⁻ allele. The letters above each genome indicate the retention of the following loci: C, *cap1*; E, *ery1*; O, *oli1*; A, *ana1*; OII, OLI2; P, *par1*. Results with only 4 *rho*⁻ strains shown; results with 10 other *rho*⁻ strains complement the above data.

locus is located within the region from 0 to 5 units on the map common to petites U4 and Y8. This is the polar region of the genome containing the *cap1* and *ery1* loci and the 21 S rRNA gene.

The order of the three loci *cap1*, *ery1* and *tsr1* within the polar region was determined by three different analyses, namely, the patterns of co-retention of the three markers in petite mutants, recombination frequencies in non-polar crosses and polarity of transmission and recombination in polar crosses. A placement for *tsr1* between *cap1* and *ery1* was first suggested from the examination of spontaneous petite strains derived from a grande strain carrying the *tsr1*⁺ allele together with five antibiotic resistance markers. All petite isolates which retained both the *cap1-r* and *ery1-r* markers also retained the *tsr1* locus, while several petites which retained *ery1-r* together with the adjoining locus *oli1* had lost the *tsr1* locus. These results indicate that *tsr1* is located on the *cap1* side of *ery1*, distal from *oli1*. Linkage of *tsr1* to *ery1* was also indicated in this analysis since the frequency of separation of these two markers was less than for any other pair of markers.

The order *cap1 tsr1 ery1* and linkage of *tsr1* to *ery1* is also shown in the results of polar and non-polar crosses (Table 1). In the non-polar cross (upper section, Table 1) linkage between *tsr1* and *ery1* is clearly shown by a recombination frequency of only 5%, while the frequencies of 10% for the *cap1 tsr1* pair and 13% for *cap1 ery1* indicate the order *cap1 tsr1 ery1*. Recombination frequencies with other markers outside the polar region are of the order of 20% and show no linkage. The polarity of transmission and recombination observed in the polar cross (lower section, Table 1) confirm the order *cap1 tsr1 ery1*. The results for the polar cross are the mean of three separate crosses, in each case the transmission of the *tsr1* marker exceeded that of *ery1*.

In summary, the several mapping procedures show that the *tsr1* locus is located between the *cap1* and *ery1* loci and is more closely linked to *ery1*. Results presented in the last section

TABLE 1

MAPPING OF tsr1⁻ MUTATION BY TRANSMISSION AND RECOMBINATION IN POLAR AND NON-POLAR CROSSES.

CROSS	TRANSMISSION (%)		RECOMBINATION FREQUENCY (%)		
			cap1/tsr1	10	
[ω^+tsr1$^-$]			cap1/ery1	13	
X			tsr1/ery1	5	
[ω^+cap1-r tsr1$^+$ ery1-r			cap1/OII	22	
ana1-r OII par1-r]			tsr1/OII	20	
			ery1/OII	19	
	cap1-s	95	cap1/tsr1	23	(7)
[ω^+tsr1$^-$]	tsr1$^-$	81	cap1/ery1	19	(166)
X	ery1-s	77	tsr1/ery1	6	
[ω^-cap1-r tsr1$^+$ ery1-r	oli1-s	58	cap1/oli1	33	(336)
oli1-r ana1-r par1-r]			tsr1/oli1	22	
			ery1/oli1	18	

Figures in brackets are polarity of recombination; for example, between cap1 and ery1 the polarity is the ratio cap1-s ery1-r/cap1-r ery1-s, where "r" and "s" indicate resistance and sensitivity to the antibiotic. For complete details of these crosses see English et al.,1977.

of this paper, relating to deletion of the *tsr1* locus and associated effects on the ω locus, provide additional evidence for this conclusion.

BIOCHEMICAL CHARACTERIZATION OF THE MUTANT STRAIN 151K-2-2C [*tsr1*$^-$].

As a result of the localization of the *tsr1* locus in a short segment of the mtDNA which includes the cistron for the 21 S rRNA component of the 50 S mitochondrial ribosomal subunit and resistance loci for antibiotics affecting mitochondrial protein synthesis, it can be expected that the *tsr1*$^-$ mutation might also affect protein synthesis by the mitochondria. Biochemical studies of the mutant has indicated this to be the case.

In all experiments comparing the biochemical properties of the mutant at the permissive (28°) and restrictive (18°) temperatures, the growth of cells was performed in glucose-limited chemostat to minimise glucose repression of mitochondrial development. Examination of the cytochrome spectra of mutant cells grown at 18° revealed the presence of only trace amounts of cytochromes $\underline{aa}_3$ and b but normal levels of cytochrome c, while all cytochromes were present at normal levels in cells grown at 28°. Table 2 shows the specific activities of cytochrome c oxidase, NADH-cytochrome c reductase and oligomycin-

ENZYME ACTIVITIES IN MITOCHONDRIA ISOLATED FROM STRAINS L410[tsr1$^+$] AND 151K-2-2C[tsr1$^-$].

STRAIN	GROWTH TEMPERATURE	ATPase	NADH-CYTOCHROME c REDUCTASE	CYTOCHROME c OXIDASE
L410 tsr1$^+$	28°	0.17 (80)	0.90	3.6
	18°	0.17 (75)	1.38	3.9
151K-2-2C tsr1$^-$	28°	0.16 (75)	1.28	3.7
	18°	0.16 (15)	0.26	0.3

Activities are expressed as umoles substrate converted/min./mg protein. Figures in brackets show the percentage of ATPase activity which is sensitive to inhibition by 20 ug oligomycin/mg protein. For details of assays see Spithill et al.,1977.

sensitive ATPase in mitochondria from mutant cells grown at the two temperatures, together with corresponding measurements obtained with mitochondria from the wild type strain. Similar values for each of the membrane bound enzyme activities are found in mutant and wild type mitochondria from cells grown at 28°, but at 18° extremely low levels of activity are observed in the mutant mitochondria. In the case of the oligomycin-sensitive ATPase, the total F_1 ATPase activity is at a normal level in mitochondria from mutant cells cultured at 18°, but the activity is insensitive to oligomycin. Since growth of the mutant at 18° leads to the simultaneous loss of a number of mitochondrial enzyme activities, all of which require mitochondrial translation products for normal function, a defect in mitochondrial protein synthesis at 18° seems most probable.

The ability of the mutant to carry out mitochondrial protein synthesis at 18° was examined with whole cells and *in vitro* with both mitochondria and isolated mitochondrial ribosomes. The mitochondrial protein synthesizing activity *in vivo* was measured as the cycloheximide-resistant incorporation of [^{3}H]-leucine into exponential phase cells. As shown in Figure 2, active mitochondrial protein synthesis occurs in both strains during the first 30 min. of incubation at 28°, comparable rates being observed with both strains irrespective of whether the cells had been grown at 18° or 28° prior to assay at 28° (results for L410 not shown). Likewise, although the rate of [^{3}H]-leucine incorporation is slower at 18° than 28°, as expected, there is no marked difference between the two strains in rate of mitochondrial protein synthesis. Indeed, the mutant strain at both 18° and 28° shows a slightly higher rate of protein synthesis than the wild type, and more noticeably, for mutant cells grown at 18° the linear rate of synthesis is maintained well after incorporation into wild type cells has ceased. It is clear that the mutant strain is capable of performing mitochondrial protein synthesis at the restrictive temperature.

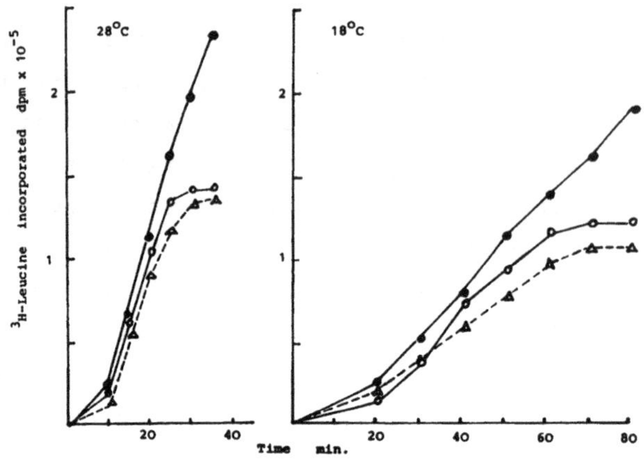

FIGURE 2. MITOCHONDRIAL PROTEIN SYNTHESIS IN CELLS OF STRAIN 151K-2-2C[*tsr1*$^-$] AND L410 [*tsr1*$^+$].
The incorporation of [^{3}H]-leucine (in the presence of cyclo-heximide at 0.6 mg/ml) into cells incubated at 28° and 18° was measured as described by Spithill *et al*, 1977. △---△ , strain L410 grown at 28° (essentially the same result is obtained for cells grown at 18°); o——o, strain 151K-2-2C grown at 28°; ●——●, strain 151K-2-2C grown at 18°.

For the measurement of *in vitro* mitochondrial protein synthesizing activity, mitochondria were isolated from the mutant and wild type strains grown at both 28° and 18°. When isolated mitochondria were assayed at 28°, the same high rates of incorporation (20 - 23 picomoles [^{14}C]-leucine incorporated/mg protein/20 min.) were obtained for both strains (Table 3), this being irrespective of whether the mitochondria were obtained from cells grown at 28° or 18°. When assayed at 18° the mitochondria from strain L410 showed a reduced rate of protein synthesis, namely 0.28 times the rate at the higher temperature. By contrast, mitochondria from the mutant showed a markedly decreased rate of amino acid incorporation at 18°, being only 0.1 times that observed at 28°. This distinct cold-sensitivity is observed in the mitochondria of strain 151K-2-2C regardless of the growth temperature for the cells.

While the results with isolated mitochondria indicate a cold-sensivity in some function involved in mitochondrial protein

TABLE 3

PROTEIN SYNTHESIS BY ISOLATED MITOCHONDRIA AND MITO-RIBOSOMES FROM STRAINS L410[tsr1$^+$] AND 151K-2-2C[tsr1$^-$].

STRAIN	GROWTH TEMPERATURE	PROTEIN SYNTHESIS ACTIVITY			
		MITOCHONDRIA			MITORIBOSOMES
		18°	28°	RATIO 18°/28°	28°
L410 tsr1$^+$	28°	5.9	21.2	0.28	12,808
151K-2-2C tsr1$^-$	28°	2.2	20.2	0.11	0
151K-2-2C tsr1$^-$	18°	2.3	23.0	0.10	0

Cells were grown at the temperatures indicated and protein synthesis assayed at 18° or 28°. Mitochondrial activities are expressed as pmoles ^{14}C-leucine incorporated/mg protein/20 min. Activities with isolated mitochondrial ribosomes are expressed as dpm ^{14}C-phenylalanine incorporated/unit A260nm/ 15 min. Ribosomes were incubated with ^{14}C-phenylalanine in the presence of poly-U and an E.coli S100 fraction. For details of ribosome isolation and assay procedures see Spithill et al.,1977.

synthesis, they provide no evidence relating to the particular component of the translation system (ribosomes, tRNA's, soluble factors) which is affected by the mutation. To distinguish a defect located in the mitochondrial ribosomes from one present in the tRNA or soluble component fractions of the translation system, the ability of purified mitochondrial ribosomes to catalyze poly-U directed polyphenylalanine synthesis in the presence of the soluble components of an E. coli S 100 fraction, was determined. Under the experimental conditions employed (15), the mitochondrial ribosomes extracted from strain L410 had an activity comparable to that of E. coli ribosomes (Table 3); the activity was completely abolished by addition of excess unlabelled phenylalanine and was dependent on the addition of poly-U and the soluble factors present in the E. coli S 100 fraction. The activity was insensitive to cycloheximide but was inhibited to greater than 90% by the specific mitochondrial protein synthesis inhibitor mikamycin. However, mitochondrial ribosomes isolated from strain 151K-2-2C grown at either 28° or 18°, and prepared concurrently with the active ribosomes from L410, were completely inactive in the translation of the poly-U message (Table 3).

During the course of a number of experiments aimed at obtaining active ribosomal preparations from the mutant strain, aliquots of the ribosomal preparations were taken and examined by sedimentation velocity in sucrose gradients. The preparations from strain L410 routinely contained the normal relative amounts of the 50 S and 37 S subunits, but the preparations from the mutant were quite variable. Thus, whilst some mitochondrial ribosomal preparations from strain 151K-2-2C contained normal levels of both subunits, other preparations showed profiles in which the 50 S subunit was either moderately or severely reduced relative to the 37 S subunit. These observations indicate there is no gross change in the synthesis or overall structure of the mitochondrial ribosomes in the mutant grown at either temperature, but rather some subtle change in the 50 S subunit which manifests itself during extraction and results in ribosomes non-functional in protein synthesis.

It is evident from the assessment of all the biochemical data that the *tsr1*$^-$ mutation effects some perturbation in the mitochondrial ribosomes, most probably the 50 S subunit, at the restrictive temperature. An effect on the 50 S subunit is consistent with the likely location of the mutation in the gene for the 21 S rRNA. Although the ribosomes catalyze protein synthesis *in vivo* at 18°, the products of this synthesis are non-functional as shown by the deficiencies in a number of enzyme complexes containing polypeptide components made on the mitochondrial ribosomes. Presumably,the change in the ribosomes at 18° affects some step in the mitochondrial protein synthesis cycle, such as correct chain initiation, fidelity of codon recognition or correct chain termination, resulting in the inability to faithfully translate mitochondrial messenger RNA. Experiments in progress are comparing the products of mitochondrial translation formed in cells at 28° and 18°, using sodium dodecylsulphate-polyacrylamide electrophoresis and immunoprecipitation procedures. Preliminary results have indicated some minor, but significant differences in the patterns of products formed at the two temperatures.

The biochemical properties found for strain 151K-2-2C resemble in several features those reported by Mason *et al.*, (6) for a strain carrying the mitochondrial mutation cs 23. The cs 23 mutation is located in the polar region of the genome and leads to the general reduction in the activity of several mitochondrial enzyme complexes and also appears to affect the level of mature 50 S mitochondrial ribosome subunits that can be extracted from the cells.

DELETION OF THE *tsr1* LOCUS IN PETITE MUTANTS AND ASSOCIATED EFFECTS ON THE ω LOCUS.

The region of mtDNA containing the *cap1* and *ery1* loci has been shown by physical measurement to be a maximum of 5 map units, equivalent to approximately 3,750 base pairs (8). Contained within this region is the major part, if not all, of the cistron coding for the 21 S rRNA, the minimum size for this gene being about 3,600 base pairs. Despite intensive investigation, it is not yet firmly established whether both the *cap1* and *ery1* loci lie within the 21 S rRNA gene (see articles by Nagley *et al.*, Sanders *et al.*, Heyting and Sanders, and Locker and Rabinowitz in ref. 16).

This laboratory has previously suggested, following analyses of the segregation of ω alleles in minority recombinants, that the ω locus may be located between the *cap1* and *ery1* loci (11). This location of ω has also been suggested by Heyting and Sanders (17) in their analysis of restriction-endonuclease fragments of petite mutants containing the RIB1-RIB3 region of mtDNA. If both the *cap1* and *ery1* loci are within the 21 S rRNA gene and ω is positioned between them, then the ω locus must also occur in the rRNA gene, a possibility which seems unlikely. The availability of the *tsr1*$^-$ mutation, which we have shown is most probably located in the rRNA gene, has prompted us to reexamine the location of ω, on the assumption that specific deletion of the *tsr1* locus might possibly have an effect on the ω locus and the properties of mitochondrial recombination.

From a grande strain of mitochondrial genotype [ω^- *cap1-r tsr1*$^+$ *ery1-r oli1-r ana1-r par1-r*] spontaneous petite mutants were selected for the retention of only the *cap1-r tsr1*$^+$ *ery1-r* markers. These petites were repeatedly subcloned to a high level of stability (>95% of subclones retaining all three markers), and then during further subcloning steps petites of genotype *cap1-r tsr1-o ery1-r* (where *tsr1-o* indicates deletion of this marker) were obtained at low frequency. It was found that each of these latter petites were very stable (>95%). As the *tsr1* locus is located between *cap1* and *ery1*, the isolates of genotype *cap1-r tsr1-o ery1-r* must have arisen from an internal deletion between the two antibiotic resistance loci.

To establish the properties of recombination of these petite mutants each was crossed to a pair of antibiotic-sensitive tester strains, one carrying the ω^+ allele and the other ω^-. The results are shown in Table 4, together with similar crosses of the parent grande strain. The petite 68.4-2 [*cap1-r tsr1*$^+$ *ery1-r*] shows similar properties of recombination as the

TABLE 4

ANALYSIS OF TRANSMISSION AND RECOMBINATION IN CROSSES INVOLVING PETITE STRAINS IN WHICH THE [tsr1$^+$] LOCUS IS EITHER RETAINED OR DELETED.

STRAIN	GENOTYPE •			TESTER STRAIN	TRANSMISSION FREQUENCY (%)		cap1/ery1 RECOMBINATION		APPARENT ω ALLELE
	cap1-r	tsr1$^+$	ery1-r		cap1-r	ery1-r	FREQUENCY (%)	POLARITY ⧧	
ω^- rho$^+$	+	+	+	ω^-	38.3	44.1	12.5	0.37	ω^-
				ω^+	21.0	47.8	26.8	0/79	
68.4-2 rho$^-$	+	+	+	ω^-	68.5	72.5	6.8	0.56	ω^-
				ω^+	4.5	61.0	37.7	1/275	
68.4-3 rho$^-$	+	o	+	ω^-	66.4	3.7	62.7	222/0	ω^+
				ω^+	5.6	5.4	8.8	1.07	
68.4-4 rho$^-$	+	o	+	ω^-	90.8	59.7	27.7	202/0	ω^+
				ω^+	29.5	25.1	13.3	1.55	
68.4-5 rho$^-$	+	o	+	ω^-	58.8	2.5	56.2	199/0	ω^+
				ω^+	8.5	6.2	12.4	1.44	

The parent strain from which the petite strains were derived has the genotype [ω^-cap1-r tsr1$^+$ ery1-r oli1-r ana1-r par1-r]. The ω^- and ω^+ tester strains carried the sensitive alleles for each of the antibiotic resistance markers. Retention or deletion of the tsr1$^+$ allele was shown in a separate cross to a strain carrying the tsr1$^-$ allele.

• "+" indicates retention of the locus in the strain and "-" the absence.

⧧ Polarity of recombination is the ratio of the frequencies of genotypes cap1-r ery1-s/cap1-s ery1-r.

grande strain from which it was derived, namely, no marked polarity of transmission or recombination in the cross to the ω^- tester, but high polarities and elevated recombination frequencies in the ω^+ tester cross. This result is typical for many petites of this genotype and it can be concluded that they carry an unaltered ω^- allele. By contrast, three petite strains (68.4-3, 68.4-4, 68.4-5), all derived from 68.4-2 by deletion of the *tsr1*$^+$ locus, show the direct reversal of these properties. Thus polar recombination properties are now observed in the cross to the ω^- tester and non-polar recombination and transmission in the cross to the ω^+ tester strain. This is most clearly seen in the case of petite 68.4-4, but is also shown by petite 68.4-3 and -5, albeit with a noticeably lower transmission of *cap1-r* and *ery1-r* in the cross to the ω^- tester. These recombination properties, namely polarity in a cross to an ω^- tester and no polarity in a cross to ω^+ are those anticipated for strains carrying the ω^+ allele.

The simplest interpretation of these observations is that the order of loci in the polar region is *cap1* ω^- *tsr1 ery1* and that the deletion event which removes the *tsr1*$^+$ locus also results in a partial deletion of the ω^- allele. This lesion in ω^- produces the ω^+ allele. This interpretation would be consistent with the suggestion of Dujon *et al.* (9) who proposed that ω^+ is related to ω^- by a macrolesion following their studies on the derivation of ω^n (neutral) mutants. Although the actual separation between *cap1* and *ery1* is not known, this mechanism would probably necessitate the deletion of a sizeable segment of mtDNA, since the ω locus is linked to *cap1* and *tsr1* to *ery1*. An alternative explanation for the apparent change in the ω alleles could be based on molecular changes occurring in mtDNA following the initial deletion event. Thus there might be selective amplification of some fragment of the retained mtDNA; for instance, repetition of some sequences located close to *cap1* relative to sequence linked to *ery1*. Dependent on the size of the repeated sequence and the degree of reiteration, a possible consequence of this repetition could be the formation of a mismatched region during pairing of the petite genome with the genome of the grande tester strain. This mismatched sequence might initiate the events resulting in polar recombination in a manner similar to that hypothesized by Perlman and Birky (18).

The resolution of these or other explanations for the observed phenomena will require further experimentation, but may provide an insight into the nature of the ω locus. Further genetic examination of these petite mutants, together with studies of their mitochondrial genomes by hybridization and restriction endonuclease analysis, are in progress in our laboratory.

SUMMARY

The mitochondrial mutation *tsr1*$^-$ confers a cold-sensitive respiratory deficient phenotype. The mutation has been shown to be located between the *cap1* (RIB1) and *ery1* (RIB3) loci, being more closely linked to *ery1*. This small segment of mtDNA includes the gene for the 21 S rRNA component of the 50 S subunit of the mitochondrial ribosome. Biochemical studies of protein synthesis by mitochondria and mitochondrial ribosomes isolated from a strain carrying the *tsr1* mutation have shown that the mitochondrial ribosomes (most probably the 50 S subunit) are affected by the mutation such that non-functional mitochondrial translation products are formed at the restrictive temperature. The genetic mapping and biochemical analyses suggest the mutation resides in the gene for the 21 S rRNA. Deletion of the *tsr1* locus from petite mutants that otherwise retain the *cap1* and *ery1* loci has been found to be associated with an apparent conversion of the ω^- allele to ω^+. This observation provides additional evidence for location of the ω locus between the *cap1* and *ery1* loci.

REFERENCES

1. Handwerker, A., Schweyen, R.J., Wolf, K. and Kaudewitz, F. Evidence for an extrakaryotic mutation affecting the maintenance of the rho factor in yeast. J. Bacteriol. 113, 1307-1310

(1973).

2. Storm, E.M. and Marmur, J. A temperature sensitive mitochondrial mutation of Saccharomyces cerevisiae. Biochem. Biophys. Res. Commun. 64, 752-759 (1975).

3. Trembath, M.K., Monk, B.C., Kellerman, G.M., and Linnane, A.W. Biogenesis of Mitochondria, 27. Genetic and biochemical analysis of a mitochondrially determined cold sensitive oligomycin resistant mutant of Saccharomyces cerevisiae with affected mitochondrial ATPase assembly. Molec. Gen. Genet. 141, 9-22 (1975).

4. Lancashire, W.E. Mitochondrial mutations conferring cold sensitivity in Saccharomyces cerevisiae. In, "Genetics and Biogenesis of Chloroplast and Mitochondria". Th. Bücher et al., eds. Elsevier/ North Holland Biomedical Press, Amsterdam. pp 481-490, (1976).

5. Bolotin-Fukuhara, M., Faye, G., and Fukuhara, H. Temperature sensitive respiratory deficient mitochondrial mutations: Isolation and genetic mapping. Molec. Gen. Genet. 152, 295-305 (1977).

6. Mason, T., Boerner, P. and Biron, C. The use of double mutant strains containing heat- and cold-sensitive mutations in studies of mitochondrial biogenesis. In, " Genetics and Biogenesis of Chloroplasts and Mitochondria", Th. Bücher et al., eds., Elsevier/ North Holland Biomedical Press, Amsterdam, (1976).

7. Groot Obbink, D.J., Hall, R.M., Linnane, A.W., Lukins, H. B., Monk, B.C., Spithill, T.W. and Trembath, M.K. Mitochondrial genes involved in the determination of mitochondrial membrane proteins. In reference 16, pp 163-173.

8. Choo, K.B., Nagley, P., Lukins, H.B. and Linnane, A.W. Biogenesis of mitochondria, 47. Refined physical map of the mitochondrial genome of Saccharomyces cerevisiae determined by analysis of an extended library of genetically and molecularly defined petite mutants. Molec. Gen. Genet. 153, 279-288 (1977).

9. Dujon, B., Bolotin-Fukuhara, M., Coen, D., Deutsch, J., Netter, P., Slonimski, P.P. and Weill, L. Mitochondrial Genetics, XI. Mutations at the mitochondrial locus ω affecting the recombination of mitochondrial genes in Saccharomyces cerevisiae. Molec. Gen. Genet. 143, 131-165 (1976).

10. Dujon, B., Slonimski, P.P. and Weill, L. Mitochondrial genetics, IX. A model for recombination and segregation of mitochondrial genomes in Saccharomyces cerevisiae. Genetics, 78, 415-437 (1974).

11. Howell, N., Hall, R.M., Linnane, A.W. and Lukins, H.B. Genetic analyses of the polarity alleles in recombinants from genetic crosses. J. Bacteriol. 119, 1063-1065 (1974).

12. English, K.J., Devenish, R.J., Oliver, C.J., Linnane, A.W. and Lukins, H.B. Biogenesis of mitochondria, 49. Identification and mapping of a new mitochondrial locus (*tsr1*) which maps within the polar region of the yeast mitochondrial genome. Molec. Gen. Genet. submitted, (1977).

13. Sriprokash, K.S., Molloy, P.L., Nagley, P., Lukins, H.B. and Linnane, A.W. Biogenesis of mitochondria, 41. Physical mapping of mitochondria genetic markers in yeast. J. Mol. Biol. 104, 485-503 (1976).

14. Linnane, A.W., Lukins, H.B., Molloy, P.L., Nagley, P., Rytka, J. Sriprokash, K.S., and Trembath, M.K. Biogenesis of mitochondria. Molecular mapping of the mitochondrial genome of yeast. Proc. Nat. Acad. Sci. U.S.A. 73, 2082-2085 (1976).

15. Spithill. T.W., English, K.J., Nagley,P. and Linnane, A.W. Biogenesis of mitochondria, 50.Biochemical characterization of a new cold-sensitive mutant of yeast with a mutation in mitochondrial DNA affecting the large subunit of the mitochondrial ribosome. Molec. Gen. Genet., submitted (1977).

16. Saccone, C. and Kroon, A.M., eds., "The Genetic Function of Mitochondria DNA", Elsevier/North Holland Biomedical Press, Amsterdam. (1976).

17. Heyting, C., and Sanders, J.P.M. The physical mapping of some genetic markers in the 21 S ribosomal region of the mitochondrial genome of yeast. In reference 16, pp 273-280.

18. Perlman, P.S. and Birky, C.W. Mitochondrial genetics in baker's yeast: A molecular mechanism for recombinational polarity and suppressiveness. Proc. Nat. Acad. Sci. U.S.A. 71, 4612-4616 (1974).

Preferential Loss of a Specific Region of Mitochondrial DNA by rho^- Mutation

H. Fukuhara and M. Wesolowski
Fondation Curie, Institut du Radium, Section de Biologie, Bâtiment 110, 91405 - ORSAY, France.

The deletion of mitochondrial DNA sequences in rho^- mutation has been studied by many laboratories to characterize the genetic and physical organization of yeast mitochondrial DNA. Although the physical deletion of various sequences can be assessed by the loss of genetic markers, the scarcity of appropriate markers has been a major limitation in this approach. When we began the deletion mapping study using rho^- mutants in order to locate the transfer and ribosomal RNA genes, it soon became apparent that the rho^- mutants are indeed a powerful tool of gene mapping. However, double or multiple deletions do occur which can introduce ambiguity in ordering the mitochondrial genes, especially when their distances are large.

In our earlier studies, we have used some forty rho^- mutants to order a dozen mitochondrial tRNA genes (1). These mutants were purified stable rho^- clones obtained after two or three subclonings following mutagenesis. Many of them have been defined with respect to a very limited number of genetic markers. Genetic structures of some of the mutants could only be explained by secondary or multiple deletions. This was later confirmed by the use of many other mitochondrial genetic markers now available, and also by restriction enzyme analysis of rho^- mitochondrial DNA (2,3).

The use of a large number of mitochondrial genetic markers has offered us an opportunity to re-examine, by a different approach, the validity of the previously proposed gene order, and to uncover some new aspects of the rho^- mutation. A grande strain was mutagenized by ethidium bromide and the fate of fifteen mitochondrial genetic markers was followed in unselected primary clones.

Mutagenesis

We used a grande strain of *Saccharomyces cerevisiae*, MH41-7B/011 (*a* ade_2 $his_1 bet_{011}$). It carries four mitochondrial drug resistance markers (C^R_{321} $E^R_{514} O^R_{145} P^R_{454}$) and a nuclear mutation bet 011, which confers resistance to ethidium bromide. When grown on a complete glucose medium at 35°C in the presence of ethidium bromide (10 μg/ml, pH 6.5), the strain produces only a small percentage of rho^- cells (less than 2 % per cell doubling). After 20 hours growth under these conditions, about 20 % of cells become rho^-, while the spontaneous rho^- production is 0.5 %. The original sensitive strain gives 100 % petites by the treatment and more than 90 % are of rho^- type. Detailed properties of the ethidium resistant mutant will be described elsewhere.

We analyzed the genetic structure of the primary petite clones generated from the strain MH41-7B/011 treated with ethidium under the conditions described above. After the mutagenesis, a total of 379 rho^- primary clones were randomly picked up and aligned on glucose complete agar plates. The presence or absence of fifteen mitochondrial loci was tested in every clone by the replica cross technique : the matrix plates carrying the spots of primary clones were replicated onto the lawns of different mit^- mutants to generate diploids, then the latter were replica-plated onto glycerol complete agar. The growth was scored after 4 days at 28°C. Any rho^- clones that can restore the growth on glycerol after the cross with a mit^- tester should contain the corresponding mit^+ locus (4). In the case of temperature-sensitive mit^- testers, the glycerol plates carrying the diploids issued from the cross were incubated at non-permissive temperature. The presence of drug resistance loci were tested by a replica cross on a drug sensitive grande strain (ILI25-10C, α ura), followed by replica plating onto various drug media (5). The mit^- mutants used are the following (see ref. 2) : tscs 983, mit 175, mit 226 (cob-1), mit 206 (oxi-3), mit 231 (oxi-2), ts 982, mit 170, mit 201 (oxi-1), M10-152 (cob-2, from Dr. Tzagoloff), M6-239 (pho 1, from Dr. Tzagoloff) and SM50-11C (tsm 8, from Drs. Schweyen and Monnerot). All these mutants can give rise to grande revertants indicating that they are not deletion mutants ; the choice of non-deletion testers was important in order to avoid a bias in the determination of marker retention frequency in rho^- clones.

Most of the primary clones show a single deletion pattern.

A complete analysis of the randomly collected 379 primary petite clones was made with regard to fifteen genetic markers. All the clones were classified according to their genotype, and the frequency of each genotype scored. The results are shown in figure 1.

At the top of the figure 1, the 15 markers are aligned circularly according to the order deduced from our previous mapping studies using purified stable rho^- strains (non-random, selected clones). Assuming that this gene order is the correct one, we can examine the deletion structure of the primary clones by means of the markers they retain. We found that, out of 379 primary clones examined, 321 (84.7 %) were compatible with single deletion structures, 23 (6.1 %) have lost all markers, and only 35 (9.2 %) showed double deletions. Any gene arrangements other than that indicated at the top of the figure 1 lead to an extremely high frequency of double deletions. In other words, the direct analysis of 379 random primary clones has led us to the same arrangement of 15 markers as that deduced from the analysis of individually purified non-random petite clones.

Deletions found in the primary clones are often small.

Among the 15 markers that we have used in this mutagenesis experiment, two linked markers, C and E, always segregated together (simultaneous loss or retention), because the number of clones examined was too small to detect their disjunction. Therefore, we may here consider them as a single block, that is, the effective number of markers was 14. Thus the number of all possible single deletion genotypes is 14 x (14 - 1) = 182. Actually 78 different single deletions were found with various frequencies (0.3 - 9.5 %).

Figure 1. Frequencies of various mitochondrial genotypes in rho⁻ primary clones.

Heavy lines represent the DNA sequences retained in the rho⁻ clones, as determined by the presence of the genetic markers indicated on the top of the diagram. 379 randomly collected rho⁻ primary clones were analyzed, and 321 clones assembled here were of single deletion type. 23 that contained no markers and 35 that showed double deletion genotypes are not represented. The frequency of each genotype is indicated to the right. The markers are arranged in a circular order.

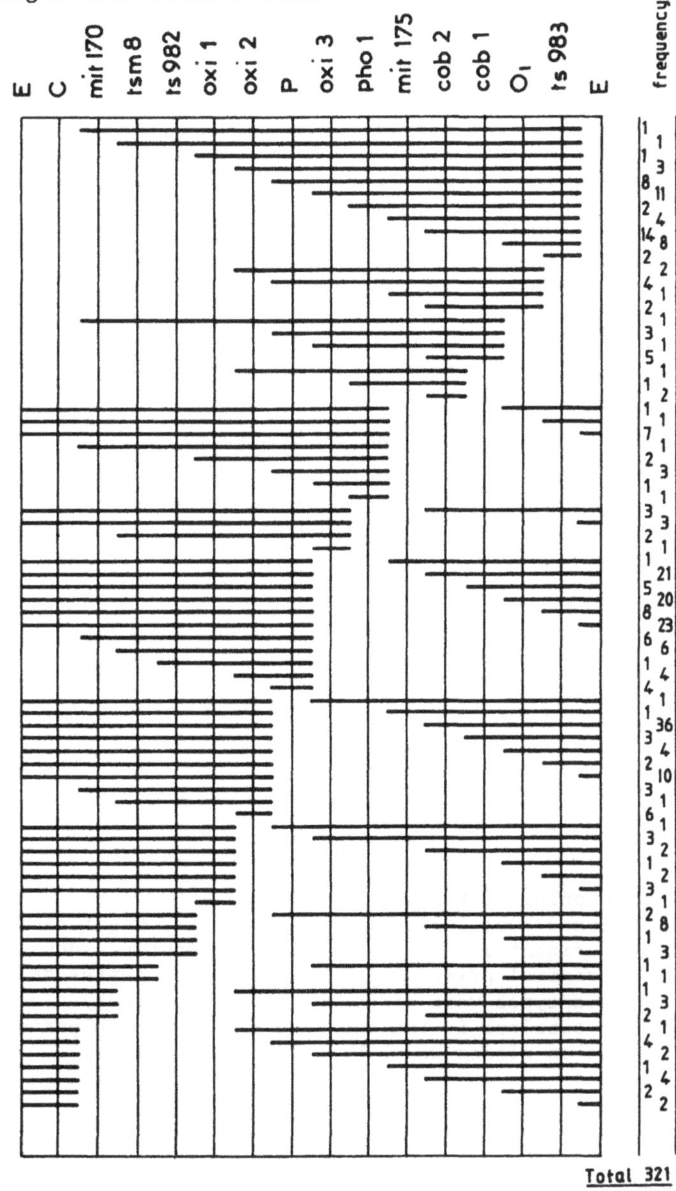

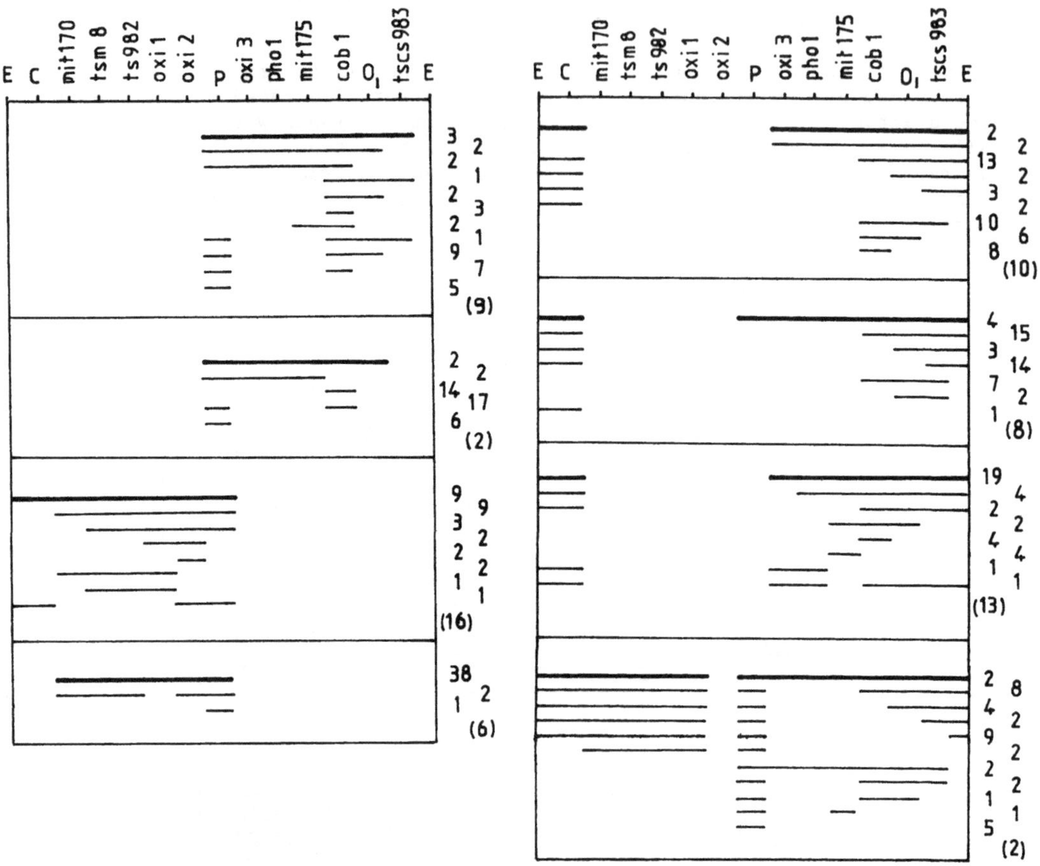

Figure 2. Subcloning of the primary clones.
The primary clones from figure 1 were subcloned. Eight examples are shown. Heavy horizontal lines stand for the genotype identical to that found in the original primary clones. Thin lines represent other subclones whose genotypes were different from the primary clone. The frequency of each subclone is indicated to the right. In parentheses are the frequencies of clones which have lost all markers.

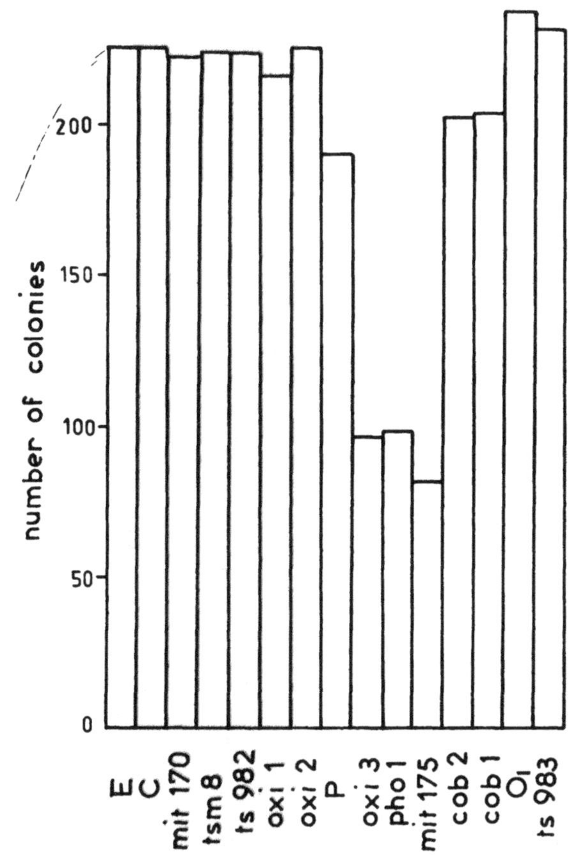

Figure 3. Unequal loss of mitochondrial genetic markers from the grande strain treated with ethidium bromide.

The number of primary clones that retained the genetic marker indicated was scored over the total of 379 rho$^-$ primary clones randomly collected after mutagenesis.

It is striking that the mitochondrial DNA sequences retained in many primary clones are apparently very large, as estimated from the number of genetic markers retained. About 50 % of the clones have retained at least 8 markers. In some cases, the deletion was represented by a loss of only one or two markers. To see whether these apparently single-deleted complex genotypes reflect the presence of large continuous genomes rather than a mixture of many fragmented sequences, nineteen primary clones were selected and subcloned. Each subclone was examined again for all genetic markers.

As shown in the examples of figure 2, we found that the genotypes assigned for the primary clones can be recovered among the secondary clones in most cases examined (18 out of 19). The frequency of such secondary clones is generally low, and most of other subclones represented much fragmented genotypes. Many double deletions appeared in these secondary clones.

The secondary clones that have retained a large part of the wild type genome are often unstable and do not survive further subclonings. However some of them have been purified in the stable form (see examples in ref.2).

The fact that the deletions in primary clones can be very small, as compared to most of the purified petite subclones may indicate that the initial event of the rho$^-$ deletion is a small deletion. Such deleted molecules seem to be more or less unstable, although some may survive for approximately 40 cell generations as seen above. Most of them will then fragment before becoming stable in the form of a repetitive DNA.

<u>A specific region of the mitochondrial DNA is preferentially lost by rho$^-$ mutation.</u>

Different regions of the mitochondrial DNA are retained in rho$^-$ clones, and the genetic markers carried by these regions can be determined by appropriate genetic crosses. It is therefore possible to examine whether the loss of different regions is statistically at random. The number of primary petite clones retaining a given marker over the total number of primary rho$^-$ clones is called <u>frequency of retention</u> of that marker. This frequency (%) is scored independently from the presence or absence of other markers. When we compared the frequency of retention of each of the fifteen genetic markers (figure 3), we found that most of the markers were retained with very similar frequencies, except those located in one region. Oxi-3, pho-1 (and O_{II} closely linked to it) and mit 175 were retained with a frequency of 23 % (± 2.4 s.d.) while all others showed a frequency of 55 % (± 3.3). The difference between these two distinct groups of frequencies is statistically highly significant ($P < 0.01$). Clearly the mitochondrial DNA sequences are not deleted in a random fashon.

Several laboratories have already studied the frequency of marker retention (6,7,8). However, the available data have dealt with only a limited number of drug resistance markers. The region including oxi-3, pho-1 and mit 175 has not been studied in detail. Although we have previously obtained some data which suggested a preferential loss of O_{II} region, convincing evidence was lacking (6). Juliani et al (9) have reported that

Table 1. Disjunction of DNA between adjacent markers.

Markers	Frequency of disjunction between adjacent markers	Survival sequence extends	
		to the left	to the right
E/C	0	0	0
C/mit 170	28	16	12
mit 170/tsm 8	16	6	10
tsm 8/ts 982	3	2	1
ts 982/oxi-1	8	14	4
oxi-1/oxi-2	31	13	18
oxi-2/P	96	67	29
P/oxi-3	123	99	24
oxi-3/(pho-1/O_{II})	13	9	4
(pho-1/O_{II})/mit 175	25	17	8
mit 175/cob-2	99	0	99
cob-2/cob-1	12	4	8
cob-1/O_I	48	10	38
O_I/tscs 983	24	9	15
tscs 983/E	106	55	51

At the disjunction between each pair of adjacent markers, the petite DNA may be rescued to the left or to the right of the disjunction point as seen in figure 1. The frequency of each orientation of rescue was scored from the data of figure 1. Left and right refer to the orientation of the gene order shown in figure 1.

the petite induction by photoactivated 3-carbethoxy-psoralen led to a preferential loss of O_{II} marker as compared to C and E markers. Recently, Moustacchi has repeated the psoralen experiment using our 15 markers and has clearly shown that the oxi-3/pho-1/mit 175 region was indeed lost with an exceptionally high frequency (Moustacchi and Fukuhara, in preparation). These experiments are the first cases in which the rho^- mutagenesis was examined with a maximal number of genetic markers to detect such a specific region.

It is to be mentioned that the phenomenon is not specifically associated to the bet 011 mutant used here. The original ethidium sensitive strain showed also a very similar pattern of marker retention. A preferential loss of any particular region of the mitochondrial DNA obviously introduces a constraint in the combination of genetic markers that can be found in petites. In fact, among the petite clones of single-deletion type, some structures are rare or absent, as can be noticed in figure 1. As this situation may be pertinent to replication or repair mechanisms, we examined how the constraint appears in individual rho^- clones.

To generate a single deletion, the wild type DNA has to be cut at two points. The frequency of disjunction between two adjacent markers, A and B, can be measured by scoring the number of clones that have lost only one of the two markers (A^+B° and $A^\circ B^+$). The frequency of disjunction may be used as a measure of the distance between the two markers, provided that the frequencies of retention of all the individual markers are the same. However, since all the markers are not lost with the same frequency as we have shown above, the estimates of distances between markers can be strongly biased in the direct measurement of disjunction frequency.

When a disjunction occurs between two adjacent markers, the survival sequence, that forms a petite DNA, may extend either to the left (say, A^+B° type) or to the right ($A^\circ B^+$). The choice between left and right survivals is not always symetrical since the frequencies of retention of A and B may not be similar. From the data of figure 1, we scored, for each interval of adjacent markers, the frequency of the "survival to the left" and the "survival to the right" (table 1). It is apparent that at some disjunction points, the survival is preferential in one direction. The disjunction between mit 175 and cob 2 represents an extreme case : all the survival sequences extend to the right, producing exclusively the petites of $(mit\ 175)^-(cob\text{-}2)^+$ type and not $(mit\ 175)^+(cob\text{-}2)^\circ$ type. The disjunction between P and oxi-3 shows also a preferential direction.

The particularity of the oxi-3/pho-1/mit 175 region also can be seen in its unstability in the primary clones, as illustrated in some examples of figure 2. This may be one of the difficulties of precisely localizing the mitochondrial 16S ribosomal RNA gene by petite deletion mapping (10, 11,12).

Supported in part by the D.G.R.S.T. contracts N° 77-7-0306 and 77-7-0307.

References

1) Fukuhara H., Bolotin-Fukuhara M., Hsu J.H. & Rabinowitz M. Molec. Gen. Genet. 145, 7-17 (1976).
2) Bolotin-Fukuhara M., Faye G. & Fukuhara H., Temperature-sensitive respiratory-deficient mitochondrial mutations : Isolation and mapping. Molec. Gen. Genet. 152, 295-305 (1977).
3) Morimoto R., Lewin A., Hsu H.J., Rabinowitz M. & Fukuhara H. Proc. Nat. Acad. Sci. (Wash.) 72, 3868-3872 (1975).
4) Slonimski P. & Tzagoloff A. Europ. J. Biochem. 61, 27-41 (1976).
5) Deutsch J., Dujon B., Netter P., Petrochilo E., Slonimski P., Bolotin-Fukuhara M. & Coen D. Genetics 76, 195-219 (1974).
6) Bolotin-Fukuhara M. & Fukuhara H. Proc. Nat. Acad. Sci. (Wash.) 73, 4608-4612 (1976).
7) Molloy P.L., Linnane A.W. & Lukins H.B. J. Bacteriol. 122, 7-18 (1975).
8) Schweyen R.J., Steyer U., Kaudewitz F., Dujon B. & Slonimski P. Molec. Gen. Genet. 146, 117-132 (1976).
9) Juliani M.H., Hixon S. & Moustacchi E. Molec. Gen. Genet. 145, 249-254 (1976).
10) Faye G., Kujawa C., Dujon B., Bolotin-Fukuhara M., Wolf K., Fukuhara H. & Slonimski P. J. Molec. Biol. 99, 203-217 (1975).
11) Faye G., Kujawa C., Fukuhara H. & Rabinowitz M. Biochem. Biophys. Res. Commun. 68, 476-482 (1976).
12) Sanders J.P.M., Heyting C., DiFranco A., Borst P. & Slonimski P. in the Genetic Function of Mitochondrial DNA, eds. Saccone C. and Kroon A.M., 1976, Elsevier/North-Holland Medical Press, Amdterdam, pp. 259-272.

PREFERENTIAL LOSS OR RETENTION OF MITOCHONDRIAL GENES IN rho$^-$ CLONES

S. Mathews, R. J. Schweyen, F. Kaudewitz
Genetisches Institut der Universität München, Maria-Ward-Str. 1a,
D-8 München 19, W. Germany.

SUMMARY

In large populations of rho$^-$ clones from different origins the segments of the mitochondrial genome retained were determined by studying the retention of genetic markers the location of which on the map is known. The <u>cob-oli1</u> segment of the genome is found to be retained most frequently. Co-retention of other markers together with these decreases with their distance on the map. A second preference of retention is found with the <u>oxi2</u> segment.

INTRODUCTION

Mitochondrial genomes of rho$^-$ petites are known to consist of a segment of the rho$^+$ mitDNA which can be reiterated to form redundant molecules of the approximate length of the grande mitDNA molecule. In the majority of rho$^-$ clones the monomeric mitDNA sequence retained is a contiguous segment of the rho$^+$ genome ranging from less than 100 base pairs up to more than 10% of the total genome(1, 2, 3).
Although any one of the genetic markers in mitDNA studied so far and any mitDNA sequence is found to be retained in a population of rho$^-$ clones, some segments are more frequently retained than others (3, 4, 5). This study intends to correlate retention frequencies of segments as determined from marker retention in large populations of rho$^-$ clones to their location on the map of the mitochondrial genome.

RESULTS AND DISCUSSION

The distribution of genome segments retained in 135 rho$^-$ clones is shown in Figure 1. A horizontal bar represents a segment of mitDNA. Its length is defined by the distance between the markers co-retained in the respective rho$^-$ petite. This presentation is justified as the length determination based on co-retention analysis of genetic markers is in good agreement with length determinations by molecular means(cf. Grosch et al., Schweyen et al., this volume). In this population of rho$^-$ clones, derived from essentially homoplasmic rho$^-$ cells, genome segments retained are fairly long, averaging about 30% of the length of mitDNA. Clones carrying non-contiguous segments of the genome appear to be rare (about 5%). The majority of rho$^-$ genome segments retains the cob-oli1 region and extends from this to the cap-oxi2 or, less frequently to the pho-oxi3 region or to both. Genomes which do not harbour the cob or oli1 markers are much less frequent. The majority of the latter genomes carries either oxi2 or par. All of the other markers studied can be retained together with either cob-oli1 markers or with par-oxi2 markers.

Genome segments retained were determined in three different populations of rho$^-$ clones derived from different yeast strains. The results of these analyses were used to calculate i) overall retention frequencies of individual markers (Figure 2, upper curve) and ii) co-retention frequencies of any one of the markers studied with the oli - cob region (Figure 2, lower curve).

Co-retention of any marker with oli1 decreases consistently with the distance between a marker regarded and oli1. This holds true for both sides, i.e. towards cob-oli2-oxi3 and towards cap-oxi1-oxi2, although the slopes are not completely symmetrical. Overall retention frequencies show a similar decrease from oli1 to oxi3; in the other part of the genome, however, their distribution is quite different.

A comparison of overall retention frequencies with co-retention frequencies reveals the retention of any marker independent of the retention of oli1. Obviously, markers in the par-cap segment have a high, markers

Figure 1 : Distribution of genome segments retained in a population of rho^- clones.

Retention of genetic markers (3) was determined in rho^- clones derived from strain RM 590-9D, tsp25, cap^r, oli^r, par^r, mit^+ (3). If two genetically linked markers were present, the genome segment flanked by these markers was assumed to be retained in a rho^- clone and alienated to the circular genetic map of the mitochondrial genome (Schweyen et al., this volume).
A horizontal bar represents one rho^- genome.

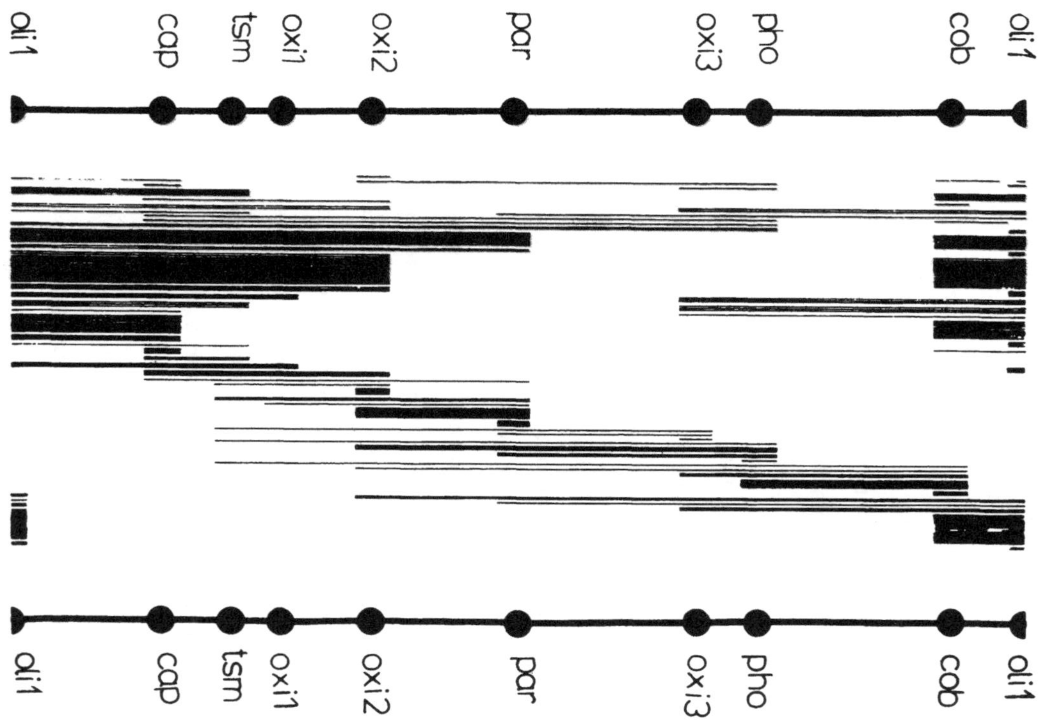

in the cob-oxi3 segment a low chance of surviving in a rho^- petite of genotype $oli1^o$.

Comparable preferences in marker retention were observed with other rho^- populations, independent of whether they were derived i) from different strains carrying mutations tsp-25, tsp-20, tsp-30 or tsm-8 (all of which cause frequent spontaneous formation of rho^- petites) or ii) from a strain grown in $MnCl_2$, boncrecic acid or FCCP in order to obtain frequent formation of rho^- petites (Weiss-Brummer et al. in prep.). The fact that other laboratories studying rho^- clones of different origin have also observed preferential loss of markers in the oli2-oxi3 region of the genome (5, and Fukuhara et al. this volume) points to a phenomenon generally observed with populations of rho^- clones.

In summary, mitDNA segments which harbour the oli1-cob region of the mitochondrial genome (or a closely linked nucleotide sequence) have a high preference in surviving in rho^- clones. A segment of second preference carries the oxi2 region. Other genes are co-retained with one or the other of these segments. If they are distant to both, e.g. oxi3, oli2, their overall retention frequency in rho^- populations is low.

Two alternative explanations for these facts may be envisaged:

i) The rho^- mutation process preferentially gives rise to the formation of some classes of rho^- genomes as compared to others; propagation of these genomes is independent of the mitDNA sequence retained.

ii) Initially formed rho^- genomes harbour every region of mitDNA with equal frequency; however, some classes of rho^- molecules have a higher chance of surviving, e.g. because they carry a mitDNA sequence giving them a preference in replication over other rho^- molecules in the same, heteroplasmic cell. When segregated, all classes have a similar chance of survival in homoplasmic cells, which would account for the fact that established rho^- clones, harbouring oxi3-pho1 segments only, do not show unusual instability.

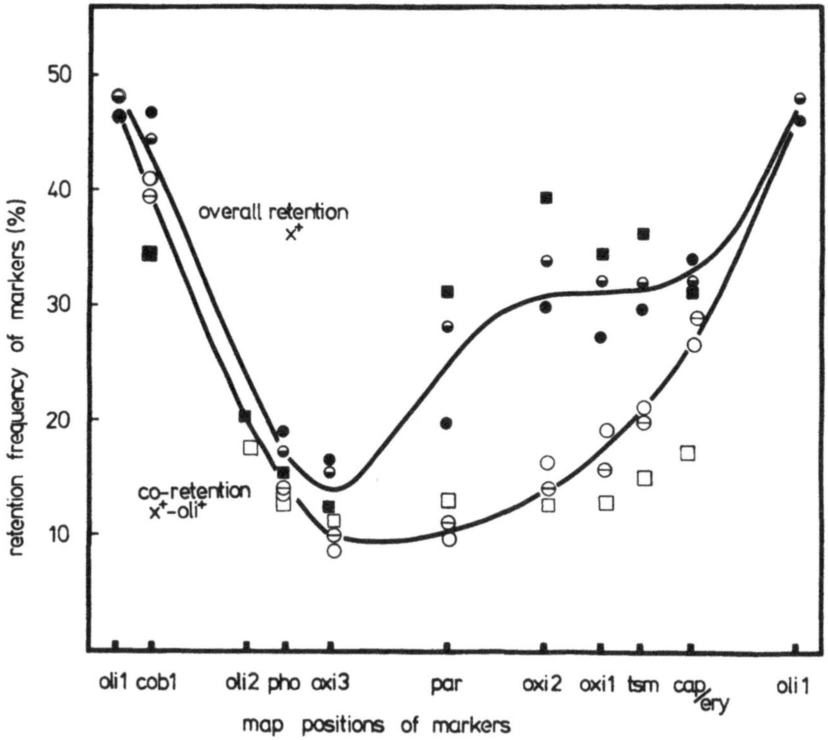

Figure 2 : Frequencies of marker retention in three populations of rho⁻ clones.

Mitochondrial genotypes of three populations of rho⁻ clones were determined (3). Based on this analysis the number of clones retaining a given marker and the number of clones retaining a given marker simultaneously with oli1 in a population were determined ("overall retention" and "co-retention" respectively).
The number of rho⁻ clones studied, their origin (3) and symbols used in this Figure are :

520 (RM 590-9D) ● ○
396 (RM 591-1D) ◒ ⊖
292 (RM 511-4A) ■ □

ACKNOWLEDGEMENTS

The authors wish to thank Dr. B. Backhaus and B. Weiss-Brummer for helpful discussion and suggestions.
This work was supported by the Deutsche Forschungsgemeinschaft.

REFERENCES

1) Faye, G., Fukuhara, H., Grandchamp, C., Lazowska, J., Michel, F., Casey, J., Getz, G. S., Locker, J., Rabinowitz, M., Bolotin-Fukuhara, M., Coen, D., Deutsch, J., Dujon, B., Netter, P., Slonimski, P. P.: Mitochondrial nucleic acids in petite colony mutants: deletion and replication of genes. Biochimie 55, 779-792 (1973)

2) Borst, P., Heyting, C., Sanders, J. P. M.: The Control of Mitochondrial DNA Synthesis in Yeast Petite Mutants. In: Genetics and Biogenesis of Chloroplasts and Mitochondria. Th. Bücher et al. eds. 1976, Elsevier/ North Holland Biomedical Press, Amsterdam, The Netherlands.

3) Schweyen, R. J., Weiss-Brummer, B., Backhaus, B., Kaudewitz, F.: The The genetic map of the mitochondrial genome in yeast: map positions of drug^r and mit$^-$ markers as revealed from population analyses of rho$^-$ clones in Saccharomyces cerevisiae. (this volume)

4) Deutsch, J., Dujon, B., Netter, P.; Petrochilo, E., Slonimski, P. P., Bolotin-Fukuhara, M., Coen, D.: Mitochondrial Genetics. VI. The petite mutation in Saccharomyces cerevisiae: Interrelations between the loss of the rho$^+$ factor and the loss of the drug resistance mitochondrial genetic markers. Genetics 76, 195-219 (1974)

5) Juliani, M.H., Hixon, S., Moustacchi, E.: Mitochondrial genetic damage induced in yeast by a photoactivated furocoumarin in combination with ethidium bromide or ultraviolet light. Molec. gen. Genet. 145, 249-254 (1976)

THE GENETIC MAP OF THE MITOCHONDRIAL GENOME INCLUDING THE FINE STRUCTURE OF COB AND OXI 3 CLUSTERS

R. J. Schweyen; B. Weiss-Brummer; B. Backhaus; F. Kaudewitz
Genetisches Institut der Universität München; Maria-Ward-Str. 1A

INTRODUCTION

Mapping of the mitochondrial genome, i.e. determination of order and distances of genetic markers, has been attempted with considerable success by three approaches which make use of the rho$^-$ petite mutants: i) "physical mapping" based on restriction analysis of rho$^+$ and rho$^-$ mitDNA (1, 2), ii) "molecular mapping" based on hybridization of rho$^+$ mitDNA to rho$^-$mitDNA or of rho$^+$ mit-cRNA to rho$^-$ mitDNA (3, 4), iii) "genetic mapping" by determination of frequencies of deletion events separating genetic markers in rho$^-$ petites (5, 6). The latter method has been shown to allow a rapid and faithful determination of the location of genes in various segments of the mitochondrial genome (7, 8). This report summarizes these studies and presents results of additional studies on the fine structure mapping of mit$^-$ mutational sites in the cob and oxi3 region of the mitochondrial genome.

RESULTS AND DISCUSSION

I. Order and distances of genes on a circular map

Genetic mapping of mitochondrial markers is based on the findings that i) spontaneously arising rho$^-$ petites retain - with few exceptions - contiguous segments of the rho$^+$ genome, ii) these segments are of various length and iii) genes of any region of the mitochondrial genome can survive in rho$^-$ clones. Frequent co-retention of two markers x and y in a population of rho$^-$ clones thus implies that these two markers are located closely together in a small region of the genome whereas frequent separation of two markers x and z, i.e. retention of one and loss of the other marker in a rho$^-$ clone (x^+z^o, x^oz^+), indicates their location in distant regions of the mitochondrial genome.

Separation frequencies of a series of genetic markers were determined in four populations of rho$^-$ clones, derived from four different rho$^+$ strains. Based on these data separation coefficients (Q) ranging from 0 to 1 were calculated. Values close to 0 indicate rare separation or " close linkage" and are assumed to correspond to small distances; values close to 1, the maximal value found, indicate frequent separation and are assumed to correspond to long distances between two markers (cf. legend to Table 1).

Q values found with pairwise combinations of any one of the markers studied and the markers cap/ery (which are hardly separated by rho$^-$ deletion) or cob2 are given in Table 1. These data show that i) highly different degrees of linkage are detected with different pairs of markers, ii) for a given pair of markers linkage relationships are similar, independent of the origin of rho$^-$ clones from different strains. The first of these findings is in favour of the assumed correlation between Q values and distances between markers, the second indicates constancy of the gene order in different yeast strains and of the pattern of rho$^-$ genomes formed.

These two sets of data suffice to define by a classical three-factor-analysis the order of the markers studied, e.g.: oli1 is localized within the segment cap-cob since both the distances oli1-cap and oli1-cob are smaller than the length of the segment cap-cob; tsm-8, however, is localized outside this segment, close to cap since i) it is more distant from cob than cap, ii) it is closely linked to cap. The order concluded on the basis of the data presented in Table 1 is : par-oxi2-oxi1-tsm8-cap-oli1-cob-oli2-pho-oxi3.

A more refined map which includes evidence for its circularity and for distances of markers is obtained when combinations of three markers neighbouring each other (x, y, z) are studied (Figure 1). Distance coefficients thus found show good additivity in any region of the genome in that the Q value for a total distance x-z is in all cases significantly higher than for either of the partial distances x-y and y-z. Therefore, the order of markers

pair of markers	separation coefficients (Q) found with rho⁻ clones from RM 590-9D	RM 591-1D	RM 511-4A	BM 1421-3C
cap/ery - tsm-8	.26	.35	.34	.21
- oxi1	.40	.50	.42	.40
- oli1	.54	.58	n.d.	.61
- oxi2	.63	.77	.60	n.d.
- cob	.73	.71	.72	.80
- par	.92	.95	.71	.79
- oli2	n.d.	n.d.	.93	n.d.
- pho1	1.0	.92	.99	n.d.
- oxi3	.99	1.0	1.0	.93
cob - oli1	.29	.26	n.d.	.34
- oli2	n.d.	n.d.	.48	n.d.
- pho1	.78	.70	.59	n.d.
- cap	.73	.71	.72	.80
- oxi3	.86	.75	.65	.80
- tsm-8	.81	.83	.99	.99
- oxi1	.86	.89	1.03	.99
- oxi2	1.0	1.05	1.0	n.d.
- par	1.03	1.01	.87	1.0

TABLE 1: Separation coefficients (Q) found between various markers and cap/ery or cob

The number of rho⁻ clones studied per population are: 520 (RM 590-9D), 395 (RM 591-1D), 292 (RM 511-4A), 312 (BM 1421-3C). The clones were derived from rho⁻ petites which were formed spontaneously with high frequency in cultures of the four strains carrying the nuclear mutation tsp-25. Origin of strains and genetic markers, isolation of rho⁻ clones and the genetic analysis of marker retention were described elsewhere (8, 10). The genetic markers studied are : cap-321, ery-221, oli1-1, oli2-144, cob2-1231, oxi3-2111, oxi2-2532, oxi1-1273, pho1-229, par-454.

Separation coefficients (Q) were calculated as described earlier (6, 8):

a) In a population of rho⁻ clones the frequencies of retention and of loss of individual markers (X^+), (Y^+), (X^o), (Y^o) and the frequencies of separation of two markers, i.e. the frequencies of phenotypes $(X^+Y^o) + (X^oY^+)$ were determined.

b) Expected frequencies of separation in case of random association - dissociation of X and Y were calculated to be:

$$(X^+Y^o)\ \text{exp.} = \frac{(X^+)\cdot(Y^o)}{N}\ ; \quad (X^oY^+)\ \text{exp.} = \frac{(X^o)\cdot(Y^+)}{N} \quad (N = \text{total clones})$$

c) The ratio $\frac{(X^+Y^o) + (X^oY^+)\ \text{obs.}}{(X^+Y^o) + (X^oY^+)\ \text{exp.}} = Q$

Q = 1 indicates a random distribution of X^+ and Y^+ in the population, Q 1 indicates preferential association of X^+ and Y^+, i.e. "linkage".

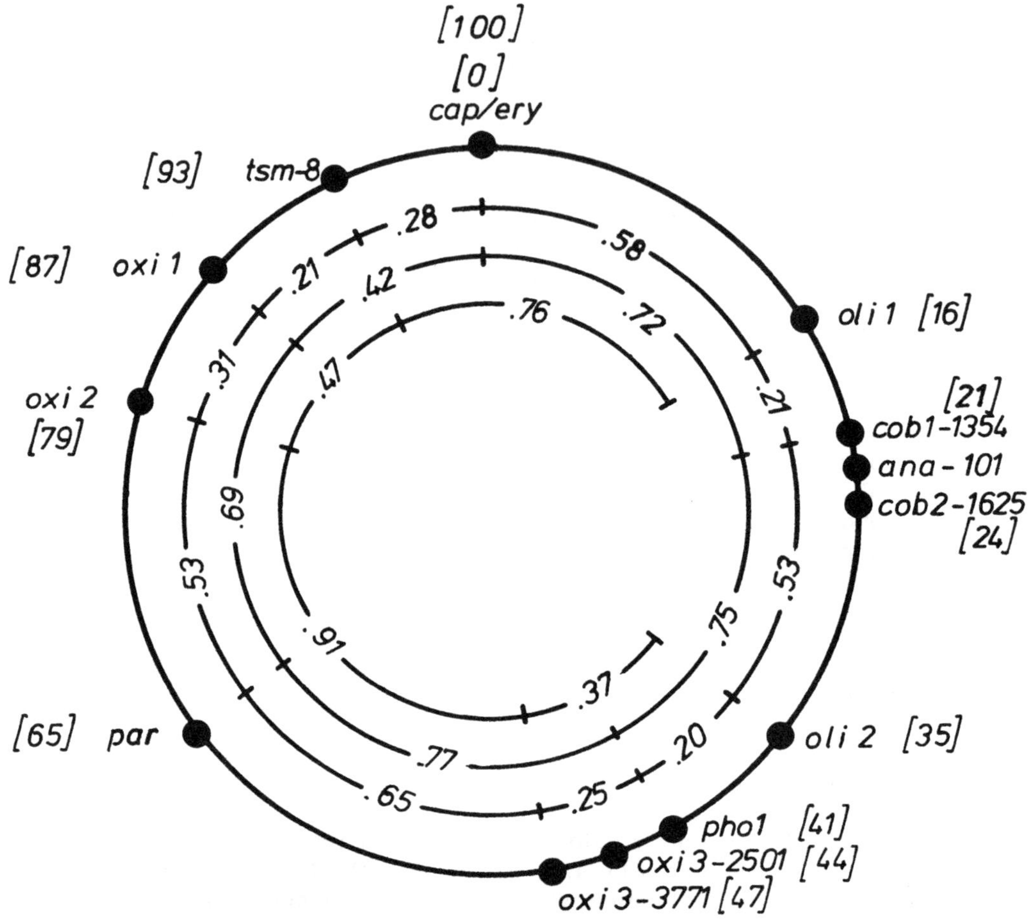

FIGURE 1: Circular map of the mitochondrial genome
Figures in the range from 0 to 1 (inside the circle) are average separation coefficients (Q) found with the respective pair of markers in the four rho⁻ populations studied (cf. Table 1). Figures in the range from 0 to 100 are positions of markers in map units. The calculations of these positions is based on the Q values found with pairs of neighbouring markers, the sum of which corresponds to 100 map units. Genetic markers studied are the same as given in Table 1 except in case of cob and oxi3 clusters where the mutational sites are indicated.

on a circular map as shown in Figure 1 is unambiguous. Furthermore, the additivity of Q values found indicates that these values are - at least to some extent - related to the physical distance of markers on the mitochondrial DNA.

This relationship is fully confirmed when positions of markers, as concluded from the data presented here (Figure 1), are compared with the positions of the respective markers as determined a) by molecular mapping based on hybridization of mitDNA from various, genetically characterized rho^- clones with rho^+ mitDNA or rho^+ mit-cRNA (3,4) or b) by physical mapping based on the analysis of restriction endonuclease digests of mitDNA from various, genetically characterized rho^- petites and of rho^+ mitDNA (2, 9).

II. Mapping of mutational sites in the cob and oxi3 region

In the cob region as well as in the oxi3 region of the genome mit^- mutations have been shown to occur at several sites which were found to be separable by recombination and - to some extent - by rho^- deletion (5, 11). The study presented here aims at mapping a series of mit^- mutations in these regions by use of the latter, i.e. the rho^- deletion analysis.

In a first step the marginal sites in the cob and in the oxi3 region were determined. Therefore, a large series of rho^- clones was isolated which had been shown to retain either oli1, which is neighbouring the cob region, or pho1, which is next to oxi3. Among these clones co-retention of all available cob and oxi3 markers were determined. Loss of all but one or retention of all but one (or a few) of the cob or oxi3 mutational sites in a rho^- clone was taken as indicative of their location most proximal or most distal to oli1 or pho1. The marginal markers thus determined were cob-1354 and cob-1625, oxi3-2501 and oxi3-3771 (or -3781, -6111, -6291). When mutants carrying these marginal markers were involved in a mapping experiment as described in the first part of this report, their distance in the cob or oxi3 region was found to be in the order of a few map units (cf. Figure 1).

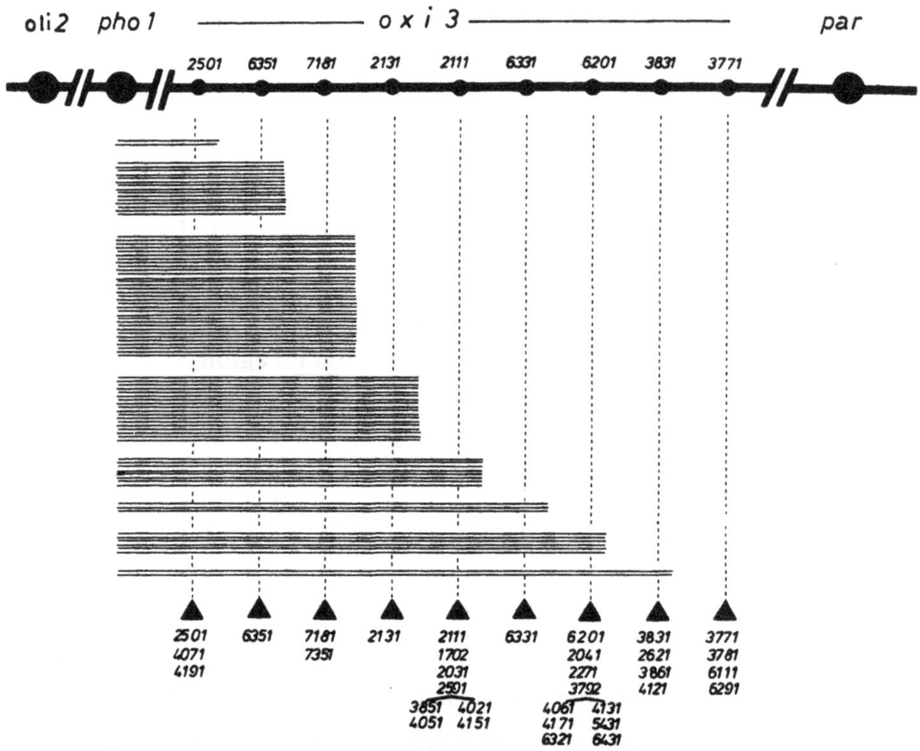

FIGURE 2: Rho$^-$ deletion map of the oxi3 region

A horizontal bar represents the genome segment retained in a rho$^-$ clone. Numbers on top of the Figure refer to mutational sites resolved, numbers at the bottom to mutations allocated to the same site.

In a second step rho^- clones were isolated which showed separation of mutational sites within the cob or oxi3 region. Therefore, rho^- clones were screened which retained only one or the other of the two marginal sites of a region. The order of other sites then was inferred from their co-retention with one of the marginal sites as shown in Figures 2 and 3.

The oxi3 region is resolved into nine different sites (Figure 2). To most of these sites several mutations were allocated which differ in phenotype (heat sensitivity, cold sensitivity, leaky). This indicates that there are more mutational sites than resolved with the rho^- clones studied so far. The map of this region probably can be refined further by recombination analyses as reported previously (5). It is remarkable that all rho^- genomes carried by this collection of rho^- clones extend from sites within the oxi3 region towards pho1-oli2, none towards par. This observation may be related to the general phenomenon of a polarity of retention discussed in the accompanying paper by Mathews et al. Alternatively, it could be explained by the assumption that one of the marginal markers (oxi3-3771) used in screening these clones is a deletion which extends far towards the par region. This possibility has to be studied further.

The cob region is resolved into twelve different sites (Figure 3). To one of these sites a large series of mutations is allocated which are of independent origin, not of identical phenotype and, thus, may represent a series of different mutational sites. In contrast to the distribution of rho^- genomes shown in Figure 2 genomes discriminating between cob markers extend to both sites of this region with about equal frequencies.
Within the cob region there are two clusters to be distinguished according to the observed frequent co-retention of sites. The majority of clones separates the sites cob-3881 and cob-2011 which may be taken as indicating that the inter-cluster - space is larger than each of the clusters itself. The cluster which is proximal to oli1 is defined as being cob1 since it contains a mutation (cob-241) which was allocated by Tzagoloff et al. (11) to the first of two cob loci. With a second mutation of the series studied by these authors and allocated to the locus cob2 more complex results were found: cob-162

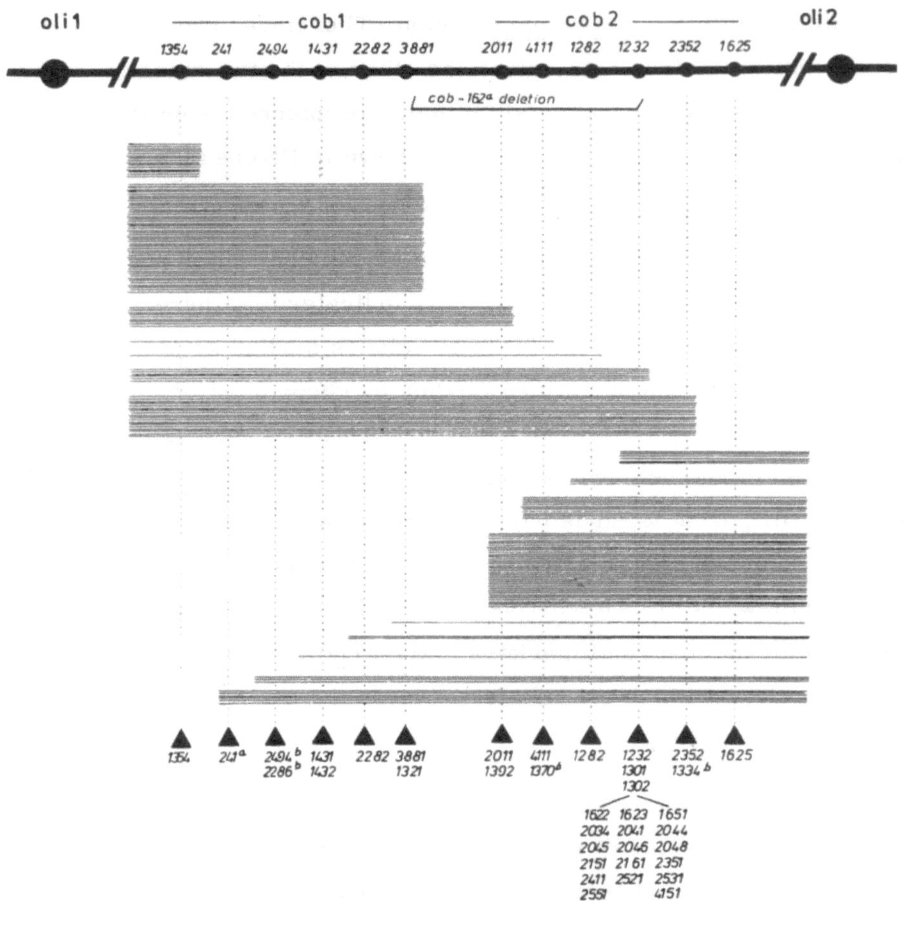

FIGURE 3: Rho$^-$ deletion map of the cob region

a) Mutants of the collection of A. Tzagoloff (original designations are M 24-241 and M 17-162)
b) Mutants of the collection of P. P. Slonimski and Z. Kotylak (original designations are G 2286, G 2494, G 1370, G 1334).

For further explanation see legend to Figure 2.

could not be separated from a series of five different sites (cob-3881, -2011, -1282, -1231). Therefore cob-162 was concluded to represent a deletion. Since in crosses rho^+ cob-162 x rho^+ cob-3881 recombinants of genotype mit^+cob^+ were obtained (unpublished), the deletion is assumed to extend from a point close to cob-3881 in the inter-cluster-space into the cob2 cluster, covering three or four of its sites. In the map of Tzagoloff et al. (11) cob-162 is a site within the cob2 cluster. This is contradictory to the findings presented here. However, according to Kotylak and Slonimski (this volume) some sites in the cob region have to be remapped which may solve the discrepancy concerning mutation cob-162.

Four of the mutations mapped by Kotylak and Slonimski (this volume) were also included in this analysis. Two of the mutations, G 2286 and G 2494 are found to map in cob1, the other two (G 1334, G 1370) in cob2, which is in agreement with a map presented by these authors.

The location of drug-resistance markers on the map is revealed from the work of Burger et al. (this volume) who allocated ana^r and fun^r markers to the map presented here. These $drug^r$ markers are localized in the oli1 distant parts of cob1 and cob2 clusters, which is consistent with the localization of similar $drug^r$ markers by Colson and Slonimski (this volume), and further illustrates the agreement between the maps of the cob region found in the two different laboratories.

ACKNOWLEDGEMENTS

The authors are grateful to Dr. P. P. Slonimski and Dr. A. Tzagoloff for providing mutant yeast strains.
This work was supported by the Deutsche Forschungsgemeinschaft.

REFERENCES

1. Morimoto, R.; Lewin, A.; Merten, S.; Rabinowitz, M.: Restriction endonuclease mapping and analysis of grande and mutant yeast mitochondrial DNA. In: Genetics and biogenesis of chloroplasts and mitochondria, Th. Bücher, W. Neupert, W. Sebald, S. Werner eds.. North-Holland Publishing Company, Amsterdam, New York, Oxford, 519-524 (1976)

2. Sanders, J.P.M.; Heyting, C.; DiFranco, A.; Borst, P.; Slonimski, P.P. The organization of genes in yeast mitochondrial DNA. In: The genetic function of mitochondrial DNA, C. Saccone and A.M. Kroon eds, North Holland Publishing Company, Amsterdam, New York, Oxford 259-272 (1976)
3. Sriprakash, K.S.; Molloy, P.L.; Nagley, P.; Lukins, H.B.; Linnane, A.W. Biogenesis of mitochondria. XLI. Physical mapping of mitochondrial genetic markers in yeast, J. Mol. Biol. 104, 485-503 (1976)
4. Grosch, G.; Doxiadis, I.; Lang, B.; Schweyen, R.J.; Kaudewitz, F: A one-step method for rho^+cRNA/rho^-DNA hybridization in mapping of the mitochondrial genome, (1977) this volume
5. Slonimski, P.P.; Tzagoloff, A.; : Localization in yeast mitochondrial DNA of mutations expressed in a deficiency of cytochrome oxidase and for coenzyme QH_2-cytochrome c reductase. Europ. J. Biochem. 61 27-41 (1976)
6. Schweyen, R.J.; Steyrer, U.; Kaudewitz, F.; Dujon, B.; Slonimski, P.P.: Mapping of mitochondrial genes in S. cerevisiae: Population and pedigree analysis of retention or loss of four genetic markers on rho^- cells. Molec. gen. Genet. , 146 131-165 (1976)
7. Schweyen, R.J.; Weiss-Brummer, B.; Backhaus, B.; Kaudewitz, F.: Localization of seven gene loci on a circular map of the mitochondrial genome of Saccharomyces cerevisiae. In: The genetic function of the mitochondrial DNA, C. Saccone and A.M. Kroon eds., North Holland Publishing Company, Amsterdam. New York, Oxford, 251-258 (1976)
8. Schweyen, R.J.; Weiss-Brummer, B.; Backhaus, B.; Kaudewitz, F.: The genetic map of the mitochondrial genome in yeast: Map positions of $drug^r$ and mit^- markers as revealed from population analyses of rho^- clones in Saccharomyces cerevisiae. Molec. gen. Genet. in press
9. Lewin, A.; Morimoto, R.; Merten, S.; Martin, N.; Berg, P.; Christianson, T.; Levens, D.; Rabinowitz : Physical mapping of mitochondrial genes and transscripts in Saccharomyces cerevisiae, (1977) this volume
10. Weiss-Brummer, B.; Mathews, S.; Schweyen, R.J.; Schnittchen, P.; Bandlow, W.; Kaudewitz, F. : in preparation
11. Tzagoloff, A.; Foury, A.; Akai, A.: Assembly of the mitochondrial membrane system XVIII. Genetic loci on mitochondrial DNA involved in cytochrome b biosynthesis. Molec. gen. Genet., 149 , 33-42 (1976)

Extrachromosomal inheritance in Schizosaccharomyces pombe

VI. Preliminary genetic and biochemical characterization of mitochondrially inherited respiratory - deficient mutants

G. Seitz, G. Lückemann, K. Wolf, F. Kaudewitz
Genetisches Institut der Universität München

M. Boutry and A. Goffeau
Laboratoire d'Enzymologie, Université de Louvain,
Louvain - La - Neuve, Belgique

INTRODUCTION

Mitochondrially inherited respiratory - deficient mutants with intact mitochondrial protein synthesis have been isolated in Saccharomyces cerevisiae as temperature - sensitive mutants (1, 2, 3) or as mutants unconditionally defective in respiration and / or oxidative phosphorylation (for review see 4). Mapping of these mit^- - mutants has already led to a detailed picture of mitochondrial DNA. In addition, biochemical studies of mutants defective in the oxi-1 locus have led to the identification of the structural gene for cytochrome oxidase subunit II (5). Recently, mit^- - mutants have also been isolated in the petite - negative yeast Schizosaccharomyces pombe by use of the mutator - properties of the antimycin - resistant mutants ana^r - 8 (6, 7) and ana^r - 14. Since loss of mitochondrial DNA seems to be lethal for petite - negative yeasts (8) it may be hypothesized that mitochondrial DNA of these yeasts harbour essential genetic information. In order to get insight into the phenomenon of petite - negativity we are trying to find out the functions which are encoded by the mitochondrial DNA.

RESULTS

1) Isolation of mutants

Respiratory - deficient mutants spontaneously occur up to 1 % in stationary phase glucose grown shake cultures. Since mutants ana^r - 8 and ana^r - 14 carry the ade-7 marker, respiratory - deficient mutants can easily be recognized by their brownish colour on adenine - limited media. Mutant colonies were isolated and at least purified twice. Mutant frequencies could be increased up to 20 - 30 % in a stationary culture by growing cells in glucose - complete - medium (GLU) containing the well - known rho^- - inducing drugs ethidium bromide (EB, 2 - 50 µg/ml) and janus green (JG, 1 - 20 µg/ml) (Table 1).

Table 1: Frequency of spontaneously arising respiratory - deficient mutants in ana^r - 8 and ana^r - 14. Mutants were grown in glucose - complete - medium to late logarithmic phase, plated out and replica - plated on glycerol - complete - medium supplemented with 0,1 % glucose.

mutant	colonies	RD colonies	% RD colonies
ana^r - 8	3426	24	0,7
ana^r - 14	3761	19	0,5

2) Genetic analysis of respiratory - deficient mutants

Respiratory - deficient mutants from the antimycin - resistant mutants ana^r - 8 and ana^r - 14 with the nuclear genotype ade 7 - 50 h^- were crossed with the respiratory - competent, antimycin - sensitive, sporulation - deficient strain mei 1 - 102 ura 1 - 171 arg 1 - 1. Diploid clones were selected on minimal medium with phloxin and patched on GLU. By replica - plating on glycerol - complete - medium enriched with 0,1 % glucose (0,1 e GLY) and 0,1 e GLY + antimycin the rates of transmission of respiratory - deficiency and antimycin - resistance were measured. To determine the reversion rate, 10^6 cells of stationary phase cultures of the different mutants were plated on 0,1 e GLY and scored after 10 days of growth. One group of mutants did not revert at all, the other to various degrees. According to their transmission rates for respiratory - deficiency and antimycin - resistance to the diploid progeny, mutants could be divided in the following groups:

- mutants showing transmission of respiratory - deficiency and antimycin - resistance,

- mutants that transmit antimycin - resistance, but fail to transmit respiratory - deficiency,
- mutants unable to transmit antimycin - resistance, but can transmit respiratory - deficiency,
- mutants lacking both transmission of respiratory - deficiency and antimycin - resistance.

Results summarized in Table 2 show that transmission rates for antimycin - resistance varies from 0 % to 53 %. With the same technique, transmission rate of antimycin - resistance of ana^r - 8 is only 20 %. Transmission of respiratory - deficiency is very low (maximum 5,4 %), which may be due to a selective disadvantage for respiratory - deficient diploids under the conditions of a standard cross.

Table 2: Transmission rates of antimycin - resistance and respiratory - deficiency in mit^- - strains derived from ana^r - 8 and ana^r - 14.

No of mit^- strains	derived from mutant	range of ana^r - transmission in %	range of RD - transmission in %	No of mit^- - strains that revert
12	ana^r - 8	0,7 - 40,3	0,2 - 1,4	3
20	ana^r - 8	0,4 - 36,2	0,0	8
3	ana^r - 8	0,0	0,2 - 0,9	3
1	ana^r - 8	0,0	0,0	1
4	ana^r - 14	5,4 - 10,9	0,5 - 5,4	1
9	ana^r - 14	2,9 - 53,0	0,0	4
2	ana^r - 14	0,0	0,0	1

On the basis of the results given in Table 2, the following conclusions can be drawn:

- mutants that revert cannot be of the rho^- or rho^o type,
- mutants showing transmission of antimycin - resistance cannot be rho^o.

For the group lacking transmission of respiratory - deficiency it is possible to decide, whether they are nuclear pet^- - mutants, rho^- or mit^- - mutants. This, however, can be achieved by m - fluorophenylalanine - induced haploidization. Haploidization of a diploid heterozygous for a dominant or recessive chromosomal mutation will always segregate into haploid clones with wild - and mutant - phenotype. Thus, a diploid with a nuclear recessive pet$^-$- mutation will segregate into respiratory - deficient and respiratory - competent haploid clones. Respiratory - competent diploids issued from a cross with a mit$^-$ or a rho$^-$ will show respiratory - competent haploid offspring only. It cannot yet be decided, whether a mutant lacking transmission of respiratory - deficiency and lacking the ability to revert, is a rho$^-$ or a mit$^-$- mutant. This can be checked by testing the mitochondrial protein synthesis in these mutants. The haploidization technique revealed extrachromosomal inheritance for all respiratory - deficient mutants. An example is given in Table 3 for haploidization of diploids issued from crosses with an extrachromosomal and a chromosomal respiratory - deficient mutant. The deviation from the 50 % : 50 % ratio of RC and RD haploid clones issued from the cross NG 19 ade 7-50 h$^-$ x mei 1 -102 ura 1-171 arg 1 - 1 is due to a selective disadvantage for respiratory - deficient clones during haploidization.

Table 3: Haploidization of respiratory - competent diploids issued from crosses of an extrachromosomal (SP 1) and a chromosomal (NG 19 respiratory - deficient mutant with mei 1 - 102 ura 1 - 171 arg 1- 1.

mutant	RC	RD	total	
SP 1	371	0	371	colonies
	100,0	0,0	100,0	%
NG 19	275	124	385	colonies
	68,9	31,1	100,0	%

3) Biochemical analysis

Respiration rates, enzyme activities, spectra, and mitochondrial protein synthesis were analyzed for 12 respiratory - deficient mutants, derived from ana^r- 14.

a) Respiration

The respiratory - deficient strains were grown in a medium containing 10 % raffinose, 2 % yeast extract, and 75 mg/ml adenine (pH 4,5). Respiration rates of stationary phase cultures were measured in a medium containing 0,1 M phthalate (4,5) and 320 mM glucose as substrate. All mutants show a residual respiration with a Q_{O2} (µl O_2/hour/mg wet weight) of about 12 % of the wild type ade 7 - 50 h^- or mutant ana^r - 14. Respiration of all mutants proved to be insensitive to antimycin, whereas the respiratory activity of the wild type is reduced to 8 %. All mutants are resistant to 1 mM KCN, but sensitive to 3 mM NaN_3. Results are summarized in Table 4.

Table 4: Respiration of whole cells of the wild type ade 7 - 50 h^-, the mutant ana^r - 14 and mit^- - strains with glucose as substrate. The effect of antimycin (AA), KCN and NaN_3 on the respiratory activity is shown.

strain	Q_{O2} $µl_{O2} \times h^{-1} \times mg^{-1}$ ww	+ AA 30 µM	+ KCN 1 mM	+ NaN_3 3 mM
wt	3,3	0,3	0,2	< 0,1
ana^r - 14	3,2	3,0	0,3	< 0,1
SP 1	0,4	0,3	0,3	0,1
SP 2	0,4	0,4	0,4	0,2
SP 3	0,5	0,4	0,3	0,2
SP 4	0,4	0,4	0,4	0,1
EB 2	0,5	0,3	0,4	0,1
EB 3	0,5	0,4	0,4	< 0,1
EB 4	0,4	0,3	0,3	< 0,1
EB 5	0,4	0,3	0,3	0,1
EB 7	0,4	0,4	0,4	0,1
EB 8	0,3	0,2	0,3	< 0,1
EB 9	0,4	0,4	0,4	< 0,1
EB 11	0,3	0,3	0,3	< 0,1

b) Low - temperature spectra

Difference spectra of mechanically prepared mitochondrial particles were recorded at the temperature of liquid CO_2. The mutants could be divided in three different spectral types:

- mutants deficient in cytochrome aa_3
- mutants with deficiency in cytochrome b
- mutants deficient in both cytochromes aa_3 and b.

The same spectral types have been found for mit^- - mutants in Saccharomyces cerevisiae, descibed as OXI -, COB -, and BOX - mutants (9). Spectra of the wild type ade 7 - 50 h^-, mutant ana^r - 14, and one representative mutant of the three classes are given in Figure 1.

Figure 1: Low temperature spectra of mitochondrial particles of the wild type ade 7 - 50 h^-, mutant ana^r - 14, and the mit^- - mutants EB 3, EB 5, and SP 3.

a) Spectral region from 530 - 570 nm, sample reduced by dithionite.

b) Spectral region from 530 - 570 nm, sample reduced by NADH (100 mM).

c) Spectral region from 580 - 620 nm, sample reduced by dithionite.

References for a, b and c were oxidized by ferricyanide (10 mM).

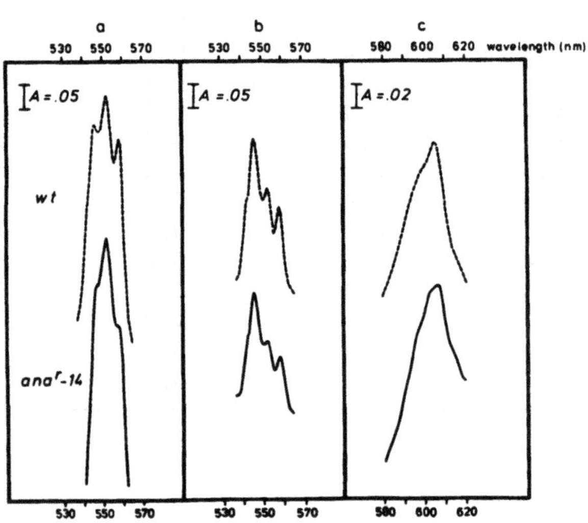

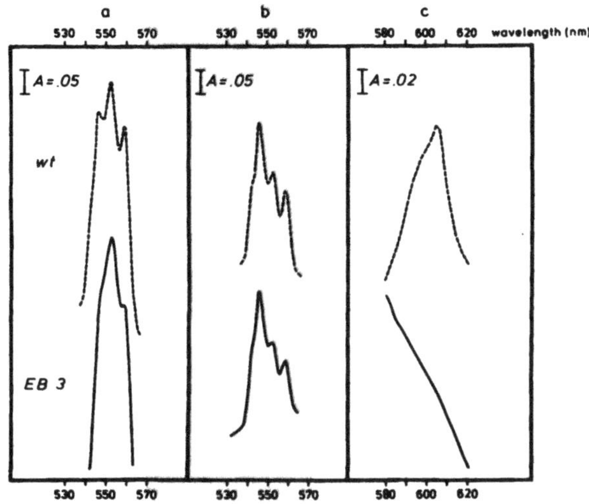
a
b
c
530 550 570
530 550 570
580 600 620 wavelength (nm)
A=.05
A=.05
A=.02
wt
EB 3
530 550 570
530 550 570
580 600 620

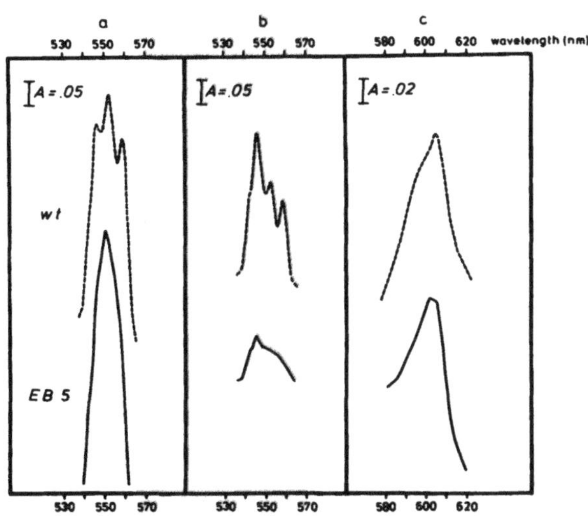
a
b
c
530 550 570
530 550 570
580 600 620 wavelength (nm)
A=.05
A=.05
A=.02
wt
EB 5
530 550 570
530 550 570
580 600 620

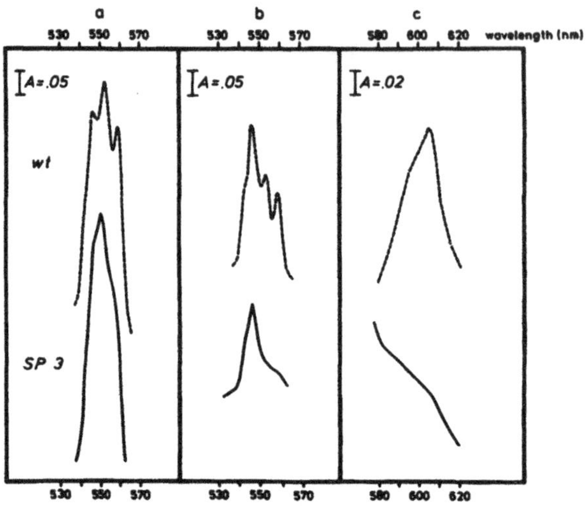

c) <u>Enzyme activities</u>

Tests on NADH - cytochrome c - reductase, cytochrome c - oxidase and ATPase were performed with mitochondrial particles (10). Enzymatic and spectral analysis are in good agreement. Lack or strong reduction of the cytochrome b peak is always correlated with a deficiency in NADH - cytochrome c - reductase. Presence of a cytochrome b peak corresponds to 17 - 33 % of the wild type activity of NADH - cytochrome c - reductase. Lack of spectral cytochromes aa_3 is also accompanied by a strongly reduced enzymatic activity. All mutants show an oligomycin - sensitive ATPase. Data are summarized in Table 5.

Table 5: Enzyme activities and results of difference spectra of mitochondria of the wild type ade 7 - 50 h^-, mutant ana^r - 14 and mit^- - strains.

strain	NADH cyt c reductase n moles cyt c x min^{-1} x mg^{-1}	+ AA 1 µg/ml	cyt c oxidase min^{-1} x mg^{-1}	ATPase µ mol P_i x min^{-1} x mg^{-1}	+ Oli 80 µg/ml	Spectra c	b	a
wt	278,6	< 0,1	3,0	0,5	0,2	+	+	+
ana^r - 14	174,4	< 0,1	2,1	0,8	0,2	+	+	+
SP 1	47,4	0,7	0,2	0,9	0,2	+	+	-
SP 4	92,2	0,9	< 0,1	0,8	0,4	+	+	-
EB 3	81,2	1,9	< 0,1	0,4	0,3	+	+	-
EB 8	46,0	1,5	< 0,1	0,4	0,3	+	+	-
SP 2	1,1	0,4	0,8	0,3	0,2	+	-	+
EB 4	2,0	0,8	3,8	0,4	0,2	+	-	+
EB 5	1,2	1,0	0,8	0,9	0,4	+	-	+
EB 7	3,3	1,3	2,2	0,8	0,3	+	-	+
EB 9	3,3	1,9	1,1	0,4	0,3	+	-	+
EB 11	10,3	2,3	0,6	0,5	0,3	+	-	+
SP 3	2,8	1,4	0,2	0,5	0,2	+	-	-
EB 2	4,1	< 0,1	0,2	0,5	0,4	+	-	-

d) Mitochondrial protein synthesis

The mutants were grown in a medium containing 10 % raffinose, 2 % yeast extract, 0,35 % $(NH_4)_2SO_4$, and 75 mg/l adenine to early stationary phase. Cells were washed, incubated in sulfate - free minimal medium and labelled with $^{35}SO_4^{--}$. Cytoplasmic protein synthesis was inhibited by cyclohexi-mide (CHI, 4 mg/ml), mitochondrial protein synthesis by chloramphenicol (CAP, 8 mg/ml). Incorporation of $^{35}SO_4^{--}$ into proteins was measured by the method of Mans and Novelli (11).

Table 6 shows that all mutants posses chloramphenicol - sensitive protein synthesis. The differences of incorporation in the mit^- - mutants may be due to different growth rates of the strains.

Table 6: Mitochondrial protein synthesis of whole cells labelled with $^{35}SO_4^{--}$.

	cpm/ml		
strain	no antibiotics	+ CHI	+ CHI + CAP
wt	83227	5550	2687
ana^r - 14	114878	5243	2816
SP 1	97431	5217	2813
SP 2	93518	6565	3384
SP 3	106128	5306	2183
EB 2	8965	2678	1074
EB 3	13900	1641	966
EB 4	97833	5178	3836
EB 5	48713	5596	2320
EB 7	5141	1238	1107
EB 8	74041	5000	1134
EB 9	11249	1604	691
EB 11	45382	2865	1632

CONCLUSIONS

In this work we have demonstrated that it is possible to isolate mitochondrially inherited respiratory - deficient mutants from the petite - negative yeast Schizosaccharomyces pombe by use of the mutator - properties of the strains ana^r - 8 and ana^r - 14. It could be established both by genetic and biochemical methods, that the mutants descibed belong to the class of mit$^-$ - mutants. In analogy to Saccharomyces cerevisiae the mutants could be assigned to the three groups OXI, COB and BOX. It remains to be tested, whether the BOX - phenotype is due to an enhanced glucose - repression of cytochrome c - oxidase, as it is the case in Saccharomyces cerevisiae (12).

Mutants that do not transmit antimycin - resistance to the diploid progeny are of special interest, since this may be due to small deletions. The possibility, however, that these mutants are not able to recombine, cannot be ruled out. Isolating respiratory - deficient mutants from a strain with several mitochondrial markers, one could pick up deletions in different regions of the genome. By this approach, essential and non - essential regions of the genome could be separated.

ACKNOLEDGEMENTS

The authors appreciate the skilful technical assistance of Miss R. Stets, Mrs. B. Sydow and M. Sorini. One of us (C. Seitz) is grateful to the European Communities for a short - term fellowship at the Laboratoire d'Enzymologie, Louvain - La - Neuve, Belgium. This work was supported by the Deutsche Forschungsgemeinschaft.

REFERENCES

(1) Handwerker, A., Schweyen, R.J., Wolf, K., Kaudewitz, F. (1973), J. Bacteriol. 113 , 1307

(2) Louie, A., Boerner, P., Biron, C.A., Mason, T.L. (1975), Federat. Proc. 34, 1556

(3) Storm, E.M., Marmur, J. (1975), Biochem. Biophys. Res. Commun. 64, 752

(4) Tzagoloff, A. (1976) In: Mitochondria, Bioenergetics, Biogenesis, and Membrane Sructure. Lester Packer and A. Gómez -Puyon eds. Academic Press inc., New York, San Francisco, London, pp 241

(5) Cabral, F., Solioz, M., Rudin, Y., Schatz, G., Clavilier, L., Slonimski, P.P. (1977) in press

(6) Wolf, K., Lang, B., Burger, G., Kaudewitz, F. (1976), Molec. gen. Genet. 144, 75

(7) Lang, B., Burger, G., Wolf, K., Bandlow, W., Kaudewitz, F. In: Genetics, Biogenesis and Bioenergetics of Mitochondria. Proceedings of a Symposium held at the Genetisches Institut der Universität München, Munich, September 11-13 (1975) (W. Bandlow, R.J. Schweyen, D.Y. Thomas, K. Wolf, F. Kaudewitz, eds) pp 379, Berlin-New York, Walter de Gruyter

(8) Bulder, C.J.E.A. (1964), Antonie van Leeuwenhoek 30, 442

(9) Tzagoloff, A., Akai, A., Needleman, R.B., Zulch, G. (1975) J. Biol. Chem. 250, 8236

(10) Tzagoloff, A., Akai, A., Needleman, R.B. (1975), J. Bacteriol. 122, 826

(11) Mans, R.J., Novelli, G.D. (1961), Arch. Biochem. Biophys. 94, 48

(12) Pajot, P., Wambier-Kluppel, M.L., Kotylak, Z., Slonimski, P.P. In: Genetics and Biogenesis of Chloroplasts and Mitochondria. Th. Bücher et al. eds. (1976), Elsevier/North-Holland Biomedical Press, Amsterdam, The Netherlands, pp. 443.

FINE STRUCTURE GENETIC MAP OF THE MITOCHONDRIAL DNA REGION CONTROLLING COENZYME QH_2-CYTOCHROME C REDUCTASE.

Z. KOTYLAK and P.P. SLONIMSKI
Institute of Microbiology, University of Wroclaw, Poland and Centre de Génétique Moléculaire du C.N.R.S. GIF sur Yvette 91190 - France.

It is now well established, that the genetic information necessary for the biosynthesis, assembly, function and regulation of the coenzyme QH_2-cytochrome c reductase and cytochrome b is mainly,if not exclusively, located in a unique region of mtDNA in yeast. Evidence comes from studies of two types of mutants, those deficient in enzyme activity and those displaying a drug resistant phenotype. In the first study performed on this subject, seven deficient mutants were localized in a cluster close to the OLI1 genetic marker (14). Shortly after, eight new deficient mutants were mapped in a vicinity close to the previous ones (7). Then, another eight similar mutants were localized in this region (15) and finally a few scattered mutants obtained in other laboratories were added to the list (cf. 1, 2). Concomitantly, some new drug resistant mutants were obtained and tentatively assigned to the same genetic region (cf. 2 and 4, for review).
This region of mtDNA is of particular interest for two reasons : i) It comprises mutations which lead to a variety of different phenotypes (4, 6, 9, 10, 14, 15), the analysis of which should provide a better understanding of structure and function of mitochondrial membranes; ii) It has been located already on the molecular map of wild type mtDNA and it may comprise or be included within, the EcoR1 restriction fragment R5 and the Hind II + III restriction fragment H5 (5, 12), the analysis of which should provide a better understanding of mtDNA sequence organization. However, a prerequisite for further studies is to establish a detailed genetic map of the region. Until now two partial maps have been published, one comprising eight mutational sites (7), the other fourteen sites (15). The mutants were obtained from two different non-isogenic strains, 777-3A and D273-10B, respectively, and the derived maps have not been compared.

The purpose of the present work is three fold : i) to construct a fine structure map of 22 mutants, out of which 14 are new ones, designated by the generic term BOX, and isolated from the strain 777-3A; ii) to re-map 8 mutants studied previously (14, 15), designated by the generic term COB, and isolated from the strain D273-10B and iii) to "synthesize", if possible a congruent genetic map of the region from the two sets of data. We will show that a unique and self-consistent genetic map can be constructed provided some internal contradictions and, we believe, erroneous conclusions of one of the previously published maps (15) are corrected. The new genetic map presented here will serve as a basis for the localization of diuron, mucidin and antimycin A resistant mutants (4), for the biochemical characterization of coenzyme QH_2-cyt.c reductase and cytochrome c oxidase deficient mutants and their revertants (10) and for the analysis of polypeptide chains coded by this genetic region (3).

I - ISOLATION AND ANALYSIS OF MUTANTS -

Fourteen new mutants, in addition to the eight described previously (7), were isolated from the strain 777-3A $op1^-$ ade after Mn^{++} mutagenesis (11). A detailed description will be published elswhere (8) and only major outlines will be given here. (Mutants isolated in Gif begin with the letter G followed by a number and mutants isolated in Wroclaw begin with the letter W followed by a number).

The $OP1^+$ ρ^+ mit^- strains of opposite mating type and appropriate nuclear auxotrophies were constructed by meiosis after crossing to the strain KL14-4A, as described in (7). Standard media and procedures were used (7, 14). Restoration tests with a collection of genetically stable and pure ρ^- clones, all derived from the strain KL14-4A, were done as described in (7, 14) after selection of prototrophic diploids on a minimal medium. Frequency of glycerol positive (presumably wild type) recombinants in ρ^+ mit_i^- x ρ^+ mit_j^- crosses was determined by the procedure originally described in (7). It is important to stress that only this type of procedure i.e. very high glucose concentration, 10%, and practically anaerobic growth conditions permits one to determine a true frequency of mit^+ recombinants which is not overshadowed by the selective advantage of mit^+ cells over the mit^- ones.

Frequency of wild type recombinants (in %) in crosses within loci

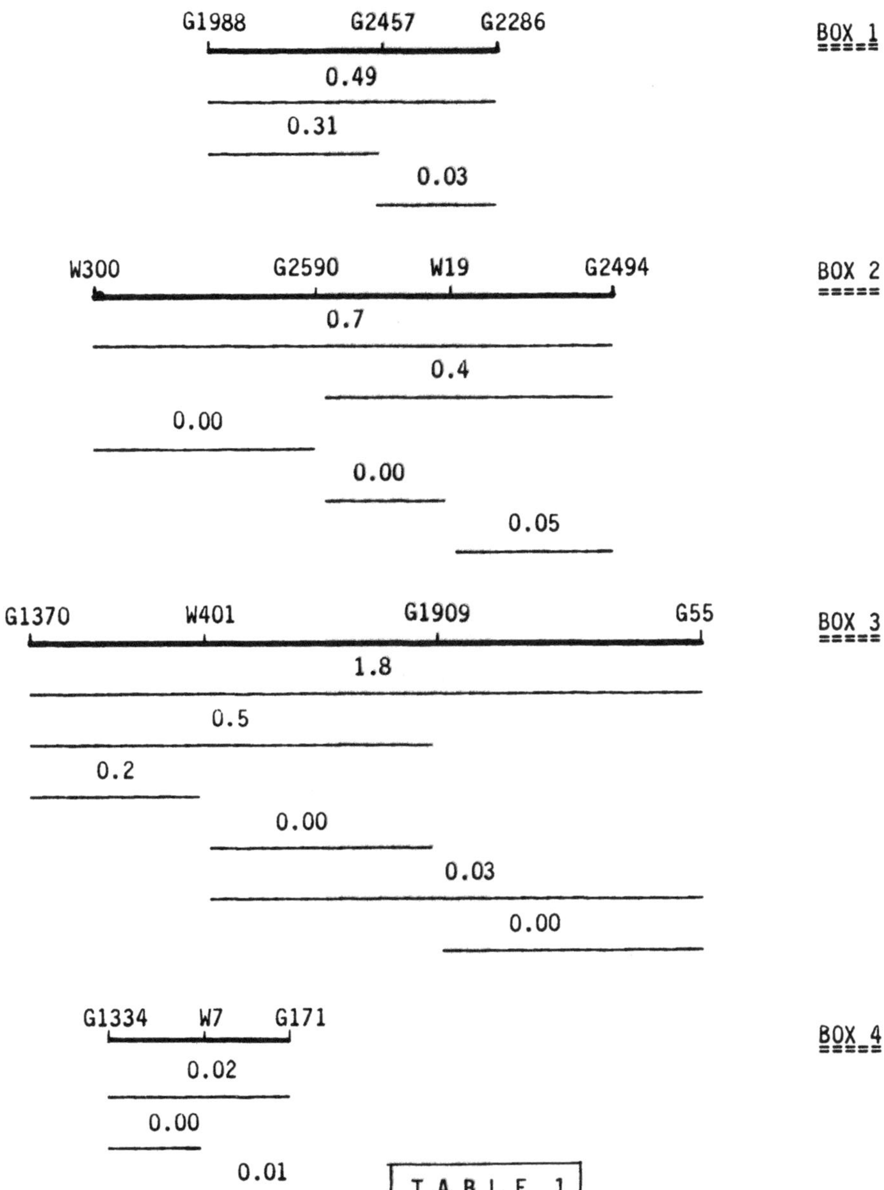

TABLE 1

II - CONSTRUCTION OF THE GENETIC MAP -

All the 22 mutants have been crossed with each other in pairwise combination and the frequency of ρ^+ mit^+ recombinants determined among the total ρ^+ progeny of each cross. There are altogether 484 combinations corresponding to 22 mutants in the α mating type and 22 mutants in the a mating type. More than 200 combinations have been examined already and we believe that they faithfully represent the total matrix of crosses for following reasons : i) reciprocal crosses i.e. a mit_i^- x α mit_j^- vs α mit_i^- x a mit_j^- give, in most cases, identical or very similar results (see ref. 9 for a numerical example); ii) we have shown that crosses are not biased in thèir genetic inputs and no polarity of recombination phenomena have been found; iii) the results are self-consistent and in excellent agreement with the restoration matrix by ρ^- testers (see below).

Table 1 shows that 14 mutants can be allocated to 4 clusters of very tightly linked mutational sites. We shall refer to such clusters as genetic loci. The four loci BOX1 to BOX4 are those described previously (7) and 6 additional mutants have been assigned to them. Some mutations may be small deletions. For instance, mutation G2590 of the locus BOX2 could be a deletion overlapping the sites W300 and W19. However, none of the mutants listed in table 1 is a large deletion covering many sites in dïfferent loci. It is apparent that a classical "map expansion" phenomenon takes places in intra-locus recombination. This is attested by the fact that frequency of recombinants between distal sites (AC) is greater than the sum of recombinant frequencies between intermediate sites (AB + BC). Numerous examples of this contention are given in table 1. Four other mutants constitute two additionnal clusters which will be referred to as BOX5 locus (mutants G706 and G392) and BOX6 locus (mutants W267 and G925). The first two are tightly linked (0.13% wild type recombinants) and the two other moderately linked (2.0% recombinants). Mutant G625 is a large deletion covering many sites. Mutant G1659 is located half way between the locus BOX1 and BOX2. It gives 1.3% recombinants on the average, when crossed to mutants of BOXI and 2.1% on the average, when crossed to mutants of BOX2 (table 2).

We have shown previously that interlocus recombinant frequencies were higher than intralocus frequencies and rather close to the upper limit of

Frequency of wild type recombinants (in %) in crosses between loci

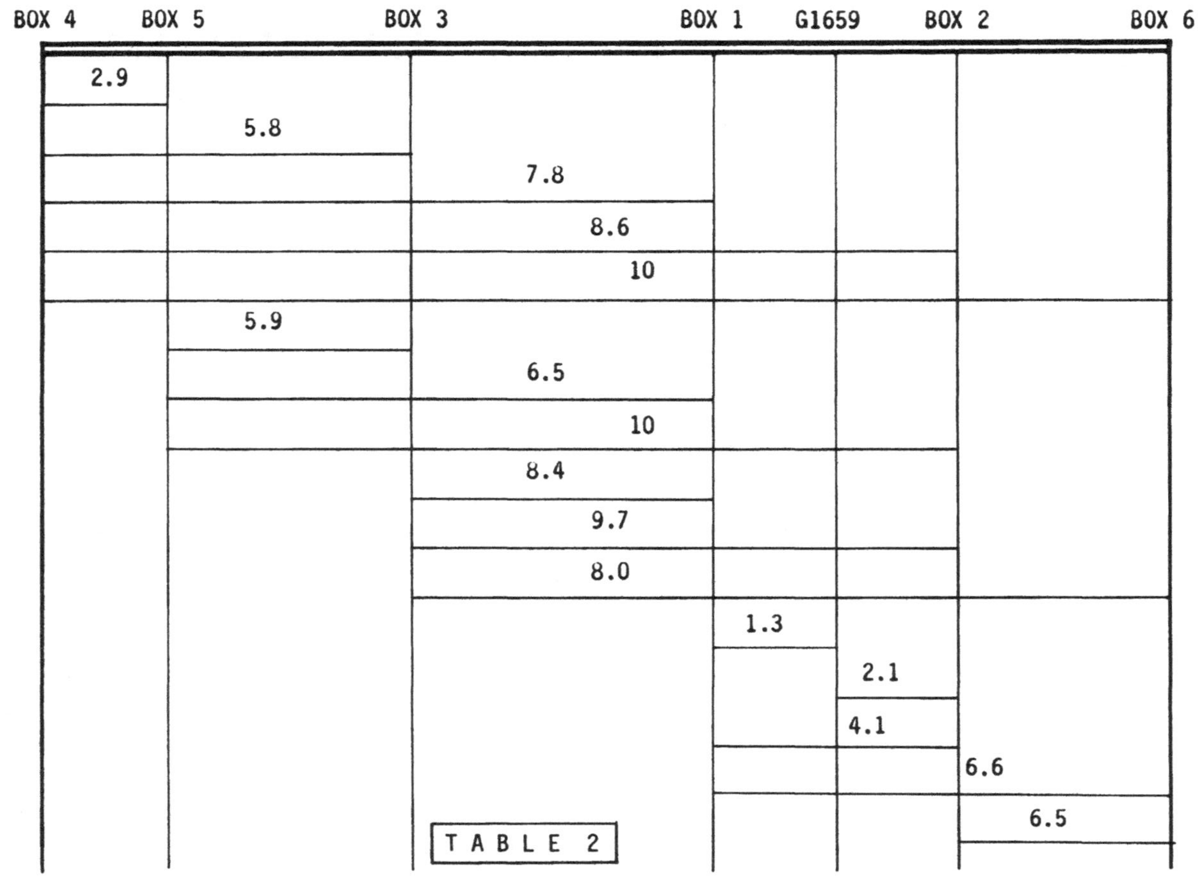

TABLE 2

TABLE 3

DELETION MAP OF CYTOCHROME C REDUCTASE REGION OF MTDNA

	G625	G839	G1334	W7	G171	G706	G392	G1370	W401	G1909	G55	W91	G1988	G2457	G2286	G1659	W300	G2590	W19	G2494	W267	G925	OLI1
Segment		SEGMENT II											SEGMENT I										
Locus BOX			4			5		3					1				2				6		
Site			1	3	2	2	1	1	3	2	4	5	1	3	2		4	2	3	1	2	1	
PM1	+	+	+	+	+	+	+	+	+	+	+	+	+	+	+	+	+	+	+	+	+	+	+
RA1	-	+	+	+	+	+	+	+	+	+	+	+	+	+	+	+	+	+	+	+	-	-	-
SC773A2B3	-	-	-	-	-	-	-	-	-	-	-	-	+	+	+	+	+	+	+	+	+	+	+
PE1	-	-	-	-	-	-	-	-	-	-	-	-	+	+	+	+	+	+	+	+	+	+	-
PI1	-	-	-	-	-	-	-	-	-	-	-	-	-	-	-	+	+	+	+	+	+	+	-
SC632A1	-	-	-	-	-	-	-	-	-	-	-	-	-	-	-	-	-	-	-	-	+	+	+
746A1	-	-	-	-	-	-	-	-	-	-	-	-	-	-	-	-	-	-	-	-	+	+	-
RD1	-	-	-	-	-	-	-	+	+	+	+	+	+	+	+	+	+	+	+	+	+	+	-
RE1	-	+	+	+	+	+	+	+	+	+	+	+	+	+	+	+	-	-	-	-	-	-	-
RC3	-	-	+	+	+	+	+	+	+	+	+	+	-	-	-	-	-	-	-	-	-	-	-
PH1	-	-	+	+	+	+	+	+	-	-	-	-	-	-	-	-	-	-	-	-	-	-	-
PK1	-	-	+	+	+	+	+	-	-	-	-	-	-	-	-	-	-	-	-	-	-	-	-
RG1	-	-	-	-	-	+	+	-	-	-	-	-	-	-	-	-	-	-	-	-	-	-	-

recombinants found in non-polar crosses (7). Table 2 confirms and extends this observation. With the exception of BOX1 and 2 which are moderately linked (cf.7) and BOX4 and 5 which are even more linked, all other combinations give 6 to 10% wild type recombinants. This would correspond to 12 to 20% total recombinants if the reciprocal mit_i^- mit_j^- recombinant class occurs with the same frequency as the mit_i^+-mit_j^+ class.
Table 3 shows the results of deletion mapping with discriminating ρ^- testers. Only 13 petite clones are shown in the table out of some 50 analyzed. These 50 discriminating petite were obtained after screening of more than 5000 ρ^- mutants. Two important conclusions can be drawn : i) the map is selfconsistent and no internal deletions appear in the retained mtDNA segments of petites covering the 22 BOX mutational sites; ii) the map is congruent with the map derived from mit_i^- x mit_j^- crosses. It is particulary rewarding to find that all sites previously allocated to the same genetic locus are always either lost together or maintained together (compare tables 1 and 3). The only exception confirms this rule : petite PH1 restores the mutation G1370 and not others at BOX3 locus. Now, G1370 is precisely located at the very edge of this locus (table 1). Furthermore, localization of G1659 is clearly demonstrated by the petite testers PI1 and RE1 in agreement with the data of table 2. As pointed out previously (13, 14), interpretation of deletion mapping by petites greatly depends not only on the nature of the retained mtDNA segments but also on the frequency with which these segments are generated. Two sites which separate frequently are interpreted as being located further apart than two sites which separate less frequently. We have shown previously that loci BOX1 and 2 were separated relatively often from loci BOX3 and 4 (7). The present work confirms and extends this observation. In 27 independant petites examined, the separation was found to be complete in the sense that all 10 sites of loci BOX3, 4, 5 were lost and all 10 sites of loci BOX1, 2, 6 were retained (like the petite PE1) or vice versa (like the petite RC3). All other combinations of loss and retention occur with frequencies much smaller.

In conclusion, the wild type mtDNA is composed of two relatively easily separable but adjacent segments I and II, the first segment comprises three genetic loci 1, 2 and 6 and ten mutational sites, the second segment includes three genetic loci 3, 4, 5 and ten sites.

TABLE 4

INCONSISTENCIES IN THE MAPPING DATA ON COB1 and COB2 LOCI OF TZAGOLOFF et al (15)

	Frequency of (+) recombinants (%)	Frequency of separating§ ρ^- clones.
COB1 x COB2	5.1*	6 out of 20
COB2 x COB2 (M9-228) (M17-162)	0.037	9 out of 20
COB2 x COB2 (M9-228) (M21-71)	0.04	8 out of 20

Data from table 3 and Fig.2 of (15). *Average from 10 crosses in the range 4.0% to 6.5%.

§A ρ^- clone which restores one but not the other mit⁻ mutation is a separating clone.

III - SIGNIFICANCE OF THE GENETIC MAP -

The map of BOX mutants, as all genetic maps, has no interest in itself except for the intellectual pleasure it gives rise to during its construction. A map acquires biological significance when it is related to function and structure of the genetic information. Research in this direction has begun (9) and is being continued (3, 4, 10) and the comparison of the map of BOX mutants with that of COB mutants should be useful.

TZAGOLOFF et al. (15) have reported that fourteen mutational sites can be allocated to two separate genetic loci COB1, COB2 oriented OLI1 - COB1 - COB2 and suggested that the structural gene of cytochrome b corresponds to COB2. Exactly the same mapping procedures, mit^- x mit^- crosses and mit^- x ρ^- crosses, were used in (7), (15) and the present work. TZAGOLOFF et al. stated that "their maps obtained by the two independant methods are in good agreement" (15).

Analysis of their data discloses an inconsistency which apparently has escaped the author's attention. This contradiction is of critical importance because it concerns the mutant M9-228, which is precisely the mutant used to draw the conclusion that COB2 is the structural gene. The contradiction is apparent in table 4. The average frequency of wild type recombinants in interlocus crosses is high and comparable to interlocus crosses in (7) and the present work. The frequency of crosses involving M9-228 with two other COB2 mutants is reported to be ca 100 times smaller. This is in a complete disagreement with the reported frequency of separation by ρ^- deletions. By this method M9-228 and the two other mutants allocated to the <u>same</u> locus appear as more distant than mutants allocated to <u>different</u> loci. This contradiction has prompted us to remap representative COB mutants, including M9-228, by crossing them with our collection of 50 discriminating ρ^- testers. Results were unambiguous. The M9-228 site is included in the segment I of mtDNA and belongs to the locus BOX1. It is rewarding to find that a unique order of 28 BOX and COB sites can be established (table 5), provided the mutant M9-228 is placed in the OLI1 <u>proximal</u> segment I of mtDNA and not in the distal one. When this is done, the two maps agree and the locus 1 of COB mutants corresponds to the segment I of BOX mutants, the locus 2 to the segment II and the structural gene of cytochrome b would be located in the

TABLE 5

OVERALL MAP OF 28 MUTATIONAL SITES BOX AND COB IN THE CYTOCHROME C REDUCTASE REGION OF MTDNA

← BOX SEGMENT I = COB LOCUS 1 →

(OLI1)—(M24·241)—(M7·40;M13·101;M8·53;G925;W267)—(M6·200)—(G2494;W19;G2590;W300)—(G1659)—(M9·228;G2286;G2457;G1988)—

—(G55)—(G1909)—(W401)—(G1370)—(G392)—(G706)—(G171;W7;G1334)—(M21·71)—(M17·162)—(G839)—

← BOX SEGMENT II = COB LOCUS 2 →

The order is established by deletion mapping with ρ^- testers (see table 3) and by recombination mapping in pairwise mutant by mutant crosses (see table 1 and 2). Both methods lead to the same order. Mutational sites allocated to different brackets are unambigously ordered. Mutational sites allocated to the same bracket cannot be oriented in respect to the flanking sites outside the bracket, although they can be ordered within the bracket in respect to each other (see table 1). Mutants underlined derive from the strain 777-3A, non underlined from D273-10B.

locus COB1 and not COB2. An important prediction can be made. If this allocation is right, one expects to find structural changes of cytochrome b polypeptide chains in mutants G1988, G2457 or G2286 included in the locus BOX1. This is the case (3).

Acknowledgments.

Z. KOTYLAK greatfully acknowledges the receipt of EMBO long term fellowship held at Gif in 1975-76. This work could not have been done without excellent technical collaboration of G. PERRODIN and J. LUTKA. The work has been supported by DGRST Grants N° 75-70748 and 77-70420.

References.

1. "The Genetic Function of Mitochondrial DNA". Eds. C. SACCONE and A.M. KROON (1976). Elsevier/North Holland Biomedical Press, Amsterdam, pp. 1-354.
2. "Genetic and Biogenesis of Chloroplasts and Mitochondria". Eds Th. BUCHER et al. (1976). Elsevier/North Holland Biomedical Press, Amsterdam, pp. 1-895.
3. CLAISSE, M., SPYRIDAKIS, A. and SLONIMSKI, P.P. (1977). This volume.
4. COLSON, A.M., & SLONIMSKI, P.P. (1977) this volume.
5. DI FRANCO, A., SANDERS, J.P.M., HEYTING, C., BORST, P. and SLONIMSKI, P.P. (1976) in "The Genetic Function of Mitochondrial DNA, Eds, C. SACCONE and A.M. KROON. Elsevier/North Holland Biomedical Press, Amsterdam, pp. 291-304.
6. FOURY, F. and TZAGOLOFF, A. (1976). Molec. Gen. Genet. 149, 43-50.
7. KOTYLAK, Z. and SLONIMSKI, P.P. (1976) in "The Genetic Function of Mitochondrial DNA". Eds C. SACCONE and A.M. KROON. Elsevier/North Holland Biomedical Press, Amsterdam pp. 143-154.
8. KOTYLAK, Z., SLONIMSKI, P.P., PERRODIN, G., and LUTY, J. (1977). Molec. Gen. Genet. submitted.
9. PAJOT, P., WAMBIER-KLUPPEL, M.L., KOTYLAK, Z., and SLONIMSKI, P.P. (1976) in "Genetic and Biogenesis of Chloroplasts and Mitochondria" Eds Th. BUCHER et al. Elsevier/North Holland Biomedical Press, Amsterdam. pp. 443-451.

10. PAJOT, P., WAMBIER-KLUPPEL, M.L. and SLONIMSKI, P.P. (1977) this volume.
11. PUTRAMENT, A., BARANOWSKA, H. and PRAZMO, W. (1973). Molec. Gen. Genet. 126, 357-366.
12. SANDERS, J.P.M., HEYTING, C., DI FRANCO, A., BORST, P. and SLONIMSKI, P.P. (1976) in "The Genetic Function of Mitochondrial DNA." Eds C. SACCONE and A.M. KROON. Elsevier/North Holland Biomedical Press, Amsterdam, pp. 259-272.
13. SCHWEYEN, R.J., STEYRER, U., KAUDEWITZ, F., DUJON, B., SLONIMSKI, P.P., (1976) Molec. Gen. Genet., 146, 117-132.
14. SLONIMSKI, P.P. and TZAGOLOFF, A. (1976). Eur. J. Biochem., 61, 27-41.
15. TZAGOLOFF, A., FOURY, F. and AKAI, A. (1976). Molec. Gen. Genet. 149, 33-42.

CYTOCHROME C REDUCTASE AND CYTOCHROME OXIDASE FORMATION IN MUTANTS AND REVERTANTS IN THE "BOX" REGION OF MITOCHONDRIAL DNA.

P. PAJOT, M.L. WAMBIER-KLUPPEL and P.P. SLONIMSKI
Centre de Génétique Moléculaire du C.N.R.S. GIF sur YVETTE,91190-FRANCE

In Bari, last year, Kotylak and Slonimski presented a new series of mitochondrial DNA mutations in Saccharomyces cerevisiae which control the biosyntheses of both cytochrome b and cytochrome aa_3. That region of mtDNA was named BOX, and, at that time, four different loci BOX1, 2, 3 and 4 were distinguished (1).

A few months later, in Munich, we presented a preliminary characterization of some of these mutants and it was shown that, whereas cytochrome b was always totally absent, in several of them, the formation of cytochrome aa_3 appeared to be particularly sensitive to catabolite repression. The degree of sensitivity to the repression was mitochondrial allele specific and identical in different nuclear contexts (2).

Since then, several other mutants have been localized in that mtDNA region and mapped by recombination and deletion (3). This communication presents a few physiological and biochemical characteristics of these mutants and of revertants originating from them.

MATERIAL AND METHODS.

The original mutant strains of the $op1^-$ series were obtained as described in (1,3); those in the $OP1^+$ nuclear context were constructed as described previously (2).

The presence of cytochromes was observed in recording the absorption spectra of fully dithionite reduced cells or mitochondria at liquid nitrogen temperature (4,5). Cells were harvested on agar plates containing 25 ml of a medium (1% yeast extract, 1% bacto peptone, 2% galactose+0.1% glucose and 0.002% adenine) after four days of growth at 28°.

Revertants were obtained by plating 10^7 to 10^8 cells issued from a single mutant clone on glycerol plates. Individual revertant clones were collected after 8 or 15 days of incubation at 28°.

Enzymic activities were measured on mitochondria prepared by mechanical breakage as described in (2).

SPECTRAL AND ENZYMIC CHARACTERISTICS OF "BOX" MUTANTS.

Table 1 summarizes a few characteristics of the BOX mutants in the two nuclear contexts,$op1^-$ and $OP1^+$, and Fig. 1 shows a few examples of absorption spectra of mitochondria from these mutants.

All the mutants listed in Table 1 are devoid of coenzyme QH_2-cytochrome c reductase activity (data not shown) and all the mutations, except two, result in the absence of the absorption bands characteristic of cytochrome b. BOX4-3 and BOX6-1 mutants contain the hemoprotein but are devoid of complex III electron transfer activity. This has been verified by measuring

TABLE 1 : LIST OF MUTANT STRAINS, SPECTRAL CHARACTERISTICS OF INTACT CELLS AND FREQUENCY OF REVERTANTS.

	Nuclear genotype : $op1^-$			Nuclear genotype : $OP1^+$					
							Revertants number of colonies		
mitochondrial genotype	strain number	cyt.b	$cyt.aa_3$	strain number	cyt.b	$cyt.aa_3$	$glycerol^+$	$glycerol^-$	frequency
	777-3A / G625	–	–	PS424	–	–	1	3×10^8	0.3×10^{-8}
	" / G839	–	–	PS420	–	–	0	1×10^9	<0.1 "
BOX 4-1	" / G1334	–	–	PS400	–	+	0	2×10^9	<0.05 "
BOX 4-2	" / G171	–	–	PS411	–	–	9 *	2×10^9	0.5 "
BOX 4-3	" / W7	+	++	PS429	+	++	38 *	2×10^9	2 "
BOX 5-2	" / G706	–	–	PS415	–	–	5	3×10^8	2 "
BOX 5-1	" / G392	–	–	PS412	ε	ε	250	3×10^8	80 "
BOX 3-1	" / G1370	–	–	PS401	–	–	130 *	1×10^7	1000 "
BOX 3-3	" / W401	–	–	PS434	–	–	0	2×10^9	<0.05 "
BOX 3-2	" / G1909	–	–	PS402	–	–	0	2×10^9	<0.05 "
BOX 3-4	" / G55	–	–	PS425	–	–	0	2×10^9	<0.05 "
BOX 1-1	" / G1988	–	+	PS403	–	++	7	5×10^8	1 "
BOX 1-2	" / G2286	–	+	PS405	–	++	6	5×10^8	1 "
BOX 1-3	" / G2457	–	++	PS406	–	++	400*	1×10^7	4000 "
	" / G1659	–	–	PS423	–	–	7	2×10^9	0.4 "
BOX 2-1	" / G2494	–	++	PS407	–	++	34	5×10^8	7 "
BOX 2-2	" / G2590	–	++	PS408	–	++	109	3×10^8	30 • "
BOX 2-3	" / W19	–	+	PS430	–	++	60 *	1.5×10^8	40 "
BOX 2-4	/ W300	–	+	PS433	–	+	2 *	2×10^9	0.1 "
BOX 6-1	" / G925	+	++	PS422	++	++	5	3×10^8	2 "
BOX 6-2	" / W267	–	++	PS432	–	++	26	2×10^9	1 "
BOX^+	777-3A	++	++	PS409	++	++			

Cytochrome spectrum recorded after growth on galactose :(–) = absent ; (ε) = barely detectable and uncertain ; (+) = about 1/3 of the wild type content : (++) = between 1/3 and the wild type content.

Glycerol positive colonies scored after 15 days of incubation at 28°C.

* Most of the glycerol positive colonies are of minute size and appear late

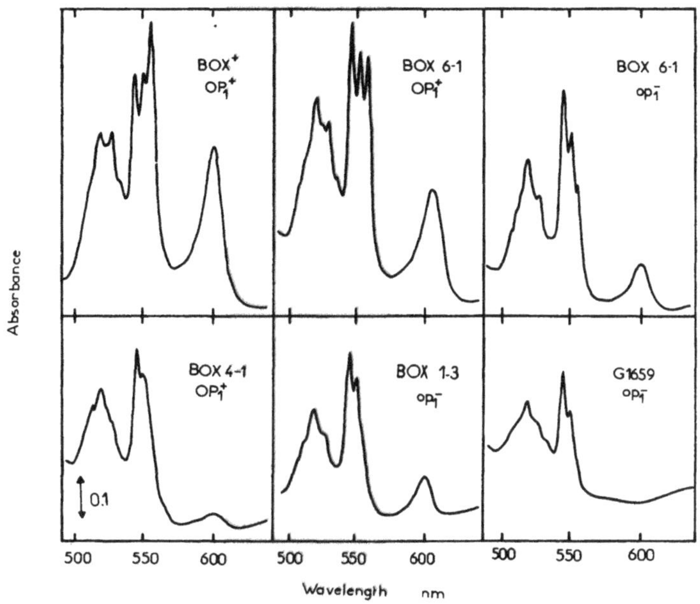

Fig.1 : Cytochrome spectra of mitochondria from various BOX mutants. Absolute spectra of fully reduced mitochondrial particles prepared after mechanical breakage of cells grown on galactose. The protein concentrations were from 23 to 27 mg/ml depending on the mutant studied. (liquid nitrogen temperature, sample thickness 2 mm).

both NADH and succinate-cytochrome c reductase activities in the presence of added coenzyme Q analog, 2,3-dimethoxy-5-methyl decyl -1,4 benzoquinone, kindly given by Dr. K. Folkers (6). For both mutants, these activities were less than 1% of those in the corresponding wild type strain. It is worth noting that the two mutations are located far apart on mtDNA but, nevertheless lead to very similar phenotypes. In both cases, the frequency of revertants is about 10^{-8} suggesting that BOX4-3 and BOX6-1 mutations are point mutations or small rearrangements

This is the first time that, by mutation, coenzyme QH_2-cytochrome c reducase activity has been completaly dissociated from spectrally observable cytochrome b. Two types of hypotheses could explain our findings:

a) Mutations modify the structure of the enzyme and abolish its activity without altering the heme binding sites of cytochrome b apoprotein itself. That would imply that there are two, relatively distant, structural genes coding for the cytochrome b apoprotein which correspond to loci BOX4 and BOX6. This, in turn, would mean that there are two structurally different polypeptide chains which constitute cytochrome b.

b) Mutations affect an unknown product of mtDNA which is not cytochrome b but is a necessary part of coenzyme QH_2-cytochrome c reductase.

It is obvious that further studies of these mutants will be of interest. When one considers the occurence of cytochrome aa_3, two distinct series of mutations appear (Table 1). Except in two mutants at locus BOX4, cytochrome oxidase is absent in the remaining seven mutations located at loci BOX4, BOX5 and BOX3. The reverse situation is observed with mutants BOX1, BOX2 and BOX6; in this segment of mtDNA, nine mutants out of ten analyzed display cytochrome aa_3 spectrum and activity. The only exception is mutant G1659 which is totally devoid of cytochrome oxidase whether studied in the $op1^-$ or the $OP1^+$ nuclear context.

It is rewarding to find that the separation into two groups of mutations corresponds exactly to the separation into two segments, II and I respectively, of mtDNA as shown by deletion mapping with defined rho^- mtDNA segments (3).

We have verified that cytochrome oxidase of mutants BOX6-1, BOX1-3, BOX2-1 and BOX4-3, in the $OP1^+$ context, has normal enzymic activity because it allows cells to oxidize lactate in conjunction with yeast lactate-cytochrome c oxidoreductase (7).

VARIOUS CLASSES OF REVERTANTS FROM "BOX" MUTANTS.

Revertants cannot be obtained in the $op1^-$ series of strains due to the nuclear mutation which prevents mit^+ cells from growing on a respirable substrate. Revertants were thus obtained from the $OP1^+$ strains by isolating glycerol positive colonies. Table 1 gives numerical data on the revertant frequency; it is apparent that this frequency depends on the nature of the original mutation. In 14 cases,revertants were found, whereas none was observed in 5 mutants; in two cases, only one or two glycerol positive colonies were observed, further investigations will be necessary to assess whether this is significant.

Essentially two criteria were used for screening various classes of revertants: the rate of growth on glycerol containing medium at the normal temperature, 28°, and the temperature dependance of the growth at three temperatures, 16, 28 and 36°.

Four classes of revertants have been distinguished:

- those displaying wild type characteristics at the three growth temperatures
- those growing very slowly, even at 28°,
- those displaying growth at high temperature and no growth at 16° (cryo-sensitive revertants),
- those growing almost normally at 16° and not at 36° (thermosensitive revertants).

Each mitochondrial BOX mutant has its own spectrum of revertants which can be characterized by the frequency and the type of revertants it gives rise to. Only a few examples will be presented here.

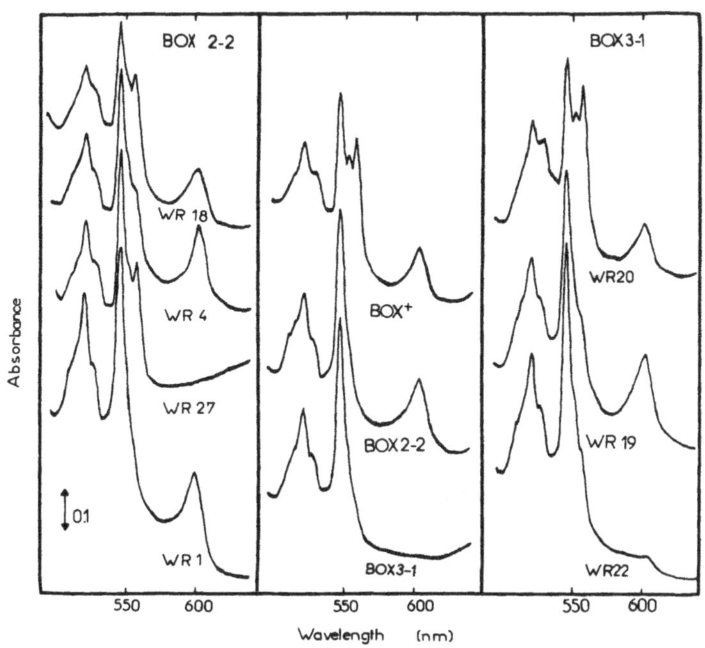

Fig.2 : Independant revertants from BOX mutants display different spectra of cytochromes.
Absolute spectra of fully dithionite-reduced cells of typical revertants derived from BOX2-2 and BOX3-1 mutants (left and right hand side of the figure). The three spectra in the central part are those of the wild type strain and the two parental mutants.Cells were grown on galactose plates. (liquid nitrogen temperature, sample thickness 1 mm).

Thermosensitive revertants originated mainly from the mutant BOX3-1; cryosensitive revertants were found rather frequently in the study of mutants BOX2-2, BOX2-3 and BOX6-1; "slow" revertants were principally obtained from mutants BOX2-2, BOX1-3, BOX5-2 and BOX6-1. Most, if not all, revertants isolated from mutants such as BOX1-2 and BOX2-1 behaved as wild-type strains as regards their growth and spectral characteristics.
A number of revertants have been analyzed for their cytochrome contents in recording the spectrum of cells grown under nonrepressing conditions. Many different spectral phenotypes appeared. Typical examples are shown in Fig. 2. Some revertants show almost complete deficiencies in cytochrome b and in cytochrome aa_3, their generation time on a respirable substate is then about 20 hours. Other revertants, such as BOX2-2/WR1, contain cytochrome oxidase and very little, if any, cytochrome b. The reverse situation is observed with the revertant BOX2-2/WR27 which displays a cytochrome b peak but synthetizes much less cytochrome oxidase in comparison with the parental mutant, grown under identical conditions.
This rapid and non exhaustive enumeration thus reveals a great variety of biochemical phenotypes, the analyses of which should be of a considerable interest. Two examples will illustrate this proposition with some more details.

TEMPERATURE DEPENDANT REVERTANTS

Revertants BOX2-2/WR4 and BOX3-1/WR19 were grown at 16, 28 and 36° on galactose and the absorption spectra of the cells were recorded (Fig. 3). It is apparent that, in these revertants, the formation of cytochrome b is quite sensitive to the temperature of growth. Two opposite effects are observed in the two types of revertants. With the cryosensitive strain WR4 the synthesis of cytochrome b is normal at 36° and abolished at 16°, the reverse is observed with the thermosensitive strain. A control experiment with the wild type strain shows the presence of that cytochrome whatever the temperature of growth.
It is of interest to note that the formation of cytochrome aa_3 does not parallel that of cytochrome b; for instance, WR19 displays a higher cytochrome aa_3 peak at 28° than at 16° and WR4 shows a higher formation of cytochrome b at 36° than at 28° but more cytochrome aa_3 at 28° than at 36°.

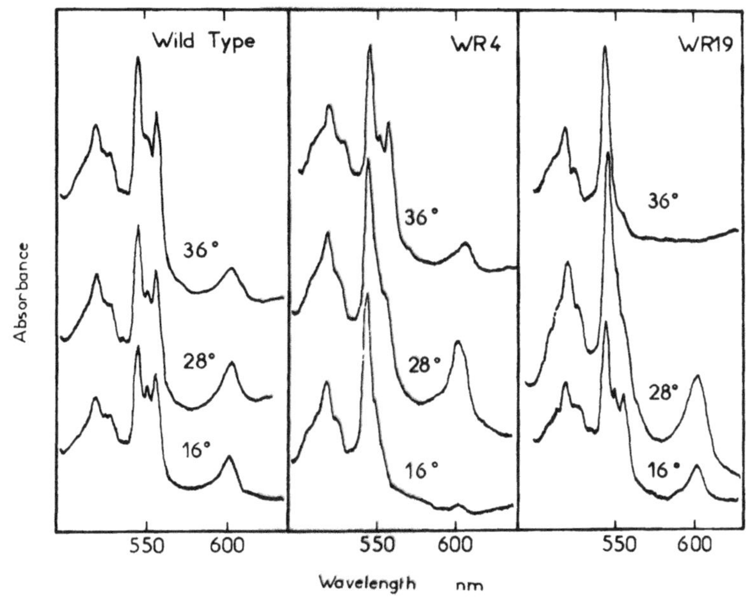

Fig. 3 : Temperature dependance of cytochrome formation in two revertants BOX2-2/WR4 and BOX3-1/WR19.
Cells were grown on solid galactose medium for 4 days at 36 and 28° and 6 days at 16° and the absolute spectra were recorded as described in Fig. 2 The wild type strain spectra are given for comparison.

We have measured the NADH cytochrome c reductase activity at two different temperatures, 28 and 36°, of mitochondria of BOX2-2/WR4, the cells having been grown at 28° and we did not find a temperature coefficient differing from that observed with wild type mitochondria. Conversely, measurements made with mitochondria from cells grown at various temperatures show that the decrease of enzymic activity parallels closely the decrease of the cytochrome content.
From these observations, it appears that the revertants BOX2-2/WR4 and BOX3-1/WRI9 revertant phenotypes reflect a temperature dependant alteration of the process of biosynthesis and/or assembly of cytochrome b rather than a modification of the structure of the protein itself.

REVERTANTS HYPERSENSITIVE TO CATABOLITE REPRESSION

An interesting observation was made in the study of BOX2-2/WR27. This revertant forms apparently a normal quantity of cytochrome b whereas the parental mutant is completely devoid of that hemoprotein; however, when this revertant is grown on galactose, it forms very little, if any, cytochrome aa_3. Furthermore, BOX2-2/WR27 can grow on a respirable substrate, but, it requires a long lag before growth starts, and then, it grows only a little slower than a wild type strain. The spectrum of harvested cells indicates the presence of cytochrome oxidase, but, if the cells grown on lactate or glycerol are inoculated to a galactose medium, the cells which are obtained are cytochrome oxidase deficient. This is illustrated in Fig. 4. The diminution of cytochrome aa_3 is likely to be due to an hypersensitivity to catabolite repression.

In conclusion, the reversion BOX2-2/WR27 has two effects: It restores cytochrome b formation and it diminishes that of cytochrome aa_3. We find

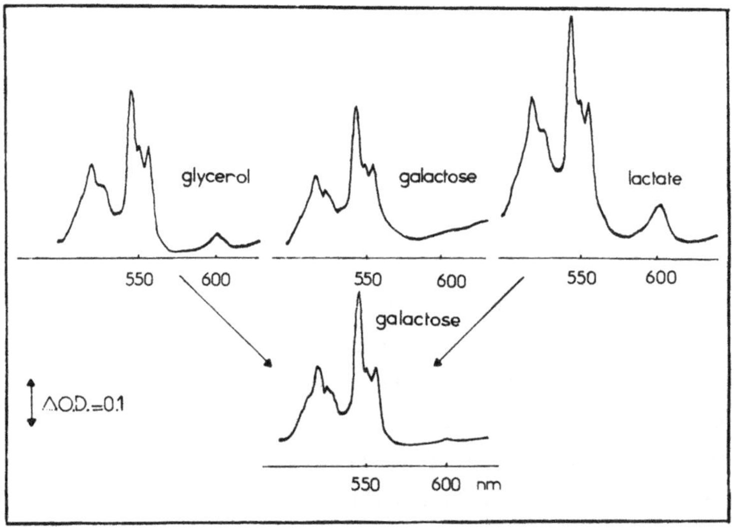

Fig. 4 : Reversibility of the repression-derepression of cytochrome oxidase formation in the revertant BOX2-2/WR27.

Cells were grown to stationary phase in liquid medium containing either lactate, glycerol or galactose and the absolute spectra of fully reduced cells are shown in the upper part of the figure. Cytochrome oxidase was found on non-fermentable substrates. In order to check the reversibility of the derepression, inocula from the cultures on glycerol and lactate were further grown on galactose. Two identical spectra were obtained with these cells, only one is shown.(liquid nitrogen temperature, sample thickness 0.5 mm)

again, in the case of this revertant, the relation between the formation of two respiratory enzymes, cytochrome c reductase and oxidase, and catabolite repression which we have previously described in the case of BOX mutants (2).

GENETIC NATURE OF REVERSIONS

It was of interest to establish the nature of the second mutation leading to restoration of the capacity to utilize a respirable substrate.

A priori, the second mutation may be due to:

-a back mutation at the same site of the mtDNA as the forward mit$^-$ mutation,

-a mitochondrial second-site mutation occuring at a site distinct from the forward mutation. The second site may be linked or unlinked to the forward site,

-a nuclear mutation which acts as a suppressor of the mitochondrial forward mutation.

In order to distinguish between these three possibilities, the following strategy was adopted. The revertants, which are diploid cells heterozygous $OP1^+/op1^-$, were sporulated and the haploid spores analyzed (revertants, which can respire, sporulate whereas the BOX mutants do not).

$op1^-$ spores can be easily distinguished because they are lethal in a glucose medium containing ethidium bromide, while $OP1^+$ spores survive as rho$^-$ cells. If the BOX^- BOX^+ reversion was due to a nuclear gene, a meiotic segregation of the active suppressor versus its inactive allele should occur. We have analyzed the meiotic $OP1^+$ progeny and found no segregation for growth on glycerol. Thus, the reversion is not due to a nuclear mutation. This was supported by the fact that ethidium bromide treatment of glycerol positive $OP1^+$ haploid progeny of revertants eliminates the transmission of the glycerol positive phenotype in crosses with the original BOX mutants.

In conclusion, revertants are due to mutations which occur in mitochondrial and not nuclear DNA. Further studies are necessary to localize the second mutation in respect to the forward one although it is already clear that the revertants which present different physiological or biochemical phenotypes cannot result from a back mutation sensu sticto at the same site.

CONCLUSION.

We would like to emphasize the interest of BOX mutants and revertants for investigating the biogenesis and function of the mitochondrial respiratory chain.

1) The two mutations, which result in the presence of cytochrome b and in a deficiency in coenzyme QH_2-cytochrome c reductase, are likely to be located in the structural genes for that electron carrier. It is satisfactory to find that mitochondrially synthesized polypeptide chain(s) of cytochrome b appear to be structurally modified by mutations localized precisely at the same loci, BOX4 and BOX6 (8), as well as drug resistance mutations (9).

2) Conversely, mitochondrial loci BOX2 and BOX3 may control the assembly processes rather than the structure of coenzyme QH_2-cytochrome c reductase itself. Evidence comes from the study of mitochondrial temperature dependant and catabolite repression hypersensitive revertants originating at those loci. Such strains may be useful for studies of the assembly of mitochondrial membranes as it was stressed by Mason et al.(10).

3) The present work on the relationships between cytochrome b and cytochrome aa_3 formation confirms and extends the previous one (2). It permits to correlate the two adjacent segments of the BOX region distinguished by genetic analysis (3) with two phenotypic characteristics. Mutations in segment I have a moderate effect on the formation of cytochrome oxidase, whereas mutations in segment II result in a more drastic effect.

4) The variety of different biochemical phenotypes which can be generated by a combination of forward and second-site mitochondrial mutations should provide an interesting basis for further investigations.

Acknowledgements.

M.L. Wambier Kluppel greatfully acknowledges the receipt of a fellowship from the Federal University of Parana (Brasil) during her leave of absence from that university. This work was supported by Grants N° 75-70748 and 77-70240 from the Délégation Générale à la Recherche Scientifique et Technique.

REFERENCES.

1. KOTYLAK, Z. and SLONIMSKI, P.P.(1976) in "The Genetic Function of Mitochondrial DNA, Eds C. SACCONE and A.M. KROON. Elsevier/North Holland Biochemical Press, Amsterdam, pp. 143-153.

2. PAJOT, P., WAMBIER-KLUPPEL, M.L., KOTYLAK, Z. and SLONIMSKI, P.P. (1976) in "Genetics and Biogenesis of Chloroplasts and Mitochondria", Eds Th. BUCHER et al. Elsevier/North Holland Biochemical Press, Amsterdam, pp. 443-451

3. KOTYLAK, Z. and SLONIMSKI, P.P. (1977) this volume.

4. CLAISSE, M.L., PéRé-AUBERT, G.A., CLAVILLIER, L.P. and SLONIMSKI, P.P. (1970) Eur. J. Biochem. 16, 430-438.

5. CLAISSE, M.L. and PAJOT,,P. (1974) Eur. J. Biochem. 49, 49-59.

6. WAN, Y.P., WILLIAMS, R.H., and FOLKERS, K. (1975) Biochem. Biophys. Res. Commun. 63, 11-15.

7. PAJOT, P. and CLAISSE, M.L. (1974) Eur. J. Biochem. 49, 275-285.

8. CLAISSE, M.L., SPYRIDAKIS, A. and SLONIMSKI, P.P. (1977), this volume.

9. COLSON, A.M. and SLONIMSKI, P.P. (1977), this volume.

10. MASON, T., BOERNER, P. and BIRON, C. (1976), in "Genetics and Biogenesis of Chloroplasts and Mitochondria" Eds. Th. Bucher et al. Elsevier/North Holland Biochemical Press, Amsterdam, pp. 239-246.

MAPPING OF DRUG-RESISTANT LOCI IN THE COENZYME QH_2-CYTOCHROME C REDUCTASE REGION OF THE MITOCHONDRIAL DNA MAP IN SACCHAROMYCES CEREVISIAE

A.M. COLSON AND P.P. SLONIMSKI
Laboratoire d'Enzymologie, Université de Louvain, 1348 Louvain-la-Neuve Belgique and Centre de Génétique Moléculaire du C.N.R.S. Gif sur Yvette 91190-France.

Diuron (1), mucidin (2), antimycin A (3) and funiculosin (4) are specific respiratory inhibitors of the coenzyme QH_2-cytochrome c reductase ($coQH_2$-cyt c red) since they block the electron flow between the cytochromes b and c_1. Two diuron-resistant loci, DIU, (5), two mucidin-resistant loci, MUC, (6,7), two antimycin-resistant loci, ANA, and one funiculosin-resistant locus, FUN, (8,9,10) were identified on the mitochondrial DNA of *Saccharomyces cerevisiae*. The resistance of the *in vitro* respiration of submitochondrial particles to diuron (5), of isolated mitochondria to mucidin (7), and insensitivity of cytochrome absorption spectrum of mitochondrial particles to funiculosin (9) indicates that, in the mutants, the sensitivity of the mitochondrial inner membrane to the inhibition of the electron flow by the drugs is diminished. Furthermore, the inhibitory effect of antimycin A on NADH oxidase is reversed by the addition of bovine serum albumin in the antimycin-resistant mutants but not in the antimycin-sensitive strains (10, 11). These observations indicate the existence of modifications in the $coQH_2$-cyt c red protein of the mutants. Two COB genetic loci (12) and six BOX genetic loci (13,14) are localized between OLI1 and OLI2 in the $coQH_2$-cyt c red region of the mt DNA. Mutants of COB and BOX phenotypes are respiratory-deficient devoid of cytochrome b spectrum and $coQH_2$-cyt c red activity. The BOX mutants display, in addition, an increased sensitivity to catabolite repression of cytochrome oxidase formation (15, 16). Diuron-resistance has been, tentatively, localized in the vicinity of COB loci (5), an antimycin-resistant mutant has been localized in the vicinity of OLI1 locus (10) and a mutant isolated as resistant to mikamycin but presently considered as resistant to antimycin A (17) has also been allocated to this region (18). In the present communication, diuron-resistant loci, mucidin-resistant loci, antimycin-resistant loci and a funiculosin-resistant locus were localized relative to BOX loci. Using deletion and recombination mapping, a map representing allelism relationships between drug-resistant loci and BOX loci in the $coQH_2$-cyt c red region of the mt DNA is proposed.

NOMENCLATURE AND ORIGIN OF STRAINS

Drug-resistant mutants

D^R_{721} and D^R_{764} are derived from IL125-2B and were described previously as *diu^r-1* for D^R_{721} and *diu^r-64* for D^R_{764} (5).

D^R_{724} is derived from D273-10B/A21.

D^R_{725}, D^R_{726}, D^R_{727} and D^R_{728} are derived from KL14-4A.

D^R_{731} and D^R_{732} are derived from D243-4A and were kindly sent to us by Dr. G. Michaelis as *Diu1* for D^R_{731} and *Diu2* for D^R_{732}.

M^R_{771} and M^R_{772} are derived from D225-5A and were kindly sent to us by Dr. J. Subik as D225-5A/3D for M^R_{771} and 101-4B for M^R_{772} (7).

Mit$^-$ mutants

The BOX mutants are derived from 777-3A and are fully described in the companion papers in this volume (14,15).

Rho$^-$ mutants

The rho$^-$ mutants are derived from KL 14-4A and are fully described in the companion paper in this volume (14).

Table 1. Identification of three groups among six diuron-resistant and two mucidin-resistant mutants in relation to BOX1, BOX2, BOX3 and BOX4 genetic loci of the $coQH_2$-cyt c red region

	BOX1		BOX2		BOX3		BOX4		
	α	a	α	a	α	a	α	a	
α M^R_{772}	X	13 (9-17)	X	16 (12-20)	X	35 (30-40)	X	15 (12-18)	Group 1
a D^R_{721}	0.3 (0-0.9)	X	14 (10-18)	X	13 (10-16)	X	16 (13-19)	X	Group 2
α D^R_{724}	X	0.4 (0.2-0.6)	X	15 (12-18)	X		X		Group 2
a D^R_{727}	0 (<0.1)	X	17 (14-20)	X	5 (4-6)	X	11 (8-14)	X	Group 2
a D^R_{725}	10 (8-12)	X	15 (12-18)		4 (3-5)	X	0.7 (0.1-1.3)	X	Group 3
a D^R_{726}	4 (3-5)	X		X	3 (2-4)	X	0.6 (0-1.2)	X	Group 3
a D^R_{728}	6 (4-8)	X	12 (10-14)	X	7 (5-9)	X	0.3 (0-0.8)	X	Group 3
α M^R_{771}	X	18 (14-22)	X	49 (43-55)	X	32 (24-40)	X	0 (<0.2)	Group 3

Crosses were performed between respiratory-competent, drug-resistant mutants ($mit^+D^RM^S$ or $mit^+D^SM^R$) and respiratory-deficient drug-sensitive mutants ($mit^-D^SM^S$). The numbers underlined represent the percent of wild type (i.e. $mit^+D^SM^S$) recombinants. The numbers given in parentheses are the 95% confidence limits which were calculated from the standard error of each percentage. Crossed out cases correspond to combinations which could not be performed because of mating type incompatibility.

Table 2. Allelism tests between various diuron-resistant and mucidin-resistant mutants

	$\alpha\ D^R_{721}$	$\alpha\ D^R_{724}$	$a\ D^R_{727}$	α or $a\ D^R_{764}$	$a\ D^R_{725}$	$a\ D^R_{726}$	$a\ D^R_{728}$	$\alpha\ M^R_{771}$	
$\alpha\ M^R_{772}$	X	X	13 (11-15)	X	13 (11-15)	6 (5-7)	7 (7-8)	17[1]	Group 1
α or $a\ D^R_{721}$		0 (<0.1)	0 (<0.4)	5 (4-6) 14 (12-16)	6 (3-9)			X	Group 2
$\alpha\ D^R_{724}$			0.1 (0-0.2)	0 (<0.2)	3.1[2]	2.2[2]	2.1[2]	X	Group 2
$a\ D^R_{727}$				9 (6-12)	X	X	X	10 (8-12)	Group 2
$\alpha\ D^R_{764}$					0 (<0.7)			X	Group 3
$a\ D^R_{725}$						X	X	0.2 (0-0.4)	Group 3
$a\ D^R_{726}$							X	0 (<0.1)	Group 3
$a\ D^R_{728}$								0.2 (0-0.4)	Group 3
								$\alpha\ M^R_{771}$	

The numbers underlined represent the percent wild type (i.e. drug-sensitive) recombinants. The numbers given in parentheses are the 95% confidence limits which were calculated from the standard error of each percentage.

[1] This result is taken from Subik et al., 1977. The percent value given is the mean value of 10 independent crosses between various strains carrying M^R_{771} and M^R_{772}.

[2] The percent value given is the mean value of 2 or 3 crosses.

Crossed out cases correspond to combinations which could not be performed because of mating type incompatibility.

RESULTS

The diploid progeny of crosses between respiratory-competent drug-resistant mutants and respiratory-deficient BOX mutants, isolated from a drug-sensitive strain was analyzed for the frequency of BOX^+ $DRUG^S$ recombinants (see table 1). Allelism or close linkage is characterized by a very low recombinant frequency while non-allelism is indicated by high recombinant frequency. Three groups of drug-resistant mutations are distinguished. In group 1, the mucidin-resistant mutation, M^R_{772}, is non-allelic to the BOX1, BOX2, BOX3 and BOX4 loci tested. Diuron-resistant mutations, D^R_{721}, D^R_{724} and D^R_{727} belong to group 2 and are allelic or closely linked to BOX1 locus and non-allelic to BOX2, BOX3, and BOX4. In group 3, the diuron-resistant mutations D^R_{725}, D^R_{726}, D^R_{728} and mucidin-resistant mutation M^R_{771} are allelic or closely linked to BOX4 locus and are non allelic to BOX1, BOX2 and BOX3. Allelism tests between durg-resistant mutants are presented in table 2. The diploid progeny of crosses between drug-resistant mutations was analyzed for the frequency of $DRUG^S$ recombinants. The results of allelism tests are in agreement with those of table 1. Mutation M^R_{772} (allocated to group 1) is non-allelic to mutations from group 2 and group 3. The mutations D^R_{721}, D^R_{724} and D^R_{727} are allelic to each other while they are non-allelic to mutations belonging to group 3. The mutations D^R_{725}, D^R_{726}, D^R_{728} and M^R_{771} (group 3) are allelic or closely linked while they are non-allelic to mutations belonging to group 1 and group 2. Mutation D^R_{764} is non-allelic to mutations D^R_{721} and D^R_{727} while it is allelic to D^R_{725}. The localization of M^R_{772} remains to be established. In table 3, deletion mapping experiments using discriminating rho^- mutants, are presented in order to allocate mutation M^R_{772} to a segment of mt DNA. The absence of diploid rho^+ cells carrying the sensitive allele of the M^R_{772} mutation in crosses with RE1, RA1 and PG1 indicates that the rho^- parental mutants utilized in those crosses dit not retain the mucidin-sensitive allele. Among the progeny of crosses with PI1, SC632/A1 and 746/A1 recombinant rho^+ diploid cells which carry the sensitive allele of M^R_{772} were obtained. It appears that M^R_{772} does not belong to the same segment of DNA as OLI1.

Table 3. Allocation of the M^R_{772} mutation in the BOX6 locus of the mitochondrial DNA by deletion mapping with discriminating rho$^-$ mutants

	M^R_{772}	BOX4	BOX5	BOX3	BOX1	BOX2	BOX6	OLI1
RE1	0 (<0.2)	+	+	+	+	-	-	-
RA1	0 (<0.2)	+	+	+	+	+	-	-
PI1	89 (86-92)	-	-	-	-	+	+	-
PG1	0 (<0.2)	+	+	+	-	-	-	-
SC632A1	48 (47-49)	-	-	-	-	-	+	+
SC773A2B31	27 (22-32)	-	-	-	-	-	+	+
746A1	7 (4-10)	-	-	-	-	-	+	-

The symbol (+) means that the mitochondrial DNA segment present in the rho$^-$ testers includes the wild type allele of the mit$^-$ mutants. The numbers underlined represent the percent of recombinant diploid rho$^+$ cells that carry the wild type (i.e. sensitive) allele of the drug-resistant mutants. The numbers given in parentheses are the percent limits which were calculated from the standard error of each percentage. The full genotype of the rho$^-$ testers used here is given in the table 3 of (14). The restauration spectrum of the rho$^-$ clone PG1 is identical to that of RC3 given in this publication.

Table 4. Allelism tests between diuron-resistant mutants from different origins

	aD^R_{731}	aD^R_{732}
D^R_{721}	<u>0</u> (0.1)	<u>7.5</u> (4.7-9.3)
D^R_{764}	<u>5.8</u> (4.3-7.3)	<u>0</u> (0.1)

The numbers underlined represent the percent wild type (i.e. drug-sensitive) recombinants. The numbers given in parentheses are the 95 % confidence limits which were calculated from the standard error of each percentage.

The presence of rho^+ M^S diploid recombinants are observed in crosses where the discriminating rho^- parent still possesses the alleles belonging to the BOX6 locus. Thus, M^R_{772} belongs to the same segment of DNA as the locus BOX6. In table 4, D^R_{731} and D^R_{732}, two diuron-resistant mutations are crossed with D^R_{721} from group 2 and D^R_{764} from group 3. The results indicate that D^R_{731} is allelic to D^R_{721} and non-allelic to D^R_{764} while D^R_{732} is allelic to D^R_{764} and non-allelic to D^R_{721}. Results of crosses obtained by Dr. G. Michaelis (personal communication) indicate that D^R_{731} is allelic or closely linked to ANA2 and that D^R_{732} belongs to ANA1. ANA1 and ANA2 are the two antimycin-resistant loci identified by Pratje and Michaelis (8,9).

DISCUSSION

As shown in fig.1, three genetic loci possessing distinct mutated allelic forms are observed. They cover a large spectrum of respiratory-deficient BOX or respiratory-competent drug-resistant phenotypes. At the first locus, diuron- or antimycin- or funiculosin-resistant or respiratory-deficient alleles are found ; at the second locus, diuron- or antimycin- or mucidin-resistant or respiratory-deficient alleles are found ; finally, at the third locus, only mucidin-resistant or respiratory-deficient alleles are observed. In conclusion, DIU1 is allelic or closely linked to ANA2, FUN1 and BOX1 ; DIU2 is allelic or closely linked to ANA1, MUC1 and BOX4 ; MUC2 is allelic or closely linked to BOX6. The high recombinant frequency observed between the three loci (13 % on the average for 33 various combinations analyzed) suggests the existence of three genes coding for three distinct gene-products. The resistance of the respiratory chain observed *in vitro* in diuron-resistant (5), in mucidin-resistant (7), in antimycin-resistant (10) and in funiculosin-resistant mutants (9) and the allelism relationships between respiratory-competent drug-resistant loci and $coQH_2$- cyt c red deficient (BOX) loci strongly suggest that each of the three drug-resistant loci codes for a structural gene-product which is essential for the normal $coQH_2$-cyt c red activity of the respiratory chain. It is important to stress that resistance to diuron, mucidin and antimycin is for each drug coded by a pair of genetic loci.

Figure 1. The map of drug-resistant and cytochrome b deficient mutants in the $coQH_2$-cyt c red region of the mitochondrial DNA

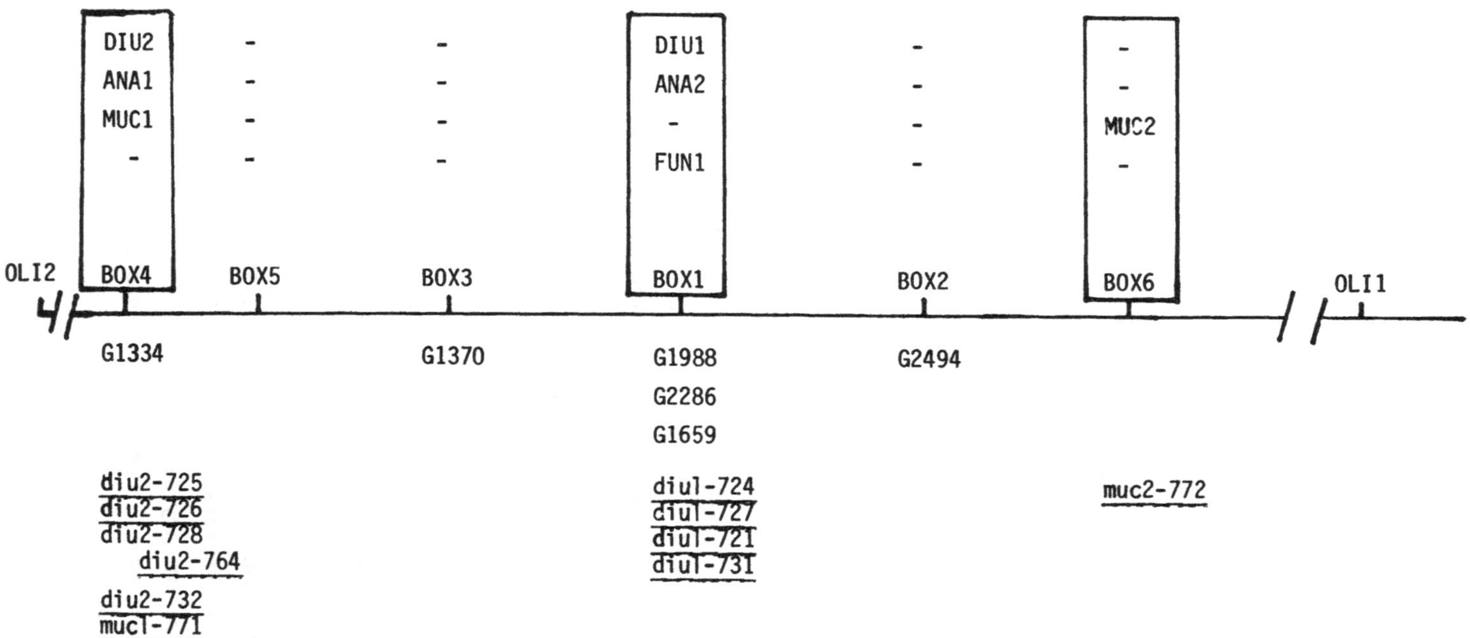

1. The names above the line correspond to the names of drug-resistant genetic loci given by Subik (7) to the mucidin-resistant, by Pratje and Michaelis (9) to antimycinA-resistant and funiculosin-resistant mutants, and in the present communication to diuron-resistant mutants, as well as to the $coQH_2$- cyt c red deficient mutants by Kotylak and Slonimski (14).
2. The names below the line correspond to the drug-resistant mutations used in the present work as well as to the respiratory-deficient BOX mutants used to establish the localization of various genetic loci. The overall order of the genetic map and details of its construction are given in (14) and biochemical characterization of BOX mutants in (16).

A mutation occuring at any one of the two genetic loci is sufficient for the expression of a resistant phenotype. In other words, any member of a gene pair overshadows, when mutated to resistance, the other non-mutated gene. Such a situation could possibly be due to gene-duplication of the drug-resistant loci. This interpetation, however, is unlikely because recessive $coQH_2$-cyt c red deficient mutations can be obtained at each drug-resistant locus. In addition, since the drug-resistant spectrum is different for each locus and partially overlapping, it seems likely that each pair of drug-resistant loci is constituted of two non-identical genetic loci. This latter explanation implies that each drug-resistant genetic locus codes for a different gene-product necessary to the $coQH_2$-cyt c red activity. Although the mechanism of resistance is still unknown it is plausible that the binding of the drugs to their specific inhibition-sensitive sites is lacking or less effective. From the genetic data, it can be postulated that two diuron-, two antimycin- and two mucidin-binding sites are carried by the $coQH_2$-cyt c red protein (5). If only one mutated genetic locus is enough for the expression of the resistant phenotype, it is likely that maximum inhibition of $coQH_2$-cyt c red is obtained only when the two binding sites are sensitive to the drug. This implies that drug-resistant mutants with one mutated locus should still bind the drug while doubly mutated mutants at the two genetic loci should no longer bind the drug. The existence of two distinct binding sites for inhibition of $coQH_2$-cyt c red has recently been predicted on the basis of kinetic data (19, 23).

The two mucidin binding sites are different from the two diuron binding sites because we have found that diuron-resistant mutants remain sensitive to mucidin and that mucidin-resistant mutants remain sensitive to diuron. This is true for mutations occuring at DIU1, DIU2, MUC1 and MUC2 genetic loci. Since DIU1 and MUC2 belong to distinct genetic loci, it is logical that their binding sites are different ; although DIU2 and MUC1 are allelic or closely linked to BOX4, one expects, however, that further refined genetic analysis of the BOX4 locus should reveal two distinct mutational sites for mucidin- and diuron-resistant alleles. We have examined whether mutants isolated as diuron-resistant present cross-resistance to antimycin A. Preliminary data show that diuron-resistant alleles from DIU2 exhibit at least partial cross-resistance to antimycin A while diuron-resistant alleles from DIU1 locus remain sensitive to antimycin A. These results suggest that at least at one locus diuron- and antimycin-binding sites overlap partially on the coQH_2-cyt c red protein (5). Funiculosin-resistance is localized at only one genetic locus (9). This specificity suggests that the phenotypic expression of this locus possesses a property which is lacking at the two other genetic loci. The differences between the genetic loci for their drug-resistant spectrum are in agreement with differences of the same drugs in their inhibitory effects on the coQH_2cyt c red activity (20). In conclusion, a complete picture of cross-resistance in a large number of mutants and refined genetic analysis within each locus should reveal possible interactions between binding sites or should establish their mutual autonomy. Cytochrome b is an essential component of coQH_2-cyt c red (see for review 21) and is obviously a good candidate for a gene product of the drug-resistant loci mapped in this communication. We show, however, that three non-allelic genetic loci confer the resistance. Thus, three structural genes would exist for cytochrome b and consequently there should exist, for the reasons discussed above, three different cytochrome b or three different polypeptide chains of cytochrome b. Although such a hypothesis is contrary to the generally accepted view (21) it is rewarding to find that a convergent body of facts points toward it. In addition to our observation on drug-resistant loci, Pajot et al. (16) conclude from comparisons between cytochrome spectra and coQH_2-cyt c red activity that BOX4 and BOX6 genetic loci would correspond to structural genes and Claisse et al (22) find modifications of polypeptide length at the BOX4, BOX1 and BOX6 genetic loci. Now, precisely these three genetic loci are allelic to the loci which modify the drug-resistance of the coenzyme QH_2-cytochrome c reductase.

ACKNOWLEDGMENTS

A.M. Colson gratefully acknowledges the receipt of a short term fellowship from the European Communities.
The authors take pleasure in thanking Professor André Goffeau for his constant interest during the progress of this research and Dr Bernard Dujon for stimulating discussions.

Dr. J. Subik and Dr. G. Michaelis are gratefully acknowledged for sending us their strains.
The work has been supported by D.G.R.S.T. grant n° 75-70748 (France) and by F.R.F.C. grant n° 2.4557.75 (Belgique).

REFERENCES.

1. Inoue,Y., Ishizuka,K. and Mitsui,S. (1967) Agr. Biol. Chem. 31, 422-427.
2. Subik,J., Behun,M. and Musilek,V. (1974) Biochem. Biophys. Res. Comm. 57, 17-22.
3. Keilin,D. and Hartree, E.J. (1955) Nature (Lond.) 176, 200-206.
4. Moser, U.K. and Walter, P. (1975) FEBS Letters 50, 279-282.
5. Colson, A.M., The Van, Luu, Convent,B., Briquet,M. and Goffeau, A. (1977) Eur.J. Biochem. 74, 521-526.
6. Subik, J. (1975) FEBS Letters 59, 273-276.
7. Subik, J., Kovacova, V. and Takacsova, G. (1977) Eur.J. Biochem. 73, 275-286.
8. Pratje, E. and Michaelis, G. (1976) in Genetics and Biogenesis of Chloroplasts and Mitochondria (Bücher, Th., Neupert, W., Sebald, W. and Werner, S., eds.) pp 467-471, Elsevier/North-Holland Biomedical Press, Amsterdam.
9. Pratje, E. and Michaelis, G. (1977) Molec.gen.Genet. 152, 167-174.
10. Lang, B., Burger, G., Bandlow, W., Kaudewitz., F. and Schweyen, R.J. (1976) in Genetics and Biogenesis of Chloroplasts and Mitochondria (Bücher, Th., Neupert, W., Sebald, W. and Werner, S., eds.) pp 461-465 Elsevier/North-Holland Biomedical Press, Amsterdam.
11. Burger, G., Lang, B., Bandlow, W., Schweyen , R.J., Backhaus, B. and Kaudewitz, F (1976) Biochem. Biophys.Res.Comm. 72, 1201-1208.
12. Tzagoloff, A., Foury F., and Akai, A. (1976) Molec.gen.Gent. 149,33-42
13. Kotylak, Z. and Slonimski, P.P. (1976) in The Genetic Function of Mitochondrial DNA (Saccone, S. and Kroon, A.M.eds) pp 143-154, Elsevier/ North-Holland Biomedical Press, Amsterdam.
14. Kotylak, Z. and Slonimski, P.P. (1977) This volume.
15. Pajot, P., Wambier-Kluppel, M.L., Kotylak, Z and Slonimski, P.P. (1976) in Genetics and Biogenesis of Chloroplasts and Mitochondrial (Bücher, Th., Neupert, W. Sebald, W. and Werner, S., eds.) pp 443-451.
16. Pajot, P., Wambier-Kluppel, M.L. and Slonimski, P.P. (1977) This volume.
17. Groot Obbink, D.J., Spithill, T.W. Maxwell, R.J. and Linnane, A.W. (1977) Molec.gen.genet. 151, 127-136.

18. Cobon, G.S., Groot Obbink, D.J., Hall, R.M., Maxwell, R., Murphy, M., Rytka, J. and Linnane, A.W. (1976) in Genetics and Biogenesis of Chloroplasts and Mitochondria (Bücher, Th., Neupert, W., Sebald, W. and Werner, S., eds.) pp 453-460.
19. Convent, B. (1977) Ph D Thesis.
20. Convent, B. and Briquet, M. in preparation
21. Weiss, H. (1976) Biochim. Biophys. Act. 456, 291-313.
22. Claisse, M., Spyridakis, A. and Slonimski, P.P. (1977) This volume.
23. Burger, G., Lang, B., Bandlow, W. and Kaudewitz, F. (1975) Biochim. Biophys. Acta 396, 187-201

Resistance to High Concentrations of Antimycin A in Saccharomyces cerevisiae : Synergistic Interaction of Nuclear and Mitochondrial Mutations

Elke Pratje and Georg Michaelis
Universität Bielefeld, Fakultät für Biologie, Postfach 8640, D-4800 Bielefeld 1, Federal Republic of Germany

Recent studies focus on the mitochondrial structural and regulatory genes taking part in the biosynthesis of the cytochrome bc_1 complex of the respiratory chain. Antimycin A, a well known inhibitor of the electron flow between cytochrome b and cytochrome c, was used to isolate resistant mutants. Such mutants were obtained with low concentrations of the antibiotic (50-100 ng/ml) in the selective medium. Two general classes of antimycin A resistant colonies have been found : mitochondrial mutants (1, 2, 3, 4) and nuclear mutants. Both types are completely sensitive to high concentrations of the drug (1 µg/ml).

1. Mitochondrial mutations :

Genetic analysis of mitochondrial mutants resistant to low concentrations of antimycin A has revealed the presence of two distinct loci, designated A_I and A_{II} (5, 6). Mutations of the A_{II} locus can be specifically selected on the basis of mitochondrial funiculosin resistance (6). The two new mitochondrial A^R loci are not closely linked, as shown by a recombination frequency varying from 8 to 21% in different crosses. Both loci are involved in the determination of component(s) of the cytochrome bc_1 complex (6). The A_I^R and A_{II}^R mutations were localized in the O_I-O_{II} region of the mitochondrial genome by deletion mapping with petite clones. In addition, the mapping relationship of mitochondrial antimycin A resistance to various cytochrome b deficient

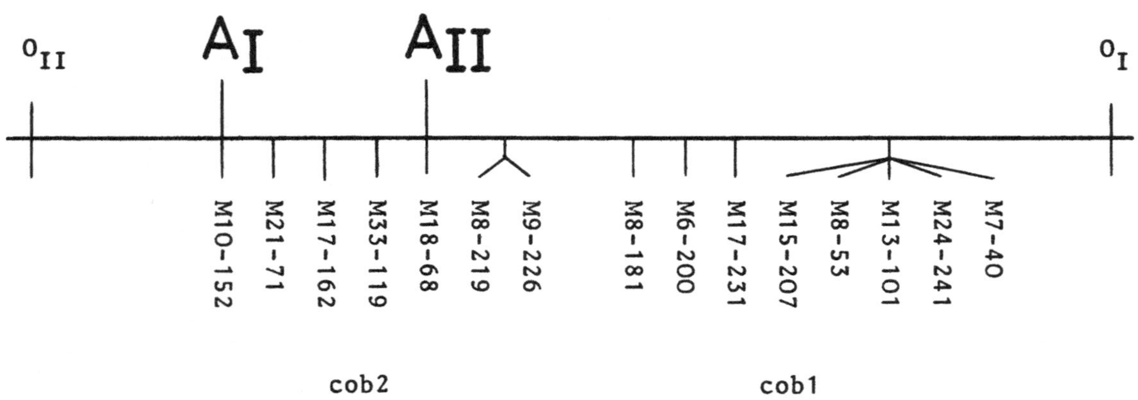

Fig. 1. Fine structure map. Arrangement of the two antimycin A resistance loci A_I and A_{II} on the mitochondrial genome of Saccharomyces cerevisiae. The order of the cob markers follows Tzagoloff et al. (7).

cob^- mutants of Tzagoloff et al. (7) has been investigated. The results are summarized in Figure 1. The genetic loci are mapped in the following order :
O_{II} - A_I - A_{II}-cob1 - O_I .

2. Nuclear mutations :

In an attempt to isolate yeast cells resistant to high concentrations of antimycin A, the mitochondrial antimycin A resistant mutants were mutagenized with $MnCl_2$ which increases the frequency of mitochondrial mutations in Saccharomyces cerevisiae (8). Colonies resistant to 5 µg antimycin A per ml occurred with a frequency of about 10^{-7} irrespective of the presence or absence of the mutagen. The high resistant colonies were replica-crossed to a petite tester which lacked all mitochondrial DNA, and the diploids were analyzed for high antimycin A resistance. According to this cross the second new resistance mutation could be characterized as a) recessive or b) either dominant or mitochondrial. Only the double mutants belonging to the latter category were analyzed further. In order to distinguish between nuclear dominant and mitochondrial mutations the strains were treated with ethidium bromide under conditions which would completely eliminate mitochondrial DNA. The resulting ρ^0 petites were characterized by two test crosses :

1. ρ^0 x A^S $\longrightarrow$ diploids, which are all sensitive. After meiosis a 2:2 segregation of low antimycin A resistance versus sensitivity was observed.
2. ρ^0 x A^R $\longrightarrow$ diploids, which are all high resistant.

These crosses demonstrate the existence of an interesting nuclear mutation conferring resistance to low concentrations of antimycin A. When such a low resistant nuclear mutant is crossed to a low resistant mitochondrial mutant, high resistant as well as sensitive diploids are produced. These results strongly suggest that the nuclear mutation is recessive in

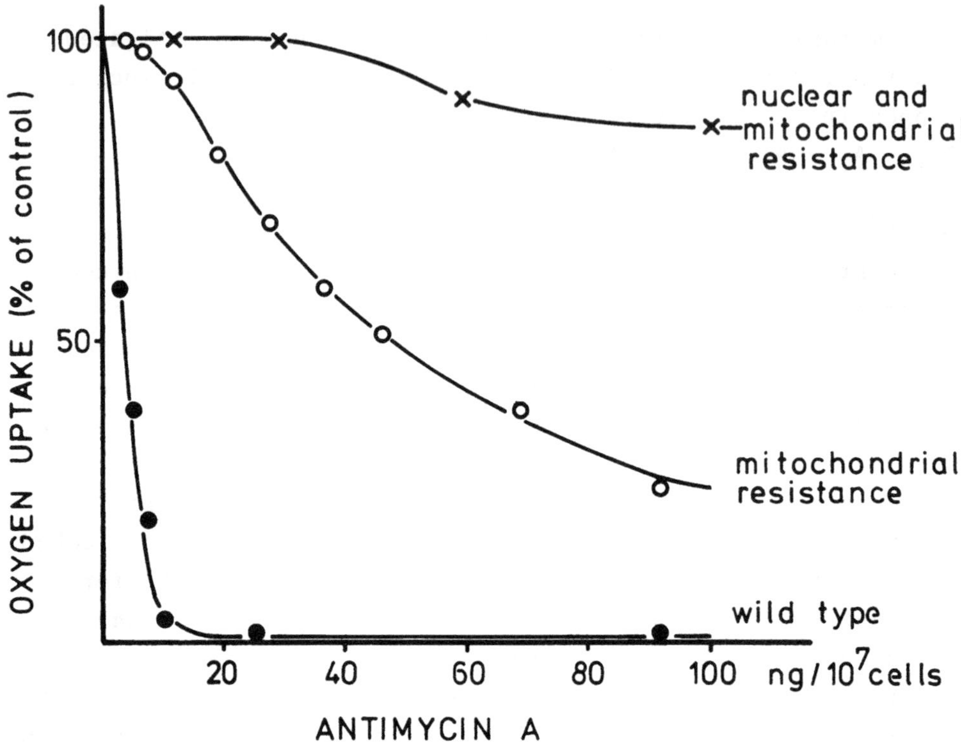

Fig. 2. Effect of antimycin A on respiration of whole cells. Oxygen uptake was measured polarographically at 28°C. The reaction vessel contained in 3 ml : 0.6 M mannitol, 40 mM potassium phosphate, 10 mM Tris-maleate (pH 6.5), 0.8% ethanol, antimycin A as indicated and yeast cells (3.6×10^7 cells/ml) grown in glucose medium to the stationary phase and washed in 0.4% sodium chloride solution. The cells were pre-incubated with antimycin A for 1 h before the activity was measured.

x - x - x double mutant D243-4A/A32B resistant to high concentrations of antimycin A
o - o - o mitochondrial mutant D8-22A/G1a (locus A_I)
● - ● - ● wild type strain D243-4A

crosses to sensitive wild type strains. However, the same nuclear mutation displays dominance of the high resistance phenotype if the mitochondrial resistance mutation is present in the cell. The high antimycin A resistance phenotype arises from a synergistic interaction of a nuclear and a mitochondrial low resistance gene.

Respiration of whole cells as a function of increasing antimycin A concentrations is illustrated in Figure 2. A sensitive wild type strain, a mitochondrial A^R mutant and a high resistant double mutant are compared.

Conclusions :

Evidence is presented that the cob2 region specifies two sites which are involved in the binding of antimycin A and funiculosin. The two mitochondrial mutations confer resistance only to low concentrations of antimycin A. High antimycin A resistance arises from a synergistic interaction of mitochondrial and nuclear resistance mutations in the cell.

Acknowledgements : We would like to thank Dr. A. Tzagoloff for providing his cob^- and petite mutants. This work was supported by the Deutsche Forschungsgemeinschaft.

References :

1) Michaelis, G.: Cytoplasmic inheritance of antimycin A resistance in Saccharomyces cerevisiae. Molec. gen. Genet. 146, 133-137 (1976)

2) Groot Obbink, D.J., Spithill, T.W., Maxwell, R.J., Linnane, A.W.: Biogenesis of mitochondria 48. Mikamycin resistance in Saccharomyces cerevisiae - A mitochondrial mutation conferring resistance to an antimycin A-like contaminant in mikamycin. Molec. gen. Genet. 151, 127-136 (1977)

3) Burger,G., Lang, B., Bandlow, W., Schweyen, R.J., Backhaus, B., Kaudewitz, F.: Antimycin resistance in Saccharomyces cerevisiae : a new mutation on the mitDNA conferring antimycin resistance on the mitochondrial respiratory chain. Biochem. biophys. Res. Commun. 72, 1201-1208 (1976)

4) Wolf, K., Burger, G., Lang, B., Kaudewitz, F.: Extrachromosomal inheritance in Schizosaccharomyces pombe. I.

Evidence for an extrakaryotically inherited mutation conferring resistance to antimycin. Molec. gen. Genet. 144, 67-73 (1976)

5) Pratje, E., Michaelis, G.: Two mitochondrial antimycin A resistance loci in Saccharomyces cerevisiae. In : (Th. Bücher, W. Neupert, W. Sebald, S. Werner, eds.), Genetics and Biogenesis of Chloroplasts and Mitochondria, pp. 467-471. Amsterdam : Elsevier/North-Holland Biomedical Press 1976

6) Pratje, E., Michaelis, G.: Allelism studies of mitochondrial mutants resistant to antimycin A or funiculosin in Saccharomyces cerevisiae. Molec. gen. Genet. 152, 167-174 (1977)

7) Tzagoloff, A., Foury, F., Akai, A.: Assembly of the mitochondrial membrane system XVIII. Genetic loci on mitochondrial DNA involved in cytochrome b biosynthesis. Molec. gen. Genet. 149, 33-42 (1976)

8) Putrament, A., Baranowska, H., Prazmo, W.: Induction by manganese of mitochondrial antibiotic resistance mutations in yeast. Molec. gen. Genet. 126, 357-366 (1973)

Note added : The classification of cob^- mutations into cob1 and cob2 follows Tzagoloff et al. (7). Slonimski and coworkers have presented evidence at this meeting that the cob^- mutations corresponding to the A_{II} locus should rather be regrouped at the terminal part of the cob1 region.

MUTATIONS TO DRUG-RESISTANCE IN THE COB REGION OF THE MITOCHONDRIAL GENOME

G. Burger, B. Lang, B. Backhaus, K. Wolf, W. Bandlow and F. Kaudewitz
Genetisches Institut der Universität München, Maria-Ward-Str.1a, D-8000 München 19, Germany

INTRODUCTION

Until now, no valid information is available concerning the nature and number of mitochondrially coded components of the bc_1-segment of the respiratory chain in yeast. To approach this problem, mitochondrially inherited mutants of *Saccharomyces cerevisiae* and *Schizosaccharomyces pombe* were studied, which are resistant to drugs blocking the electron transfer between cytochrome *b* and c_1. Mapping of these mutations and enzymatic analysis of the respiratory chain of the mutants were carried out in order to relate functional alterations of the bc_1-segment to structural genes on the mitochondrial genome.

RESULTS

Mitochondrially inherited mutants of *Saccharomyces cerevisiae* and *Schizosaccharomyces pombe* were isolated and characterized, being resistant to the inhibitors of the bc_1-segment of the respiratory chain: antimycin, funiculosin and diuron.

1. Biochemical analysis

In order to test whether the resistance is due to transport or metabolic phenomena or due to functional alterations in the respiratory chain caused by a mutation in a structural gene, the efficiency of the inhibitors on the NADH oxidase activity of mitochondrial particles was determined (e.g. Fig.1).

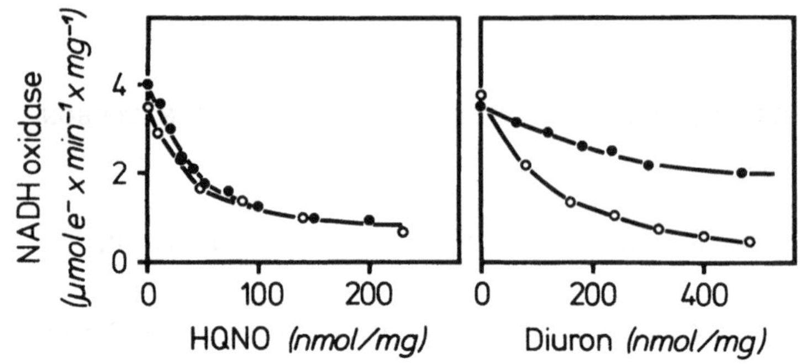

Fig. 1 Inhibition of the NADH oxidase of the mutant diu^{r}-301 (●-●) and the wild type (o-o) of Sch.pombe by HQNO and diuron. The mitochondrial particles were prepared by a mechanical procedure (5). The respiration of mitochondrial particles was measured with a clark type electrode as described earlier (6).

The data concerning growth and respiration in the presence of the inhibitors are summarized in Table I.

Three classes of mutants were obtained with respect to the biochemical analysis:

a) Mutants, in vitro resistant only to that inhibitor applied in the selection procedure
   - fun^{r}-209 (S.c.)
   - ana^{r}-104 (S.c.)
   - diu^{r}-301 (Sch.p.)

b) in vitro resistant mutants being also resistant to inhibitors in the bc_1-segment, not applied in the selection procedure
   - ana^{r}-101 (S.c.) is in vitro resistant to HQNO and diuron. The resistance to antimycin could be demonstrated in the presence of either ubiquinone-3 or bovine serum albumin, both reducing the sensitivity to the inhibitor (7).
   - ana^{r}-8 and ana^{r}-14 (both Sch.p.) are in vitro resistant to antimycin and funiculosin.

TABLE I Resistance of mutants to inhibitors in the bc$_1$-segment

		in vivo resistance			*in vitro* resistance			
		anti-mycin	funicu-losin	diuron	anti-mycin	funicu-losin	diuron	HQNO
Saccharomyces cerevisiae	ana^r-101	+++	-	+++	+++	-	+++	+++
	ana^r-104	++	-	-	+	-	-	-
	fun^r-209	-	++	-	-	++	-	-
	fun^r-214	-	++	-	-	-	-	-
Schizosaccharo-myces pombe	ana^r-8	++++	++	-	++++	++	-	-
	ana^r-14	++++	+++	--	++++	+++	--	-
	diu^r-301	-	-	++	-	-	++	-

(+) ....(++++)/(-), (--): degree of resistance/sensitivity compared with the wild type.

Test of *in vivo* resistance: growth of cells on glycerol.

Test of *in vitro* resistence: NADH oxidase activity of mitochondrial particles.

c) - fun^r-214 (S.c.) is *in vivo* resistant, but *in vitro* sensitive to funiculosin. Because no alteration of the bc_1-segment was found with the applied biochemical methods it is conceivable that the efficiency of the inhibitor in the mutant is lowered by chemical modification or a restricted permeability of the membrane for the drug.

These results allow to conclude that the resistance of type a) and b) is obviously based on an alteration of the function of the respiratory chain of mitochondrial particles. The content of cytochrome *b* and the respiratory activities of all mutants analyzed are the same in wild type and mutant cells. The loss of a component or an altered regulation of the cob gene products thus cannot account for the resistance.

The occurrence of type b) mutants allows to localize resistance markers - although not being analyzed genetically - in a known mutational cluster. E.g. the resistance of ana^r-101 (S.c.) to antimycin as well as to diuron localizes a mutational site for diuron resistance in this gene cluster for antimycin resistance. Very likely it is identical with one of the gene clusters for diuron resistance found by others (8).

2. Genetical analysis

In *Saccharomyces cerevisiae*, ana^r and fun^r mutational sites were localized by deletion mapping. Four mutants which were derived from $ana^s fun^s$ strains and carry deletions in the *cob* region (ρ^- H211, ρ^- G311, ρ^- J31, $\rho^+ mit^-$ cob-162; cf. Fig.2) were crossed with strains carrying one or the other of the resistance mutations. Occurrence or absence of $drug^s$ recombinants in the progeny of these crosses indicate that the analyzed $drug^r$ sites map in a region retained or deleted in the tester strains, respectively.

Results given in Table 2 reveal that all but one drug^r mutational site (fun^r-214) are located within a mitDNA sequence, common to ρ^- clones G311 and J31. According to the fine structure map of Schweyen et al. (this volume) this is that part of the cob1 cluster which is more distant from oli1. fun^r-214, however, is located in the cob2cluster, proximal to oli2, as is revealed from the fact that it is retained in two ρ^- clones (H211, J31), covering the cob2 cluster, and also in the mit$^-$ mutant cob-162 which is deleted in the oli2 distant part of the cob2 cluster.

A more precise mapping of ana^r-101 is based on the analysis of a population of ρ^- clones: out of 163 ρ^- clones none was found to separate ana^r-101 from mit$^-$ 3881 (Schweyen et al., this volume). Therefore, ana^r-101 is localized at the same site as cob-3881 at the oli-distant end of cob1.

Localization of ana^r-101 and ana^r-104 by recombination analysis with mit$^-$ mutants of the cob-region (e.g. 3881, 2281, 1431) failed since recombinant frequencies were found to be similar to those between ana and oli1 or cap markers: ana^r mutational sites and mit$^-$ sites in the cob region appear to be genetically unlinked.

TABLE II: Recombination of ana^r and fun^r mutants with different tester strains

drug^r mutant	deletion mutant ρ^- H211	ρ^- G311	ρ^- J311	cob-162
ana^r-101	-	+	+	n.d.
ana^r-104	-	+	+	+
fun^r-209	-	+	+	+
fun^r-210	-	+	+	+
fun^r-211	-	+	+	+
fun^r-213	-	+	+	+
fun^r-214	+	-	+	+

Drug-resistance mutants, originated from wild type KL 14-4A (a his, trp; ω^+ cap^r-321, oli^r-1, par^r-454) were crossed with either three individual tsp-induced ρ^- clones (Schweyen et al., this volume) or with mutant mit$^-$ cob 162 showing a small deletion in the cob-region (a mutant from A. Tzagoloff, originally designated M17-162). Among the progeny of these crosses the occurrence of drug sensitive cells was determined indicating recombination between the mutant and the tester strain.

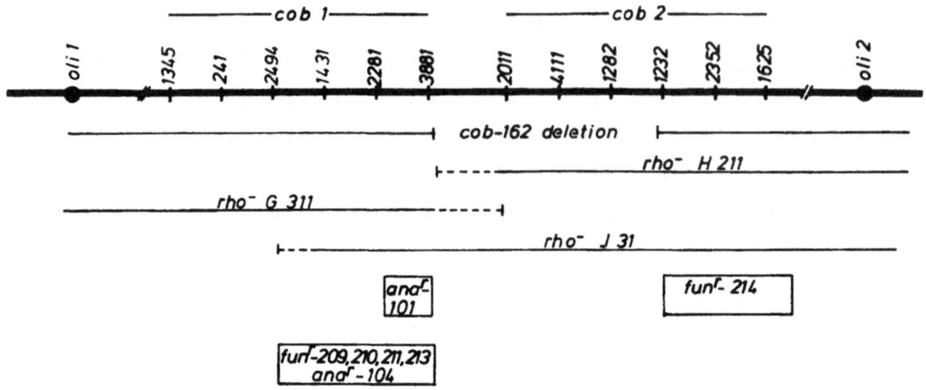

Fig. 2 Map of drug resistance markers in the cob region of the mitochondrial genome.

This figure presents the comparison of the fine structure map of the cob-region (Schweyen et al., this volume) and the results of deletion analysis shown in Table II.

In *Schizosaccharomyces pombe*, three different mutational sites on the mitDNA were separated by random spore analysis. The recombination frequencies between ana^r-8, ana^r-14 and diu^r-301 range from 0.06 to 2% (Wolf et al., this volume).

DISCUSSION

By rho$^-$ deletion mapping the *cob*-region of the mitDNA of *Saccharomyces cerevisiae* can be separated into 2 defined regions: *cob1* and *cob2* (R.J. Schweyen, this volume). Drug resistance markers are found in both *cob1* and *cob2*. The sites ana^r-101, ana^r-104 and fun^r-209 are clustered in the region of *cob1* harbouring 6 mutational sites probably including the ana^r-2 and diu^r-1 markers described by Colson and Slonimski (8). *In vitro* resistance (i.e. resistance of mitochondrial particles) of these mutants indicates an alteration of a polypeptide in the bc_1-segment caused by the mutation in a structural gene in *cob1*.

In vitro resistance of mutant fun^r-214 was not found. Its mutational site in *cob2* presumably is localized in the diu^r-2/ana^r-1 cluster described by others (8). Because *in vitro* resistance conferred by mutations in this cluster was shown earlier (9), also fun^r-214 may be a mutational site in a structural gene.

The location of $drug^r$ mutational sites in 2 clusters which are rather distant from each other as revealed by their frequent separation in rho^- mutants, may indicate that the clusters code for 2 structural genes. The question is then which polypeptides are the gene products of these genes. It was shown that cytochrome b isolated from Neurospora crassa (1) and cytochrome b in an enzymatically inactive bc_1 preparation from yeast (2) is synthesized on mitochondrial ribosomes. Presumably, it is coded for by mitDNA. Although 2 slightly different apoproteins of cytochrome b could be separated (1), it was impossible to decide whether they are coded for by one or by two genes.

In Saccharomyces cerevisiae the special analysis reveals at least 3 different spectral forms of cytochrome b with absorption maxima at about 566, 564 and 562 nm, at room temperature. An additional absorption maximum at 557 nm is found in Schizosaccharomyces pombe (cf. 10). The fact that in the mitochondrial mutant ana^r-8 of Sch.p. one of the b absorption maxima is shifted to the red (3) also points to the involvement of (a) mitochondrial gene(s) in the synthesis of cytochrome b apoprotein. However, the identity of the antimycin binding component with one of the b-type cytochromes becomes questionable by the finding of Das Gupta and Rieske (4) that an antimycin analogue can be covalently attached to a polypeptide of molecular weight 11 500 carrying no heme. It was previously shown (3,7) that the antimycin resistant mutants described here bind the drug less firmly. Provided that the antimycin binding protein of beef heart also occurs in yeast cells, it would be possible that this protein is changed in the antimycin resistant mutants. Therefore, this "antimycin binding protein" is a further candidate for a protein subunit of the bc_1 segment, which may be a mitochondrially coded gene product.

Finally, in Schizosaccharomyces pombe a stoichiometry of 2 different specific antimycin binding sites per 3 b-cytochromes has been proposed (6). It may represent a particular property of this fission yeast. A correlation with the genetical data is not possible because the arrangement of the 3 mutational sites ana^r-8, ana^r-14 and diu^r-301 in the cob-region of Schizosaccharomyces pombe is still unknown.

Our interest is to analyze whether the different forms of cytochrome b are different gene products coded for by mitDNA, or are modified polypeptides belonging to one or two genes only, and whether the antimycin binding protein of molecular weight 11 500 found in beef heart (4) also exists in yeast and is coded for by a structural gene of the cob-region.

ACKNOWLEDGEMENT

The authors are very much obliged to R.J. Schweyen for providing rho^- tester strains and for many helpful discussions; to G. Dörr for excellent technical assistance, and to H. Bruckner for biochemical data concerning ana^r-104. This work was supported by Deutsche Forschungsgemeinschaft.

REFERENCES

1. Weiss, H. and Ziganke, B. (1974) Eur.J.Biochem. 41, 63-71

2. Katan, J. (1975) Eur.J.Biochem. 43, 237-244

3. Lang, B. et al. (1975) Molec.gen.Genet. 137, 353-363

4. Das Gupta, U. and Rieske, J.S. (1973) Biochem.Biophys.Res. Commun. 54, 1247-1254

5. Lang, B. et al. (1977) Anal.Biochem. 77, 110-121

6. Burger, G. et al. (1976), Biochim.Biophys.Acta 396, 187-201

7. Burger, G. et al. (1976), Biochem.Biophys.Res.Commun. 72, 1201-1207

8. Colson, A.M. and Slonimski, P.P., this volume

9. Colson, A.M. et al. (1977) Eur.J.Bioch. 74, 521-526

10. Lang, B. (1977) Thesis

THE PHYSICAL MAP OF YEAST MITOCHONDRIAL DNA ANNO 1977[†]

(A REVIEW)

P.BORST, J.L.BOS, L.A.GRIVELL, G.S.P.GROOT, C.HEYTING, A.F.M.MOORMAN, J.P.M.SANDERS, J.L.TALEN, C.F.VAN KREIJL AND G.J.B.VAN OMMEN

Section for Medical Enzymology and Molecular Biology, Laboratory of Biochemistry, University of Amsterdam, Eerste Constantijn Huygensstraat 20, Amsterdam (The Netherlands)

CONTENTS

1. Introduction
2. The variability of yeast mtDNA
3. Maps of yeast mtDNAs, often used in biochemical experiments
4. The structure of petite mtDNA
5. Mapping genetic markers on the fragment map
6. Mapping rRNAs and 4S RNA
7. A preliminary map of transcripts
8. Mapping genes for mitochondrial proteins
9. Mitochondrial promotors, poly(A), precursor RNAs and primary protein products
10. The 21S rRNA region and the nature of C, E and ω
11. Genes in the C-P span and the nature and map position of P
12. The OXI-3 region
13. Genes in the E to OLI-2 span
14. AT-rich segments in yeast mtDNA and the Bernardi model for genes and spacers
15. Petite mutant RD1A and the AT-rich segments in yeast mtDNA
16. Concluding remarks

---

† Presented in abbreviated form by J.P.M.Sanders.

1. Introduction

The mtDNA of yeast is a circular duplex of 25 μm; its molecular weight is 5 x 10^7 and it contains about 75 000 bp (1). We have constructed a physical map of this DNA and used this to learn more about genes on mtDNA. In this paper we summarize the main results obtained up till June 1977.

2. The variability of yeast mtDNA

The restriction endonucleases HpaII (or its isoschizomer HapII) and HaeIII cleave yeast mtDNA in about 100 fragments (2,3). When the size distribution of fragments of mtDNAs from different yeast strains was compared, large differences were observed and some of the mtDNAs had hardly any fragment in common. Bernardi and co-workers have attributed this to multiple insertions/deletions involving putative "spacer segments" between genes (see section 14). To analyse this phenomenon further, Sanders *et al*. (4-6) have constructed independent physical maps of mtDNAs from three yeast strains, *Saccharomyces carlsbergensis* (C), *Saccharomyces cerevisiae* JS1-3D (J) and *Saccharomyces cerevisiae* KL14-4A (K). The circular maps are shown in Fig. 1; in Fig. 2 the linearized maps are compared. This comparison leads to three main conclusions:

1. The overall gene order in the three mtDNAs is the same, notwithstanding numerous small rearrangements and several large insertions/deletions. This follows from the overall similarity in the size distribution of large fragments obtained with several restriction endonucleases and it has been rigorously proven by hybridization experiments that demonstrate the sequence homology of fragments at corresponding map positions.

2. When homologous fragments of the three mtDNAs in the 1000-4000 bp range are compared, small length differences corresponding to 25-50 bp are found (6). In some of the larger HindII+III and EcoRI fragments no differences could be detected, but this may be due to the cancelling out of several small changes in opposite direction, as suggested by Bernardi. Our

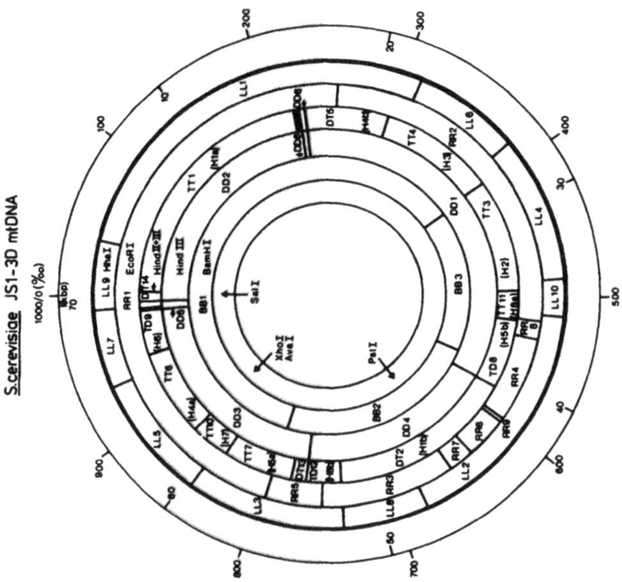
S.cerevisiae JS1-3D mtDNA

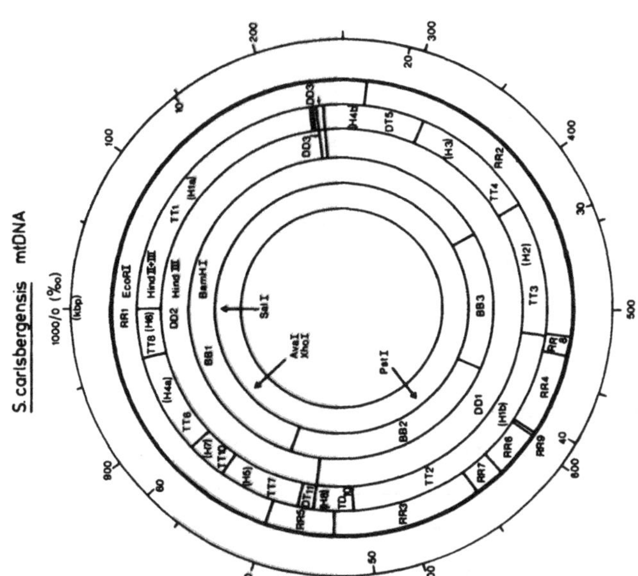
S.carlsbergensis mtDNA

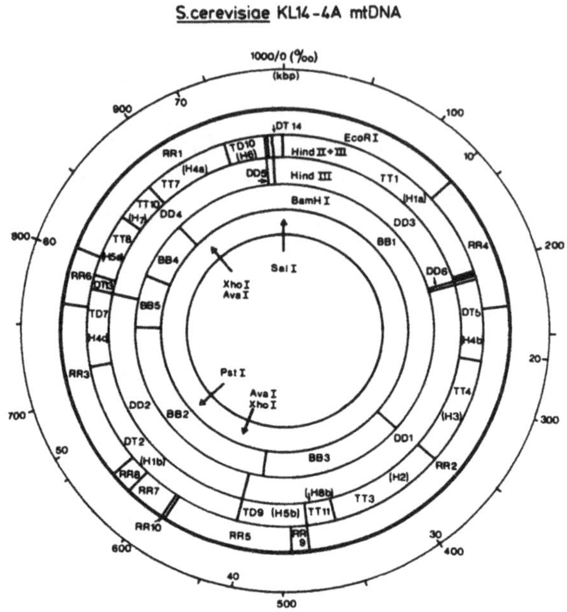

Fig. 1. The physical maps of mtDNA from three Saccharomyces strains. The DNAs were digested with restriction endonucleases and the maps were constructed as described by Sanders et al. (4-6). Standard abbreviations are used for restriction endonucleases, i.e. endonucleases HindII+III is a mixture of enzymes II and III from Haemophilus influenzae, serotype d (7). Fragments are designated by a two-letter code (see ref. 6), which specifies the endonuclease recognition sites on both ends. The order of site identification is in a clock-wise direction and each fragment is given a number, which indicates in order of decreasing length, its position in the first fragmentation pattern in which it is observed.

results are, therefore, compatible with Bernardi's conclusion that insertions/deletions frequently occur in all major segments of mtDNA. We find that most of these insertions/deletions are small.

3. In addition to the small rearrangements, also large deletions/insertions are present (5,6). Relative to S. carlsbergensis mtDNA, JS1-3D mtDNA contains three large insertions (900, 1000 and 1500 bp; differences I, VI and IV in Fig. 5) and KL14-4A mtDNA five large insertions (additionally 2600 and 3000 bp; differences II and III in Fig. 5). These insertions account for the differences in length of the three mtDNAs, which is 68 000 bp for S. carlsbergensis, 70 750 bp for JS1-3D and 75 750 bp for KL14-4A.

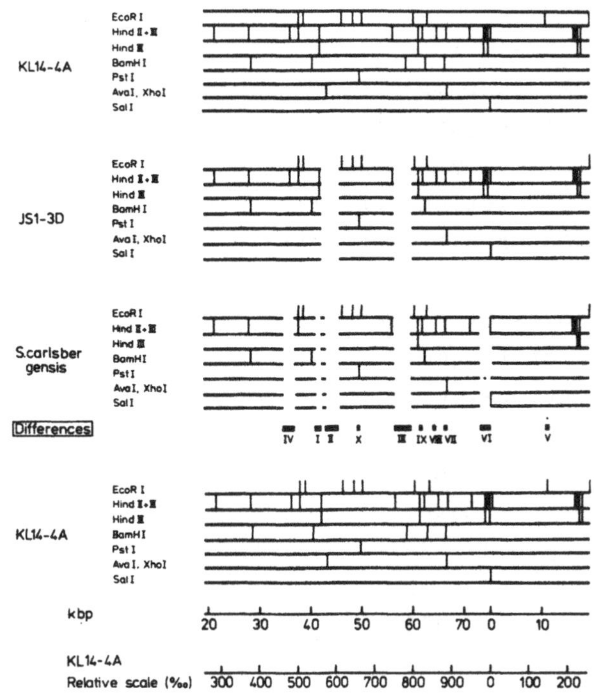

Fig. 2. Comparison of the physical maps of mtDNA from three Saccharomyces strains. The maps presented in Fig. 1 were linearized by cutting at the EcoRI site within DT5 (H4b) and aligning sequences considered homologous. From ref. 6.

The nature of the "extra sequences" present in insertions has been studied by Meijlink, Heyting and Sanders. They made complementary RNA (cRNA) of the endonuclease HhaI fragment LL10 of JS1-3D mtDNA (see Fig. 1), which covers most of the insert IV, present both in JS1-3D and KL14-4A mtDNA. This cRNA did not hybridize significantly with S. carlsbergensis mtDNA or with any other fragment than LL10 in the JS1-3D HhaI digest (unpublished result). This suggests that the sequence of insert IV (in so far as it is represented in the cRNA probe) is not present in S. carlsbergensis mtDNA nor elsewhere in JS1-3D mt-DNA. It may have come from another yeast mtDNA in a similar fashion as the transmission of an ω^+ allele into ω^- mtDNA entails the transfer of insert VI (see below). Whether the other "extra sequences" in KL14-4A represent duplications or contain "extra" information absent from S. carlsbergensis mtDNA remains to be determined.

Whereas the small insertions are spread more or less randomly over the DNA, four major ones are clustered in one quarter of the genome. This quarter also contains the large (dA.dT) clusters identified by hybridization with poly(U) (5,8).

The multiple deletions and insertions complicate the comparisons of mtDNAs of different strains and make an unambiguous designation of map segments difficult. Corresponding fragments often differ somewhat in length, the larger insertions change the size order of fragments and, because of the clustering of the large insertions, homologous segments do not have the same position on circular maps even if the maps are normalized to a percentage of total length scale. For instance, in bp the "right half" of the three maps in Fig. 1 is nearly identical, the 36 000 bp span TT1 (H1a)-TT3 (H2) differing less than 200 bp in the three mtDNAs. However, it covers 520 map units in S. carlsbergensis and 458 in KL14-4A. The resulting distortion is readily seen from a comparison of Fig. 5A (KL14-4A) with Fig. 7 (S. carlsbergensis). The shorter distance between the two rRNA cistrons appears larger in S. carlsbergensis mtDNA, whereas in reality it is 1000 bp smaller. The most informative comparison, therefore, is that presented in Fig. 2, in which equivalent cleavage sites are put at equivalent map positions, whereas deletions are indicated by gaps.

3. Maps of yeast mtDNAs, often used in biochemical experiments

For the practice of gene mapping the apparent conservation of gene order in yeast mtDNA has the obvious advantage that the maps now available will probably apply with minor adjustments to all yeast strains. In fact, most of the biochemical experiments done on yeast mtDNA can be related already to one of the DNAs mapped. Most of the work of Slonimski and Fukuhara and their co-workers has been done with derivatives of KL14-4A. Rabinowitz and co-workers have made a restriction fragment map of mtDNA from their S. cerevisiae strain MH41-7B (9), which differs from the KL14-4A map in one point only and this minor difference may even be due to a mapping error, as suggested by Sanders et al. (6). The S. cerevisiae strain D273-10B, used by Tzagoloff and co-workers, is nearly identical to JS1-3D (6,10) and most of the experiments in this laboratory have been done with JS1-3D or S. carlsbergensis mtDNA. Furthermore, the Hind-II+III fragment pattern of the mtDNAs of many other strains has been analysed and most of these closely resemble that of JS1-3D or KL14-4A mtDNA (6). No restriction fragment analysis has been published for the mtDNA from the yeast strains used by Linnane and co-workers. This mtDNA has been mapped, however, using petite mutant mtDNA and on the resulting physical map the order and relative distance of genetic markers and ribosomal RNAs (rRNAs) is about the same as in our strains (see below). We infer that the physical map is similar to those presented in Fig. 1.

4. The structure of petite mtDNA

Current ideas about the origin of the altered mtDNA in cytoplasmic petite mutants are summarized in Fig. 3. In the primary mutagenic event the mtDNA somehow fragments; some of these fragments replicate and after a process of compensatory amplification the cell ends up with the same amount of mtDNA it started out with. In theory, therefore, petite mtDNAs are segments of wild-type mtDNA purified by nature. It is clearly of importance to verify if this idyllic picture is correct and how extensively it is disturbed by the secondary rearrangements and deletions also included in the scheme of Fig. 3. This can be done by the detailed mapping of the wild-type DNA segments present in the petite mtDNA.

Our analysis of a series of petites isolated by Heyting or by the Gif group has shown (12,13; Heyting, C. and Talen, J.L., unpublished) that they can be roughly sub-divided into three types:

1. Some rho^- mtDNAs appear to contain a single, continuous, amplified segment of rho^+ mtDNA without secondary rearrangements. The repeat units are arranged head to tail. Treatment

A b c d e f g h i j k l m n o p q r s t u A

↓ Fragmentation

A b c d e f g h i j k + l m n o p q + r + s t u

↓ Illegitimate recombination with replicator and compensatory amplification

A b c d e f g h i j k b c d e f g h i j k A

\+

A r r r r r r r r r r r r r r r r r r r r A

\+

A b A b A b A b A b A b A b A b A b A b A

etc.

↓ Cloning

A b c d e f g h i j k b c d e f g h i j k A

↓ Rearrangements, secondary deletions and compensatory amplification

A b c d f g g'f'd'c'b'b c d f g g'f'd'c'b'A

↑↓ Excision of monomer repeats

b c d f g g'f'd'c'b'b

Fig. 3. Speculative model for the origin of petite mutants (from ref. 11). A, Replicator sequence; underlined terminal letters indicate circularity, i.e. in the top line A is joined to b and to u.

of such mtDNAs with restriction endonucleases yields a set of fragments that are adjacent in the rho$^+$ mtDNA plus one new "junction" fragment. All fragments are present in stoicheiometric amounts. It is difficult to prove rigorously that a petite is an ideal one, especially when the repeated segment is long. F41 of Heyting and Sanders (12) is an apparent example; RD1A (see below) comes close to it.

2. Most of the rho$^-$ mtDNAs analysed contain a continuous rho$^+$ mtDNA segment, but with secondary rearrangements within this segment. Digests of such mtDNAs contain more than one "junction" fragment and fragments may not be present in stoicheiometric amounts. To prove the nature of the "junction" fragments, the sequences contained in it must be characterized by hybridization with rho$^+$ mtDNA fragments or by other means.

3. In the third type of petite, mtDNA sequences that are far apart in the rho$^+$ genome have been joined. Fortunately, such mutants are rare and their isolation may result from the simultaneous selection for markers far apart on the rho$^+$ DNA.

From our analysis we conclude that petites of class 1 and most of the petites of class 2 can be used for gene purification and gene mapping, if the rho$^+$ mtDNA sequences contained in them are characterized.

The secondary rearrangements found in class 2 mutants are of interest, because they may arise by the same mechanism that produces insertions/deletions in the rho$^+$ mtDNA. Moreover, such mutants may give information on the mechanism of the intriguing amplification process, which may also be of interest for an understanding of the regulation of copy number and copy sequence in other multi-gene families. A detailed analysis of rearranged petites in the region of the large rRNA cistron, made by C.Heyting and J.L.Talen (unpublished results), will be presented elsewhere.

5. Mapping genetic markers on the fragment map

By genetic crosses one can identify the mitochondrial genetic markers present in a petite mutant. These markers lie in the mtDNA segment retained in the petite. By mapping this segment on the wild-type mtDNA, the approximate location of the markers on the fragment map can be determined. The principle is illustrated in Fig. 4. This fragment-marker co-retention analysis has been applied to many genetic markers and the results are presented in Fig. 5A. Fig. 5B gives the results obtained by Linnane and co-workers with a different mapping procedure, but also using petite mutants. In general the maps in Figs 5A and 5B are in good agreement, the only significant discrepancy being at the P locus. We return to this discrepancy below.

In the case of mit$^-$ mutants, in which the biosynthesis of

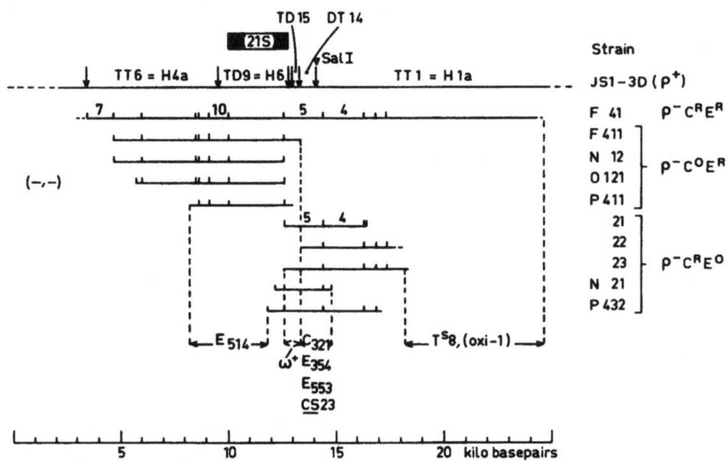

Fig. 4. Fine-structure map of the 21S rRNA region of yeast mtDNA, analysed by means of restriction enzyme digests of petite mtDNAs. The upper line represents part of the rho$^+$ mtDNA. Arrows indicate recognition sites of the endonucleases HindII+III and the triangle the single SalI site (see Fig.1). The lower lines represent the mtDNAs of petites carrying genetic markers of the 21S rRNA region with the arrangement of their HapII fragments. The vertical bars are the recognition sites for endonuclease HapII. The most probable position of the gene for 21S rRNA is indicated (but see also section 9 and Fig. 9). Each marker is located on the DNA segment common to all petite mutants able to transmit the marker in genetic crosses. The position of OXI-1 is uncertain; F41 was found to transmit the wild-type allele of this locus in early experiments (12), but not more recently (see Fig. 5). The markers belong to the following loci: E_{514} = E = RIB-3; C_{321} = C = RIB-1; E_{354} = RIB-2; E_{553} = RIB-2; cs23 (see ref. 14) = probably RIB-2. Based on ref. 12 and unpublished results of C.Heyting and J.L.Talen.

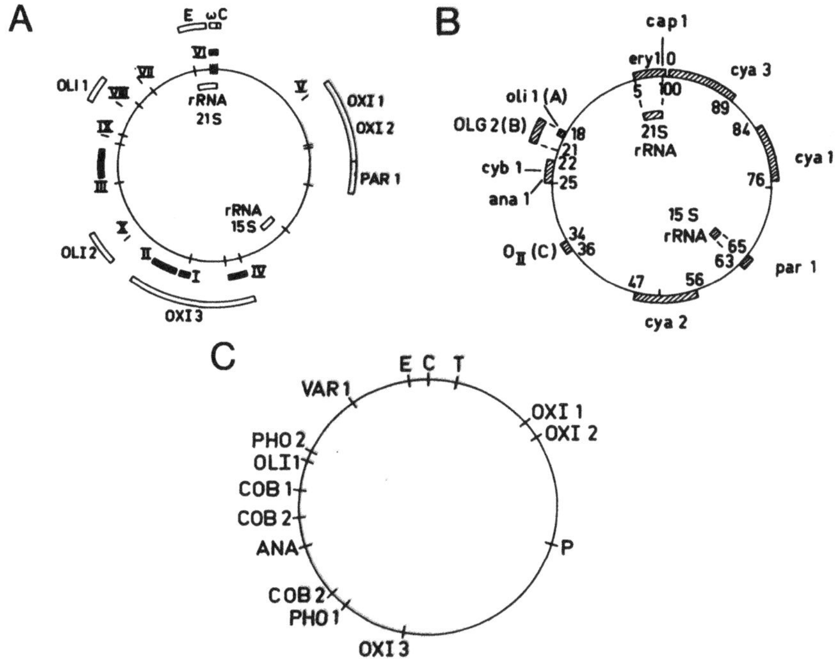

Fig. 5. The location of genetic loci on the physical and genetic maps of yeast mtDNA. A. Location of genetic loci on the physical map of mtDNA from S. cerevisiae KL14-4A. The central solid line represents the DNA with the recognition sites for endonucleases HindII+III indicated by cross lines. The orientation of the map is the same as in Fig. 1. The inner ring gives the position of the two rRNA genes (4) and of 4S RNA genes (see Fig. 7). The roman numerals and black boxes indicate the position of the major differences between the three mtDNAs (see Fig. 2). In the outer ring the position of genetic loci is indicated by open blocks. E (RIB-3), C (RIB-1), P (PAR-1) and OLI are loci for erythromycin, chloramphenicol, paromomycin and oligomycin resistance; OXI-1 to 3 are three loci involved in the biosynthesis of cytochrome c oxidase; ω is a locus affecting polarity of recombination in the 21S region; T (TSM-1) is a locus affecting the stability of rho^{+} mtDNA. The data for E, C, T and ω (Fig. 4), P (13) and OLI-2 (15) were obtained by restriction analysis of genetically-marked petites, as illustrated in Fig. 4. The position of OXI-3 is taken from Fig. 6. We have inferred the approximate position of OXI-1 and 2 from the petite analysis by Faye et al. (66) in conjunction with the unpublished observation by L.A.Grivell that petite mutant F41 (see Fig. 4) cannot restore mit^{-} mutations in the OXI-1 locus. B. Location of genetic loci on the physical map of mtDNA from S. cerevisiae strain 770-7B (16). Ery 1, cap 1, par 1 and ana 1 are loci for erythromycin, chloramphenicol, paromomycin and antimycin resistance, respectively; oli 1, OLG 2 and O_{II} are three loci for oligomycin resistance (OLG 2 is equivalent to OLI-1 and O_{II} to OLI-2 in maps A and C). cya 1-3 are loci involved in the biosynthesis of cytochrome c oxidase, cyb

only one or two of the mitochondrially-synthesized enzymes is affected, a second mapping procedure is available. Some of the mit$^-$ mutants are deletions and these can be recognized in the fragment pattern of the DNA and mapped as first shown by Rabinowitz et al. (20). The principle is illustrated in the map of the OXI-3 region, given in Fig. 6. This procedure has been applied to other types of mit$^-$ mutants, but no mappable deletions have yet been found (20; Grivell, L.A., unpublished).

Comparison of the physical maps in Figs 5A and 5B with the genetic map in Fig. 5C shows a gratifying correspondence. Both

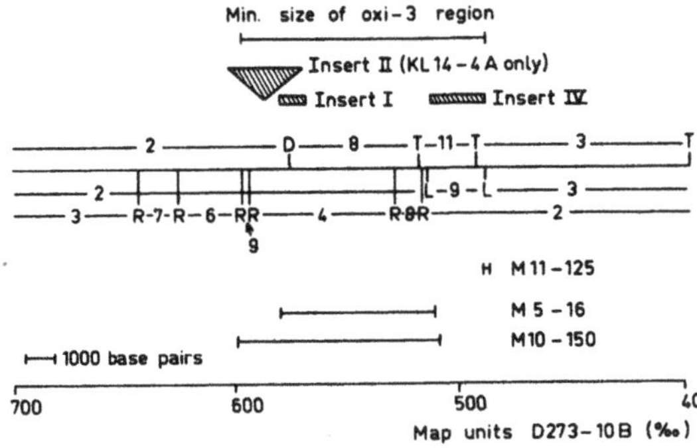

Fig. 6. A fine-structure map of the OXI-3 region in yeast mtDNA. The map shows a segment of mtDNA from strain D273-10B (identical to JS1-3D in this region), the map position of three mit$^-$ mutants, carrying deletions, and the positions of the known insertions in this region (see Fig. 2). The minimal size of the OXI-3 region follows from the existence of OXI-3 mutants outside the region covered by the three deletion mutants mapped here. From ref. 10.

Legend Fig. 5 - continued.

1 in the biosynthesis of cytochrome b. C. Location of loci on the genetic map of S. cerevisiae. PHO-1 and 2 are loci involved in the biosynthesis of the ATPase complex; COB-1 and 2 are loci involved in the biosynthesis of cytochrome b; ANA is an antimycin-resistance locus. The other abbreviations are the same as in map A. The data for C, T, P, OLI and ANA are taken from ref. 17, the position of E from ref. 4, the position of VAR-1 from ref. 18 and that of the other markers from ref. 19. The map given in ref. 19 has been adjusted to allow superposition of OLI-1 on both maps. The relative positions of C, T, P, OLI-1 and ANA have been determined by an analysis of the frequency of co-retention of genetic markers in petite mutants and are more accurate than those of the other markers.

the order and the relative distance of the markers is the same within the accuracy of the maps. We conclude from these results (cf. ref. 4) that the two main techniques used for constructing the genetic map - determination of recombination frequencies and co-retention of markers in petite mutants - yield the approximate real physical distance of the markers studied, at least when the markers are far apart. This implies that there are not one or two major hot-spots for recombination or strongly-preferred sites for deletions leading to petite mutants in yeast mtDNA.

6. Mapping rRNAs and 4S RNA

By straightforward hybridization experiments it is possible to determine the positions of mitochondrial RNAs on the physical map. The position of rRNAs and 4S RNAs on the HindII+III map in Fig. 7 shows three points of interest:

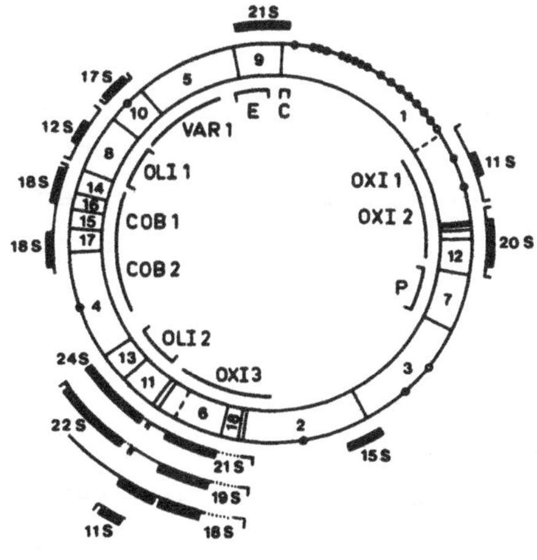

Fig. 7. A transcription map of yeast mtRNA. The circular map gives the fragments obtained by digesting S. carlsbergensis mtDNA with a mixture of endonucleases HindII+III and EcoRI. The dotted lines indicate extra EcoRI or HindIII recognition sites present in KL14-4A mtDNA, that have been used to get more precise map positions for some RNA species. The black dots represent 4S RNAs, the two open dots the two met-tRNAs. The outer circle gives the position of larger mtRNAs on the map; the inner circle gives the position of genetic loci, taken from Fig. 5A. The position of COB and VAR-1 has not been determined by physical mapping and is deduced from the genetic map in Fig. 5C. From ref. 21.

1. The two rRNA cistrons are far apart, nearly 30 000 bp. There is no precedent for this in nature; even all mtDNAs studied, including those of man, fly, toad, mouse, rat and *Neurospora crassa*, have the two cistrons adjacent. We do not know why yeast mitochondria should be an exception in this respect.

2. The number of tRNA cistrons is around 26, significantly less than the 32 tRNAs required for reading all codons according to Crick's wobble hypothesis (22). It is of interest that about the same number of "spots" is found when total 4S RNA in mitochondrial preparations is analysed by two-dimensional chromatography (see references in ref. 23). The possibility that mitochondria would disregard Crick and read 61 codons with 26 tRNAs will be considered elsewhere.

3. The 4S RNAs (probably all or nearly all tRNAs) are largely clustered in a segment of 9000 bp on the largest HindII+III fragment. Analysis of the HapII fragments from this region (see ref. 23) has shown the presence of one 460 bp fragment that contains at least five 4S RNA genes, leaving only about 80 bp for other (regulatory) sequences. If the 4S RNA genes are transcribed separately these 80 bp must include at least four promotor and termination sequences. The possibility that these five genes are transcribed into a single precursor RNA is, therefore, an attractive possibility, although we have not yet been able to find this precursor.

7. A preliminary map of transcripts

When yeast mtRNA is fractionated on an acrylamide gel, the electropherogram is dominated by the two rRNA peaks at 21S and 15S and the 4S peak. Nevertheless, discrete other minor components are reproducibly present with S_E values of 24, 22, 19, 18, 12 and 11S (see Fig. 2 of ref. 21). Such minor components should be derived from all over the genome, because Rabinowitz *et al*. have shown that at very high R_ot values around 50% of yeast mtDNA can be converted into a DNA-RNA hybrid.

To map these minor components, Van Ommen and Groot have fractionated labelled yeast mtRNA on polyacrylamide-agarose gels under conditions that minimize aggregation (cf. ref. 21); sliced up the gel, eluted the RNA from each slice and hybridized each fraction separately against the EcoRI x HindII+III fragments of mtDNA by the Southern technique. The results are presented elsewhere in this volume (21) and summarized in Fig. 7.

There are three points of interest in this preliminary map of transcripts:

1. Specific transcripts have been found that roughly correspond to most of the known genetic markers, i.e. an 11S transcript is found in the OXI-1 region, a 20S transcript in the OXI-2 region, etc. P and T are the only clear exceptions and we return to these below. Neither the mapping of genetic markers, nor the mapping of transcripts is as yet precise enough to say that the transcript covers the gene in which the genetic marker is present.

2. In the OLI-2 to OXI-3 region there are several RNA size classes that hybridize in an overlapping fashion with the same fragments. This suggests that long precursor RNAs undergo stepwise processing to yield discrete products.

3. The transcripts detected cover about 45% of one strand of mtDNA. There is considerable uncertainty in this estimate because the size of the transcripts has been calculated from their mobilities relative to the rRNAs in non-denaturing gels. It is clear, however, that there are large "empty" spaces on the mtDNA, where we can not find stable transcripts.

8. Mapping genes for mitochondrial proteins

Given a map and the means to identify translation products, it should be simple to map the structural genes for mitochondrial proteins on mtDNA. This can be done either by taking defined segments of mtDNA (e.g. restriction fragments) and making the protein(s) corresponding to this segment in a coupled transcription-translation system, or by translating mitochondrial mRNAs that have been mapped by hybridization. Prime candidates for proteins specified by mtDNA are the proteins made

TABLE 1

PROTEINS SYNTHESIZED ON MITOCHONDRIAL RIBOSOMES IN YEAST (see refs 24,25)

	Molecular weight ($\times 10^{-4}$)	Daltons duplex DNA required for coding ($\times 10^{-6}$)
Cytochrome oxidase subunits	4	0.8
(3 out of 7 present)	3	0.6
	2	0.4
ATPase complex	3*	0.6
(4 out of 9 present)	2	0.4
	1	0.2
	0.8*	0.16
Cytochrome $\underline{bc}_1$ complex	3	0.6
(1 out of 7 present)		
Small ribosomal subunit	3.5	0.7
(1 out of 22 present)		
	Total:	4.46

* Not made on mitochondrial ribosomes in *Neurospora* (26).

on mitochondrial ribosomes and these are listed in Table I. Unfortunately, protein mapping has run into two formidable obstacles:

1. The products made are hydrophobic (like the products of mitochondrial protein synthesis *in vivo*) and tend to aggregate even in sodium dodecylsulphate and urea. They also stick to any carrier protein added and to the proteins of the protein-synthesizing system used to translate the RNA. In initial experiments in several laboratories this pitfall was not realized and hence claims were made for the biosynthesis of full-length subunits of cytochrome oxidase in the *in vitro* system. In all these experiments cytochrome oxidase was added as carrier or marker in gels and we now know that an unambiguous interpretation of such experiments is impossible (see refs 27-29).

2. The great majority of the products made *in vitro* has a molecular weight below 10 000; none of the larger polypeptides co-migrates with the *in vivo* products of mitochondrial protein

synthesis and they may all be aggregation artefacts (27,28). This holds both for RNA-directed and DNA-directed in vitro systems and for all in vitro protein-synthesizing systems tested, including several modifications of systems from Escherichia coli, wheat germ, rabbit reticulocytes or Xenopus oocytes (29). The mtRNA is extracted without apparent damage from mitochondria that do synthesize full-length mitochondrial peptides, even after isolation from the cell. Moreover, the isolated RNA acts as a very effective template for amino acid incorporation, especially in the system from E. coli. The absence of large products is, therefore, surprising. Since all these in vitro protein-synthesizing systems are non-mitochondrial, they may all be unable to properly recognize the initiation signals on mitochondrial mRNAs. It remains unusual, however, that no long products are made at all. The possibility that in vivo the mitochondrial translation machinery makes small peptides that are covalently linked later or jumps loops in mRNAs, is unattractive but cannot be ruled out.

Fortunately, however, fragments of mitochondrial proteins contain specific antigenic determinants and by the use of double-antibody precipitation techniques and rigorously controlled procedures these can be identified in the peptide products made in vitro. The results obtained thusfar are presented in detail elsewhere (29,30) and they are summarized in Fig. 8.

Fig. 8a shows the ability of mtRNA, fractionated into discrete size classes by sucrose gradient centrifugation, to stimulate the incorporation of (^{35}S)methionine into acid-insoluble material in an E. coli protein-synthesizing system. mRNA activity parallels the A_{260nm} profile, even though RNA from purified ribosomes is completely inactive (27). When the products are analysed by specific immunoprecipitation, the results are quite different, however. The mRNA for antigenic determinants of cytochrome c oxidase is found in a sharp peak at 11S (Fig. 8b), just behind the 12S RNA in the A_{260nm} profile, the mRNA activity for antigenic determinants of the ATPase complex is mainly found at 6S (Fig. 8c).

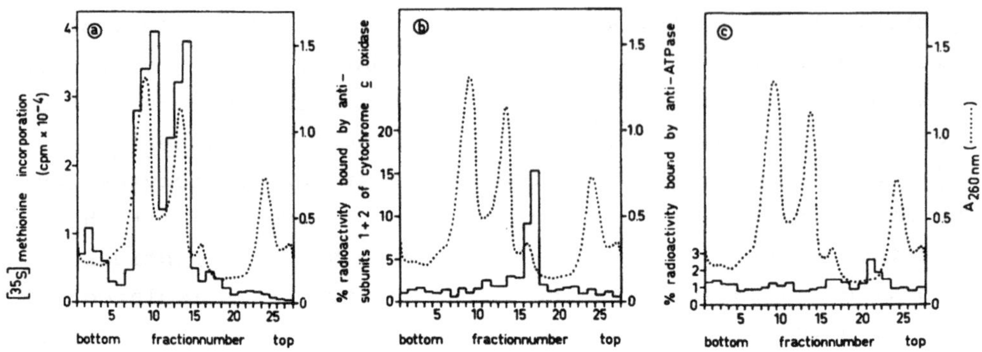

Fig. 8. Identification of mRNAs for subunits of cytochrome c oxidase and the ATPase complex. 100 µg mtRNA from S. carlsbergensis was fractionated by sedimentation through isokinetic sucrose gradients containing 3% formaldehyde. RNA of each fraction was assayed for the ability to stimulate an E. coli extract to incorporate (^{35}S)methionine into acid-insoluble material (a). Aliquots of the product were assayed in double immunoprecipitation tests for binding to antibodies raised against subunits 1 + 2 of cytochrome c oxidase (b) or antibodies raised against the ATPase complex (c). See refs 29 and 30 for experimental details.

The RNA species in the 11-12S region have been resolved by re-sedimentation through sucrose gradients and by gel electrophoresis and their transcriptional origin has been determined. The major RNA, sedimenting at 12S, migrates at 12 S_E through gels and maps on the HindII+III x EcoRI fragment on which OLI-1 lies (Fig. 7). We attribute the marginal mRNA activity for ATPase in the 12S region to this mRNA. We do not know whether the immunoprecipitation is so low because this RNA is poorly translated or because the antibody has a low affinity for the antigenic determinants made.

The minor 11S RNA usually splits in gels into a diffuse 11 S_E double band. The major component from this band hybridizes at 2.30 o'clock in Fig. 7 and could correspond to the OXI-1 locus; the minor component hybridizes at 7 o'clock and could correspond to the OXI-3 locus. The relative contribution of these RNAs to the in vitro synthesis of peptide material that reacts with the antibody directed against subunits I and II of oxidase, is not known.

The map location of the 6S ATPase mRNA is not yet known.

Hybridization experiments with RNA from this region are complicated by the presence of degradation products of larger RNAs.

The coupled transcription-translation system programmed with mtDNA has been used previously to prove that the structural genes for oxidase subunits I and II (and possibly III) are in mtDNA. By programming this system with purified restriction fragments of mtDNA we are locating these structural genes on the mtDNA. Although the preliminary results are compatible with the conclusion that the OXI-1, 2 and 3 loci are the structural genes for subunits I-III of oxidase, the lack of mono-specific antibodies and cross-contamination of restriction fragments has precluded a precise localization of each individual gene on the map (see ref. 10).

9. Mitochondrial promotors, poly(A), precursor RNAs and primary protein products

mtDNA transcription in HeLa cells is symmetric and yields full-length transcripts of both strands, presumably each starting at a single promotor. These transcripts are rapidly processed to yield rRNAs, tRNAs and a spectrum of other RNAs, some of which contain a poly(A) segment of about 55 nucleotides (see ref. 31).

The nature of the primary transcript and its processing in yeast is less clear and in part controversial. Polyacrylamide gel analysis of yeast mtRNA labelled with a 4-min ^{32}P pulse, shows a heterogeneous distribution of ^{32}P throughout the gel with considerable radioactivity migrating slower than 21S RNA (32). This suggests that the primary transcript could be a very long RNA that is more rapidly processed than the pulse time. The results in Fig. 7 also indicate that RNA processing occurs. Definitive proof that the large transcripts are completely free of aggregates is still lacking, however.

A considerable fraction of symmetric transcripts was found in yeast mtRNA by Hendler et al. (20,33). In excess RNA hybridizations they could convert 46% of mtDNA into a DNA-RNA hybrid before self-annealing their RNA preparations and this value decreased to 29% after self-annealing. The symmetric transcripts are present at very low concentrations because in the pulse-labelled RNA of Van Ommen (unpublished) the amount of duplex RNA after self-annealing was less than 2%.

On the basis of these results it is still possible that in wild-type yeast both strands of the mtDNA are completely transcribed from one specific promotor followed by rapid processing of the primary transcripts. This becomes less likely, however, from the finding by Faye and co-workers (34,35) that petite mutants, that have conserved the 21S rRNA gene but not the 15S rRNA gene, still make apparently normal 21S rRNA, whereas petites that only contain the 15S rRNA gene still make 15S rRNA. This is not easily compatible with a model in which RNA synthesis on each strand always starts from a single promotor.

The presence of poly(A) in yeast mtRNA is still controversial. After we had shown that no poly(A) tract >30 nucleotides was detectable (32), Hendler et al. (36) reported the isolation of an RNA fraction with a short 3'-terminal poly(A) segment of 20-30 nucleotides. This poly(A)-RNA was preferentially labelled in pulse-labelling experiments and was found to direct the synthesis of peptide material in a protein-synthesizing system from E. coli (37). This led to the conclusion that yeast mitochondrial mRNA contains poly(A) (37). Experiments in this and Rabinowitz's laboratory failed to substantiate this claim, however, and no significant difference was found in the ability to stimulate protein synthesis between RNA fractions that do and that do not bind to poly(U)-Sepharose (Grivell, L.A. and Moorman, A.F.M., unpublished; 33). In more recent experiments Moorman and Grivell (unpublished) have shown that the mitochondrial 11S RNA that shows a high specific oxidase mRNA activity (Fig. 8b) does not significantly bind to poly(U)-Sepharose under the conditions used by Hendler et al. (36) to isolate poly(A)-RNA. We conclude that even a short poly(A) segment is not an essential feature of yeast mitochondrial mRNAs. It seems possible that the binding of a small fraction of yeast mtRNA to poly(U)-Sepharose is a non-specific effect, resulting from the high A content of yeast mtRNA.

The translation of mitochondrial mRNAs in vitro has not yet yielded products in the size range found for the true products of mitochondrial protein synthesis (see section 8). Hence, it is not certain whether these proteins are made as such or in the form of long precursors. The finding of F-met as N-terminal amino acid on some known products of mitochondrial protein synthesis in animal mitochondria (38; Muijsers, A.O., personal communication), makes it unlikely that processing at the N-terminal end is a general phenomenon. Secondary modification of proteins may occur in mitochondria, however, as Weiss and co-workers (39) have obtained evidence for two forms of cytochrome b, which appear to be largely identical in sequence.

10. The 21S rRNA region and the nature of C, E and ω

Heyting and Sanders (12) have reported earlier that the map positions of genetic markers for chloramphenicol (C_{321}, RIB-1 locus) and erythromycin (E_{514}, RIB-3 locus and E_{354}, RIB-2 locus) resistance overlap the rRNA cistron. The mapping was not precise enough, however, to decide whether these loci are just within or just outside the rRNA cistron. The polarity locus ω was placed between E_{514} and C_{321}.

Further analysis of this region by Heyting and co-workers (6,12; Heyting, C. and Talen, J.L., unpublished) has uncovered the surprising situation summarized in Fig. 9. Two points are noteworthy:

1. Compared with ω^+ mtDNA, the ω^- mtDNA always lacks the small Hind fragments between TT1 (H1a) and TD9 (H6) and always has a TD9 (H6) fragment which is about 50 nucleotides shorter. S. carlsbergensis mtDNA behaves as ω^- in this respect. DNA segments on both sides of this large insertion hybridize with rRNA, irrespective whether this RNA is isolated from highly purified ribosomal subunits or taken from the 21S region of gels or gradients.

2. Two of the small HindII+III fragments in the ω^+ mtDNA do

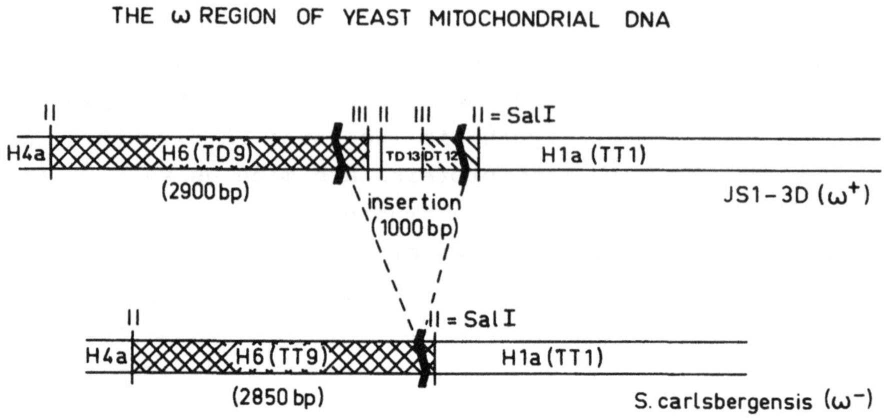

Fig. 9. The ω region of yeast mtDNA. The maps show the HindII+III fragments in the region of the 21S rRNA cistron (the SalI site is at the 0 map position in Fig. 1). Roman II and III indicate recognition sites for HindII and III, respectively. Fragments hybridizing with rRNA are hatched over their entire length, although we do not know which part of the fragment hybridizes. Cross-hatched fragments hybridize more strongly than hatched fragments. Based on results of Heyting and co-workers (6,12).

not hybridize with homologous 21S rRNA even though the surrounding fragments do.

Trivial explanations for these results, to be presented in detail elsewhere by Heyting and co-workers, have been ruled out and two possibilities remain, both interesting:

a) There is a partial duplication of the rRNA cistron and this accounts for the hybridization of fragments to the right of ω in Fig. 9. If this is so, ω is outside the rRNA cistron and so is C_{321} (and E_{354} which is closely associated with C). The implication is that C_{321} and E_{354} cannot be mutations in the 21S RNA and they must lie in an as yet unidentified mitochondrial gene. Since all mitochondrial proteins of the large ribosomal subunit are apparently made on cell-sap ribosomes (see ref. 40) and since the secondary modification of rRNA is minimal (41), this new gene could either specify an enzyme that modifies a ribosomal protein or a 5S RNA equivalent.

The former alternative is most unattractive: even though an alteration in a protein-modifying enzyme could theoretically lead to a chloramphenicol-resistant ribosome, we do not see how another alteration in the very same enzyme could result in ribosomes that are resistant to erythromycin. Compared with this objection, the absence of a large transcript hybridizing in this region (see Fig. 7) is a minor worry. The 5S RNA alternative is not unreasonable, however. Although there is no 5S RNA in yeast (42), there could be a 5S RNA equivalent of the size of tRNA. In fact, one 4S RNA gene is present in the HapII fragment that overlaps the position of C_{321} (23).

b) Mature rRNA in yeast mitochondria contains segments that are not adjacent in the mtDNA, as recently reported for some adenovirus RNAs (43). Covalent coupling of these segments could either be due to a jumping RNA polymerase, to the synthesis of two separate RNA fragments with subsequent ligation or to the synthesis of a large precursor followed by internal processing and ligation. According to this interpretation, C_{321} and E_{354} could still be mutations in the rRNA cistron and the difference between ω^+ and ω^- would be an insert between the two "parts" of the rRNA cistron.

In recent hybridization experiments (Heyting, C. and Bos, J.L., unpublished), weak but specific homology has been found between segments of the ω^+ insert and HindII+III fragment H6 (TT8 in S. carlsbergensis, TD9 in JS1-3D and TD10 in KL14-4A), which carries the bulk of the rRNA cistron. This supports alternative a), but it remains to be seen whether this duplication can completely account for the hybridization with rRNA, shown in Fig. 9.

The genetic analysis of the three different alleles of ω - ω^+, ω^- and ω^n - has suggested that these alleles differ by macro-lesions rather than point mutations (44). Polarity effects are thought to arise from a gene conversion process starting at ω and converting the alleles on the ω^- strand to those on the ω^+ strand. Perlman and Birky (45) have proposed that the ω^- allele is an insertion, which is attacked by an endonuclease in the ω^-/ω^+ heteroduplex.

Our results largely agree with these proposals. They show

that the ω^+ marker covers the 1000 bp insert found in the ω^+ mtDNA and since the complete insert is present in all ω^+ strains and none of the ω^- strains tested, it appears that the ω^+ genotype requires the complete 1000 bp insertion. Rather than an insertion, as proposed by Perlman and Birky (45), the ω^- allele is a deletion, however.

The polarity effects elicited by ω are not merely due to the presence of a large insertion, because no polarity is found in crosses of ω^+ and ω^n strains, even though the restriction fragment pattern of ω^n mtDNA is indistinguishable from that of ω^- mtDNA (Heyting, C., unpublished). Whether polarity is induced by the other large insertions in mtDNA, *i.e.* in the OXI-3 region and in the COB region, remains to be verified.

11. Genes in the C-P span and the nature and map position of P

The C-P span contains tRNA genes, T, OXI-1 and 2. Schatz and co-workers (46) have obtained strong evidence that OXI-1 is the structural gene for subunit II of cytochrome *c* oxidase. Since subunits I, II and III of this enzyme are made on mitochondrial ribosomes (Table I) and since there is circumstantial evidence to link the OXI-3 locus to subunit I (see below), OXI-2 could code for subunit III. The 11S and 20S RNAs mapped in the OXI-1,2 region (Fig. 7) could be the mRNAs for these subunits.

The nature of T is unknown. Mutation *tsm*-8 at this locus leads to rapid induction of cytoplasmic petite mutants at the restrictive temperature and recent results suggest that a shut-off of transcription is one of the first effects observed after raising the temperature (47). There is no major transcript detectable that could correspond to T (compare Figs 5 and 7) and if T codes for a protein, it is a minor one making its mRNA or the protein itself undetectable in labelling experiments. It is unclear why the maintenance of rho$^+$ mtDNA would require continued mtRNA synthesis (or protein synthesis, see ref. 48), whereas the continued orderly replication of rho$^-$ mtDNA does not.

The position of the P locus is controversial. Faye et al. (49) studied a series of petites that had retained their marker P_{454} and found that some of these petites had lost the 15S rRNA cistron. From preliminary mapping experiments they deduced the order C-15S-P rather than C-P-15S, as shown in Fig. 5. Subsequent precise mapping of P_{454} on the restriction fragment map, using the petites of the Gif group, led to the conclusion (4) that the correct order is C-P-15S and that the distance between P and 15S is at least 6000 bp, i.e. nearly 10% of the physical map.

As shown in Fig. 5B, a different map position for P has been published by Linnane and co-workers. They find that in their set of (rather complex) petites the 15S RNA and P are always co-ordinately lost and they map both loci at the same position within two map units (1500 bp), as shown in Fig. 5B. This position is identical to that found for 15S RNA in KL14-4A where it straddles the HindII+III site between TT3 and TT4 at 365%o, which is equivalent to 63.5% in Linnane's terminology.

Two possibilities have been considered by Choo et al. (16) to account for this discrepancy:

1. The sequence variations between wild-type mtDNAs may affect the distance between P and the 15S rRNA cistron. Although this cannot be ruled out, we consider it unlikely because this segment of the map has been found to be free of major insertions in the mtDNAs studied thusfar.

2. The paromomycin resistance markers used by us and in Monash are known to have a different origin and might represent different loci.

As a third explanation, the possibility of a mapping error cannot be excluded. The mapping of genetic markers, using petites, is based on certain assumptions that we have specified previously (4). These assumptions can only be verified by fine-structure mapping of an overlapping set of petite mtDNAs on the wild-type mtDNA to rigorously exclude the possibility that all petites used share a common internal deletion or rearrangement in their DNA. This has only been done up till now by Heyting for the region of 21S rRNA, but not for the P region by us or by Choo et al.

Paromomycin resistance is known in bacteria to result from an alteration in the small subunit of the ribosome. There is circumstantial evidence that also in yeast the resistance is due to an alteration in the ribosome (Grivell, L.A., unpublished). If there is a P locus overlapping the 15S rRNA cistron, paromomycin resistance could be simply explained by an alteration in the base sequence of 15S rRNA. However, if a P locus exists at the position deduced by DiFranco et al. (13), mutations in this locus must affect the sensitivity to paromomycin in a different way. Three possibilities can be mentioned:

1. P is the structural gene for the single protein of the mitochondrial ribosome that is made within the mitochondrion itself (40). This appears to be ruled out by recent work of Groot et al. (50) who have shown that this protein co-migrates in sodium dodecylsulphate gels with VAR-1, a product of mitochondrial protein synthesis, specified by a gene that maps between E and OLI-1 (18).

2. P codes for an enzyme that modifies the RNA or a protein of the small subunit of the mitochondrial ribosome. The 15S RNA only contains a single modified nucleotide (41) making the former alternative unlikely. A problem with both alternatives is that there is no clearly detectable transcript of the fragment which harbours P (Fig. 7).

3. P codes for F-met-tRNA. Since the main action of paromomycin is inhibition of the start of translation, a modified F-met-tRNA might conceivably mitigate inhibition. Martin and Rabinowitz (51) have shown that their met-$tRNA_{II}$ maps close to P and from the results of Martin et al. (52) we infer that this is the formylatable met-$tRNA_F$, presumably used in the initiation of mitochondrial protein synthesis. To verify the position of this tRNA on the restriction map, Moorman and Van Ommen (unpublished result) have loaded mitochondrial tRNA with (^{35}S)-methionine and hybridized the resulting (^{35}S)met-tRNA with the HindII+III fragments of mtDNA (JS1-3D). Only fragments TT1 (H1a)

and TT4 (H3) hybridized and not DT5 (H4b) known to contain the P locus (see Fig. 7). One could argue, however, that a genetic marker that lies close to a restriction site might be mapped on the wrong side of this site, especially if the gene is only 80 nucleotides long, as is the case for a tRNA. In the case of P, the two petite mtDNAs used by DiFranco et al. (13) have a repeat length of >500 bp and if P were excentrically located in both repeats, it is possible that the repeats would be found to hybridize with DT5 and that the small hybridization with TT4 would be missed. Although admittedly far-fetched, this possibility should be checked by a more precise mapping of the relative positions of P and the met-$tRNA_{II}$ gene, as the other present alternative for a function of a P locus lying on DT5 is not attractive either.

The hybridization experiments with met-tRNA also provide new information on the relative order of markers in the P region. Bolotin-Fukuhara et al. (53) and Martin and Rabinowitz (51) present this order (clock-wise on the map going from C to OLI-2, cf. Fig. 5) as met-$tRNA_I$, OXI-1, (val-tRNA, met-$tRNA_{II}$, OXI-2), pro-tRNA, trp-tRNA, P. Van Ommen et al. (23) have found 2-3 4S RNA genes in the P to OLI-2 region and they have pointed out that the mapping results of Fukuhara and Rabinowitz and co-workers do not exclude that some of the tRNA genes linked to P and the 15S rRNA gene are actually located in the P to OLI-2 span. This is now confirmed for met-$tRNA_{II}$ and it could also hold for pro-tRNA and trp-tRNA, if our mapping of P is correct.

12. The OXI-3 region

The interesting situation in the OXI-3 region is summarized in Fig. 6. Mutations in this locus cover a region which is at least 8800 bp in strain JS1-3D (or D273-10B) and even 2600 bp larger in strain KL14-4A, because insert II lies within this region. There is circumstantial evidence (10) that OXI-3 is the structural gene for subunit I of cytochrome oxidase, which has a molecular weight of approximately 40 000, equivalent to 1200 bp of DNA. We are presently testing two explanations to account for this large discrepancy between locus size and peptide specified:

i) The locus codes for two proteins, both involved in oxidase synthesis. There is evidence (see section 11) that subunits II and III of cytochrome oxidase are specified by the OXI-1 and 2 loci, respectively, and the second protein in the OXI-3 region would, therefore, have to be a protein which modifies subunit I.

ii) The OXI-3 region is transcribed into a giant pro-mRNA that has to be processed to be read. Mutations in the section of the RNA that does not contain the structural gene could affect processing and prevent translation (a variant of this explanation is that the long RNA has to fold properly for reading). This hypothesis provides an explanation for the unexpected finding that in this region at least three large, discrete transcripts are found, one of which appears to be over 3000 nucleotides long (Fig. 7). It would also explain why messenger activity for cytochrome oxidase is not limited to the three size classes expected from the known sizes of the three large peptides of cytochrome oxidase that are made on mitochondrial ribosomes. If internal processing of the pro-mRNA would occur (see section 10), segments of the structural gene could even be far apart on the genome. A precedent for such an odd arrangement is provided by the recent discovery (Jeffreys, A.J. and Flavell, R.A., personal communication) that the structural gene for the β subunit of rabbit haemoglobin contains a 600 bp insert in the middle of the peptide-specifying sequence.

13. Genes in the E to OLI-2 span

The E to OLI-2 span of the map contains a multiplicity of genetic markers that can be divided into three groups: VAR-1, ATPase mutants and $\underline{bc}_1$ mutants.

1. VAR-1 is defined as a polymorphic protein made on mitochondrial ribosomes (18). This protein is probably identical with Groot's ribosome-associated protein (see section 11) and it maps between E and OLI-1 (Fig. 5C). The only large transcript

in this region maps on fragment dDRT10 (Fig. 7) and this may be the location of the structural gene for this protein.

2. Mutants that affect the ATPase complex include mit$^-$ mutants PHO-1 and 2 (see ref. 19) and three oligomycin resistance loci (see ref. 54). Mutations in OLI-1 (O_I, class B mutants), affect the aggregation behaviour of subunit 9 (the proteolipid with a molecular weight of 7500) (54,55). It is, therefore, likely that OLI-1 is the structural gene for this subunit. A second oligomycin locus (OLI-3, class A mutant) is closely linked to OLI-1 and Groot Obbink et al. (54) have obtained evidence that mutations in this locus affect ATPase subunit 6. Mutations in PHO-1 and 2 and OLI-2 have not been linked with alterations in specific ATPase subunits, two more of which are thought to be made on mitochondrial ribosomes in yeast (Table I). Our working hypothesis is that the 6S ATPase mRNA specifies the small proteolipid subunit, whereas the 12S ATPase mRNA could correspond to OLI-3 (or OLI-3 + OLI-1). The amino acid sequence of the small subunit 9 has been determined (Sebald, W., personal communication). Since OLI-1 has been localized on a 1400 bp HapII fragment (13), it should now be possible by sequence analysis to locate the structural gene for subunit 9 and to identify and sequence the transcripts corresponding to this gene.

3. This group includes two loci for antimycin resistance, A_I and A_{II} (56), one for funiculosin resistance (17,56), two for diuron resistance (57), two for mucidin resistance (58) and two classes of mit$^-$ mutants COB-1 and 2 (19). This multiplicity of genetic markers contrasts with the finding that only one of the seven polypeptides of the bc_1 complex is made on mitochondrial ribosomes, the cytochrome b subunit (59). The resistance to various antibiotics may affect the same gene, e.g. one class of funiculosin-resistant mutants maps at the A_{II} locus (56) and mutation to diuron resistance also affects antimycin resistance (57). Most of the antibiotic-resistant mutants tested are closely linked to the COB loci and could be in a single structural gene for cytochrome b. It is disturbing, however, that the

recombination frequency between COB-1 and 2 is so high (>10%; see ref. 19) and that a mitochondrial antimycin-resistant mutant has been isolated that shows no linkage to COB in recombination studies (17). This would require very high rates of recombination within a single b gene. Similar explanations for this apparent discrepancy between the (genetic) size of the gene and the size of the polypeptide specified can be invoked here as we have done already for the OXI-3 region. With cytochrome b the additional possibility exists, however, that the two similar b peptides observed by Weiss et al. (39) in Neurospora, are derived from different genes rather than from one gene.

14. AT-rich segments in yeast mtDNA and the Bernardi model for genes and spacers

With a mole percent G+C of 18%, yeast mtDNA is one of the most AT-rich functional DNAs in nature. From early nearest-neighbour analyses it was already clear that the G's and C's of yeast mtDNA are highly clustered. Detailed analysis of the distribution of AT-rich and GC-rich segments in this DNA by Bernardi and co-workers has led to the model shown in Fig. 10.

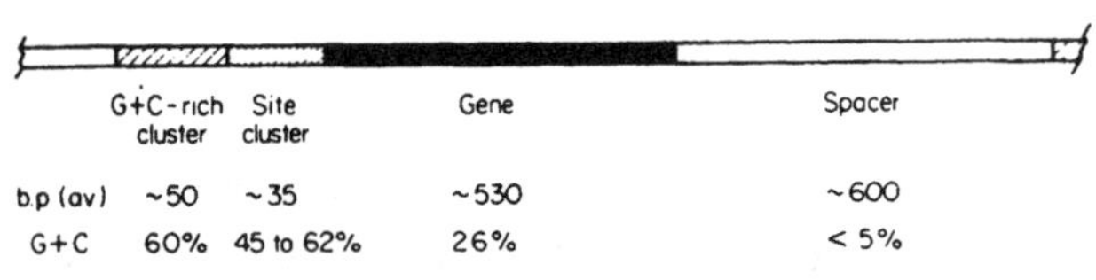

Fig. 10. The Bernardi model for the sequence organization of yeast mtDNA. From ref. 60.

Bernardi and co-workers (2,3,60) have identified four sequence elements:

a) AT-rich stretches (GC <5%) with an average molecular weight of 4 x 10^5; these make up about 50% of the DNA and are called "spacers".

b) Intermediate GC-rich stretches (GC varying from 22% to 65%) of about the same size as the AT-rich stretches and called "genes"; these also take up about 50% of the DNA.

c) Site clusters; these are short elements containing several recognition sites for the restriction endonucleases HaeIII (GGCC) and HpaII (CCGG).

d) GC-rich clusters; these are also short and characterized by the absence of any of the tetranucleotides GGCC, CCGG and GCGC.

There are about 70 copies of each of the four sequence elements and Bernardi has suggested that these may be arranged as shown in Fig. 10. The alternation of "genes" and "spacers" is supported by several lines of evidence, the positioning of the other two elements - that are thought to have some regulatory function (e.g. promotor) - is more arbitrary. Of the five implications, derived by Bernardi from this model, we quote three verbatim (60):

1. Transcription of the mitochondrial genome of yeast yields monocistronic messages; in agreement with this idea, there is no evidence for spacer transcripts in wild-type yeast mitochondria.

2. The mitochondrial genome of yeast shows an eukaryotic and not a prokaryotic type of organization in being an interspersed system of repetitive sequences and genes transcribed monocistronically.

3. Recombination phenomena taking advantage of the high level of sequence homology in the spacers and in the regulatory segments can be expected.

Our experiments bear on two of these predictions. First, it is clear from Fig. 7 that the first prediction derived from Bernardi's model does not fit our data. There are several large transcripts that should span more than one gene-spacer unit and Van Ommen et al. (23) have shown that not every tRNA gene is preceded by a HapII site. Second, our experiments do not support the conclusion that yeast mtDNA shows an eukaryotic type of organization in being an interspersed system of repetitive sequences and genes, transcribed monocistronically. Hollenberg et al. (1) showed already in 1969 that yeast mtDNA renatures as a single species with a complexity commensurate with its size. Although anomalies may arise when renaturation is studied with long DNA rather than the sonicated DNA used by Hollenberg et al., no evidence was ever found for reiterated sequences in DNA-DNA renaturation experiments. Since reiterated sequences in nuclear DNA are defined as sequences showing a higher rate of renaturation than single-copy DNA, yeast mtDNA does not contain repetitive sequences in the usual sense of the words. Of course, the statistical chance of finding an ATA trimer in a DNA segment consisting of 100% AT, is 8 times greater than in a segment containing 50% AT, but there is no evidence that the various AT-rich segments in mtDNA have more homology than expected on a statistical basis. This is borne out by the results with petite mutant RD1A presented in the next section.

Although the "eukaryotic, repetitive" character of yeast mtDNA is unproven in our opinion, the underlying observations of Bernardi and co-workers remain undisputed and require an explanation. The three main unusual features are the large fraction of very AT-rich sequences, the multiple rearrangements and the apparent large excess of DNA over that required to specify known genes. Two explanations can be considered:

1. The AT-rich sequences are "spacers" that are not transcribed and that provide quasi-homologous sequences for unequal crossing-over events. This is the explanation advanced by Prunell et al. (2) and they have proposed that this is used as a means for rapid genome evolution, because it allows duplication, translocation, inversion and deletion of genes without affecting the nucleotide sequence of the gene itself. Such processes are generally thought to play an important role in the evolution of eukaryotic nuclear DNA. A major problem with this interpretation is that there is no evidence that mtDNA is rapidly evolving or effectively adapting to new demands as far as its gene products are concerned. Even the gene order is apparently strictly conserved. Under such circumstances faulty recombination in spacer sequences should be a nuisance rather than an advantage and one would expect the DNA to get rid of

its spacers.

2. Control of gene expression is not (only) at the level of transcription, but (also) at the level of RNA processing. Processing occurs at AT-rich sites in the RNA, i.e. transcripts of the "spacers" of Bernardi. In this concept the insertions and deletions found in mtDNA are the unfortunate consequence of a processing system requiring AU-rich segments which may undergo faulty recombination, leading either to small insertions and deletions which do not interfere with processing, or to formation of petites.

The reason why such a system has evolved could be in the unusual ability of the yeast cell to shut off mitochondrial biogenesis, including the biosynthesis of some - but not all - mitochondrial proteins. In HeLa cell mitochondria, transcription of both DNA strands is complete and the differential control of the steady state concentration of individual RNA species is, therefore, exclusively at the processing level. There is circumstantial evidence mentioned in section 9 that more than two promotors are present on yeast mtDNA, but transcription control might still be limited. In this situation an increased demand for regulation might only be met by increased sophistication of the processing, requiring increasing segments of non-conserved RNA. Since the eukaryotic nucleus appears to rely heavily on RNA processing in regulating gene expression and since the enzymes required for processing the mtRNAs are probably coded in nuclear DNA, one could argue that the nucleus will choose for more elaborate processing if it has to provide a more complex regulation of gene expression in mitochondria, as required in glucose repression or anaerobiosis. In this sense, the yeast mitochondrial genetic system could still be considered a mini-replica of the eukaryotic nuclear system, although for other reasons than Bernardi has brought forward.

It is clear that hypothesis (2) can be critically tested by the analysis of RNA synthesis and processing, made possible by the map of mtDNA now available. These experiments are in progress.

15. Petite mutant RD1A and the AT-rich segments in yeast mtDNA

RD1A is a so-called "low-density" petite mutant, *i.e.* a mutant that contains mtDNA with an unusually low buoyant density in CsCl. This mtDNA has been extensively studied in Amsterdam, since Hollenberg isolated RD1A in 1968. It consists for more than 95% of perfect tandem repeats of a 68-nucleotide sequence in head-to-tail arrangement (61,62). The unusual base sequence of this repeat is shown in Fig. 11. The DNA contains only 3 mole percent G+C and the two GC base pairs per repeat are adjacent, the remainder of the repeat consisting of a run of 66 consecutive AT pairs. Since the repeat hybridizes with wild-type mtDNA (63,64), it represents an AT-rich segment of mtDNA that can be simply obtained in large amounts. Mol (64) and Van Kreijl and Bos (unpublished experiments) have used this petite mtDNA to study the divergence and evolutionary stability of the AT-rich segments in wild-type mtDNA.

The first question to be answered is whether the 68-nucleotide segment is a perfect copy of a corresponding segment of the mtDNA in the wild-type *S. cerevisiae* strain H1 from which RD1A is derived. Using DNA-cRNA hybridization we have found that the T_m for the heteroduplex of H1 mtDNA and cRNA of RD1A mtDNA is identical to the T_m of the homoduplex of RD1A mtDNA and cRNA of one repeat length. This suggests that the RD1A sequence is present as single repeat(s) in the wild-type mtDNA and that no copying errors have been made during amplification of the wild-type segment. Unfortunately, the sensitivity of the melting analysis is limited with respect to differences in chain length and if we set the accuracy of our T_m analysis at $\pm$ 1°C a mismatch up to 10 bp at one end (or 5 bp at both ends) would not be observed.

Next we have hybridized cRNA of RD1A mtDNA with the mtDNAs from *S. carlsbergensis* and the *S. cerevisiae* strains JS1-3D and KL14-4A, which show considerable differences in restriction fragment patterns with strain H1 mtDNA and among each other (Fig. 1 and ref. 6). In all cases positive hybridization was observed and the T_m's of the heteroduplex were identical either

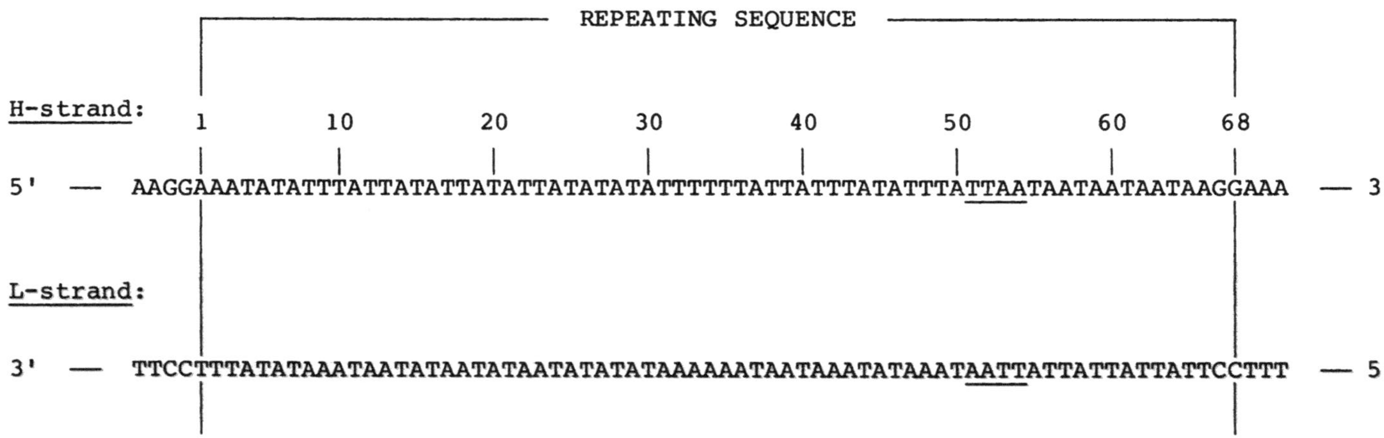

Fig. 11. The nucleotide sequence of one repeat of the repetitive mtDNA from petite mutant RD1A. From ref. 61.

with or without subsequent ribonuclease treatment. Similar results have been obtained in DNA-DNA hybridization experiments with mtDNA of S. cerevisiae strain IL-46 (64). This proves that this very AT-rich segment is faithfully conserved in yeast mtDNA.

Finally, we have studied the number of copies of the RD1A repeat in the wild-type mtDNA. Quantitative hybridization analysis showed that there are less than four copies of the RD1A sequence present in the wild-type DNA (63). In mapping experiments using wild-type mtDNAs of KL14-4A and JS1-3D, only one restriction fragment (TT1) hybridizes with cRNA of RD1A mtDNA. Moreover, on KL14-4A mtDNA the repeat is present in the smaller segment shared by TT1 and RR4 (RT6 at 134-205‰) which comprises about 7% of the wild-type genome (see Fig. 1). This suggests that a single copy of the RD1A repeat is present at a unique position in the mtDNA.

Our results provide evidence for the individuality and conservation of at least one AT-rich segment in yeast mtDNA. The function of this segment is unknown. There are 11S and 20S transcripts that hybridize with the same fragment as RD1A cRNA (see Fig. 7), but these transcripts cover only part of the DNA fragment and we do not know whether they contain the RD1A sequence. The H-strand transcript of RD1A mtDNA has stop codons in all three reading frames, the L-strand transcript only in one and the remaining reading frames could code for a protein segment containing only a very limited set of mostly hydrophobic amino acids. It does not seem likely, therefore, that the RD1A sequence is part of a structural gene for a mitochondrial protein. More precise mapping of the sequences around the repeat in the wild-type mtDNA should give more information on this point and also on the process that leads to formation of petite mtDNA.

An interesting aspect of RD1A mtDNA is that it contains small amounts of an unusual pyrimidine of unknown structure (61, 64). Although this has not yet been detected in wild-type mtDNA, the fact that the enzymes to make it are present in mitochondria suggests that it should have an essential function, possibly in mtDNA replication - the process that is still running smoothly in petites.

16. Concluding remarks

Although absorbing and often intellectually pleasing, the construction of physical maps of DNA tends to become an end in itself, a disease of young adults transmitted by middle-aged professors. It is, therefore, pleasing to see that in our case the mapping is yielding interesting information about the biology of mtDNA in several areas:

1. Mapping has confirmed that small rearrangements occur all over the map and established that nonetheless gene order is strictly conserved. By mapping we have identified five large inserts; one of these is the ω^+ allele, other large inserts lie within the OXI-3 region. In one case at least, a large insert represents a new sequence not present elsewhere in mtDNA.

2. Mapping has precisely defined the rho^+ sequences retained in rho^- mtDNA. This information is essential for the use of rho^- mtDNAs as amplified, purified segments of rho^+ mtDNA for marker localization and sequence analysis. It will also be helpful in elucidating the mechanism of mtDNA recombination and the generation of petite mutants.

3. Mapping has provided precise physical distances between known markers on mtDNA, allowing a correlation between genetic markers, transcripts and proteins. Although this correlation is only tentative in most cases, the methods are now available to precisely map genes for major proteins and the transcripts used for making these proteins. In this way the regions where control sequences in DNA are located will soon be known.

Mapping has also identified "empty" spaces on mtDNA where no major transcripts are found. The function of these spaces is unclear. They might be "spacers", they might be transcribed into non-conserved transcripts or they might contain structural genes for regulatory proteins or proteins involved in mitochondrial protein synthesis. If such a protein is present at low concentration, it would not be detected as product of mitochondrial protein synthesis, nor would its mRNA be found in hybridization experiments. We have outlined a biochemical (mapping) approach for the detection of genes for minor proteins elsewhere (65). Temperature-sensitive mutants will also be helpful.

4. Mapping has alerted us to the possibility that RNA processing could be of major importance in yeast mitochondria, that some of the mutants known might affect RNA processing rather than the structural gene sequence and that internal processing might join gene segments that are not adjacent in the genome at the RNA level.

5. Finally, the physical map is an essential step in the sequence analysis of strategic segments of mtDNA. This analysis is indispensable for a detailed understanding of the function of mtDNA and it should rapidly answer some of the more pressing questions raised in this review.

Acknowledgements

We are indebted to numerous students and colleagues (specified in other papers from this laboratory) for experimental results, advice, yeast strains, mutants, restriction endonucleases, antigens and antibodies. This work was supported in part by a grant from The Netherlands Foundation for Chemical Research (SON) with financial aid from The Netherlands Organization for the Advancement of Pure Research (ZWO).

References

1 Hollenberg, C.P., Borst, P. and Van Bruggen, E.F.J. (1970) Mitochondrial DNA. V. A 25-μ closed circular duplex DNA molecule in wild-type mitochondria. Structure and genetic complexity. Biochim.Biophys.Acta 209, 1-15.

2 Prunell, A., Kopecka, H., Strauss, F. and Bernardi, G. (1977) The mitochondrial genome of wild-type yeast cells. V. Genome evolution. J.Mol. Biol. 110, 17-52.

3 Prunell, A. and Bernardi, G. (1977) The mitochondrial genome of wild-type yeast cells. VI. Genome organization. J.Mol.Biol. 110, 53-74.

4 Sanders, J.P.M., Heyting, C., DiFranco, A., Borst, P. and Slonimski, P.P. (1976) The organization of genes in yeast mitochondrial DNA. In: The Genetic Function of Mitochondrial DNA (Saccone, C. and Kroon, A.M., Eds), North-Holland, Amsterdam, pp. 259-272.

5 Sanders, J.P.M., Heyting, C. and Borst, P. (1976) The variability of the mitochondrial genome of Saccharomyces strains. In: Genetics and Biogenesis of Chloroplasts and Mitochondria (Bücher, Th., Neupert, W., Sebald, W. and Werner, S., Eds), North-Holland, Amsterdam, pp. 511-517.

6 Sanders, J.P.M., Heyting, C., Verbeet, M.Ph., Meijlink, F.C.P.W. and Borst, P. (1977) The organization of genes in yeast mitochondrial DNA. III. Comparison of the physical maps of the mitochondrial DNAs from three wild-type Saccharomyces strains. Mol.Gen.Gen., submitted.

7 Smith, H.O. and Nathans, D. (1973) A suggested nomenclature for bacterial host modification and restriction systems and their enzymes. J.Mol. Biol. 81, 419-423.

8 Sanders, J.P.M. and Borst, P. (1977) The organization of genes in yeast mitochondrial DNA. IV. Analysis of (dA.dT) clusters in yeast mitochondrial DNA by poly(U)-Sephadex chromatography. Mol.Gen.Gen., submitted.

9 Morimoto, R., Lewin, A. and Rabinowitz, M. (1977) Restriction cleavage map of mtDNA from the yeast Saccharomyces cerevisiae. Nucl.Acids Res. 4, 2331-2361.

10 Grivell, L.A. and Moorman, A.F.M. (1977) A structural analysis of the oxi-3 region on yeast mtDNA. This volume.

11 Borst, P., Heyting, C. and Sanders, J.P.M. (1976) The control of mitochondrial DNA synthesis in yeast petite mutants. In: idem ref. 5, pp. 525-533.

12 Heyting, C. and Sanders, J.P.M. (1976) The physical mapping of some genetic markers in the 21S ribosomal region of the mitochondrial DNA of yeast. In: idem ref. 4, pp. 273-280.

13 DiFranco, A., Sanders, J.P.M., Heyting, C., Borst, P. and Slonimski, P.P. (1976) Restriction enzyme analysis and physical mapping of mitochondrial DNA from petite mutants, carrying a genetic marker for oligomycin or paromomycin resistance. In: idem ref. 4, pp. 291-304.

14 Mason, T., Boerner, P. and Biron, C. (1976) The use of double mutant strains containing both heat- and cold-sensitive mutations in studies of mitochondrial biogenesis. In: idem ref. 5, pp. 239-246.

15 Morimoto, R., Lewin, A., Merten, S. and Rabinowitz, M. (1976) Restriction endonuclease mapping and analysis of grande and mutant yeast mitochondrial DNA. In: idem ref. 5, pp. 519-524.

16 Choo, K.-B., Nagley, P., Lukins, H.B. and Linnane, A.W. (1977) Biogenesis of mitochondria. 47. Mol.Gen.Gen. 153, 279-288.

17 Lang, B., Burger, G., Bandlow, W., Kaudewitz, F. and Schweyen, R.J. (1976) Antimycin- and funiculosin-resistant mutants in Saccharomyces cerevisiae: New markers on the mitochondrial DNA. In: idem ref. 5, pp. 461-465.

18 Douglas, M.C., Strausberg, R.L., Perlman, P.S. and Butow, R.A. (1976) Genetic analysis of mitochondrial polymorphic proteins in yeast. In: idem ref. 5, pp. 435-442.

19 Tzagoloff, A., Foury, F. and Akai, A. (1976) Genetic determination of mitochondrial cytochrome b. In: idem ref. 5, pp. 419-426.

20 Rabinowitz, M., Jakovcic, S., Martin, N., Hendler, F., Halbreich, A., Lewin, A. and Morimoto, R. (1976) Transcription and organization of yeast mitochondrial DNA. In: idem ref. 4, pp. 219-230.

21 Van Ommen, G.J.B. and Groot, G.S.P. (1977) A transcription map of yeast mitochondrial DNA. This volume.

22 Crick, F.H.C. (1966) Codon-anticodon pairing: The wobble hypothesis. J.Mol.Biol. 19, 548-555.

23 Van Ommen, G.J.B., Groot, G.S.P. and Borst, P. (1977) Fine-structure physical mapping of 4S RNA genes on mitochondrial DNA of Saccharomyces cerevisiae. Mol.Gen.Gen., in press.

24 Schatz, G. and Mason, T.L. (1974) The biosynthesis of mitochondrial proteins. Ann.Rev.Biochem. 43, 51-87.

25 Borst, P., Sanders, J.P.M. and Heyting, C. (1976) Genes on Saccharomyces mitochondrial DNA. In: Genetics, Biogenesis and Bioenergetics of Mitochondria (Bandlow, W., Schweyen, R.J., Thomas, D.Y., Wolf, K. and Kaudewitz, F., Eds), De Gruyter, Berlin, pp. 85-97.

26 Jackl, G. and Sebald, W. (1975) Identification of two products of mitochondrial protein synthesis associated with mitochondrial adenosine triphosphate from Neurospora crassa. Europ.J.Biochem. 54, 97-106.

27 Moorman, A.F.M., Grivell, L.A., Katan, M.B., Van Ommen, G.J.B. and Meiland, C.M. (1976) Translation of mitochondrial messenger RNAs in vitro as an approach to the identification of gene products of yeast mitochondrial DNA. In: idem ref. 25, pp. 241-258.

28 Moorman, A.F.M. and Grivell, L.A. (1976) Gene identification by coupled transcription-translation of yeast mitochondrial DNA in vitro. In: idem ref. 5, pp. 539-546.

29 Moorman, A.F.M., Verkley, F.N., Asselbergs, F.A.M. and Grivell, L.A. (1977) Yeast mitochondrial messenger RNA is not correctly translated in heterologous cell-free systems. This volume.

30 Moorman, A.F.M., Grivell, L.A., Lamie, F. and Smits, H.L. (1977) Identification of mitochondrial gene products by DNA-directed protein synthesis in vitro. Biochim.Biophys.Acta, submitted.

31 Attardi, G., Amalric, F., Ching, E., Costantino, P., Gelfand, R. and Lynch, D. (1976) Information content and gene mapping of mitochondrial DNA from HeLa cells. In: idem ref. 4, pp. 37-46.

32 Groot, G.S.P., Flavell, R.A., Van Ommen, G.J.B. and Grivell, L.A. (1974) Yeast mitochondrial RNA does not contain poly(A). Nature 252, 167-169.

33 Hendler, F., Halbreich, A., Jakovcic, S., Patzer, J., Merten, S. and Rabinowitz, M. (1976) Characterization and translation of yeast mitochondrial RNA. In: idem ref. 5, pp. 679-684.

34 Faye, G., Kujawa, C. and Fukuhara, H. (1974) Physical and genetic organization of petite and grande yeast mitochondrial DNA. J.Mol.Biol. 88, 185-203.

35 Faye, G. (1976) Mitochondrial ribosomal subunits from Saccharomyces cerevisiae rho⁻ mutants. FEBS Letters 69, 167-170.

36 Hendler, F.J., Padmanaban, G., Patzer, J., Ryan, R. and Rabinowitz, M. (1975) Yeast mitochondrial RNA contains a short polyadenylic acid segment. Nature 258, 357-359.

37 Padmanaban, G., Hendler, F., Patzer, J., Ryan, R. and Rabinowitz, M. (1975) Translation of RNA that contains polyadenylate from yeast mitochondria in an Escherichia coli ribosomal system. Proc.Natl.Acad.Sci.U.S. 72, 4293-4297.

38 Buse, G. and Steffens, G.J. (1976) Analysis of the polypeptide chains of cytochrome oxidase from beef heart mitochondria. In: idem ref. 5, pp. 189-194.

39 Weiss, H., Ziganke, B. and Kolb, H.J. (1976) Subunit structure and biogenesis of cytochrome b from Neurospora crassa mitochondria. In: idem ref. 25, pp. 289-302.

40 Groot, G.S.P. (1974) The biosynthesis of mitochondrial ribosomes in Saccharomyces cerevisiae. In: The Biogenesis of Mitochondria (Kroon, A.M. and Saccone, C., Eds), Academic Press, New York, pp. 443-452.

41 Klootwijk, J., Grivell, L.A. and Klein, I. (1975) Minimal post-transcriptional modification of yeast mitochondrial ribosomal RNA. J.Mol.Biol. 97, 337-350.

42 Grivell, L.A. and Reijnders, L. (1974) Yeast mitochondrial ribosomes. Postępy Mikrobiologii 13, 67-80.

43 Sambrook, J. (1977) Adenovirus amazes at Cold Spring Harbor. Nature 268, 101-104.
44 Dujon, B., Bolotin-Fukuhara, M., Coen, D., Deutsch, J., Netter, P., Slonimski, P.P. and Weill, L. (1976) Mitochondrial genetics. Mol.Gen. Gen. 143, 131-165.
45 Perlman, P.S. and Birky Jr., C.W. (1974) Mitochondrial genetics in bakers' yeast: A molecular mechanism for recombinational polarity and suppressiveness. Proc.Natl.Acad.Sci.U.S. 71, 4612-4616.
46 Cabral, F., Solioz, M., Rudin, Y., Schatz, G., Clavilier, L. and Slonimski, P.P. (1977) Identification of the structural gene for yeast cytochrome c oxidase subunit II on mitochondrial DNA. J.Biol.Chem., submitted.
47 Bandlow, W., Metzke, R., Klein, A., Kotzias, K., Doxiadis, I., Bechmann, H., Schweyen, R.J. and Kaudewitz, F. (1977) Macromolecular synthesis and energy level in a mitochondrial conditional yeast mutant, tsm-8. Europ. J.Biochem. 76, 373-382.
48 Williamson, D.H., Maroudas, N.G. and Wilkie, D. (1971) Induction of the cytoplasmic petite mutation in Saccharomyces cerevisiae by the antibacterial antibiotics erythromycin and chloramphenicol. Mol.Gen.Gen. 111, 209-223.
49 Faye, G., Kujawa, C., Fukuhara, H. and Rabinowitz, M. (1976) Mapping of the mitochondrial 16S ribosomal RNA gene and its expression in the cytoplasmic petite mutants of Saccharomyces cerevisiae. Biochem.Biophys.Res. Commun. 68, 476-482.
50 Groot, G.S.P., Grivell, L.A., Van Harten-Loosbroek, N., Kreike, J., Moorman, A.F.M. and Van Ommen, G.J.B. (1977) The role of the mitochondrial genetic system in the biogenesis of the mitochondrial inner membrane. In: Structure and Function of Energy-Transducing Membranes (Van Dam, K. and Van Gelder, B.F., Eds), North-Holland, Amsterdam, in press.
51 Martin, N.C. and Rabinowitz, M. (1976) Transfer RNAs of yeast mitochondria. In: idem ref. 5, pp. 749-754.
52 Martin, R., Schneller, J.M., Stahl, A.J.C. and Dirheimer, G. (1976) Isoacceptor tRNA species in yeast mitochondria - methionine and formylmethionine specific tRNAs coded by mitochondrial DNA. In: idem ref. 5, pp. 755-758.
53 Bolotin-Fukuhara, M., Faye, G. and Fukuhara, H. (1977) Temperature-sensitive respiratory-deficient mitochondrial mutations: Isolation and genetic mapping. Mol.Gen.Gen. 152, 295-305.
54 Groot Obbink, D.J., Hall, R.M., Linnane, A.W., Lukins, H.B., Monk, B.C., Spithill, T.W. and Trembath, M.K. (1976) Mitochondrial genes involved in the determination of mitochondrial membrane proteins. In: idem ref. 4, pp. 163-173.
55 Tzagoloff, A., Akai, A. and Foury, F. (1976) Assembly of the mitochondrial membrane system. XVI. Modified form of the ATPase proteolipid in oligomycin-resistant mutants of Saccharomyces cerevisiae. FEBS Letters 65, 391-395.
56 Pratje, E. and Michaelis, G. (1976) Two mitochondrial antimycin A resistance loci in Saccharomyces cerevisiae. In: idem ref. 5, pp. 467-471.
57 Colson, A.M., The Van, L., Convent, B., Briquet, M. and Goffeau, A. (1977) Mitochondrial heredity of resistance to 3-(3,4-dichlorophenyl)-1, 1-dimethylurea, an inhibitor of cytochrome b oxidation, in Saccharomyces cerevisiae. Europ.J.Biochem. 74, 521-526.
58 Šubík, J., Kováčová, V. and Takácsová, G. (1977) Mucidin resistance in

yeast. Europ.J.Biochem. 73, 275-286.

59 Katan, M.B., Van Harten-Loosbroek, N. and Groot, G.S.P. (1976) The cytochrome bc_1 complex of yeast mitochondria. II. Site of translation of the polypeptides in vivo. Europ.J.Biochem. 70, 409-417.

60 Bernardi, G., Prunell, A., Fonty, G., Kopecka, H. and Strauss, F. (1976) The mitochondrial genome of yeast: Organization, evolution and the petite mutation. In: idem ref. 4, pp. 185-198.

61 Van Kreijl, C.F., Borst, P., Flavell, R.A. and Hollenberg, C.P. (1972) Pyrimidine tract analysis of mtDNA of a "low-density" petite mutant of yeast. Biochim.Biophys.Acta 277, 61-70.

62 Van Kreijl, C.F. and Bos, J.L. (1977) The repeating nucleotide sequence in the repetitive mitochondrial DNA from a "low-density" petite mutant of yeast. Nucl.Acids Res. 4, 2369-2388.

63 Sanders, J.P.M., Flavell, R.A., Borst, P. and Mol, J.N.M. (1973) Nature of the base sequence conserved in the mitochondrial DNA of a low-density petite. Biochim.Biophys.Acta 312, 441-457.

64 Mol, J.N.M. (1976) Ph.D. Thesis, The Analysis of (dA.dT) Clusters in DNAs, Krips-Repro, Meppel.

65 Borst, P., Sanders, J.P.M. and Heyting, C. (1978) Biochemical methods to locate genes on the physical map of yeast mitochondrial DNA. In: Methods in Enzymology (Fleischer, S. and Packer, L., Eds), Academic Press, New York, in press.

66 Faye, G., Bolotin-Fukuhara, M. and Fukuhara, H. (1976) Mitochondrial mutations that affect mitochondrial transfer ribonucleic acid in Saccharomyces cerevisiae. In: idem ref. 5, pp. 547-555.

PHYSICAL CHARACTERIZATION OF THE DIFFERENCE BETWEEN YEAST mtDNA ALLELES ω^+ AND ω^-.

C. JACQ, C. KUJAWA, C. GRANDCHAMP and P. NETTER
Centre de Génétique Moléculaire du C.N.R.S. Gif-sur-Yvette, 91190 - FRANCE.

INTRODUCTION

The molecular mechanisms responsible for genetic recombination in yeast mitochondria are poorly understood. However, a peculiar phenomenon of recombination has been very well described (1, 2, 3, 12), principally in our lab on the yeast Saccharomyces cerevisiae. This phenomenon is the polarity of recombination specified by the locus ω and affecting the three loci RI, RII, RIII also called RibI, RibII and RibIII (14). In a first attempt to tackle the molecular aspect of this phenomenon (4) the genetic recombination properties of mtDNA molecules were correlated with the structure of the DNA itself as revealed by the melting block analysis. In this article we will present a more straightforward analysis of this problem. We will first establish a precise localization of the genetic loci RI, RII and RIII and we will then compare the physical structure of mtDNA of strains carrying the mitochondrial ω^+ allele to that of strains carrying the ω^- allele, using the restriction enzyme analysis technique. It is now clear that the physical localization of genetic markers can be carried out using the petite methodology (7, 11, 15). The characterization of the difference between the ω^+ and ω^- alleles is more delicate. Since an ω^+ strain has not yet been obtained from an ω^- strain and reciprocally, it is clear that a straightforward establishment of the difference is not easy to reach. In other terms, the fact that it is not possible to compare two strains differing only by their ω allele will raise major objections concerning the interpretation of the results. Our plan to overcome this difficulty will be to establish a physical difference between an ω^+ and an ω^- strain and then to test the correlation between this difference and the ω type of a strain, in as many cases as possible.

DERIVATION OF STRAINS

The petite strains studied were derived from three grande strains. The construction of IL8 ($\rho^+\omega^+C^R_{321}\ E^R_{514}$) has already been described and results from the cross : 55R5-3C ($\rho^+\omega^+C^R_{321}$) x DP1-1B ($\rho^+\ \omega^-E^R_{514}$) (2).

The two grande strains KL263 (ω^+) and KL479 (ω^-) have been constructed in order to be as similar as possible except for their ω allele. KL263 ($\rho^+\omega^+C^R_{324}E^R_{553}S^R_{352}$) has been isolated in the cross : DP1-1B/E^R_{553} x 55R5-3C/S^R_{352} and KL479 ($\rho^+\omega^-$ $C^R_{324}E^R_{553}S^R_{352}$) has been obtained from the cross : KL263-2B x 55R5-3C/S^R_{352}.

The following table presents the properties of the different petite clones used in this paper.

Name of petite clone	Abreviated name	ω	RI	RII	RIII	Origin
IL8-8C/R53	R53	+	C^R_{321}	E^S	E^R_{514}	(2)
IL8-8C/R53/S3	S3	O	O	O	E^R_{514}	(10)
IL8-8C/E41/S1	S1	O	O	O	E^R_{514}	(10)
KL263-2B/A12	A12	+	C^R_{324}	E^R_{553}	S^R_{352}	this work
KL263-2B/B922/B8	B8	+	C^R_{324}	E^R_{553}	O	"
KL263-2B/C23	C23	O	C^R_{324}	E^R_{553}	O	"
KL263-2B/B922/CO	CO	O	C^R_{324}	E^R_{553}	O	"
KL479-2B/C51	C51	-	C^R_{324}	E^R_{553}	S^R_{352}	"
KL479-2B/A72	A72	O	O	O	S^R_{352}	"
KL479-2B/C721	C721	-	C^R_{324}	E^R_{553}	S^R_{352}	"
KL479-2B/D41	D41	-	C^R_{324}	O	S^R_{352}	"

Table 1

All the petite clones were tested for their genetic stability and were found to have retained their original drug resistance markers in at least 80 % of the cells constituting the culture from which the mtDNA was extracted. Determination of the ω allele of clones was carried out by the triploïd methodology using an ω^n tester and the full details of this procedure will be published later.

RESTRICTION MAP OF THE ω^+ SEQUENCE

Clones listed in Table 1 represent a selection of the strains actually examined. We did not take into account petite clones whose repeat unit was too long or those whose structure was complex (i.e. due to double deletion, partial amplification..). Having carried out this severe selection we have used the classical methodology to establish the restriction map. One must notice that some difficulties due to the known random degradation of mtDNA (11) were easily overcome because of the very small size (few thousands of base pairs, fig. 1) of repeating units. However, because of the nicks and gaps present in mtDNA we dit not use the end labelling technique to order the restriction fragments. In our case, double digestions, redigestions of eluted fragments and partial digestions have proved to be sufficient to construct a restriction map from fragments generated by Hae III, Alu I, Hind III and Hha. Such a map is represented in fig. 2. We will not present here all the arguments to establish the map, this will be published later. The general outlines of our strategy were the following ones. First we selected very deleted petite clones for the RIII region on one hand, and for the RI/RII region on the other hand. From these very short repeat units it was easy to order the restriction fragments. We had already done this for RIII region (10) and we could confirm these early results with the petite clone A72. Considering the RI/RII region, the petite clones B8, C23 and CO were very useful. The mtDNA repeat unit of CO was slightly shorter than that of C23 which, in turn, was shorter

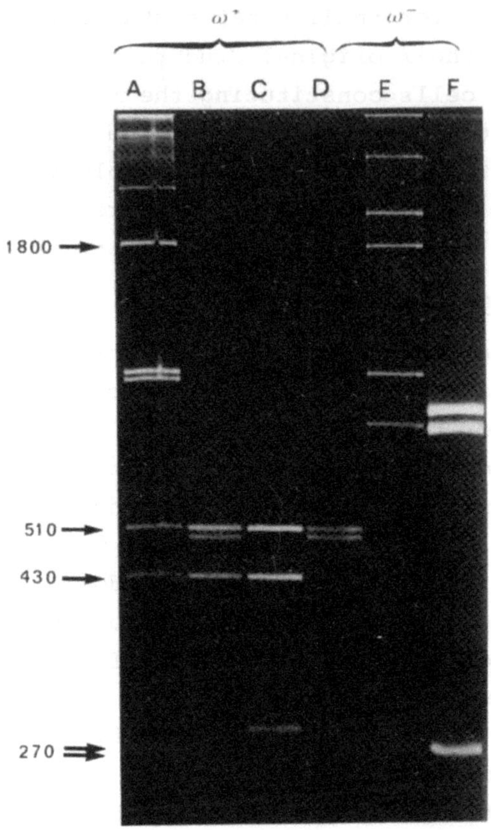

Fig. 1 : Polyacrylamide gel electrophoresis of Alu I digests of different petite mtDNAs.

All the techniques : purification of mtDNAs, conditions of enzyme action, electrophoresis and elution from acrylamide gel have already been described (9) (10). The mtDNAs digest presented in the columns A,B,C and D are purified from petite subclones derived from two ω^+ grande strains and have kept the following loci : (A). R53 (RI,RII,RIII)-(B) B8(RI,RII)-(C) C23(RI,RII)-(D) CO(RI, RII).
The other petite clones come from an ω^- grande strain : (E) C721 (RI,RII,RIII)-(F) D41(RI,RIII).

than that from B8. This gradual deletion was very convenient to link the restriction fragments together. The complete restriction map of this region was then obtained by studying the structure of mtDNA from R53 which had kept the loci RI, RII and RIII. For that purpose a particular feature of the sequence was very useful : a large part of the sequence overlapping the two regions appeared to be devoid of Hae III sites, this large Hae III fragment could be eluted from petites R53 and A12 and analyzed with other enzymes.

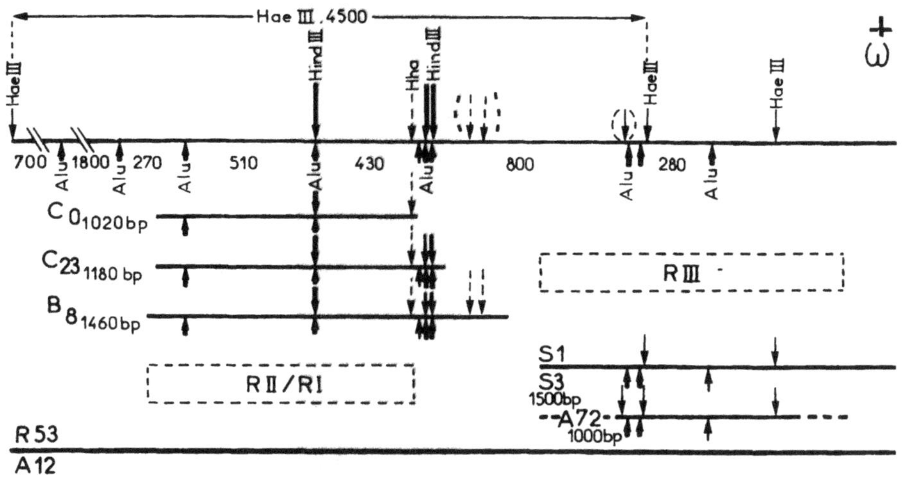

Fig. 2 : Restriction map of the ω^+ sequence.

The upper line represents the part of the rho^+ mtDNA whose map has been deduced from the restriction endonuclease analysis of the different petite mtDNAs. All the petite mtDNAs analyzed, except S1, appeared to be arranged in a tandem way.
The arrows between brackets indicate sites -for Hha or for Hae III- which have only been found in some petites depending on their origin from non-isomitochondrial "grandes" (see Table 1). CO, C23 and B8 have been used to map the region RI/RII ; these loci being necessarily localized in the sequence CO. Similarly RIII is necessarily localized within S1, S3 and cannot be situated beyond the boundaries of A72 symbolized by broken lines.

RESTRICTION MAP OF THE ω^- SEQUENCE

Having established the ω^+ sequence, the ω^- sequence was easy to understand (fig. 3). Fig. 1 represents the difference between the two alleles ω^+ and ω^-. The two Alu I fragments, 510 and 430 base pairs long, which had been shown to be closely linked to RI/RII region in the ω^+ sequence (fig. 2), appeared to be systematically absent in the ω^- petites we examined. A quantitative estimation of this difference was obtained by characterizing the large Hae III fragment from the ω^- sequence. After elution and redigestion this fragment from the petite clone C721 (fig. 6), was shown to be deleted

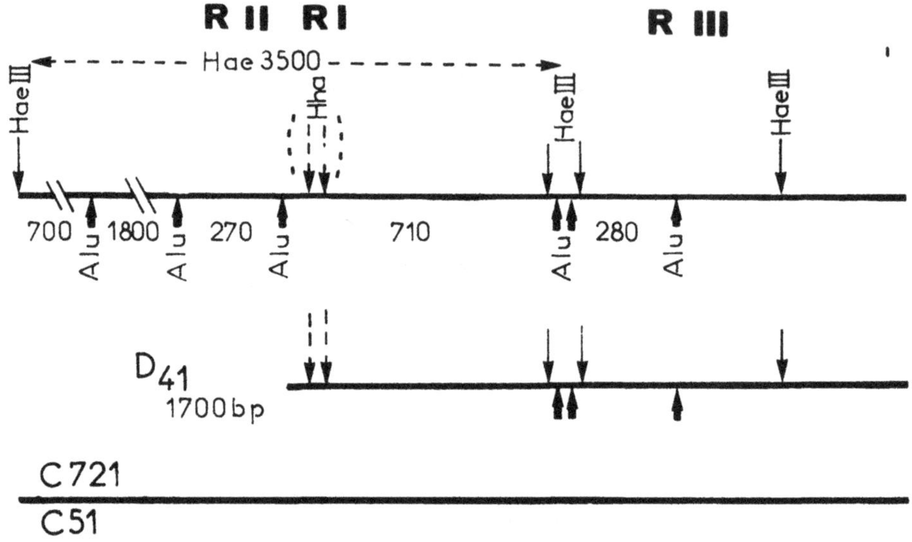

Fig. 3 : Restriction map of the ω^- sequence.

Three petite strains have been used to establish the ω^- sequence : C721 (RI, RII, RIII), C51 (RI, RII, RIII) and D41 (RI, RIII).

of 1000 base pairs comparatively to ω^+ sequence. Furthermore the Alu digest of this fragment allowed us to localize precisely this difference, which henceforth will be referred to as Δ, on the restriction map (fig. 4). An immediate consequence of this result is that it is possible to localize very accurately the loci RII and RI. These loci are not in the sequence Δ because this sequence is absent in the ω^- strains which carry them. On the other hand RI and RII have to be located in the sequence CO (fig. 2), they are thus necessarily localized in the 200 base pairs at the left end of the sequence CO (compare fig. 2 and fig. 4).
Another argument would allow us to order RI and RII on this segment. One of the petites examined, D41, was deleted for

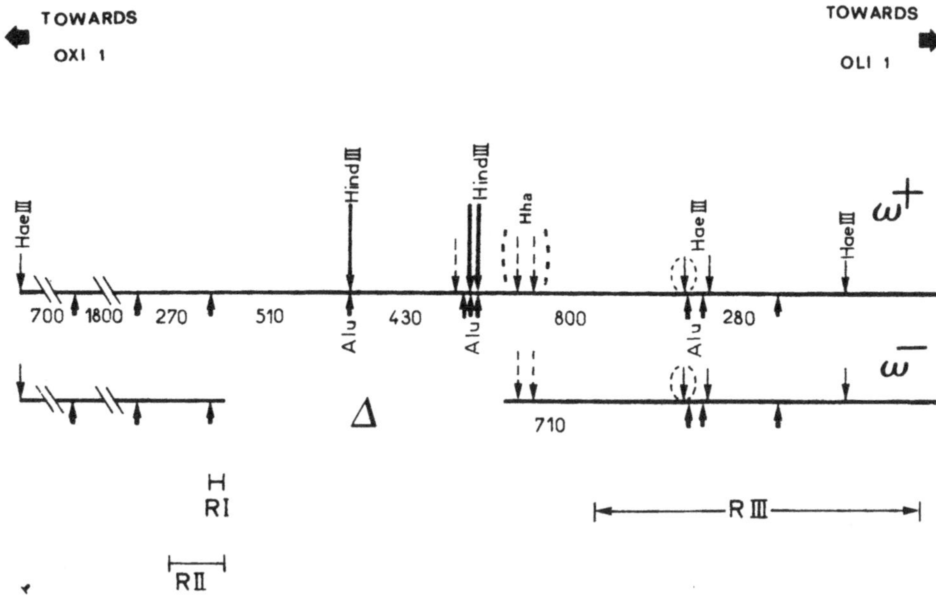

Fig. 4 : Comparison of the two sequences associated with the alleles ω^+ and ω^-.

The two restriction maps previously established are compared. The gap in the ω^- sequence was drawn to allow us to align the corresponding homologous sequence outside the Δ region. One has to keep in mind that instead of this gap there is a junction between the two regions. This junction being represented by a sequence which should not exist in the ω^+ sequence. The sequence called Δ is the segment specifically present in the ω^+ sequence. From this map the physical distance between the loci RI/RII and RIII should be comprised between 300 and 1500 base pairs in the ω^- strains and between 1300 and 2500 base pairs in the ω^+ strains.

RII and had kept RI and RIII ; its possible localization is shown (fig. 3). If this result is correct, it would mean that RI would be localized on the very edge of the sequence Δ . This point will be discussed later.

HETERODUPLEX ANALYSIS

In order to verify the restriction map we attempted an heteroduplex analysis of the difference between the two ω^+

and ω^- alleles. The large Hae III fragment overlapping the sequence Δ appeared to be a good candidate for such a study. Fig. 5 presents one of the molecules obtained after hybridization of an Hae III fragment 4500 purified from the ω^+ petite R53 and an Hae III fragment 3500 obtained from the ω^- petite C721. About fifteen such molecules were observed. The values of the different parameters evaluated by electron microscopy can be compared to those previously established by restriction mapping studies (Table 2).

	A	B(Δ)	C
Electron microscopy	0.86 μ	0.39 μ	0.26 μ
Electron microscopy (1 $\mu \equiv$ 3000 b.p.)	2580 b.p.	1170 b.p.	780 b.p.
Restriction analysis	2780 b.p.	1000 b.p.	710 b.p.

Table 2

Taking into account the low accuracy of the two methods and the usual discrepancy found between evaluations by electron microscopy and restriction analysis, one can see that the two sets of values are in very good agreement. These results confirm our restriction map of this region and establish the sequence homology outside the Δ segment.

IS THE DIFFERENCE BETWEEN ω^+ AND ω^- DIRECTLY DETECTABLE ON THE DIGESTION PATTERN OF GRANDE mtDNA ?

One must keep in mind that above results were obtained using mtDNAs only from petite clones. We have thus to wonder how faithfully the sequence established reflects the sequence of the grande mtDNA. In our case the fact that several independently obtained mutants share a common sequence is

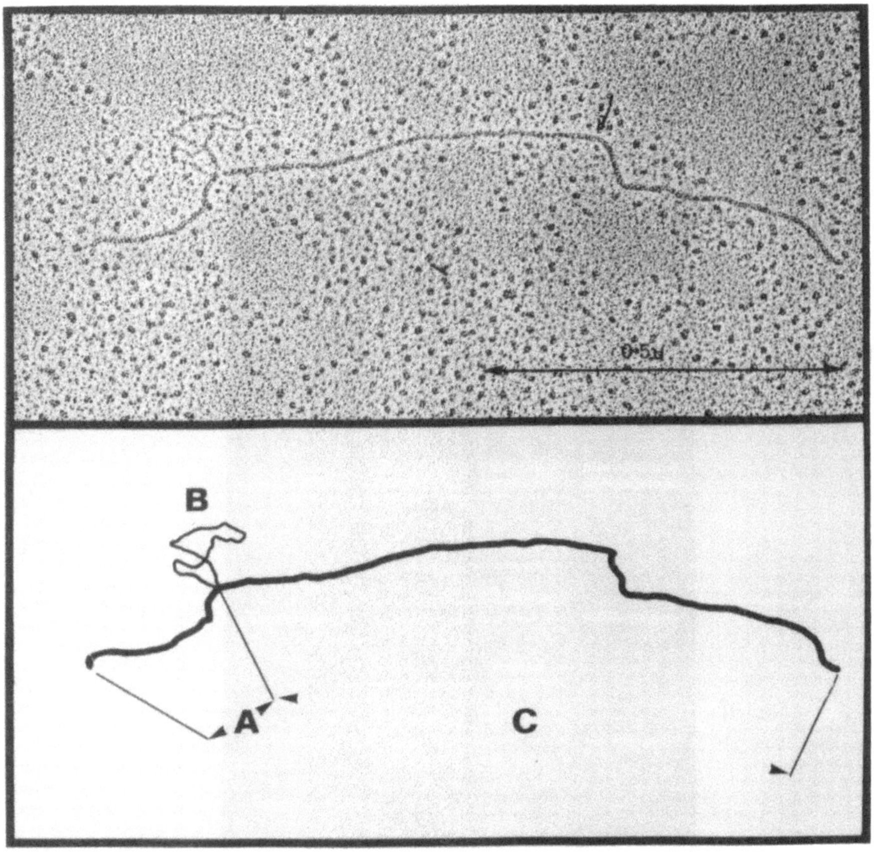

Fig. 5 : Heteroduplex between the large Hae III restriction fragments ω^+ and ω^-.

Equal amounts of purified restriction fragments : Hae 4500 (fig. 2) from R53 (ω^+) and Hae 3500 (fig. 3) from C721 (ω^-) were mixed, denaturated by heating at 80° C for 30' in NaCl 20 mM EDTA 5 mM and renatured in 50 % formamide pH 8.5 at room temperature for 1 h 30 at a concentration of about 1 ug/ml. About 15 such heteroduplexes were obtained and measured (see text). Portion B corresponds to the sequence Δ and does not show any secondary structure.

an argument in favor of the reliability of the results. In order to check these results one can look for the presence of the large Hae III fragment, overlapping the sequence Δ , in the Hae III digests of grande mtDNA. In the fig. 6 the

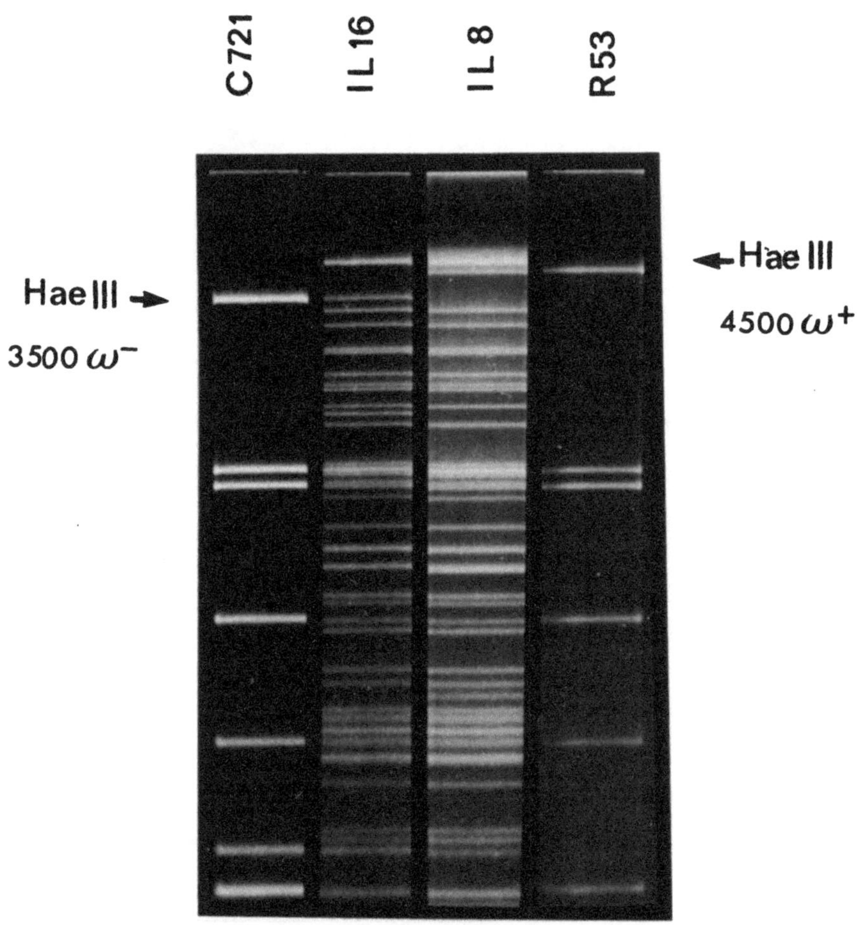

Fig. 6 : Polyacrylamide gel electrophoresis of Hae III digests of different mtDNAs.

The Hae III fragments associated with the ω^+ and ω^- alleles are the larger fragments of the Hae III digests of mtDNA respectively from R53 and C721. The grande strain IL16 shows the expected fragment for the ω^- sequence but not that of the ω^+ sequence and the opposite result is obtained for IL8 strain which is ω^+.

longer Hae III fragment of the petite is clearly apparent in the pattern of the grande IL8. Similary, IL16 (ρ^+ ω^-) which is related (see above) to KL479-2B/C721 (ρ^- ω^-) (2) share in common the Hae III fragment 3500 characteristic of the ω^- sequence. Moreover no Hae III fragment 4500 can be seen on the IL16 pattern and no fragment corresponding to the ω^- sequence is visible in the IL8 digest. In other terms it was possible to detect the difference between ω^+ and ω^- directly on Hae III digest of grande mtDNA. We thus had the possibility of extending this observation to other grande strains either ω^+ or ω^-. We have examined seven grande mtDNAs whose ω allele had been characterized in the Gif laboratory and, as far as we know, are all from independent origins and are from different laboratories (D6 from Dr. D. Wilkie, K10 Dr. A. Bevan, M12 from Dr. F. Kaudewitz, 706 from Dr. Gunge and N123 originated from Dr. R. Mortimer and was handed to us by Dr. E. Moustacchi). We asked the question : do they possess the Hae III fragment characteristic of their ω allele ? The results of this inquiry are shown in Table 3.

	Strains	Presence of Hae III 4500	Presence of Hae III 3500	Presence of Hae III 4000
ω^+ strains	IL8-8C or KL263-2B	+	-	-
	D6	+	-	-
	K10	+	-	-
ω^- strains	IL16 or KL479-2B	-	+	-
	M12	-	+	-
	706	-	-	+
	N123	-	-	+

Table 3 : Presence of the characteristic Hae III fragment in the digestion patterns of grande mtDNA.
IL8-8C and KL263-2B have been put together because they both derived from DP1-1B, similary IL16 and KL479-2B have kept ω^- sequence of 55R5-3C.

It turned out that all the ω^+ strains of different origins we looked at, had an Hae fragment corresponding in size to the insertion of the sequence Δ. For the ω^- strains two classes of results were found. One class (IL16, M12) possesses the expected Hae fragment but the other class (706, N123) had none of these two fragments and had a new fragment (4000 b.p. long). In order to examine this new class we made petite clones from the grande strain 706. As expected these petite clones, which had been selected to keep the ω region, had the Hae fragment 4000. The redigestion of this fragment by Alu showed us that the pattern was similar to the other ω^- patterns except for one fragment which had been shown to be located at the left end of the fragment Hae 3500. In other terms, this new class of ω^- sequence did not differ from the first one around the Δ region but had only lost an Hae III site on the left end of the large Hae fragment.

In conclusion, this difference could be seen in three unrelated ω^+ strains and four unrelated ω^- strains. The consistency of results between yeast strains interbred in the Gif laboratory with those obtained from other laboratories confirms the observation that the difference between the ω^+ and the ω^- alleles can be correlated with the sequence Δ .

DISCUSSION

ORDER OF THE FOUR LOCI RI, RII, RIII, ω .

One of the clearest results of this study is that it leads to an unambigous order for the loci RI, RII, RIII and ω. From the physical mapping studies, ω- or the sequence Δ related to ω^+- appears to be localized between the groups RI/RII and RIII. Moreover consistency requires that the order be RII RI ω RIII, this deduction is however based on the structure of a single petite (D41). Nevertheless if we take into account the results on the polarity of recombination (1,2,6) which demonstrate unambiguously that the origin of polarity is within the ω locus, or intimately linked to that locus, the unescapable conclusion is that the order must be RII, RI, ω , RIII. This order receives additional support from the fact that ω^n mutations are frequently associated with mutations at the RI locus (5).This result seems to be in disagreement with the previous genetic studies of Netter et al. which proposed the order ω-RI-RII-RIII, however, as it has already been pointed out by others (6), these genetic results are not incompatible with the order RII-RI-ω-RIII. Furthermore, this order has also been suggested by other recombination studies (8) and has been proposed in a preliminary fashion after physical analysis (7). It is worthy of notice that such an order for these loci involves a bidirectional mechanism for the conversion process (6).

POSITION OF THE 23S rRNA STRUCTURAL GENE RELATIVELY TO THE GENETIC MARKERS.

If the locus RIII is likely to be in the 23S rRNA gene, the question has still to be answered for the loci RI and RII. This point is all the more exciting now that the inclusion of RI and RII in the 23S rRNA gene would mean that the Δ sequence is inserted within the ribosomal gene of the ω^+ strains. Such a situation cannot be ruled out at this point since a similar case has been shown to occur in some of the 28S rRNA genes of Drosophila melanogaster (13,16). However, in this last example the genes carrying the insertion have not been shown to be transcribed : in mitochondria the 23S rRNA gene is transcribed. Our opinion today is that if RI and RII are located in the large RNA gene, this structural gene does not stretch for more than 270 base pairs to the right of the Δ sequence, this has been suggested by preliminary hybridization studies on restriction fragments. Nevertheless it is clear that more experiments should be carried out.

STRUCTURE AND FUNCTION OF Δ .

We have clearly established in this work that the ω^+ allele is associated with the presence in yeast mtDNA of the sequence Δ whereas the ω^- allele is deleted for this sequence. Δ is a well defined segment of DNA which is 1000 ± 100 base pairs long. A striking feature of this sequence is the cluster of restriction sites at its right end (fig. 4). Three Alu I sites are present on a segment which is 150 base pairs long, such a high frequency of restriction sites might be due to a palindromic structure of this part of Δ .

Other insertions/deletions have already been shown to occur in yeast mtDNA (15) from different strains. However, these insertions/deletions which are clustered pratically at the opposite of Δ on mtDNA genome, are heterogeneous in size, ranging from 500 up to 3500 base pairs.

The function of Δ is far from being clear. From this work we can say that the cross between a strain which has Δ and a strain which is deleted for Δ leads to a high polarity of recombination between markers surrounding Δ . In the conversion model (6,) which involves heteroduplex formation between ω^+ and ω^- sequences the dissymetrical initiation of conversion may be due either to the specific presence of the Δ sequence or simply to the fact that any insertion at this point creates a specific structure recognizable by enzymes. The sequence generated by the deletion of Δ in ω^- strains (or the insertion point of Δ), may be as important for this process as the Δ sequence itself. In this respect the study of mutants at the mitochondrial locus ω is very promising. Only one type of such mutants is available so far. With these ω^n mutants, derived from ω^- strains, the ω^+/ω^n or ω^-/ω^n pairings do not specifically initiate the non reciprocal act (5). Moreover it has been suggested that these mutations are deletions near the RI locus. It would be informative to know wether these deletions, if any, affect the insertion point in the ω^- sequence. This work is now in progress.

ACKNOWLEDGEMENTS

The authors are indebted to Prof. P.P. Slonimski for his continued interest and support. We also thank B. Dujon and F. Michel for helpful discussionsand R. Lang for critical reading of this manuscript. Last not least, we are grateful to E. Petrochilo for many generous discussions. This work was supported by a grant D.G.R.S.T. n° 75-70748.

REFERENCES

1. BOLOTIN, M., COEN, D., DEUTSCH, J., DUJON, B., NETTER, P., PETROCHILO, E., SLONIMSKI, P.P. (1971) Bull. Inst. Pasteur, 69, 215-239.

2. COEN, D., DEUTSCH, J., NETTER, P., PETROCHILO, E. and SLONIMSKI, P.P., (1970) Symp. Soc. Experiment. Biol., 24, 449-496 London Cambridge University Press.

3. DUJON, B., SLONIMSKI, P.P., WEILL, L. (1974) Genetics, 78, 415-437.

4. DUJON, B. and MICHEL, F. (1976) in "The Genetic Function of Mitochondrial DNA", Eds C. SACCONE and A.M. KROON. Elsevier/ North Holland Biomedical Press, Amsterdam, pp. 175-184.

5. DUJON, B., BOLOTIN-FUKUHARA, M., COEN, D., DEUTSCH, J., NETTER, P., SLONIMSKI, P.P., WEILL, L. (1976) Mol. Gen. Genet., 143, 131-165.

6. DUJON, B. (1976) PhD thesis, Université Pierre et Marie Curie, Paris.

7. HEYTING, C. and SANDERS, J.M.P. (1976) in "The Genetic Function of Mitochondrial DNA" Eds C. SACCONE and A.M. KROON Elsevier/North Holland Biomedical Press, Amsterdam, pp. 273-280.

8. HOWELL, N., HALL, R.M., LINNANE, A.W., LUKINS, H.B. (1974) J. Bacteriol., 119, 1063-1065.

9. LAZOWSKA, J. and SLONIMSKI, P.P. (1976) Mol. Gen. Genet. 146, 61-78.

10. LAZOWSKA, J., JACQ, C., CEBRAT, S., SLONIMSKI, P.P. (1976) In "The Genetic Function of Mitochondrial DNA" Eds C. SACCONE and A.M. KROON. Elsevier/North Holland, Biomedical Press, Amsterdam, pp. 325-335.

11. MORIMOTO, R., LEWIN, A., RABINOWITZ, M. (1977) Nucleic Acid Res., 4, 2331-2351.

12. NETTER, P., PETROCHILO, E., SLONIMSKI, P.P., BOLOTIN-FUKUHARA, M., COEN, D., DEUTSCH, J., DUJON, B. (1974) Genetics, 78, 1063-1100.

13. PELLEGRINI, M., MANNING, J., DAVIDSON, N. (1977) Cell, 10, 213-224.

14. PLISCHKE, M.E., VON BORSTEL, R.C., MORTIMER, R.K., COHN, W.E. in The Third Edition of Handbook of Biochemistry and Molecular Biology, edited by G.D. FASMA.

15. SANDERS, J.P.M., HEYTING, C., DI FRANCO, A., BORST, P. and SLONIMSKI, P.P. (1976) in "The Genetic Function of Mitochondrial DNA" Eds C. SACCONE and A.M. KROON. Elsevier/North Holland, Biomedical Press, Amsterdam, pp. 259-271.

16. WHITE, L.R. and HOGNESS, D.S. (1977) Cell, 10, 177-192.

Physical Mapping of Mitochondrial Genes and Transcripts in *Saccharomyces cerevisiae*

Alfred Lewin, Rick Morimoto, Sylvie Merten, Nancy Martin, Patricia Berg, Thomas Christianson, David Levens, and Murray Rabinowitz
Departments of Biology, Medicine and Biochemistry and The Franklin McLean Memorial Research Institute, University of Chicago, Chicago, Illinois 60637 USA

Our laboratory has been interested in the physical mapping of mitochondrial genes and transcripts using the restriction map we have prepared for the mt-DNA from Saccharomyces cerevisiae(1). Specifically, we have used restriction maps generated by eight restriction endonucleases for the mtDNA from several grande strains, and a variety of analytic techniques to map ant^R, mit^-, and syn^- markers as well as the mitochondrial tRNA and rRNA genes. First, we have characterized a series of cytoplasmic petites by restriction analysis as retaining single uninterrupted segments of grande mtDNA and have used the maps of these single deletion petites to localize the genetic markers they contain. Second, we have directly localized the sequences encoding 14S and 21S ribosomal RNA and tRNAs by hybridization of ^{125}I labeled RNA to restriction fragments. These data have been confirmed by electron microscopic measurement of the spacing of ribosomal RNA R-loops. Finally, we have transcribed ^3H-labeled cRNA from mtDNA of petites with single genetic markers to provide hybridization probes for restriction fragments containing these antibiotic resistance markers. Mapping by petite deletion analysis (2,3,4,5) and by RNA hybridization to restriction fragments (6,7) has also been successfully applied in other laboratories, and differences in both methodology and results will be discussed later. These mapping studies have not only helped us understand the architecture of the mitochondrial genome in both petite and grande strains, but have also provided impetus for new inquiries concerning the transcription of mitochondrial RNA. These investigations have employed, on one hand, the isolation of an RNA polymerase complex from mitochondria, and the cloning of restriction fragments of mtDNA on the other. These subjects will be treated at the end of this presentation.

The mitochondrial DNA from cytoplasmic petite mutants of yeast has been shown to contain variable sized deletions with stable retention and amplification of the retained sequences by renaturation kinetics (8,9), DNA-DNA and RNA-DNA hybridization (10,11,12,13), genetic examination of resistance markers (14,15), electron microscopic analysis (16,17), and by restriction analysis (18,19,20). Because these deletions can be extensive and can occur in any region of the mitochondrial genome, petites have been useful for mapping antibiotic resistance markers which are unlinked by recombination analysis (2,3,4). Further, deletion mapping using collections of petites characterized with respect to the ant^R markers has then been used to determine the arrangement of tRNA (21,22,23), and rRNA genes (24,25), and of mit^- and syn^- loci by recombination (26,27,28). To be rigorous deletion mapping must employ petites whose mtDNA consists of single deletions of homogeneous sequence composition. However, abnormal marker retention patterns (4) and tRNA hybridization patterns (21) indicate the presence of internal deletions in some petite clones with stable marker transmission. Also, analysis of mtDNA by restriction endonucleases (18,19) and by electron microscopy (16,17) reveals internal deletions and sequence heterogeneity in some petites which appear to be single deletions on the basis of marker retention and tRNA hybridization. Consequently we have selected only those genetically marked petite strains which are single deletions to employ in our deletion mapping.

Using the restriction fragment maps that we have determined for the mtDNA of the grande strains, we can analyze the restriction pattern of petite mtDNAs to detect internal deletions and sequence diversity (Figure 1). The grande map is set into 100 units with the single Sal I site as the origin. Since a deletion of grande mtDNA results in the loss of some restriction fragments together with the appearance of a "new" fragment comprised of the fusion of the sequences bordering the deletion, in each digest single deletions are indicated by the loss of some fragments and the presence of only a single fragments which does not comigrate with any grande fragment, while multiple deletions display more than one "new" fragment. Since in certain cases "new" bands may migrate similarly to grande bands, or multiple deletions can occur between the same two restriction sites, we have employed several enzymes and enzyme combinations in order to discriminate between multiple and single deletion petites.

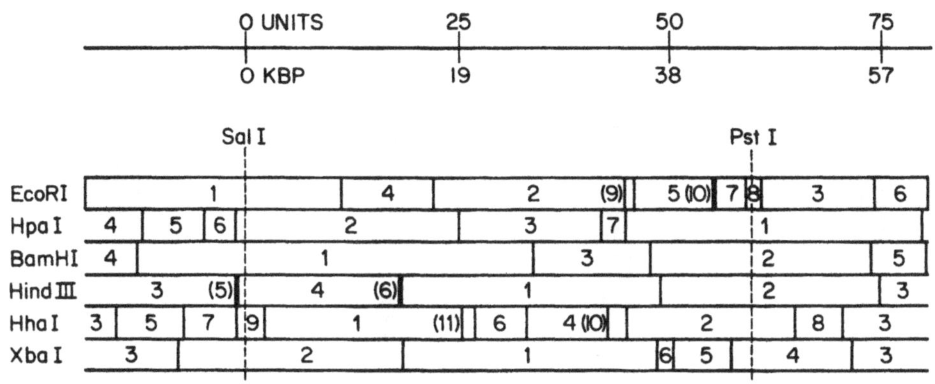

figure 1. Restriction endonuclease map of grande strain MH41-7B mtDNA.

Further, since the resolution of this technique for detecting internal deletions depends on the number of restriction sites recognized, we have employed enzymes such as Hae III, Hpa II and Alu I, each of which cleaves the mtDNA of our grande stains more than 80 times. These endonucleases are particularly useful for characterizing those petites which retain only one restriction site with the other restriction activities. Figure 2 shows the mtDNA of petite 0_I0_{II} cleaved by an enzyme used for mapping, Eco Rl, and by Hpa II . In both cases most of the DNA comigrates with grande fragments and only one differently migrating fragment is formed. Petite P27, in contrast, (figure 3), shows two extra bands both with mapping endonucleases (Bam HI/Hpa I) and with Hae III. Careful examination of these gels also reveals substoichiometric fragments which are caused by sequence heterogeneity of the mitochondrial DNA.

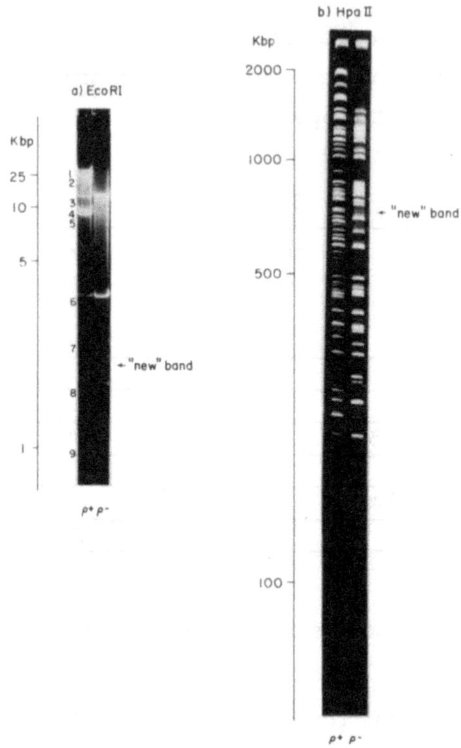

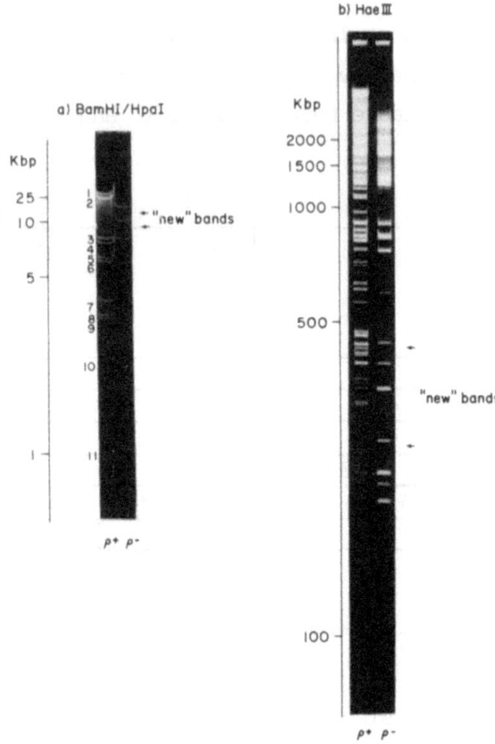

figure 2. Restriction digests of mtDNA from grande and single deletion petite O_IO_{II}. Fragments in the petite digest not comigrating with grande fragments are indicated with arrows.

figure 3. Restriction digests of mtDNA from grande and double deletion petite P_{27}.

Another criterion used to establish that a petite is derived by single deletion is the comparison of petite restriction fragments with the grande fragment map (figure 4). Petites which are single deletions show only one gap relative to the fragment map of grande mtDNA. Multiply deleted petites, analogously, show internal deletions when aligned with the grande restriction map. The restriction patterns of a few petites, however, cannot be arranged with the grande map in a colinear fashion, and are thus difficult to diagram. We have used hybridization of cRNA made to these petites to determine their origin on the grande genome.

Comparison of the restriction maps of different single deletion petite strains that retain or lose markers enables deletion mapping of ant^R, mit^-

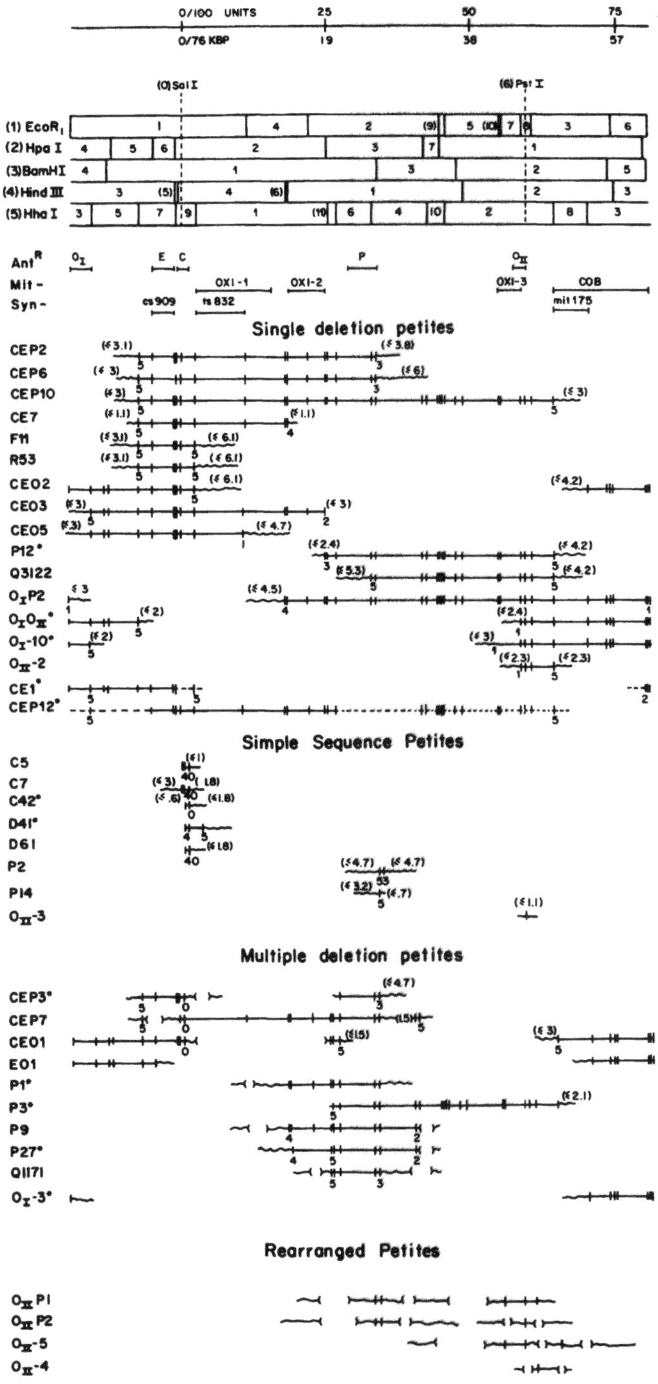

figure 4. Linear fragment maps of petite mt-DNAs aligned with the grande restriction map. Cross-hatching lines indicate restriction sites, with endonucleases producing the terminal restriction site in petites designated in a subscript. Extension of the "new" fragment is indicated by wavy lines; the number in parentheses indicating maximal possible length in kb. Map positions of ant^R, mit$^-$ and syn$^-$ markers are shown between the grande and petite fragment maps. Asterisks near strain names indicate sequence heterogeneity.

and syn⁻ mutations. The mapping of O_I^R by this technique is illustrated by the restriction maps of a few petites that transmit the O_I marker (Figure 5). Petite O_IP2 mtDNA extends less than 3.8 map units (1 map unit = 1% of the 76 kb mtDNA) past the Eco R1 site at 80.8 units on the restriction map. CEO_I-3 extends less than 3.8 units from the Hha I site at 84.6

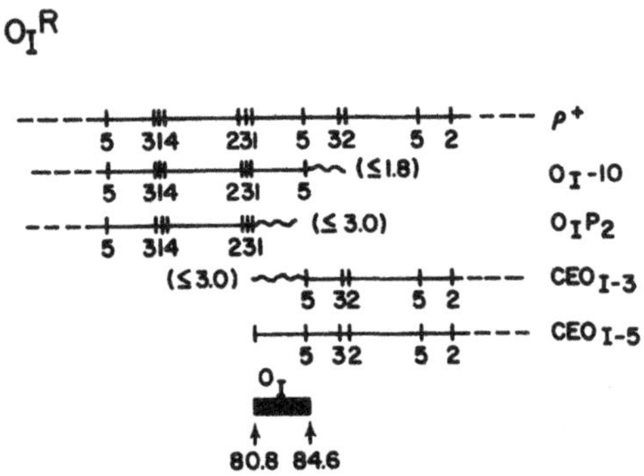

figure 5. Detail of deletion mapping of the O_I resistance marker.

units. The O_I locus is thus placed between 80.8 and 84.6 on the map, a region of 3 kb in length. Maps of other O_I containing petites (O_I-10; CEO_I-5) are consistent with this localization but do not make it more precise. Using similar comparisons of petite restriction patterns, we have localized the O_I, O_{II}, P, C and E antibiotic resistance markers, the OXI-1, OXI-2, OXI-3, and COB mit⁻²⁶ loci, and the syn⁻ mutations CS909, ts832 and mit170 (Bolotin-Fukuhara, unpublished data).

We have also combined hybridization and the petite restriction analysis to more precisely locate antibiotic resistance markers.For these experiments we have selected petites with single genetic markers that retain only short segments of the grande mitochondrial DNA. These are then used as templates for the synthesis of cRNA using labeled RNA precursors and E. coli RNA polymerase. This cRNA probe is then hybridized to grande restriction fragments immobilized on nitrocellulose filters, in order to localize the ant^R markers transmitted by the petites to segments of the physical map. Figure 6 demonstrates the hybridization patterns for cRNA from several single ant^R petite strains. These data, plus information on the size of the petite gen-

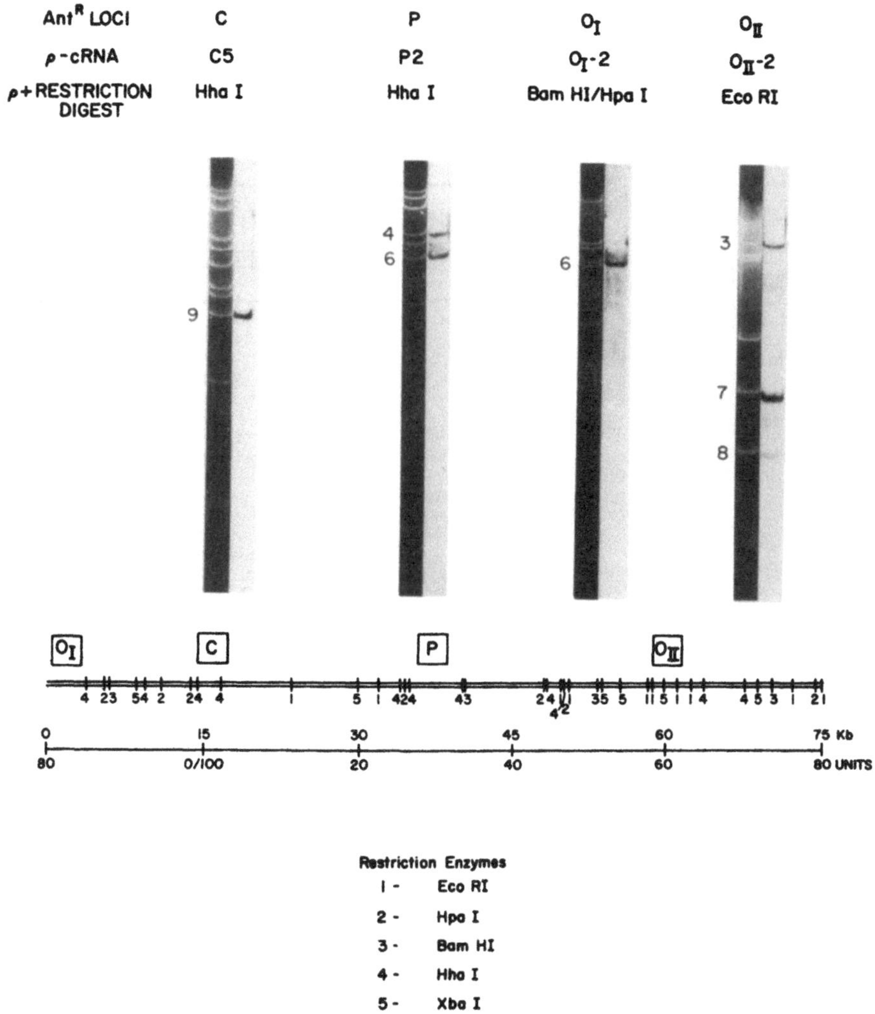

figure 6. Hybridization of ^{3}H labeled cRNA made on templates of mtDNA from single marker petites, with mtDNA restriction fragments. Autoradiograms are compared with fluorescent gel patterns. Derived map positions for individual markers is shown.

omes and the restriction sites contained therein, as well as the hybridization patterns of other single marker petite cRNAs, permits precise localization for the C^R, P^R, O_{II}^R, O_I^R and E^R markers. For example, cRNA transcribed from petite C_5 mtDNA hybridizes primarily to Hha I fragment 9. Be-

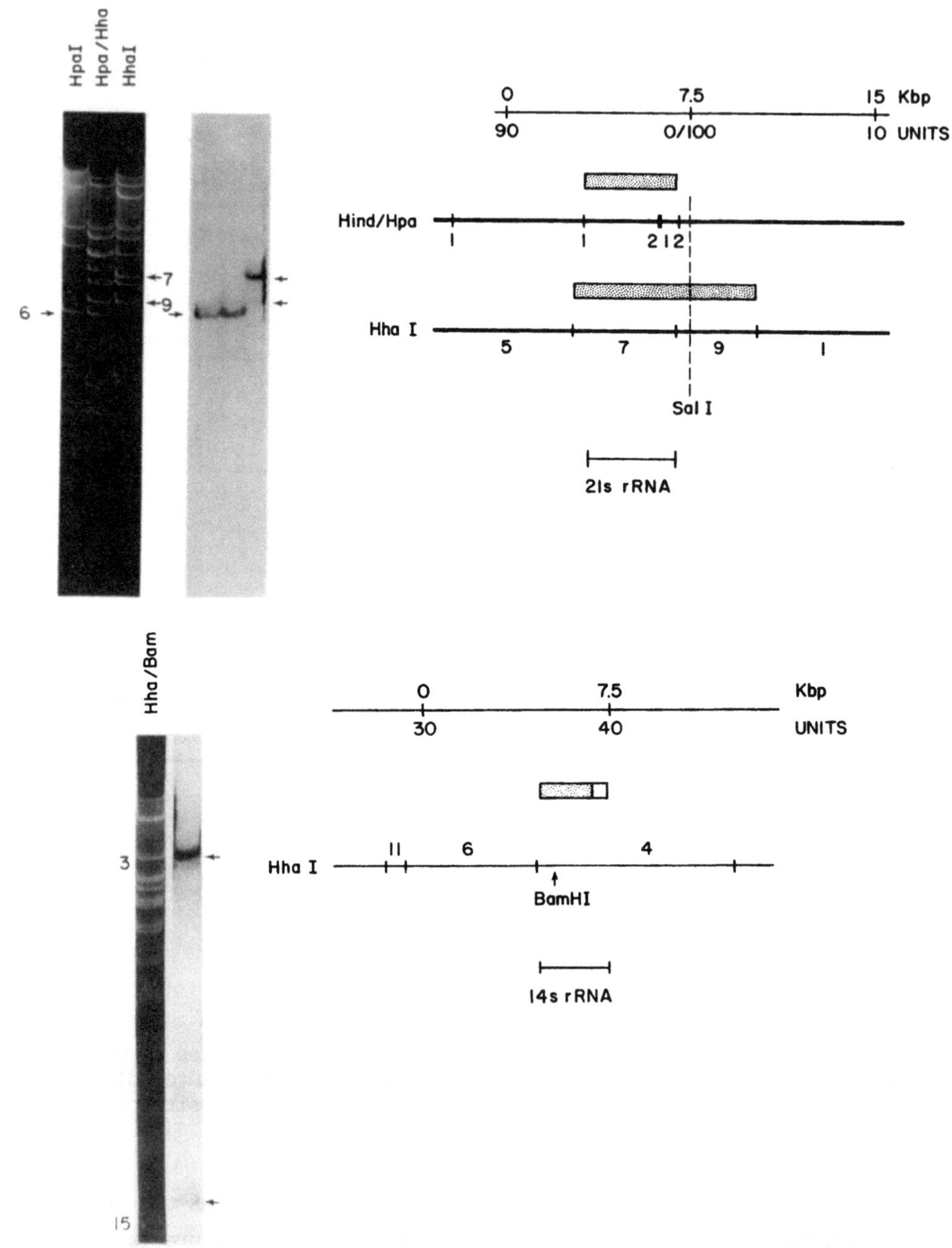

figure 7. Hybridization of ^{125}I labeled 14S and 21S rRNAs to restriction fragments, with map details indicating consequent localizations of rRNA species.

cause this petite has a unit genome of less than 2.0 kb which contains single Hha I, Hind III and Sal I sites, we can place the chloramphenicol resistance locus between 99.3 and 1.4 units on the physical map. In most instances, the mapping of genes using cRNA hybridization in conjunction with restriction mapping of petites is more precise than mapping by restriction analysis alone.

A similar method for localization of genes on restriction fragments is illustrated by Figure 7 which shows the hybridization of ^{125}I labeled mitochondrial rRNA to restriction fragments immobilized on nitrocellulose filters (2,9). The fluorescent gel pattern is shown next to the autoradiogram of the hybrids. Large ribosomal RNA (21S) hybridizes predominantly to Hpa I fragments 6 and to Hha I fragments 7 and 9. Since Hpa I band 6 is smaller than the 3.6 kb 21S rRNA, this transcript must overlap about 400 nucleotides into Hpa I band 2, placing the 21S RNA between 94.5 and 99 units. The 14S rRNA is mapped to left half of Hha I band 4 by a Bam HI/Hha I double digest, which localizes this RNA between 32.7 and 36.7 units on the restriction map. Approximately one third of the genome (33 units) lies between the two rRNA species, as previously demonstrated by hybridization to restriction fragments (6) and to petite mtDNA (24,25). We have also demonstrated the wide separation of the rRNA genes (30) using the R-loop technique (31) (Figure 8). In this method, the homologous DNA strand is dis-

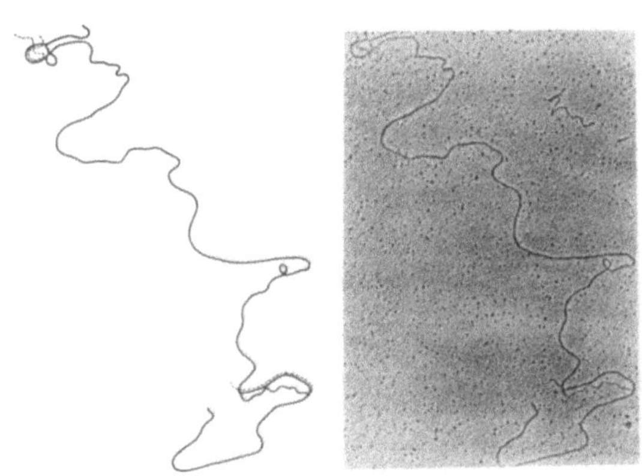

figure 8. Electron micrograph of R-loop structures produced by rRNA hybridization with mtDNA.

placed by RNA during hybridization in conditions favoring the formation of RNA/DNA duplexes. Electron microscopic measurement of the loop structures thus produced indicates that they are separated by one third of the mitochondrial genome in one direction and by two thirds in the other. ^{125}I labeled mitochondrial tRNA hybridized chiefly to two regions of the restriction map (Figure 9). The first is between the Hpa I site at 99 units

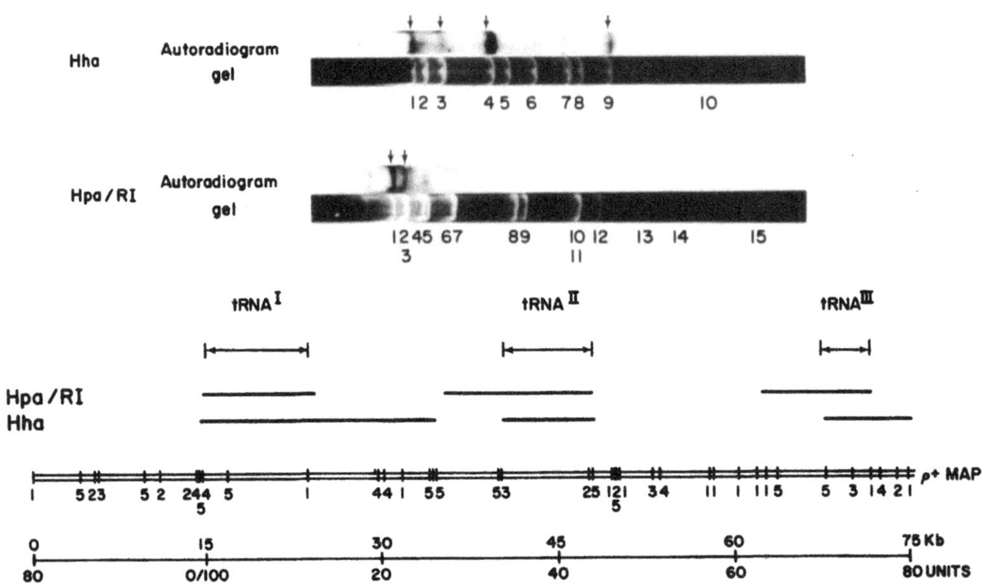

figure 9. Hybridization ^{125}I labeled tRNA to restriction fragments, with derived map positions.

and the Eco Rl site at 11.5 units, and the second is between the Bam HI site at 34 units and the Hha I site at 44 units. There is a gap of 22.5 units between these two regions in which no tRNA hybridization was detected. A third region of hybridization was demonstrated in the region bordered by the Eco Rl site at 70 units on the left and the Hha I site at 76 on the right. These hybridization data are consistent with our mapping of mitochondrial tRNAs based on the hybridization of acylated tRNAs to petite mtDNA (Figure 10). Taken together (Figure 11), it is probable that most of

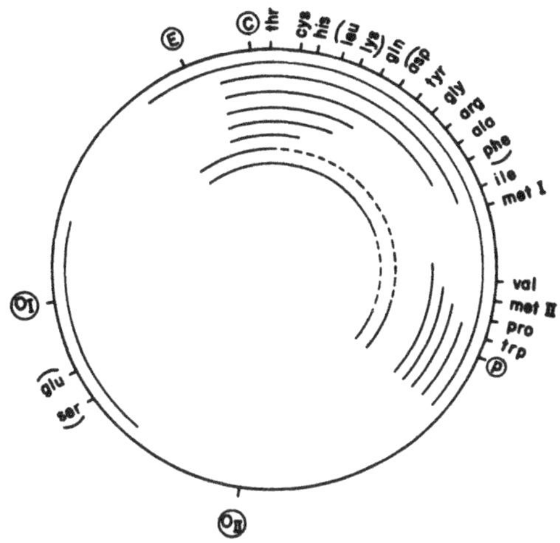

figure 10. Mapping of mitochondrial tRNAs based on the hybridization of acylated tRNAs to various petite strains.

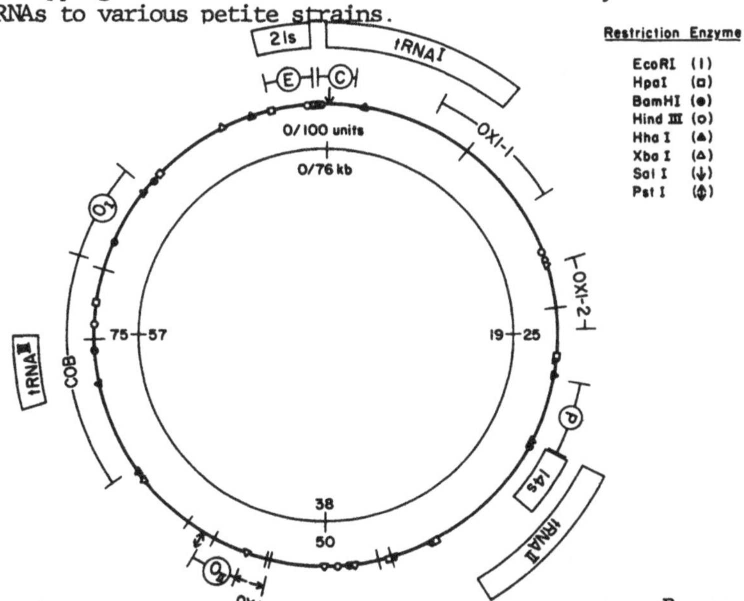

figure 11. Circularized restriction map show locations of ant^R and mit^- markers, sequences complementary to 14S and 21S rRNA and regions hybridizing mitochondrial tRNAs.

the 14 tRNAs between $tRNA^{thr}$ and $tRNA^{met\ I}$ map with the first group of tRNAs, the four tRNAs acylated with val, methionine, proline and tryptophan map in a second group between 34 and 44 units, and the tRNAs which acylate with glutamic acid and serine map between O_I and O_{II}. Figure 11 summarizes our marker and transcript mapping by showing the locations of ant^R and mit^- genes and the coding regions for mitochondrial tRNAs with reference to the circular physical map.

Our mapping of ant^R markers is similar to that produced in other laboratories (Figure 12). The mapping of C, E and 21S rRNA is almost identical in

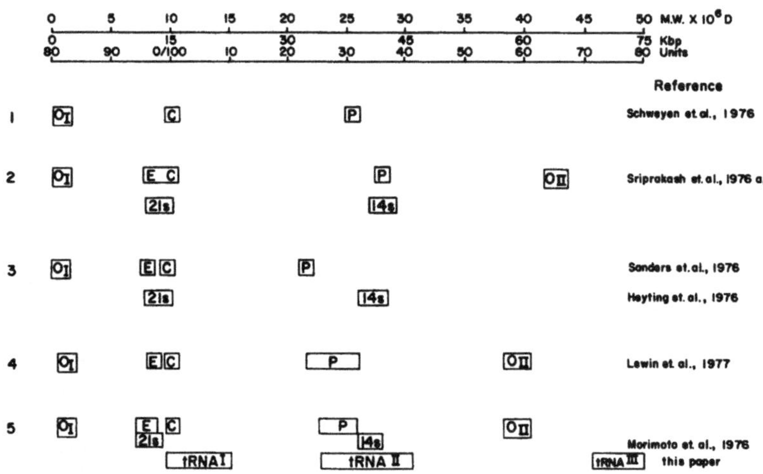

figure 12. Comparison of the map positions of ant^R markers and rRNAs between various laboratories.

the various maps. We have located C near the terminus and perhaps outside of the large ribosomal RNA gene in contrast to Sriprakash et al. (5) and Sanders et al. (7) whose localization of C overlaps with that of 21S rRNA. The greatest differences between the various maps are in the positioning of P^R relative to C^R and to the 14S rRNA. We have placed P^R adjacent to and probably outside of the DNA sequences encoding the 14S rRNA, while Sriprakash et al. (5) show P^R and 14S to be overlapping. Sanders et al. (7) place the P marker much closer to C than either Schweyen et al. (3) or we do (30 units). It is notable that the distance between the 21S and 14S rRNA genes (33-37% of the genome) is constant in all the maps, but that the spacing from C^R to P^R is variable. The probable explanation is that the C^R to P^R distance actually is different between grande strains and that the P^R and 14S rRNA genes do not overlap.

Another aspect of our current research is the cloning of individual mtDNA restriction fragments in E. coli and to attempt to identify RNA and perhaps protein products. Because it is possible to immunochemically identify mitochondrial peptides of the cytochrome oxidase and oligomycin sensitive ATPase complexes, and to characterize transcripts by hybridization, by electrophoretic mobility and, in the case of tRNA, by acylation reactions, we should be able to detect the expression and processing of mitochondrial gene products in E. coli if they occur. Furthermore, cloned DNA fragments provide pure mtDNA segments which display greater structural integrity than the DNA obtained directly from yeast mitochondria. This is useful for heteroduplex analysis and for the sequencing of the mtDNA. Using the plasmid vector pMB9, we have cloned four small EcoR1 fragments, ranging in size from 3.5 to 0.9 kb, from the grande strain 19d (Figure 13). These inser-

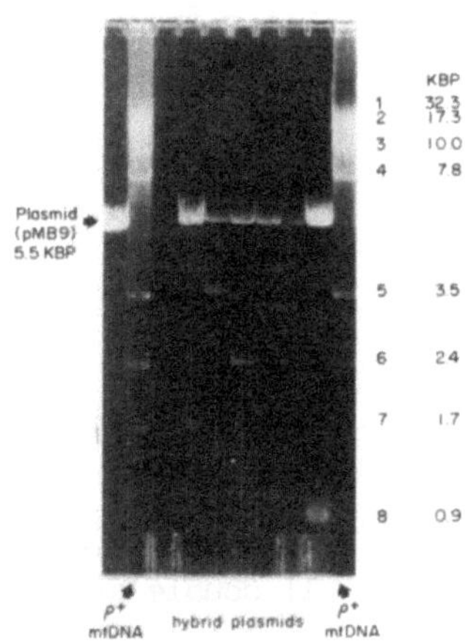

figure 13. Eco R1 cleavage products of recombitant plasmids compared to Eco R1 digest of mtDNA and plasmid pMB9.

tions are stable, because following long term culture, E. coli host strain C600 still produces faithful replicas of the cloned mitochondrial DNA according to restriction digests. The inserted fragments are identical with mitochondrial DNA, because cRNA made to chimeric plasmids hybridizes to the appropriate fragment in the grande mtDNA digest (figure 14), and because Hpa II digests of the cloned fragments compared with the original fragments isolated from yeast have the same restriction patterns. Figure 15 shows the Hpa II cleavage products of EcoR1 band seven, which is under particular

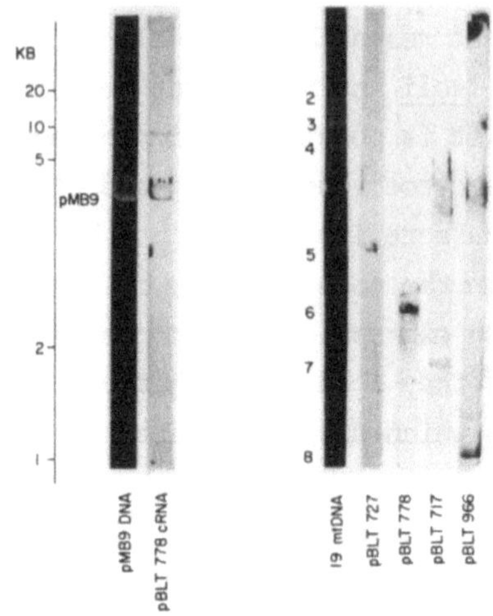

figure 14. Hybridizations of cRNA transcribed from chimeric plasmids to plasmid pMB9 and mtDNA EcoR1 restriction fragments.

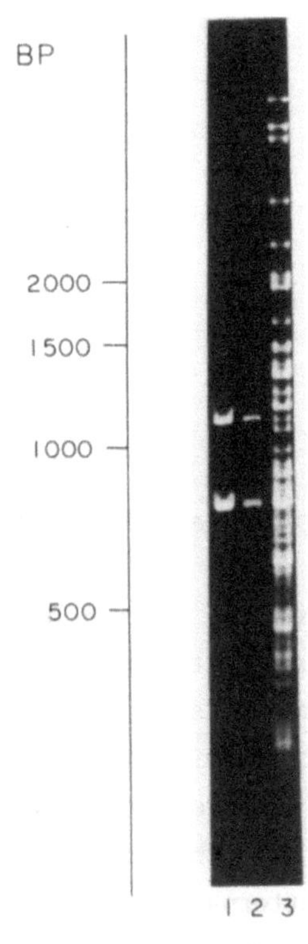

figure 15. Hpa II cleavage products of EcoR1 band 7 isolated from plasmid pBLT 778 (1) compared with the Hpa II cleavage products of band seven derived from yeast mtDNA (2) and the Eco R1/Hpa II double digestion of grande mtDNA (3).

scrutiny, since it may encode the largest mitochondrial subunit of the cytochrome oxidase complex. Accordingly, we are using E. coli minicells to characterize the RNA transcripts made from a hybrid plasmid containing band seven, and attempting to identify translation products with antigenic similarity to the cytochrome oxidase subunit.

To localize promoter regions on the mitochondrial genome, a transcriptionally active complex has been isolated from mitochondrial lysates by Sepharose 4B chromatography. At this stage in the purification, the RNA polymerase activity in this complex is not dependent on exogenously added template. ^{32}P-UTP labeled RNA made from this complex hybridizes with yeast mtDNA, but does not hybridize to excess yeast nuclear DNA or to other heterologous DNAs (Table 1). Significantly, no annealing is detected when

Nucleic Acid in Excess	RNA Transcribed from RNA Polymerase Complex % RNA Hybridized	cRNA Transcribed from mtDNA by E. Coli RNA Polymerase % RNA Hybridized
Yeast mtDNA	48	60
Yeast Nuclear DNA	0.4	0.4
E. coli DNA	0.2	0.4
Salmon Sperm DNA	0.3	n.d.
Yeast Mitochondrial RNA	0.5	29

Table I. Hybridization of in vitro synthesized RNAs.

RNA from this complex is incubated with excess yeast mtRNA, indicating that the RNA is transcribed from the same strand as the in vivo synthesized RNA and that transcription is asymmetric. As controls, we have employed RNA transcribed from mtDNA by E. coli RNA polymerase. This cRNA hybridizes 60% with mtDNA and anneals to excess mtRNA to a level of almost 30%. This indicates that E. coli RNA polymerase synthesizes RNA from both strands of the mtDNA, unlike the mitochondrial polymerase complex. When we hybridize labeled RNA made in vitro from this complex to restriction fragments immobilized on nitrocellulose filters, hybridization was found primarily with two regions of the restriction map (adjacent Hha I band 7 and 9 and Hha I band 4), precisely the regions encoding the 21S and 14S rRNAs (figure 16).

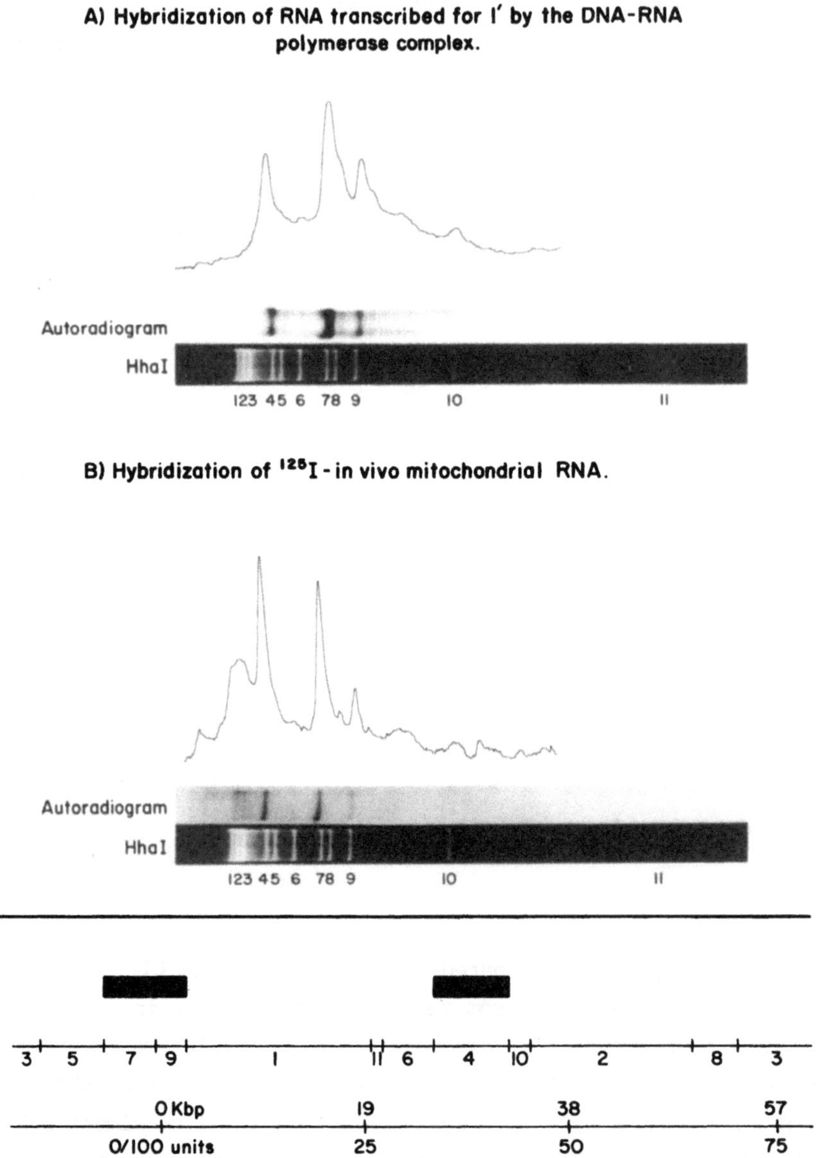

figure 16. Comparison of hybridization patterns of RNA isolated from (a) mtDNA-RNA polymerase complex, incorporating ^{32}P-UTP in vitro; (b) ^{125}I labeled mtRNA synthesized in vivo. RNAs were hybridized with Hha I restriction digests of grande mtDNA. Map positions of primary phybridizations is shown.

Other fragments were also hybridized to a substantially lower level. This profile of hybridization to restriction fragments was also obtained when ^{125}I labeled mtRNA made in vivo was hybridized. This suggests that the transcription obtained in vitro accounts for the pattern of mtRNA transcription in vivo. To exclude the possibility that the total genome could be transcribed followed by a processing step, we have studied a time course of transcription from the polymerase-mtDNA complex in vitro. With 1, 5 and 20 minutes of incorporation, the level of RNA transcription increases, but the pattern of hybridization to restriction fragments does not change. Taken together, these data suggest that the mitochondrial DNA is transcribed from at least two widely spaced promoters and that transcription is asymmetrical.

In conclusion, detailed restriction enzyme analysis has enabled us to select a series of petites with single deletions of the mtDNA and to characterize the various kinds of molecular heterogeneity present in petite strains. By using only single deletion petites for mapping as well as hybridization techniques employing cRNA, rRNA and tRNA, we have precisely mapped antibiotic resistance markers, mit⁻ loci, syn⁻ mutations, ribosomal RNAs and transfer RNAs within the mitochondrial genome. Further, we have succesfully cloned four EcoR1 fragments of mtDNA in E. coli and are in the process of analyzing their transcription and translation in minicells. Finally, we have isolated a mitochondrial RNA polymerase complex which transcribes the mitochondrial DNA in vitro specifically and asymmetrically.

This study was supported in part by Grants HL 09172 and HL04442 from the National Institutes of Health and the Louis Block Fund of the University of Chicago. The Franklin McLean Memorial Research Institute is operated by the University of Chicago for the U.S. Energy Research and Development Administration under Contract EY-76-C-02-0069. R.M. was supported by USPHS Training Grant T32-GM07197 and A.L., by USPHS Training Grant 5-T01-GM00090-19.

REFERENCES

1. Morimoto, R, Lewin, A., and Rabinowitz, M: Nucleic Acid Res. 7, 2331-2351, 1977.
2. Molloy, P.L., Linnane, A.W. and Lukins H.B.: J. Bacteriology 122, 7-18, 1975.
3. Schweyen, R.J., Steyer, A., Kaudewitz, F., Dujon, B., and Slonimski, P.P.: Mol. Gen. Genet. 146, 131-165, 1976.
4. Bolotin-Fukuhara, M., and H. Fukuhara: Proc. Nat. Acad. Sci. (USA) 73: 4608-4612, 1976.
5. Sriprakash, K.S., Molloy, P.L., Nagley, P., Lukins, H.B., and Linnane, A.W.: J. Mol. Biol. 104, 485-503, 1976.
6. Sanders, J.P.M., Heyting, C., and Borst, P.: Biochem. Biophys. Res. Commun. 65, 699-707, 1975.
7. Sanders, J.P.M., Heyting, C., and Borst, P.: In Genetics and Biogenesis of Chloroplasts and Mitochondria, T. Bucher, W. Neupert, W. Sebald, and S. Werner (eds.). Amsterdam: Elsevier/North-Holland Biomedical Publishing, pp. 511-517, 1976.
8. Locker, J., Rabinowitz, M., and Getz, G.S.: J. Mol. Biol. 88, 489-502, 1974.
9. Michel, F., Lazowska, J., Faye, G., Fukuhara, H., and Slonimski, P.P.: J. Mol. Biol. 85, 411-431, 1974.
10. Gordon, P., and Rabinowitz, M. Biochemistry 12, 116-123, 1973.
11. Gordon, P., Casey, J., and Rabinowitz, M.: Biochemistry 13, 1067-1075, 1974.
12. Lazowska, J., Michel, F., Faye, G., Fukuhara, H., and Slonimski, P.P.: J. Mol. Biol. 85, 393-410, 1974.
13. Casey, J., Gordon, P. and Rabinowitz, M.: Biochemistry 13, 1059-1063, 1974.
14. Linnane, A.W., Saunders, G., Gingold, E., and Lukins, H.: Proc. Nat. Acad. Sci.(USA) 59, 903-910, 1968.
15. Faye, G., Fukuhara, H., Grandchamp, C., Lazowska, J., Michel, F., Casey, J., Getz, G.S., Locker, J., Rabinowitz, M., Bolotin-Fukuhara, M., Coen, D., Deutsch, J., Dujon, B., Netter, P., and Slonimski, P.P.: Biochemie 55, 779-792, 1973.
16. Locker, J., Rabinowtiz, M.,and Getz, G.S.: Proc.Nat. Acad. Sci. (USA) 71, 1366-1370, 1974.

17. Lazowska, J., and Slonimski, P.P.: Molec. Gen. Genet. 146, 61-78, 1976.
18. Morimoto, R., Lewin, A., Hsu, H.-J., Rabinowitz, M., and Fukuhara, H.: Proc. Nat. Acad. Sci. (USA) 72, 3868-3872, 1975.
19. Morimoto, R., Lewin, A., Merten, S., and Rabinowitz, M.: In Genetics and Biogenesis of Chloroplasts and Mitochondria . T. Bucher, W. Neupert, W. Sebald and S. Werner (eds.). Amsterdam: Elsevier/North-Holland Publishing, 1976, pp. 519-533.
20. Heyting, C., and Sanders, J.P.M.: In The Genetic Function of Mitochondrial DNA, C. Saccone and A.M. Kroon (eds.). Amsterdam: Elsevier/ North Holland Publishing, 1976, pp. 273-280.
21. Fukuhara, H., Bolotin-Fukuhara, M., Hsu, H.-J., and Rabinowitz, M.: Mol. Gen. Genet. 145, 7-17, 1976.
22. Martin, N., and Rabinowitz, M: in Genetics and Biogenesis of Chloroplasts and Mitochondria, T. Bucher, W. Neupert, W. Sebald, and S. Werner (eds.). Amsterdam: Elsevier/North Holland Publishing, 1976, pp. 749-754.
23. Martin, N., Rabinowtiz, M., and Fukuhara, H.: Biochemistry, 1977 (In press).
24. Faye, G., Kujawa, C. Dujon, B., Bolotin-Fukuhara, M., Wolf, K., Fukuhara, M., and Slonimski, P.P.: J. Mol. Biol. 99, 203-217, 1975.
25. Faye, G., Kujawa, C., Fukuhara, H., and Rabinowitz, M.: Biochem. Biophys. Res. Comm. 68, 476-482, 1976a .
26. Slonimski, P.P. and Tzagoloff, A.: Eur. J. Biochem. 61, 27-41, 1976.
27. Bolotin-Fukuhara, M., Faye, G., and Fukuhara, H.: In The Genetic Function of Mitochondrial DNA, C. Saccone and A.M. Kroon (eds.). Amsterdam: Elsevier/North-Holland Publishing, 1976, pp. 243-250.
28. Faye, G., Bolotin-Fukuhara, M., and Fukuhara, H.: In Genetics and Biogenesis of Chloroplasts and Mitochondria, T. Bucher, W. Neupert, W. Sebald, and S. Werner(eds.). Amsterdam: Elsevier/North-Holland Biomedical Press, 1976, pp. 547-556.
29. Southern, E.M.: J. Mol. Biol. 98, 503-517, 1975b.
30. Merten, S., and Rabinowitz, M.: Fed. Proc. 36, 662, 1977.
31. Thomas, M., White, R.L., and Davis, R.W.: Proc. Nat. Acad. Sci. (USA) 73, 2294-2298, 1976.

PROPERTIES AND GENETIC LOCALIZATION OF MITOCHONDRIAL TRANSFER RNAs OF *NEUROSPORA CRASSA*

P. Terpstra*, H. de Vries and A.M. Kroon
Laboratory of Physiological Chemistry, State University, Bloemsingel 10, Groningen, The Netherlands

INTRODUCTION

The mitochondrial DNA (mtDNA) from the ascomycete *Neurospora crassa* consists of double-stranded covalently closed circular molecules with a homogeneous length of 19 µm. Attempts to isolate this mtDNA as intact circular molecules on a preparative scale have failed so far. We have isolated mtDNA from *Neurospora crassa* strain E5256 as a collection of linear fragments. Incubation with the restriction endonucleases Eco RI and Bam HI produced 12 and 5 specific fragments respectively. Analysis of the partial cleavage products allowed us to determine the sequence of the fragments of both enzymes. The 5 Bam HI recognition sites were located on the Eco RI fragments by isolation of the individual Eco RI fragments and redigestion of these fragments with endonuclease Bam HI. In this way two overlapping cleavage maps could be constructed. In addition we determined the position and order on the map of 9 out of the more than 30 Hind III restriction fragments.

The cleavage map provided a framework for the localization of the mitochondrial rRNA genes. From DNA-RNA hybridization experiments we have concluded that the mtDNA of *Neurospora crassa*, like other mtDNAs, contains one gene for the large (24S) and one for the small (17S) mitochondrial ribosomal RNA. The genes are localized in the region of the mtDNA which is relatively rich in G+C. The distance separating the two genes is about 900 basepairs. The results of these studies have been published in detail recently (1 - 4) and are consistent with the existence of a ribosomal RNA precursor, which has been described previously by Kuryama and Luck (5). Similar

---

* Present address: Dept. of Biochemistry, University of Texas Health Science Center, Dallas, Texas, USA.

althcugh not completely identical data have been presented by Küntzel and his coworkers (6 - 8). In this paper we will concentrate on our more recent studies on mitochondrial tRNAs and their genetic localization.

MITOCHONDRIAL TRANSFER RNAs OF *NEUROSPORA CRASSA*

Mitochondrial tRNAs from *Neurospora* have, up till now, not been used for hybridization studies with mtDNA, but are otherwise well characterized. Aminoacylation has been reported for 15 aminoacids (9 - 11). Twelve aminoacylated tRNAs can be readily discriminated from cytoplasmic tRNAs by chromatography (10). The occurrence of F-met-tRNA (11) and the absence of a fluorescent "Y" base (12) in *Neurospora* mitochondrial tRNA recalls the resemblance of mitochondrial to prokaryote protein synthesis so often mentioned.

In collaboration with the groups of Stahl and Dirheimer, De Vries has recently extended the characterization of *Neurospora* tRNA to base composition analysis, one- and twodimensional gel electrophoresis and chromatography on RPC5 (13). The G+C content of total mt-tRNA of *Neurospora* is about 43%, as compared to 60% for the cytoplasmic tRNA. The methylated nucleoside content is low: 2.5% as compared to 5.5% for cytoplasmic tRNA. Small amounts of Ψ, T, hU, t^6A, m^1G, m^2G and m^2_2G are present; m^1A and m^7G are absent, whereas the minimal amounts of m^5C and I present in the mt-tRNA preparations can be ascribed to tRNAs of cytoplasmic origin. Neither the eukaryotic "Y" base, nor the prokaryotic s^4U could be detected. On twodimensional polyacrylamide gel electropherograms about 25 species were separated. One species for phenylalanine, two for leucine and two for methionine could be located. Besides the 25 strong spots, which were of about equal intensity, the gels contained a small number of weak spots, representing either cytoplasmic contaminants or minor species of mt-tRNA. Coelectrophoresis of mitochondrial tRNAs from *Neurospora crassa* and from *Saccharomyces cerevisiae* shows that these two fungi contain completely different sets of mitochondrial tRNAs. The two mitochondrial tRNAs hardly cross-hybridized to the mitochondrial DNA of the two species. In spite of the low degree of sequence homology between the mitochondrial tRNAs of the two species, the mitochondrial ligases of both species appear to effectively charge the mt-tRNAs irrespective of their origin.

THE TRANSFER RNA GENES ON MITOCHONDRIAL DNA

Mitochondrial DNA contains genes for several tRNAs. Saturation hybridization with total mitochondrial 4S RNA has led to estimates of 20-25 genes in the case of yeast mtDNA (14, 15). Studies with individual tRNA species reveal that yeast mtDNA-coded tRNA accepts at least 18 different amino acids (16). In view of the question if "import" of cytoplasmic tRNAs is necessary or occurs, it was considered worthwhile to undertake experiments to locate and to determine the number of tRNA genes on *Neurospora* mtDNA. Primarily we have used the same approach as for the localization of the rRNA genes (4).

Hybridization of iodinated tRNA to stripfilters with different sets of restriction fragments is shown in Fig. 1. We have estimated the number of tRNA genes per fragment making the following assumptions:

1) The total number of tRNA genes on *Neurospora* mtDNA is about 25. This number is based on the observation of the 25 strong spots in the two dimensional gel of *N. crassa* tRNA (13). If the tRNAs represented by the weak spots are also mtDNA-coded, then they should contribute little to the amount of radioactivity bound per filter. 2) The 25 tRNAs are present in about equal amounts. This is in first approximation a reasonable assumption, because, as already mentioned, the 25 tRNA spots on the two-dimensional gel have about the same intensity. 3) The specific activity of all tRNAs after iodination is the same. The hybridization conditions are similar for each tRNA and the stoichiometry of the DNA fragments remains unaltered (*cf.* ref. 4).

Assumption 1 is necessary to calculate the p units presented in Table I. A consistent minimum percentage of 3 - 6% of the total radioactivity per filter was found in 12 fragments, leading to a p unit of 1. Assumption 2 and 3 are necessary to equate one p unit with one tRNA gene. The distribution of the p units or tRNAs on the cleavage map of the mtDNA (Fig. 2) has been calculated in the following way.

Hybridization with a Bam HI digest (Fig. 1.A) is not very informative because B1 is long and radioactivity in B2 cannot be counted separately from B1. There is one tRNA on B3 and this is confirmed by the hybridization of

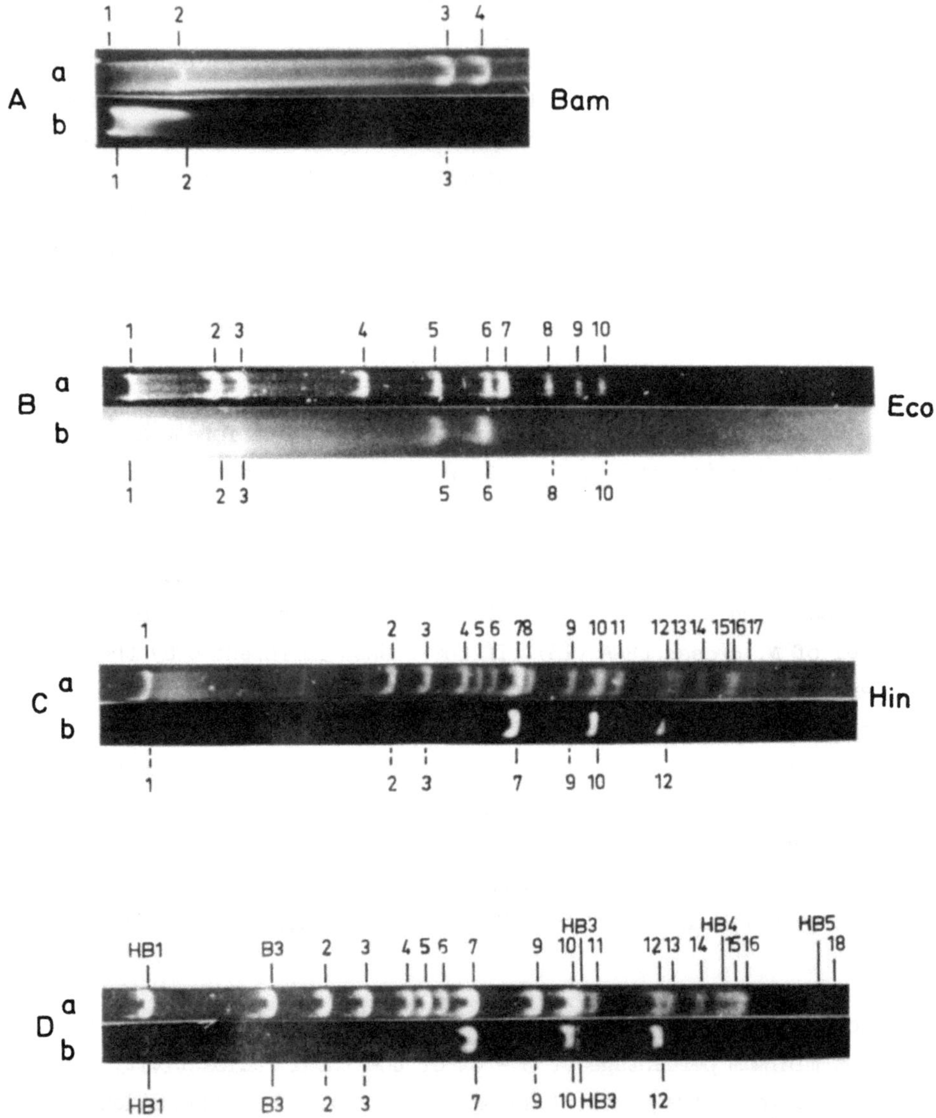

Fig. 1. *Hybridization of mitochondrial tRNA with restriction fragments of mtDNA.* Mitochondrial tRNA was isolated as described (13) and iodinated to a specific activity of 5×10^6 cpm/µg (4). RNA input in the hybridization mixture was 4 µg, together with a 75 fold excess of unlabelled 24S + 17S rRNA. mtDNA input per digest: 3 µg. a) photo of the gel. b) autoradiogram. The broken lines indicate bands in the autoradiograms that are too faint to be reproduced. A) Hybridization with a Bam HI digest. B) Hybridization with an Eco RI digest. C) Hybridization with a Hind III digest. D) Hybridization with a double digest of Hind III plus Bam HI.

TABLE I

QUANTITATION OF STRIPFILTER HYBRIDIZATION (FIG. 1)

Bands, visible in the autoradiogram, were cut and counted. After subtraction of background values, the total amount of radioactivity per filter was fixed at 100%. P = percentage hybridization per band, divided per 4 and rounded off to the nearest integer value. Each p unit therefore represents approximately 4% hybridization (see text). For the length of the restriction fragments see ref. 3.

Bam HI fragments Fig. 1.A.b	p	Eco RI fragments Fig. 1.B.b	p	Hind III fragments Fig. 1.C.b	p	HB fragments Fig. 1.D.b	p
B1 + B2	24	E1	21	H1	2	HB1	1
B3	1	E2	1	H2	1	B3 = HB2	1
		E3	1	H3	1	H2	1
		E5	1	H7	10	H3	<0.5
		E6	1	H9	1	H7	10
		E8	<0.5	H10	6	H9	1
		E10	<0.5	H12	4	H10 + HB3	7
						H12	4

tRNA with a double digest of Hind III plus Bam HI (Fig. 1.D). Hybridization with the Eco RI fragments (Fig. 1.B). locates 21 tRNAs in E1. One tRNA is present on E2 and H3 and one is present on E3 and H2. This is in accord with the data obtained by digestion of E2 and E3 with Hind III (Table II) and makes it unlikely that the fragments H7c and H10b contain a tRNA gene. Therefore, all counts bound at the positions of the Hind III fragments 7 and 10 (Fig. 1.C and D) are to be ascribed to H7a, H7b and H10a, all contained in E1. Because E5 is labelled and B4 is not, we have tentatively placed a tRNA gene on the 200 bp fragment between the Bam HI and Eco RI cleavage sites. This has to be verified by hybridization of tRNA to an Eco RI - Bam HI double digest. Fragment E8 and E10 are very faintly, although reproducibly, labelled. We speculate that they contain a tRNA gene for one of the tRNAs present in low amounts in the mitochondria. In a hybridization experiment of tRNA with an Eco RI digest, reported earlier (17), hybridization with E2, E3 and E5 was not detected. This may have been caused by the much higher background hybridization in that experiment. Hybridization

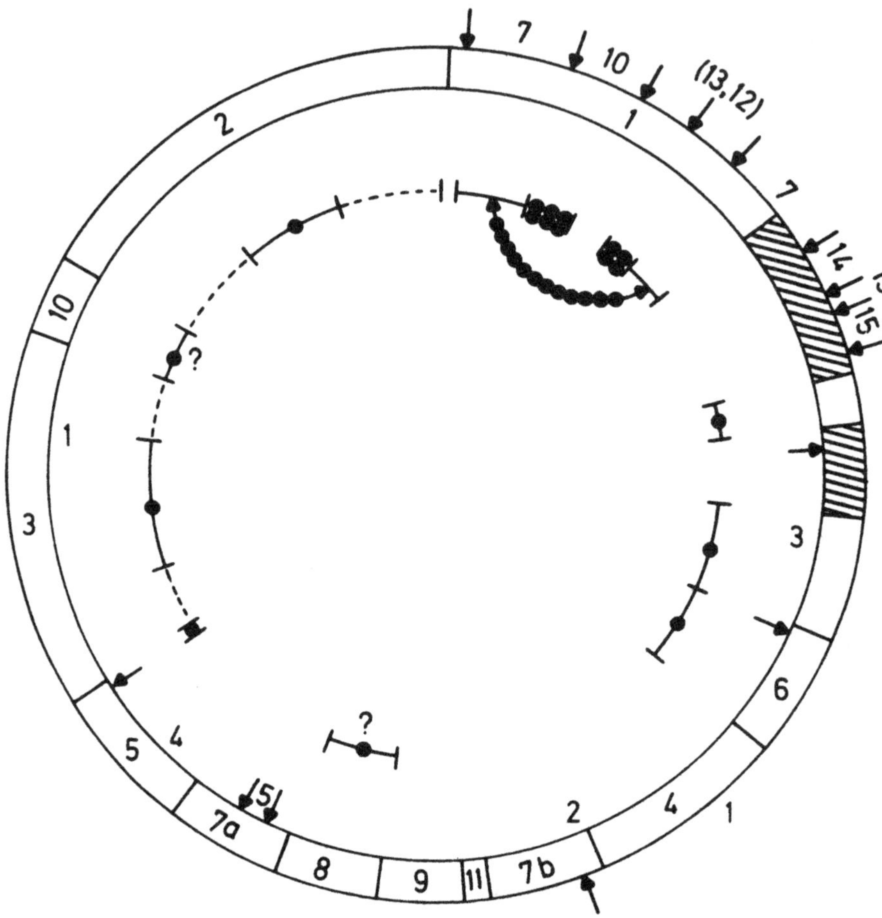

Fig. 2. *Location of rRNA and tRNA genes on the restriction fragment map of* Neurospora crassa *mtDNA.* The map combines our previous results (1-4) with those reported here. *Outer circle band:* Eco RI cleavage sites are drawn with solid lines within the circle. Hind III cleavage sites are indicated with arrows outside the circle. Bam H1 cleavage sites are indicated with arrows inside the circle. The hatched regions indicate the position of the 24S and 17S rRNA. *Inner circle:* position of tRNA genes. Each dot represents one p unit (Table I). The dots with question marks are less than one p unit. One dot has been tentatively equated with one tRNA gene (see text). The localization precision is limited by the fragment size, indicated with solid lines. The broken lines indicate that the position of fragment H3 in E2 and H2 in E3 is not known (*cf.* Table II).

TABLE II

TERMINAL FRAGMENTS, OBSERVED AFTER DIGESTION OF E1, E2 AND E3 WITH ENDONUCLEASE HIND III

Eco RI fragments E1, E2 and E3 were extracted from an agarose gel and redigested with Hind III. Electrophoresis was done in a 0.7% agarose gel, with a complete Hind III digest as marker. The discrepancy between observed and expected total length is due to the faintness of bands smaller than H10.

E1 x Hind III fragments	fragment length	E2 x Hind III fragments	fragment length	E3 x Hind III fragments	fragment length
EH1	6,450	H3	3,500	H2	3,900
H7a	2,470	H10b	1,800	H7c	2,470
H7b	2,470				
H10a	1,800				
H12	1,300				
H13	1,250				
H14	1,100				
H15	980				
E1H2	600				
H19	420				
total length	18,890		5,300		6,370
exp. length	18,900		10,200		8,700

with a Hind III digest (Fig. 1.C) shows that H7, H10 and H12 are strongly labelled. Contamination with rRNA is absent, because we could not detect any hybridization with H14 and H15. Therefore the 20 tRNAs, calculated to be present on H7, H10 and H12, can be positioned on E1 next to the 24S rRNA gene (Fig. 2). There are two fragments H7 in this region and it cannot be decided whether the tRNA genes on H7 are located on one or on both of the fragments. Fragments H12 and H13 can still be interchanged. The one tRNA gene on H9 cannot be placed on the map, but could be identical to a tRNA on E5, E8 or E10. Hybridization with a Hind III - Bam HI double digest (Fig. 1.D). confirms the localization of tRNA genes, deduced from Fig. 1.C and shows in addition that fragment HB3 is labelled. This places a tRNA gene in between the two rRNA genes.

It should be stressed once more that the total number of tRNA genes and the number of tRNA genes per fragment are speculative. Future work on the localization of all individual amino-acylated tRNAs will show how much the tRNA gene map of Fig. 2 will have to be corrected in a quantitative aspect. Nevertheless some important general aspects of the tRNA gene organization on *N. crassa* mtDNA are already apparent.

1) The majority of the tRNA genes are present in a region of maximally 15% genome length. This is a strong clustering of genes with respect to the total genome length and has not been observed before in other mtDNAs. For yeast mtDNA 19 individual tRNAs have been mapped (16). Their genes are between the two rRNA genes on a stretch of about 30% of the genome. On animal mtDNA, the tRNA genes are spread over the entire genome (18) but it should be noted that the animal mitochondrial genome is 4 - 5 times smaller than *N. crassa* or yeast mtDNA. The distribution of the tRNA genes on animal mtDNAs does not seem to be random, because they are in nearly the same position with respect to the rRNA genes in different organisms. This suggests a strong conservation of at least the order of other genes during the evolution of animal mtDNA.

2) There is a tRNA gene localized in the gap between the two rRNA genes. A 4S gene in between the two mitochondrial rRNA genes has also been found on animal mtDNAs (*cf.* ref. 18) and recently also in between the rRNA genes of *E. coli* DNA (19). Although this may point to a similarity in gene order between bacterial and mitochondrial genes, it remains to be proven that the 4S RNA in between the two rRNA genes of mtDNA has a tRNA function.

FURTHER AND FUTURE STUDIES

Our goal is to answer the following questions: 1. How many different tRNA species are used in mitochondrial protein synthesis, are these species coded for on mitochondrial DNA and is the presence of isoacceptors consistent with the use of all codons in mitochondrial messenger RNA; 2. How are the genes for tRNAs organized in mtDNA, are they transcribed separately or as parts of larger primary transcripts?

Ad 1. The characterization of the tRNAs and the hybridization studies described in this paper lead to the lower limit of 25 tRNAs being present in and coded for by the mitochondria. They do not give the final information as to origin and function of the minor species noticed in the two-dimensional gels (13). Since import of cytoplasmic tRNA is considered very unlikely (20), one has to conclude that either these minor species are of importance or that the number of 25 tRNAs suffices for decoding the mitochondrial messages. If the conclusion, provisionally reached above, that mt-tRNAs do not contain I is correct, the decoding potential is limited anyhow. As further approaches we plan to perform plateau hybridization studies with total mt-tRNA and have started stripfilter hybridization experiments with total mt-tRNA specifically charged with one labelled aminoacid. In the case of leucine we have come upon two iso-acceptors in this way, one on Hind III fragment H7 and one on H10. This is in accord with our chromatographic and electrophoretic data (13). It should be noted that 2 isoacceptors for leucine are not sufficient for decoding the 6 leucine codons unless hitherto unknown wobbling is allowed. It might also be that not all codons occur in mtDNA. Although some problems of charging the tRNA with each of the aminoacids have still to be solved, the latter approach looks promising.

Ad 2. We already mentioned that at least 20 of the mt-tRNAs are clustered in a region of 15% of the mtDNA. But 15% is still a stretch of 9000 basepairs, about twice the length of the phage ΦX 174. The length of this cluster may, however, well be overestimated, since the region contains two fragments of equal length H7a and H7b which cannot be distinguished. Furthermore the order of two fragments has not been settled (H12 and H13). We have now looked for other restriction enzymes suitable for cleavage of the mtDNA in a way that at the one hand stripfilter hybridizations with tRNA can be performed (fragments not too small) and at the other hand specific cleavage sites in the region of the tRNA gene-cluster are present sufficiently close together to delimit the cluster more precisely. From the enzymes tested Hind II and Hha I look promising in these respects. Hind II gives 15 fragments of respectively about 10200, 8200, 7100, 5400, 4500, 4400, 4300, 4175, 3400, 2575, 2300, 1350, 800, 500 and 400 basepairs. For HhaI 19 cleavage sites were detected and fragments of about 14000, 7200, 4600, 4400, 3350, 2950, 2750 (2x), 2250, 1975, 1925, 1600, 1550, 1500, 1300, 1225, 800, 700 and 500 basepairs. Most of the larger restriction fragments of both enzymes are cleaved by either Bam H1 or Eco RI and are, therefore, most likely not an integral part of the tRNA gene-cluster, which lies on the part of the mtDNA shared by the fragments E1 and B1 (*cf.* ref. 3). The complete fragment maps for Hha I and Hind II and their positions relative to the Eco RI and Bam HI maps, are presently under construction. It is hoped that a suitable fragment will be obtained for *in vitro* transcription studies with respect to promoter and termination sites and for basesequence analysis.

ACKNOWLEDGEMENTS

These studies were aided in part by a grant from the Netherlands Foundation for Chemical Research (SON) with financial support of the Netherlands Organization for Advancement of Pure Research (ZWO). We wish to thank Hennie Bakker, Marijke Holtrop and Jenny de Jonge for their skilful cooperation in the various experiments.

REFERENCES,

1. Terpstra, P., Holtrop, M., Kroon, A.M.: Restriction fragment map of *Neurospora crassa*, in: Saccone, C. and Kroon, A.M. (Eds.): The genetic function of mitochondrial DNA, Elsevier/North-Holland Biomedical Press, Amsterdam, pp. 111-118 (1976).
2. Terpstra, P., Holtrop, M, Kroon, A.M.: Heterogeneous base distribution in mitochondrial DNA of *Neurospora crassa*, Nucleic Acids Res., 4, 129-139 (1977).
3. Terpstra, P., Holtrop, M., Kroon, A.M.: A complete cleavage map of *Neurospora crassa* mtDNA obtained with endonucleases Eco RI and Bam HI, Biochim. Biophys. Acta, 475, 571-588 (1977).
4. Terpstra, P., Holtrop, M., Kroon, A.M.: The ribosomal RNA genes on *Neurospora crassa* mitochondrial DNA are adjacent, Biochim. Biophys. Acta, 478, 146-155 (1977).
5. Kuriyama, Y., Luck, D.J.L.: Ribosomal RNA synthesis in mitochondria of *Neurospora crassa*, J. Mol. Biol., 73, 425-437 (1973).
6. Bernard, U., Pühler, A., Mayer, F., Küntzel, H.: Denaturation map of the circular mitochondrial genome of *Neurospora crassa*, Biochim. Biophys. Acta, 402, 270-278 (1975).
7. Bernard, U., Pühler, A., Küntzel, H.: Physical map of circular mitochondrial DNA from *Neurospora crassa*, FEBS Lett., 60, 119-121 (1975).
8. Bernard, U., Goldthwaite, C., Küntzel, H.: Physical map of *Neurospo-*

ra crassa mitochondrial DNA and its transcription unit for ribosomal RNA, Nucleic Acids Res., 3, 3103-3108 (1976).

9. Barnett, W.E., Brown, D.H.: Mitochondrial transfer ribonucleic acids, Proc. Nat. Acad. Sci. U.S.A., 57, 452-458 (1967).
10. Epler, J.L.: The mitochondrial and cytoplasmic transfer ribonucleic acids of *Neurospora crassa*, Biochemistry, 8, 2285-2290 (1969).
11. Epler, J.L., Shugart, L.R., Barnett, W.E.: N-formyl methionyl transfer ribonucleic acid in mitochondria from *Neurospora*, Biochemistry, 9, 3575-3579 (1970).
12. Fairfield, S.A., Barnett, W.E.: On the similarities between the tRNAs of organelles and prokaryotes, Proc. Nat. Acad. Sci. U.S.A., 68, 2972-2976 (1971).
13. De Vries, H., Schneller, J.M., Martin, R.P, Dirheimer, G., Stahl, A.J. C.: *Neurospora crassa* mitochondrial transfer RNAs. Their comparison with yeast mitochondrial tRNAs, Nucleic Acids Res., submitted (1977).
14. Reijnders, L., Borst, P.: The number of 4-S RNA genes on yeast mitochondrial DNA, Biochem. Biophys. Res. Commun., 47, 126-133 (1972).
15. Schneller, J.M., Faye, G., Kujawa, C., Stahl, A.J.C.: Number of genes and base composition of mitochondrial tRNA from *Saccharomyces cerevisiae*, Nucleic Acids Res., 2, 831-838 (1975).
16. Martin, N.C., Rabinowitz, M.: Transfer RNAs of yeast mitochondria, in: Bücher, T., Neupert, W., Sebald, W., Werner, S. (Eds.), Genetics and Biogenesis of Chloroplasts and Mitochondria, Elsevier/North-Holland Biomedical Press, Amsterdam, pp. 749-754 (1976).
17. Kroon, A.M., Terpstra, P., Holtrop, M., De Vries, H., Van den Bogert, C., De Jonge, J., Agsteribbe, E.: The mitochondrial RNAs of *Neurospora crassa*: their function in translation and relation to the mitochondrial genome, in: Bücher, T., Neupert, W., Sebald, W., Werner, S. (Eds.): Genetics and Biogenesis of Chloroplasts and Mitochondria, Elsevier/ North-Holland Biochemical Press, Amsterdam, pp. 685-696 (1976).
18. Saccone, C., Pepe, G., Bakker, H., Kroon, A.M.: The Genetic organization of rat-liver mitochondrial DNA, this volume.
19. Lund, E., Dahlberg, J.E.: Spacer transfer RNAs in ribosomal RNA transcripts of *E. coli*: processing of 30S ribosomal RNA *in vitro*, Cell, 11, 247-262 (1977).

20. Martin, R.P., Schneller, J.M., Stahl, A.J.C., Dirheimer, G.: Study of yeast mitochondrial tRNAs by two-dimensional polyacrylamide gel electrophoresis: characterization of isoaccepting species and search for inported cytoplasmic tRNAs, Nucleic Acids Res., submitted (1977).

THE GENETIC ORGANIZATION OF RAT LIVER MITOCHONDRIAL DNA

Cecilia Saccone[1], Gabriella Pepe[1], Hennie Bakker[2] and Ab Kroon[2]
Laboratory of Biochemistry, University of Bari, Via Amendola 165/A, Bari, Italy[1] and Laboratory of Physiological Chemistry, State University, Bloemsingel 10, Groningen, The Netherlands[2]

INTRODUCTION

Mitochondrial DNA (mtDNA) is a double-stranded, usually circular molecule containing genes for ribosomal-, transfer- and messenger RNA. Animal mtDNA has a molecular weight of about 10.10^6; this is equivalent to 15000 base-pairs. It represents the smallest replication unit known at present that contains genetic information for these three classes of RNAs, which are characteristic for all autonomously living cells. However, owing to its low molecular weight, the potential information to be expressed is sufficient for only a limited number of products which together with a far greater number of mitochondrial components encoded in nuclear DNA are assembled to give functional mitochondria within the cell. The building up of mitochondria thus demands a careful interplay between the nucleocytoplasmic and mitochondrial genetic systems which are under control of each other. Therefore, the study of the genetic organisation of mtDNA is particularly interesting not only to elucidate the function of the mitochondrial gene products but also to shed light at the molecular level upon the possible mechanisms of fundamental regulatory processes in eukaryotic cells.

The rather limited size of animal mtDNA and the circumstance that it can be easily obtained undegraded, uncontaminated and in reasonable amounts, offer an unique opportunity to tackle the problem of its genetic organization at the molecular level. Genetic maps of animal mtDNA have been constructed on the basis of electronmicroscopical observations (1, 2) and more recently also with the aid of restriction endoncleases (3, 4, 5). This paper deals with our experiments on the genetic organization of rat-liver mtDNA.

THE ENDONUCLEASE CLEAVAGE MAP OF RAT-LIVER MITOCHONDRIAL DNA

We have perviously reported our initial studies on the restriction endonuclease cleavage map of rat-liver mtDNA and on the localization of the genes for the 16S and 12S rRNAs (7, 8). The conclusion was reached that the two ribosomal RNA genes were not adjacent but about 4000 basepairs apart. This result was unexpected because the two rRNA genes are separated by only about 150 basepairs on mtDNA from *Drosophila melanogaster* (2), *Xenopus laevis* (2) and HeLa cells (5). We have recently reinvestigated this point by the following experimental approaches: a) analysis of the length of restriction fragments of partially and completely digested mtDNA. We used slabgels of different composition also allowing the detection of small restriction fragments; b) analysis of double digests of total mtDNA and of the fragment patterns obtained after digestion of isolated restriction fragments with a second endonuclease; c) electronmicroscopical length measurements of the fragments; d) identification of the fragments of complete single and double digestions and of partially digested fragments containing the base sequences complementary to the mitochondrial rRNAs, using the stripfilter hybridization technique.

The endonucleases Eco RI, Hind III, Bam HI and Hap II gave 8, 6, 2 and 11 restriction fragments respectively. The two smallest Eco RI fragments of about 125 and 100 basepairs were not previously detected. This led us to misinterpretation of the composition of the partial digestion products and to a wrong arrangement of the fragments on the map. Using very highly labelled rRNAs we could further show unequivocally that the 16S rRNA overlapped the cleavage site between the Hind III fragments A and B and that this site was localized on Eco RI fragment D. With in addition the data of the Hap II digestion and the electronmicroscopical observations we have constructed the restriction fragment map with 25 well-defined cleavage sites. On this map the ribosomal RNA genes appear to be adjacent; the origin of replication is located in the vicinity of the small ribosomal RNA gene and the direction of replication is distant from this gene. We confine ourselves to this brief summary of these results her, because all details have been published recently (10). The data are incorporated in Fig. 3 of this paper.

TABLE I

LENGTH OF THE RESTRICTION ENZYME FRAGMENTS OF RAT-LIVER MITOCHONDRIAL DNA OBTAINED WITH THE ENDONUCLEASES Bam HI, Hpa I, Hha I AND Hind II

mtDNA was digested with appropriate amounts of the various enzymes at 37 °C and pH 7.4 for 2 h or longer. The medium contained 50 mM NaCl, 6 mM Tris HCl, 6 mM $MgCl_2$, 6 mM mercaptoethanol, 100 µg/ml gelatine for the incubations with Hind II or Hha I and 10 mM Tris HCl, 10 mM $MgCl_2$, 6 mM KCl, 1 mM dithiotreitol, 100 µg/ml gelatine in the case of Hpa I. Fragments were separated on 0.7% agarose slabgels; the Eco RI fragments of mtDNA from *Neurospora crassa* and rat-liver were used for calibration. All other experimental details are described elsewhere (10). Non-standard abbreviations: B = Bam HI; HII = Hind II.

Bam HI fragments			Hpa I fragments		
	$M \times 10^{-6}$	Basepairs		$M \times 10^{-6}$	Basepairs
BA	6.65	10 000	HpaA	8.31	12 500
BB	2.96	4 450	HpaB	1.53	2 300
	9.61	14 450		9.84	14 800

Hha I fragments			Hind II fragments		
	$M \times 10^{-6}$	Basepairs		$M \times 10^{-6}$	Basepairs
HhaA	5.52	8 300	HIIA	4.59	6 900
HhaB	2.59	3 900	HIIB	3.26	4 900
HhaC	1.26	1 900	HIIC	1.43	2 150
HhaD	0.52	780	HIID	0.42	630
	9.89	14 880	HIIE	0.17	250
				9.87	14 830

For purposes of fine mapping of genes and gene-structure analysis at the level of basesequences a number of 25 cleavage sites is of course not sufficient for a molecule of about 15000 basepairs. We have, therefore, focussed our attention on further cleavage of the mtDNA with the restriction endonucleases HpaI, HhaI and Hind II. Hpa I recognizing the sequence 5'GTTAAC3'

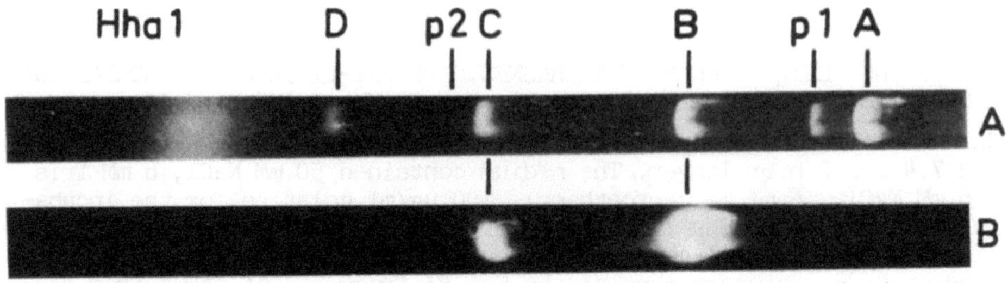

Fig. 1. *Hybridization of mitochondrial rRNA with HhaI restriction fragments of mtDNA.* Specific activity of the ^{125}I-iodinated rRNA was 5.10^{6} cpm/μg. The input was 10^{6} cpm and about 20% of the input counts were bound to the filter; roughly 90% of these to fragment HhaB and 10% to fragment HhaC. The length of the fragments is given in Table I. All further experimental details are described elsewhere (10).

cuts the DNA into 2 fragments, HhaI specific for the sequence 5'GCGC3' gives 4 fragments and Hind II with the recognition sequence of 5'GTPyPuAC3' has 5 cleavage sites. The lengths of the fragments obtained with these enzymes is given in Table I. The order of the fragments was estimated by analysis of double digests with various other enzymes. The cleavage sites are indicated in Fig. 3. The HpaI cleavage sites do not completely coincide with some of the Hind II sites as one would expect on the basis of the recognition sequences. Due to a shortage in HpaI a double digest of the two enzymes was not prepared. The distance on our map between the HpaI cleavage sites and the nearest Hind II cleavage site is, however, less than 1% of the genome length and may well reflect the inherent inaccuracies of the methods used. For the Hha digest stripfilter hybridization with total mitochondrial rRNA revealed that the sequences complementary to the rRNAs are present on the Hha I fragments B and C, which are, therefore, adjacent. This experiment is illustrated in Fig. 1, which shows the UV-photograph of the original gelslot aligned with the autoradiogram of the stripfilter after hybridization with ^{125}I-labelled rRNA.

IDENTIFICATION OF THE REGIONS CONTAINING TRANSFER RNA GENES

To identify the fragments containing sequences complementary to mitochondrial tRNAs we used the stripfilter hybridization technique (11). The extraction of the tRNA fraction from rat-liver mitochondria presented, however, several difficulties because the fraction obtained was always contaminated with other RNAs, particularly fragments of the mitochondrial ribo-

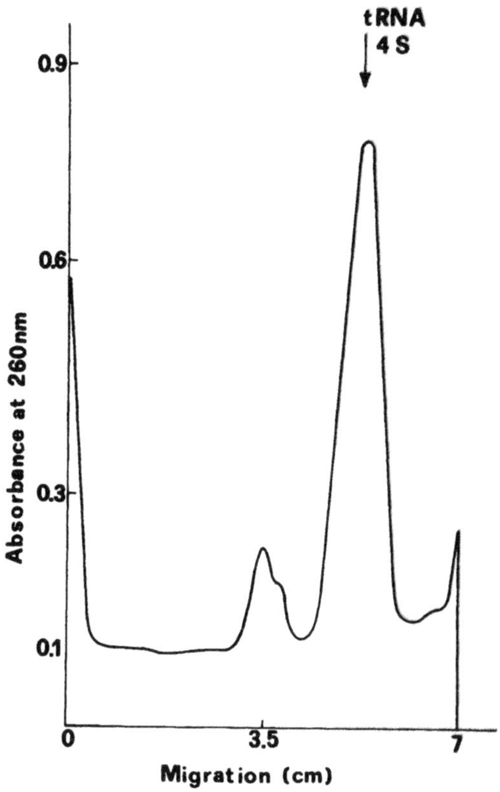

Fig. 2. *Purification of the 4S RNA fraction of rat-liver mitochondrial tRNA by electrophoresis on a 2.5% polyacrylamide gel.* The sharp and symmetrical peak was eluted from the gel, but appeared to be contaminated with rRNA fragments to such an extent that the addition of high amounts of competitor RNA in the hybridization experiments remained necessary.

somal RNAs. We used the method Buck and Nass (12) for tRNA isolation in the experiments presented here. After ethanol precipitation tRNA was iodinated with ^{125}I as described elsewhere (10). For some preparations after extraction the tRNA fraction was further purified by gel electrophoresis in 2.5 - 2.7% polyacrylamide and the peak (Fig. 2) eluted from the gel was used for iodination and hybridization. Even in this case, however, the fraction seemed to be contaminated with degradation products of larger RNA species, so that the hybridization experiments had always to be carried out in the presence of competitor ribosomal RNA. In lower eukaryotes the extraction of tRNA from mitochondria does not present these inconveniences to this extent. This is probably due to the presence in the rat-liver mitochondrial fraction of very stubborn nucleases giving rise to digestion products which are extracted with the fraction containing tRNAs.

Table II gives the quantitative data of the hybridizations of ^{125}I-labelled tRNA with the fragments obtained after complete digestion of mtDNA with Eco RI, Hind III, Hap II or Hha I and after double digestion with Hind III x Bam HI. The results are expressed as percentage of the total radioactivity bound to the fragments on the stripfilters. The minimum percentage found to any fragment varied around 5% in all experiments. The number of tRNA genes on each of the fragments has been provisionally calculated taking this 5% hybridization as corresponding to one tRNA gene. A total number of between 16 and 23 tRNA genes is obtained in this way. This figure is in good agreement with previous observations (1, 2, 5, 10). For the calculations it was assumed that the tRNAs are present in about equal amounts and that iodine-label is equally distributed over the tRNAs.

Taking into account the overlap of the various restriction fragments we have located the tRNA genes on the physical map of the mtDNA. This is shown in Fig. 3. The tRNA genes are scattered on the genome, leaving only a few areas suitable for coding the polypeptides of the mitochondrial translation products. We have placed one of the tRNA genes between the 16S and 12S rRNA genes, because the Hap II fragment E consistently binds the labelled tRNAs in the presence of a large excess of competitor RNA and because the Hind III fragment A binds more counts than the Eco RI fragment A. By extending this type of experiments to digests of other restriction enzymes we hope to delimit the tRNA genes sufficiently precisely to enable analysis of their base sequence.

THE HETEROGENEITY OF RAT-LIVER MITOCHONDRIAL DNA

The question if mtDNA within one mitochondrion or one cell is heterogeneous in base sequence has been raised many times. By denaturation-renaturation experiments gross heterogeneity could be excluded. Restriction analysis offers another tool to this problem, especially if endonucleases are used for which the number of cleavage sites is relatively high. Heterogeneity between different strains of *Saccharomyces cerevisiae* has been shown in this way (13). In our experiments with the restriction enzyme

TABLE II

THE LOCALIZATION OF MITOCHONDRIAL TRANSFER RNA GENES ON VARIOUS RESTRICTION FRAGMENTS OF RAT-LIVER MITOCHONDRIAL DNA

Stripfilter hybridization of mitochondrial tRNA with restriction fragments was performed as described elsewhere (10). Specific radioactivity of the [^{125}I]iodinated tRNA was about 10-50.10^6 cpm/µg in the various experiments. The input was 5.10^6 counts per filter in the presence of a 50 - 70 fold excess of competitor rRNA in the various experiments. The abbreviations used for Eco RI and Hind III are Eco and HIII respectively. All further abbreviations and the designations of the fragments are the same as in Table I and ref. 10.

fragment	[^{125}I]- mt-tRNA bound		number of tRNA genes	fragment	[^{125}I]-mt-tRNA bound		number of tRNA genes
	counts/ min	% of total			counts/ min	% of total	
EcoA	4114	36	7-8	HhaA	1953	61	12-13
EcoB	3226	28	5-6	HhaB	573	18	3-4
EcoC	2894	24	4-5	HhaC	506	16	3-4
EcoD	435	4	0-1	HhaD	152	5	1
EcoE	379	3	0-1				
EcoF	444	4	0-1	HIIIA	1067	41	8-9
				HIIIB	882	34	6-7
HapA	2611	27	5-6	HIIIC	503	19	3-4
HapB	2508	26	5-6	HIIID	141	5	1
HapC	1312	14	2-3				
HapD	778	8	1-2	HIIIB	3293	37	7-8
HapE	346	4	0-1	HB1	2490	28	5-6
HapF	915	10	2	HB2	1591	18	3-4
HapG	522	5	1	HIIID	1222	14	2-3
HapH	626	6	1-2	HB3	338	4	1

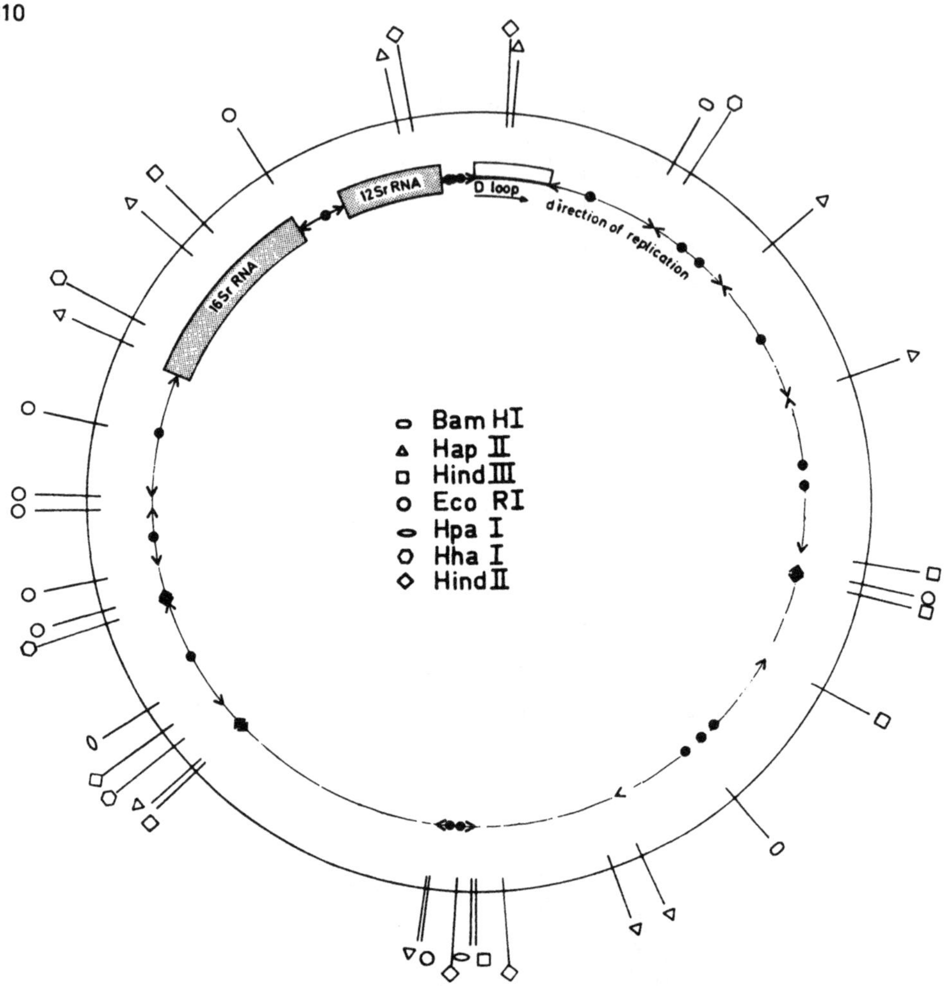

Fig. 3. *Restriction fragment map of rat-liver mitochondrial DNA and the positions of a number of genetic markers.* The outer circle represents the cleavage sites for the restriction endonucleases Bam H1, Hpa I, Eco RI, Hind II, Hind III, Hha I and Hap II. The length of the fragments is given partly in ref. 10 and partly in Table I of this paper. The positions of the cleavage sites were estimated by taking the sum of the molecular weights of the fragments of each enzyme as 100%. The length of the fragments as % of this sum was transposed to degrees of arc and placed on the circle in the right order, the direction of replication being clockwise. The cleavage site between the Eco RI fragments D and A was taken as the reference point for those sets of fragments for which we have not ascertained the origin of replication experimentally. Due to slightly unequal recoveries for the various enzymes the 100% values represent different total molecular weights in terms of basepairs (*cf.* ref. 10 and Table I). The inner circle shows the position of the rRNA genes and of the D-loop and the direction of replication (see also ref. 10). The areas in which transfer RNAs are located are indicated with the arrows; the provisional number of transfer RNAs corresponds with the number of black circles in each area. For further details see text.

Hha I, we noticed the presence in small amounts of two "partial" restriction products that could not be reconciled with any combination of the end fragments obtained. They did not contain the fragments Hha B or C as can be concluded from the hybridization experiment shown in Fig. 1. The sum of their molecular weights is equal to that of Hha A. The fragments were resistent to prolonged digestion with excess enzyme. Moreover, the relative amounts of these "partials" was not the same in different mtDNA preparations and in some preparations they were even absent. This is shown in Fig. 4, lanes A-D. The results together strongly suggested the presence of an extra Hha cleavage site in part of the mtDNA molecules. Comparing our results to those of Parker and Watson (9), it became clear that this odd cleavage pattern for Hha I is exactly the same as that found by these authors for their mtDNA from *Rattus norvegicus*. The most likely explanation is, that our Wistar strain (*Rattus rattus*) is not genetically pure. This observation further explains the results obtained previously (7, 8) with the enzyme Eco RI. We obtained two "partials", coinciding with the real partial digestion products containing the Eco RI fragments A plus B and D plus E. We explained this observation by assuming that the cleavage sites between these fragments were more resistent to digestion than the others. More likely, however, is the presence of two types of mtDNA in our preparations also at that time. The relative amounts of the Hha I fragments p1 and p2 and of the Eco RI fragments p1 and p2 (see Fig. 4, lane D and E) is indeed striking. Proof for this explanation gives the experiment of Fig. 4, lane F. Eco RI fragment p1 was isolated and digested with Hha I; three fragments were obtained indicating that Eco p1 consists of a homogeneous population of molecules all containing the extra Hha I cleavage site, adding a further ressemblance to *Rattus norvegicus* mtDNA (9). We have so far not been able to detect heterogeneity of mtDNA prepared from a single rat liver. This is in agreement with other studies, showing individual homogeneity in combination with intraspecific heterogeneity for animal mtDNA (14-16), but apparently different from the situation in higher plants (17). In our opinion the differences in the results of Parker and Watson (9) and ours have to be related wholly to the intraspecific differences between *Rattus rattus* and *norvegicus*, since the length of all fragments can be explained *vice versa* by the presence or absence of an extra cleavage site. The situation is different for the discrepancy in the Hha I maps of Koike

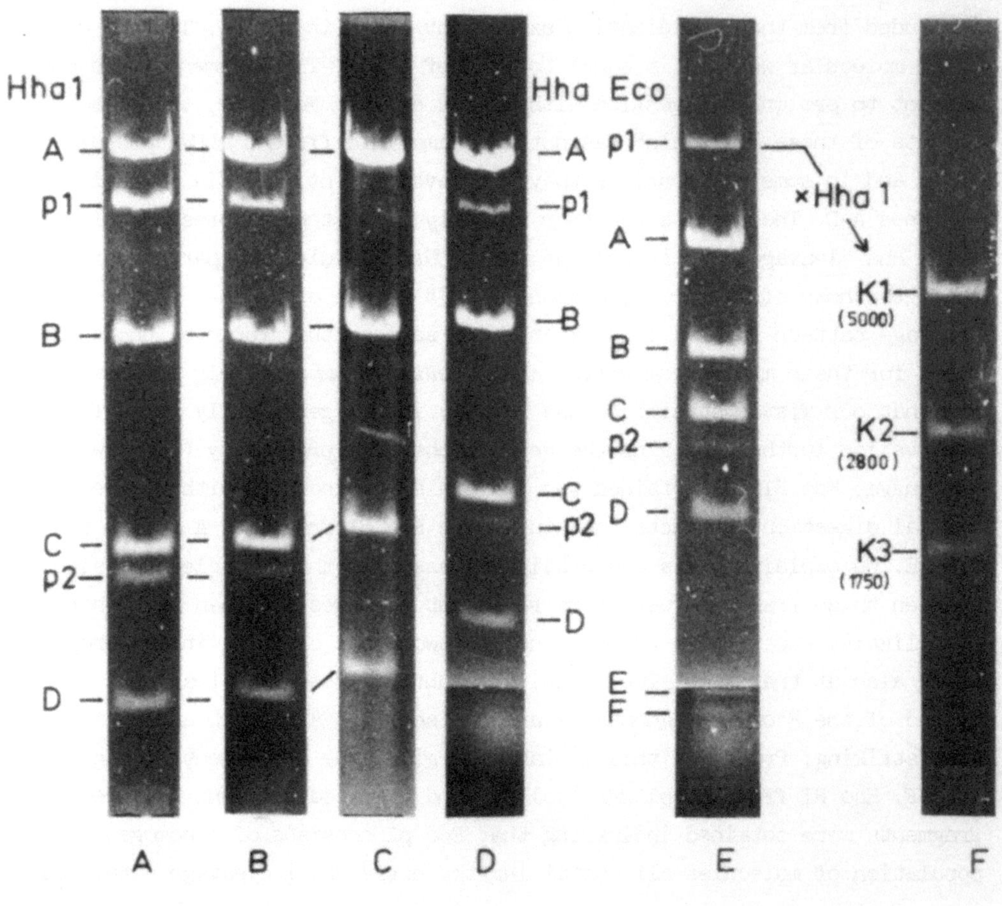

Fig. 4. *Slabgel electrophoresis of restriction fragments of mtDNA.* The composition of the slabgels, the conditions of electrophoresis and visualization of the fragments are described (8, 10). Lanes A, B, C and D are complete digests of different mtDNA preparations, treated with the endonuclease Hha I under the conditions described in the legend to Table I. The fragments indicated p1 and p2 are now considered endfragments (see text). The lanes A and B are photographs of two slots of the same gel, lane C and lane D are photographs of slots of two other gels. Lane E gives the fragment pattern of an Eco RI digest from the same gel as D. Also the fragments Eco p1 and Eco p2 are now considered endfragments. The fragment Eco p1 was isolated on a preparative scale and digested with Hha I. The three fragments obtained in this digest can be seen on the photograph of lane F; they are indicated k1, k2 and k3, their molecular weight in basepairs is given between brackets. For further details see text.

et al. (6) and ours; the Hha I fragment order BADC is most likely the correct one, since it is confirmed by the hybridization experiment shown in Fig. 1.

ACKNOWLEDGEMENTS

These studies were supported in part by grants from the Consiglio Nazionale delle Ricerche (CNR), Italy to CS and from the Netherlands Foundation for Chemical Research (SON) with financial aid of the Netherlands Organization of Pure Research (ZWO) to AK. We wish to thank Karin van Wijk for preparing the manuscript.

REFERENCES

1. Wu, M., Davidson, N., Attardi, G., Aloni, Y.: Expression of the mitochondrial genome in HeLa cells. XIV The relative positions of the 4S RNA genes and of the ribosomal RNA genes in mitochondrial DNA, J. Mol. Biol., 71, 81-93 (1972).
2. Dawid, I.B., Klukas, C.K., Ohi, S., Ramirez, J.L., Upholt, W.B.: Stucture and evolution of animal mitochondrial DNA, in: Saccone, C., Kroon, A.M. (Eds.), The genetic function of mitochondrial DNA, Elsevier/North-Holland Biomedical Press, Amsterdam, pp. 3-12 (1976).
3. Robberson, D.L., Clayton, D.A., Morrow, J.F.: Cleavage of replicating forms of mitochondrial DNA by Eco RI endonuclease, Proc. Nat. Acad. Sci. U.S.A., 71, 4447-4451 (1974).
4. Brown, W.M., Vinograd, J.: Restriction endonuclease cleavage map of animal mitochondrial DNAs, Proc. Nat. Acad. Sci. U.S.A., 71, 4617-4621 (1974).
5. Attardi, G., Albring, M., Amalric, F., Gelfand, R., Griffith, J., Lynch, D., Merkel, C., Murphy, W., Ojala, D.: Organization and expression of the mitochondrial genome in HeLa cells, in: Bücher, T., Neupert, W., Sebald, W., Werner, S.: Genetics and Biogenesis of Chloroplasts and Mitochondria, Elsevier/Noth-Holland Biomedical Press, Amsterdam, pp. 573-585 (1976).
6. Koike, K., Kobayashi, M., Tanaka, S., Mizusawa, H.: Physical map and replication of rat mitochondrial DNA, in: Bücher, T., Neupert, W., Sebald, W., Werner, S.: Genetics and Biogenesis of Chloroplasts and Mitochondria, Elsevier/North-Holland Biomedical Press, Amsterdam,

pp. 593-596 (1976).

7. Saccone, C., Pepe, G., Cantatore, P., Terpstra, P., Kroon, A.M.: Mapping of the transcription products of rat-liver mitochondria by hybridization: in: Saccone, C., Kroon, A.M. (Eds.), Function of mitochondrial DNA, Elsevier/North-Holland Biomedical Press, Amsterdam, pp. 27-36 (1976).
8. Kroon, A.M., Bakker, H., Holtrop, M., Terpstra, P.: The restriction endonuclease cleavage map of rat-liver mitochondrial DNA, Biochim. Biophys. Acta, 474, pp. 61-68 (1977).
9. Parker, R.C., Watson, R.M.: Restriction endonuclease cleavage maps of rat and mouse mitochondrial DNAs, Nucleic Acids Res., 4, 1291-1299 (1977).
10. Kroon, A.M., Pepe, G., Bakker, H., Holtrop, M., Bollen, J.E., Van Bruggen, E.F.J., Cantatore, P., Terpsta, P., Saccone, C.: The restriction fragment map of rat-liver mitochondrial DNA, Biochim. Biophys. Acta, 478, 128-145 (1977).
11. Southern, E.M.: Detection of specific sequences among DNA fragments separated by gel electrophoresis, J. Mol. Biol., 98, 503-517 (1975).
12. Buck, C.A., Nass, M.M.K.: Differences between mitochondrial and cytoplasmic transfer RNA and aminoacyl transfer RNA synthetases from rat-liver, Proc. Nat. Acad. Sci. U.S.A., 60, 1045-1052 (1968).
13. Bernardi, G., Prunell, A., Fonty, G., Kopecka, H., Strauss, F.: The mitochondrial genome of yeast: in: Saccone, C., Kroon, A.M. (Eds.): The genetic function of mitochondrial DNA, Elsevier/North-Holland Biomedical Press, Amsterdam, pp. 185-198 (1976).
14. Potter, S.S., Newbold, J.E., Hutchinson, C.A. III, Edgell, M.H.: Specific cleavage analysis of mammalian mitochondrial DNA, Proc. Nat. Acad. Sci. U.S.A., 72, 4496-4500 (1975).
15. Fauron, C.M.R., Wolstenholme, D.R.: Structural heterogeneity of mitochondrial DNA molecules within the genus *Drosophila*, Proc. Nat. Acad. Sci. U.S.A., 73, 3623-3627 (1976).
16. Upholt, W.B., Dawid, I.B., Functional organization and evolution of animal mitochondrial DNA, in: Bücher, T., Neupert, W., Sebald, W., Werner, S. (Eds.): Genetics and Biogenesis of Chloroplasts and Mitochondria, Elsevier/North-Holland Biomedical Press, Amsterdam, pp 587-592 (1976).

17. Quetier, F., Vedel, F.: Heterogenous populations of mitochondrial DNA molecules in higher plants, Nature, 268, 365-368 (1977).

GENETIC, PHYSICAL AND BIOCHEMICAL ANALYSIS OF A MITOCHONDRIAL GENE.

R. A. Butow[1], R. D. Vincent[2], R. L. Strausberg[1], E. Zanders[1], and P. S. Perlman[2,3], Department of Biochemistry, The University of Texas Health Science Center at Dallas[1], and Developmental Biology Program[2], and Department of Genetics[3], Ohio State University.

INTRODUCTION

Many laboratories are actively engaged in the detailed characterization of the genes present on yeast mitochondrial DNA. Important contributions have been made recently by the isolation of mit$^-$ mutants (1), syn$^-$ mutants (2,3) and conditional mutants (3-5). A new direction of research on this problem is the construction of a physical map of yeast mtDNA including the genetic loci currently available. The earliest physical mapping data were obtained by the correlation of genetic studies with biochemical analysis of mtDNA from petite mutants (deletion mapping) (6-9). More recently, studies with restriction endonucleases using mtDNA from wild-type, petite and mit$^-$ strains has refined the preliminary physical map (7,9-11).

Recently we identified a new locus on mtDNA, termed *var1* (12). The phenotype of mutant alleles at this locus is assessed by measuring the molecular weight of the protein product of this gene. Strains with different alleles of *var1* produce var1 proteins of different molecular weights. Analysis of many wild-type strains from our collection has revealed a number of different allelic forms of *var1*. Using two of these alleles, originally termed *var1*^{f} and *var1*^{s}, we applied the procedures of recombination and petite deletion and mapping and mapped the *var1* locus to the region between *ery* (*RIBIII*) and *oli1* (13,14).

In this communication we describe further our characterization of the varl gene and gene product. We initially set out to learn whether the differences between varl proteins are due to additions or deletions in the structural gene primary sequence. To this end we have analyzed the protein in greater detail by peptide analysis and found that significant differences exist in the primary sequence of the allelic forms. We have mapped the gene physically using restriction endonuclease analysis of mtDNA isolated from wild-type strains having different varl alleles. Our initial observation was that a restriction fragment, mapping between ery and olil is larger in $varl^{s}$ than in $varl^{f}$ strains. Deletion mapping coupled with restriction analysis of the petite mtDNA confirms the localization in a manner independent of protein size differences.

We have also developed a new method, termed zygotic gene rescue, capable of assessing the presence of intact structural genes on petite mtDNA (15). This method also supports our physical localization of the varl gene and should be of further use in locating other structural genes.

Finally, during our studies of different varl alleles, we have made a new and unexpected observation: when crosses are performed between strains having either of the two largest forms and strains having the smallest form of varl polypeptide, some diploid progeny cells are obtained with a new form of varl, different from that of either parent. The new forms have been shown to be recombinants and in some crosses occur at frequencies approaching recombination values for unlinked markers (about 20%).

Our results strongly suggest that the structural genes for different varl alleles are of different size and thus the protein differences are explained by deletions/insertions within the varl gene. The fingerprinting and recombination data are the first steps toward a fine structure analysis of the varl gene, the function of which remains to be elucidated.

RESULTS AND DISCUSSION

Variability of var1

Fig. 1 shows a composite gel of $^{35}SO_4^{=}$-labeled mitochondrial translation products from six yeast strains. It is apparent that all six strains show a unique mobility for var1 polypeptide ranging in apparent molecular weight from 46,500 for the largest (5DSS) to 42,100 for the smallest (1-1/1). The apparent molecular weights for all of the strains shown in Fig. 1 are summarized in Table 1.

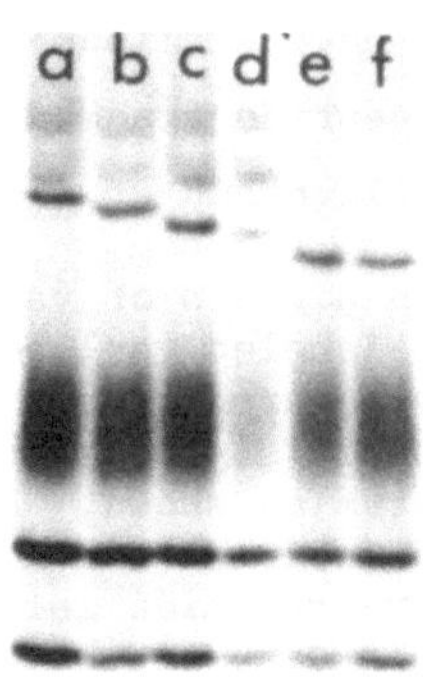

Fig. 1. Allelic forms of var1 gene product. Cell labeling procedures and analysis of mitochondrial translation products on SDS/polyacrylamide slab gels have been previously described (12). The strains displayed are: a) 5DSS; b) D6; c) 55R5-3C/221; d) NP1-3D; e) 4I-6/161 and f) 1-1/1.

TABLE 1

APPARENT MOLECULAR WEIGHTS FOR VAR1 POLYPEPTIDE:CORRELATION WITH A SPECIFIC RESTRICTION ENDONUCLEASE FRAGMENT

STRAIN	APPARENT PROTEIN MW	RESTRICTION FRAGMENT MW X 10^{-6}	
		Hinc II Band 10	Hinc II Band 10 + BamI[a]
5DSS	46,500	1.79	1.375
D6	45,800	1.795	--
55R5-3C/221	44,900	1.77	1.36
NP1-3D	44,400	1.755	--
4I-6/161	42,400	1.725	1.33
1-1/1	42,100	1.720	1.33

a. These values are for the large MW BamI fragment of band 10

One important question obviously concerns the molecular basis for the different var1 forms. Experiments designed to determine if the mobility differences are the result of post-translational modification reactions, for example, variable glycosylation of the polypeptide, were all negative. In addition, analysis of the mobilities of the different var1 species as a function of acrylamide concentration did not show anomolous migration behavior. Thus, it does not appear that some strain dependent chemical modification of the var1 polypeptide is responsible for the observed differences in mobility.

To investigate the properties of var1 further, we have examined the polypeptide in four different strains by analysis of peptide fragments according to the procedure recently described by Cleveland et al. (16).

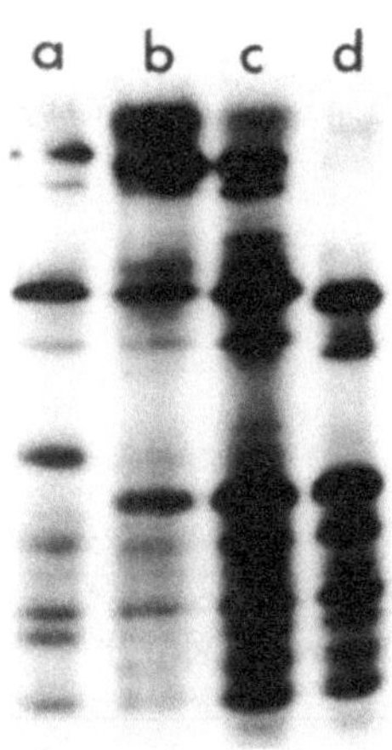

Fig. 2. Papain digests of var1 polypeptide. $SO_4^{=}$-labeled var1 polypeptide was located by autoradiography of slab gels of mitochondrial translation products of the strains shown. Digestion of the labeled material and analysis on 15% SDS/polyacrylamide gels was carried out according to the procedure of Cleveland et al. (16).
a) 5DSS; b) 55R5-3C/221; c) NP1-3D; d) 1-1/1.

The data shown in Fig. 2 allow us to draw several conclusions:

1. The pattern of peptides for var1 obtained by papain digestion shows considerable similarity between the strains; thus there is a degree of homology in their primary structure.

2. Mobility differences are readily apparent between the strains for some of the higher molecular weight bands. These most probably represent

partial digestion products and indicate differences in the absolute number of amino acid residues for a given fragment among the various varl species.

3. The presence and absence of bands can be detected in a comparison of the fragment pattern indicating unique cleavage sites.

Although in this experiment the comparisons are limited to sulfur-containing peptides, it is evident that while there exists sequence homology between the varl species, differences at the level of the primary structure nevertheless exist. Taken together, these results suggest that the mobility differences between the varl polypeptides are the result of amino acid additions or deletions.

Physical Mapping:Analysis of ρ^+strains

In light of the above results, we may consider the possibility that the genes for the largest and smallest varl proteins should differ by roughly 72,000 daltons. In principle such a difference, as well as differences between varl genes more similar in size, could be detected using restriction endonuclease analysis of mtDNA if enzymes yielding fairly small fragments in the region of interest (ery-oli1) are used.

We have analyzed the restriction patterns of mtDNA from more than 20 wild-type strains from our collection using six restriction enzymes (EcoR1, Hinc II, Hind III, Bam I, Sal and Pst). The first strain analyzed in detail was 41-6/161 and its restriction map is shown in Fig. 3. The map closely resembles that of strain MH41-7B already published by other workers (7). Although significant differences in restriction patterns exist among the strains analyzed below, it has been possible to identify homologous fragments in the region of interest from all strains shown. Data from other laboratories (7,11) (and our own deletion mapping data) show that ery (RIBIII) maps in Hinc II band 7 (3.76×10^6 daltons) whereas oli1 maps in Hinc II band 8 (3.04×10^6 daltons); thus, Hinc II band 10 (1.725×10^6 daltons) is an attractive candidate for the location for varl.

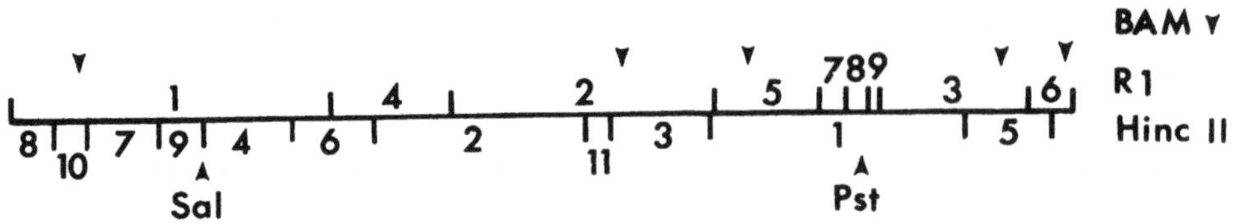

Fig. 3. Restriction endonuclease map of mtDNA from strain 4I-6/161. A physical map was constructed using restriction endonucleases EcoRI, Hinc II, Bam I, Sal I and Pst.

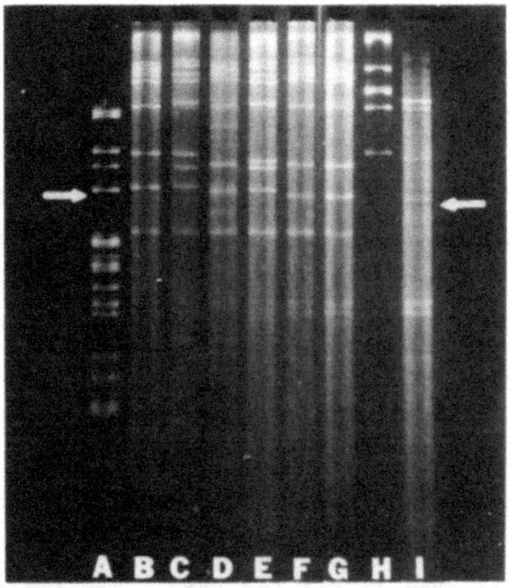

Fig. 4. Hinc II restriction endonuclease pattern of mtDNA. Restricted samples were analyzed on gels containing 2% agarose. The largest restriction fragments are not resolved in this gel but are regularly seen on gels with a lower concentration of agarose. A) phage lamda DNA (Hinc II); B) 5DSS mtDNA; C) D6 mtDNA; D) 55R5-3C/221 mtDNA; E) NPI-3D mtDNA; F) 4I-6/161 mtDNA; G) 1-1/1 mtDNA; H) phage lamda DNA (EcoRI); I) 4I-6/161 nuclear DNA. Samples were arranged in correspondence with decreasing size of the varl protein (Fig. 1). The region of band 10 (B-G) is indicated by the arrows.

Figure 4 represents Hinc II patterns of mtDNA isolated from the six wild-type strains whose varl alleles are defined in Fig. 1 and Table 1. These samples are analyzed on 2% agarose gels to enhance resolution of the three small bands of interest; the two largest fragments (1 and 2) are not resolved in this gel but are routinely observed in gels with lower agarose concentrations (see data in Mahler et al., this volume). The data are displayed in order of decreasing varl size. It is clear that the band 10 homologs (using strain 4I-6/161 to define the band numbers) vary in molecular weight (see also Table 1) and that there is a reasonably good correlation between the order of molecular weights of varl polypeptide and band 10. These data have been extended by analysis of mtDNA from a number of other wild-type strains with different varl alleles (not shown). For the six strains shown here, few differences in mobility are noted for band 7, whereas band 8 varies in size as does band 10 except for strains 55R5-3C/221 and 5DSS. The many other differences in these restriction patterns are not relevant to this analysis because they are located outside of the region of interest. These results are compatible with the conclusion that the regions of differences between different varl alleles are located within Hinc II band 10, a DNA segment of $1.7\text{-}1.8 \times 10^6$ daltons.

As shown in Fig. 3, there is a Bam I site in the Hinc II band 10 (of the standard strain, 4I-6/161) that splits band 10 into fragments of 1.33×10^6 and 0.39×10^6 daltons. Since four of the strains shown have this Bam I site we examined the molecular weight of the largest Hinc II-Bam I piece of band 10 in the cases of strains 5DSS, 55R5-3C/221, 4I-6/161 and 1-1/1 (Fig. 5; the absence of the Bam I site in strains D6 and NP1-3D is also shown in this Fig.). As is summarized in Table 1, the large derived fragment retains most of the size differences observed with Hinc II alone. These data indicate that at least a portion of the varl gene (including some allelic differences) is located to the left of the Bam site in fragment 10. Further resolution must await fine structure analysis using multi-site enzymes and such studies are now in progress.

An additional experiment using wild-type strains is shown in Fig. 6. A cross was performed between strains 55R5-3C/221 (varl [44.9]) and 4I-6/19 (varl [42.4]), a sister spore of 4I-6/161. A recombinant (strain EOP 88II-3) of genotype cap^s ery^r $oli1^r$ par^r varl (44.9) was isolated and its mtDNA analyzed using Hinc II (and other enzymes). To generate this strain, crossovers between varl and oli1 and between cap and par were required (as a

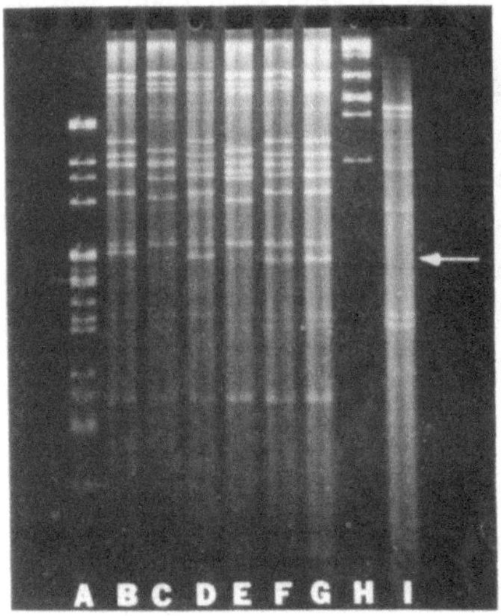

Fig. 5. Hinc II + Bam I restriction endonuclease pattern of mtDNA. Restricted samples were analyzed on gels containing 2% agarose. A) phage lamda DNA (Hinc II only); B) 5DSS mtDNA; C) D6 mtDNA: D) 55R5-3C/221 mtDNA; E) NP1-3D mtDNA; F) 4I-6/161 mtDNA; G) 1-1/1 mtDNA; H) phage lamda DNA I) 4I-6/161 nuclear DNA. The region of the large Bam I fragment of band 10 is indicated by the arrow.

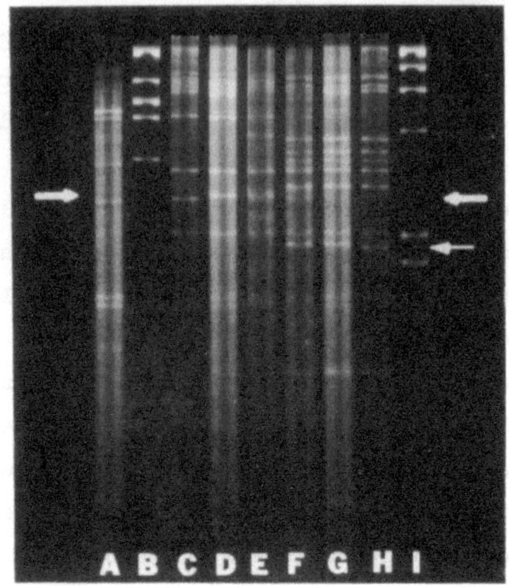

Fig. 6. Restriction endonculease pattern of strains 4I-6/161, 55R5-3C/221 and EOP 88II-3 recombinant. Samples were restricted with Hinc II (A,C,D, and E), Hinc II + Bam I (F,G, and H), EcoRI (B) or Hinc III (I) and electrophoresed through 2% agarose gels. A)4I-6/161 nuclear DNA; B) phage lamda DNA; C) 4I-6/161 mtDNA; D) EOP 88-II-3 mtDNA; E) 55R5-3C/221 mtDNA; F) 55R5-3C/221 mtDNA; G) EOP 88-II-3 mtDNA; H) 4I-6/161 mtDNA; I) phage lamda DNA. The bold arrows denote the Hinc II band region and the thin arrow denotes the large Bam I fragment of band 10.

minimum). The recombinant has band 10 identical to that of the var1 (44.9) parent while most other bands are identical to those of strain 4I-6/19, as expected (Fig. 6). In our collection of analyzed strains, we have made several other observations of this sort which support the conclusion that var1 and Hinc II fragment 10 are inherited as a unit.

Physical Mapping:Deletion Mapping Using Petites

As a confirmation of the above results, we examined mtDNA from several sets of petites of known mitochondrial genotype. We first established the presence or absence of varl by recombination analysis. Fig. 7 summarizes the results for petites derived from wild-type strain AT3. Petite AT3ρ-4 has the genotype cap^r ery^r oli^r par^s and varl (42.4) and retains Hinc II bands 7, 9 and 10 (intact) and a portion of band 8 (plus other fragments outside the region of interest). The other three petite shown are derived directly from ATP3ρ-4 by the isolation of spontaneous segregants:

i) O^r1-1/2 retains $oli1^r$ and varl but has lost cap^r and ery^r
ii) CE-1 retains cap^r, ery^r and varl but has lost oli^r
iii) CE-4 retains cap^r and ery^r but has lost varl and oli^r

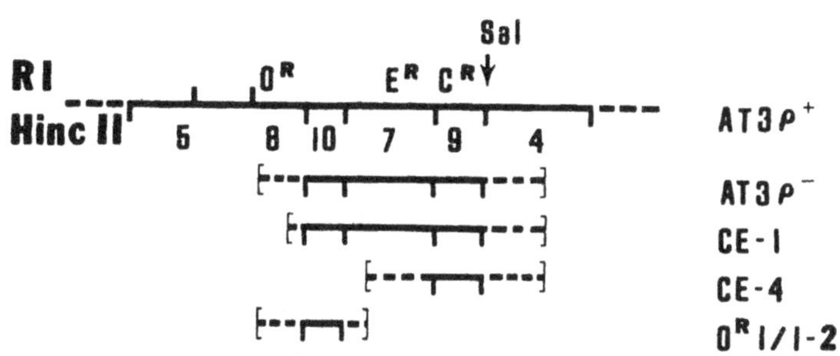

Fig. 7. Deletion mapping of varl. Petites were isolated from strain AT3, analyzed for retention or loss of all drug and mit loci and then mtDNA analyzed using restriction endonucleases.

These data eliminate most of band 7 (see O^r1-1/2) and band 8 (see CE-1) from consideration as the location of varl. A similar conclusion was reached using two other sets of petites of genotypes oli^r varl (42.4) and oli^r var^o derived from the wild-type strains 4I-6/19 and 41-6/161 both of which have the same varl allele as AT3. Using this deletion mapping protocol, our conclusions about the location of var1 is not dependent on the allelic form of var1 and thus lend support to the conclusion drawn from the analysis of wild-type strains having different var1 alleles.

Analysis by Zygotic Gene Rescue

As a further extension of the genetic and physical mapping studies of the var1 gene, we have employed a new procedure, termed zygotic gene rescue, to detect the presence of intact structural genes (and probably regulatory elements, as well) on the petite mtDNA (15). This procedure is based on the notion that genes retained on mtDNA of petite strains can be expressed in zygotes of a $\rho^+ \times \rho^-$ cross. The validity of this notion has been confirmed by analysis of the var1 product in zygotes of crosses between petites, which we demonstrated by transmission analysis to carry at least a portion of the var1 gene, and a wild-type tester with a different allelic form of var1. After 4-5 hours of mating, the var1 product characteristic of the petite appears in zygotes in amounts which cannot be accounted for exclusively by recombination; that is, it appears to be expressed by the ρ^+ mitochondrial transcription and translation apparatus.

The zygotic gene rescue technique has now been applied in the physical mapping of the var1 gene because it allows us to determine if the petite var1 gene is intact. The series of petites derived from strain AT3 described above was analyzed by the zygotic gene rescue procedure. The results shown in Fig. 8 are consistent with the proposed relationship between Hinc II band 10 and var1: petites AT3ρ^-4, CE-1 and 0^r 1/1-2 retain band 10 and show rescue of the var1 product characteristic of the parental wild-type strain, whereas petite CE-4 does not have band 10 and does not show rescue of the var1 product. Further application of this procedure, together with fine structure physical mapping of the var1 region, should enable us to define more precisely the physical limits of the var1 gene.

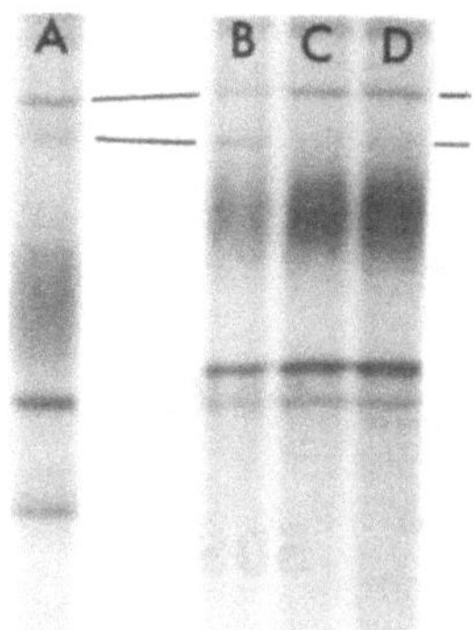

Fig. 8. Zygotic gene rescue of petites derived of strain AT3ρ+. The protocol for zygotic gene rescue has been described (15). The lines in the Fig. indicate the var1 species of the ρ+ tester (upper line) and the var1 species characteristic of strain AT3ρ+ (lower line). A) AT3ρ-; B) CE-1ρ-; C) CE4-ρ-; and D) O^r 1-1/1 ρ-.

Recombination and the Formation of New Var1 Alleles

In our original studies we analyzed only strains taken from our collection without regard to their pedigrees. Recently we have performed crosses between strains having different var1 alleles and have made the unexpected observation that some progeny of such crosses exhibit a form of var1 different from that of both parents.

The first new form of var1 obtained as a result of a cross is var1 (46.5), strain 5DSS. Restriction patterns of the two parent strains (55R-3C/221 and D6) and the new form have already been presented (Figs. 4 and 5). Insight to the mode of origin of this var1 allele can be obtained by a careful analysis of the molecular weights of the Hinc II bands flanking band 10 and the band 10 Bam I site. This is illustrated in Fig. 9. It is clear that 5DSS mtDNA is recombinant and that a crossover (or gene conversion) event occurred within the right side of band 10 (as drawn); that is, between the region of difference between D6 band 10 and 55R5-3C/221 band 10 and the Bam I site. Although we have only preliminary fingerprinting data on these strains, it is noteworthy that the new var1 allele was obtained in a strain that is related to its two parents by a recombination event in band 10.

In most of the crosses examined thus far, the recombinant forms of var1 are generally intermediate in molecular weight with respect to the parental types, and for a given cross, generally only a unique intermediate species is apparent.

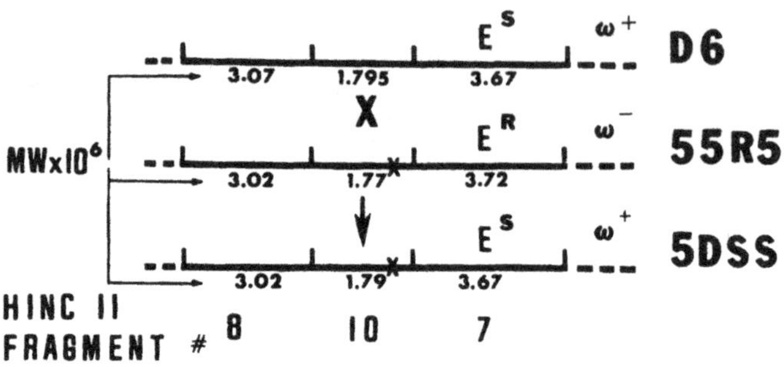

Fig. 9. Strain 5DSS is recombinant in Hinc II band 10. The Fig. illustrates the ery-olil region of the physical map of strain 5DSS and its two parent strains D6 and 55R5-3C/221. Note that 5DSS has the Bam I site in Hinc II band 10 from strain 55R5-3C/221 and the ery marker in band 7 from strain D6. Band 8 in 5DSS has the size of band 8 in 55R5-3C/221.

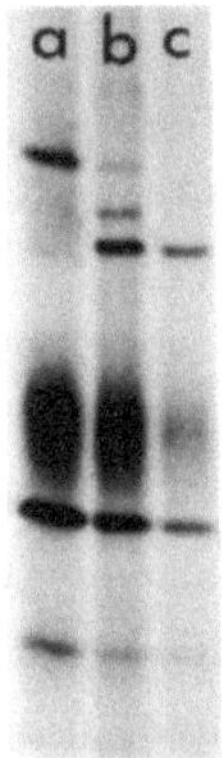

Fig. 10. Non-parental form of var1 polypeptide. Mitochondrial translation product from random diploids from the cross between strain 5DSS and 4I-6/19 were analyzed as described (12). A) 5DSS; B) random diploid population, and C) 4I-6/19.

Fig. 10 shown an autoradiograph of a gel of mitochondrial translation products in two haploid strains and in random diploids (total progeny analysis) from a cross between the two strains. It can be seen that the diploid progeny include three distinct var1 forms: the two parental types and a form of intermediate molecular weight. Quantitative analysis of the relative amounts of the different var1 forms reveals that the var1 intermediate represents roughly 20% of the total var1 population. Such a high rate of apparent intragenic recombination (see below) is unexpected since experiments from many laboratories have demonstrated that the recombination frequency between two unlinked mitochondrial loci is on the order of 20%.

We have also observed that the amount of intragenic recombination appears related directly to the absolute difference between the molecular weights of the parental var1 species; the larger the difference, the greater the frequency of var1 recombination. Table 2 shows the results of several crosses involving different allelic forms of the var1 gene. The proportion of recombinant var1 polypeptide among the random diploid progeny is seen to increase as the molecular weight difference between the two parental var1 forms increases. This unexpected high frequency of recombination of the var1 locus provides a satisfying explanation for the different mobility forms of var1 polypeptide seen in yeast stocks. While we have not observed any selection advantage to cells carrying the different forms of var1 thus far, it will be of considerable interest to understand the function of a protein which apparently can tolerate relatively dramatic alterations in its primary structure.

TABLE 2

TRANSMISSION OF VAR1

Cross	Var1 Transmission, % of Total Diploid Progeny		
	s	f	I
1. 5DSS, ω^+ var1^s (46.5) X 4I-6/19, ω^- var1^f (42.4)	16	66	18
2. D6, ω^+ var1^s (45.8) X 41I6/161, ω^- var1^f (42.4)	63	31	6
3. D6, ω^+ var1^s (45.8) X D-243, ω^+ var1^s (42.4)	54	40	6
4. 55R5-3C/221, ω^- var1^s (44.9) X 4I-6/19, ω^+ var1^f (42.4)	43.5	56	0.5

In each cross, the strain is given underlined followed by the relevant mitochondrial genotype. For simplicity, the mobility of the var1 polypeptide is indicated as s for the larger molecular weight form and f for the smaller molecular weight form. The absolute molecular weights X 10^{-3} are indicated in parenthesis. I represents a var1 recombinant with mobility falling in between the parental types.

Since the _var1_ gene maps in a region of the mitochondrial genome not far removed from ω, it is conceivable that the observed intragenic recombination may be controlled by this locus. However, we observe roughly the same amount of var1 intragenic recombination regardless of the configuration of the cross, with respect to ω; thus we do not believe _var1_ recombination to be under the control of this locus.

The segregation of the _var1_ gene in crosses is also of interest. Studies with antibiotic resistance loci indicate that mitochondrial genes segregate rapidly; that is, a heteroplasmic zygote produces diploid progeny which are all homoplasmic within about 20 generations. In the case of _var1_ rapid segregation seems to occur only when the two parental var1 species are present in a cell. On the other hand, in crosses where a high proportion of an intermediate form of var1 is produced, segregation to pure mitochondrial genotypes requires, in some cases, more than 60 generations. A more thorough analysis of this phenomenon, particularly with regard to flanking marker distribution, should provide insight into the mechanism of segregation of mitochondrial genes.

We have analyzed mtDNA from two diploids issued from the cross between 5DSS and 4I-6/19 having non-parental forms of var1 (see Fig. 10). In Fig. 11 are presented Hinc II (C-F) and Hinc II + Bam I (H-K) restriction patterns using mtDNA from the two diploids and the two parental strains. Unexpectedly, although the two diploids make the same size non-parental var1 protein, the Hinc II band 10 of diploid 2-1 resembles that of 5DSS while that of diploid 5-1 resemble that of 4I-6/19 (Fig. 11, C-F). In the Hinc II + Bam I double digestion (Fig. 11, H-K), strains 2-1 and 5-1 have the same size large Bam I fragment of band 10 and it is intermediate in size between that of the two parental strains. It is clear that the large and small Bam I fragments of band 10 are non-parental and size differences in the small Bam I fragment of band 10 do not correlate with the var1 protein molecular weight. Our analysis of these two non-parental strains, in agreement with our analysis of the construction of strain 5DSS (Fig. 9), shows that modifications in band 10 correlate with modification in the _var1_ gene product.

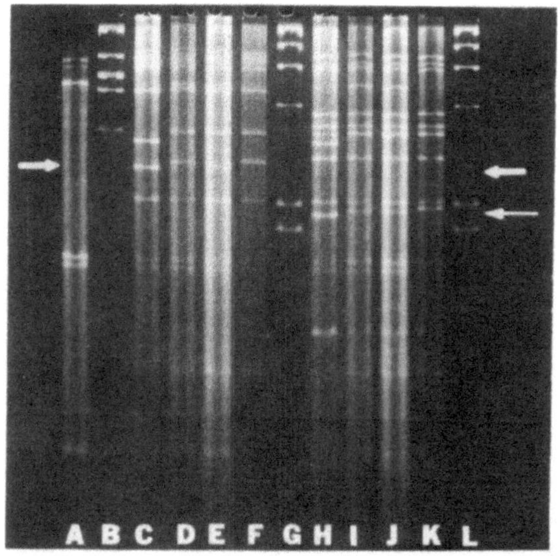

Fig. 11. Hinc II and Bam I restriction endonuclease pattern of mtDNA. Restricted samples were analyzed on gels containing 2% agarose. Note that diploid strains 2-1 and 5-1 have the same var1 forms which are intermediate in mobility from their parents. A) nuclease DNA; B) phage lamda (EcoRI); C) 4I-6/161; D) diploid 2-1; E) diploid 5-1; F) 5DSS; G) phage lamda, Hinc III; H) 4I-6/161; I) diploid 2-1; J) diploid 5-1; K) 5DSS and L) phage lamda (Hind III). C-F are Hinc II digests and H-K are Hinc II + Bam I digests.

Models for the Origin of Non-Parental Var1 Genes

Fig. 12 shows schematic representations of the generation of various non-parental forms of var1 through intragenic recombination. These tentative models depict heteroduplex pairing events between molecules which contain a large and small form of the var1 gene. For simplicity, forms of the var1 gene are shown to differ by additions and deletions of sequences corresponding to N and C terminal portions of the polypeptide; however, similar arguments discussed below would apply if the additions and deletions occur within the structural gene and are separated by regions of homology.

By reciprocal crossing over, forms of the var1 polypeptide which are of a size intermediate between the parentals, or larger than the parental forms, occur only if at least two non-homologous portions exist as shown here at

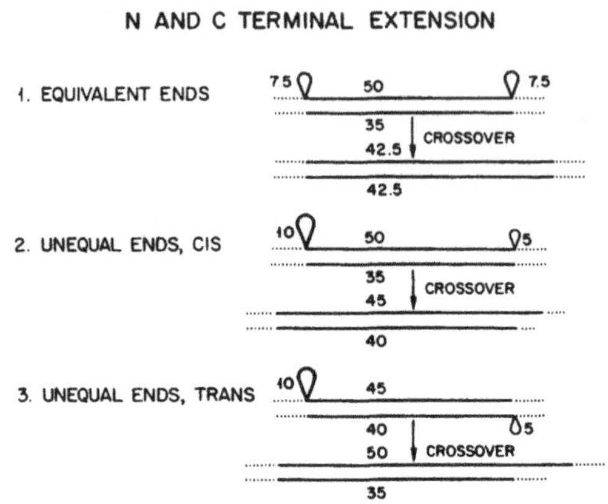

Fig. 12. Models for generation of recombinant forms of var1. Hypothetical heteroduplex pairing is shown for allelic forms of the <u>var1</u> gene; lengths indicated have been arbitarily chosen. Following a reciprocal crossover, the recombinant products with predicted gene lengths are shown below each arrow.

both the N and C termini; no forms of var1 of intermediate size would be generated with a single added or deleted portion whether that region of non-homology occurred somewhere in the middle of the gene or at either end. The model also shows that if the terminal extensions or deletions are equal in size, only a <u>single</u> non-parental form of intermediate size is generated by a crossover occurring anywhere within the gene; equal ends in the heteroduplex would give rise to <u>two</u> intermediate forms of unique size by a reciprocal recombination event. The models make a number of predictions which can be tested:

1. Heteroduplex analysis of the <u>var1</u> segment of the genome for strains which in a cross, generate unique, non-parental species should show at least two looped-out regions.

2. If only a single unique form of var1 is generated in a cross, then the looped-out regions should be equal in size; otherwise some process of gene conversion or non-reciprocal recombination must be considered to account for the appearance of a unique, non-parental form of var1.

3. Peptide analysis of parental and intermediate, recombinant forms of var1 polypeptide could, in principle, distinguish between reciprocal crossing-over events and gene conversion. The latter has been suggested previously as a general mechanism for recombination of mitochondrial genes (17).

While it is clear that much more data must be collected to understand both the recombination and segregation properties of the var1 locus, our ability to detect easily the polymorphic forms of this interesting protein should be of considerable value in probing the organization and behavior of mitochondrial genes in general.

ACKNOWLEDGMENTS

The expert technical assistance of Ms. Mary Ensign, Robbin Palmer, and Jill Ackerson is gratefully acknowledged. This work was supported by NIH Grant GM 22837 (Subcontracts to RAB and PSP), NIH Grant GM 22525 and a Grant from The Robert A. Welch Foundation to RAB. RDV was supported by a Training Grant from NIH-NICHHD. EZ is a Robert A. Welch Foundation Fellow.

REFERENCES

1. Tzagoloff, A., Akai, A., Needleman, R. B.: Assembly of the Mitochondrial membrane system:Isolation of nuclear and cytoplasmic mutants of Saccharomyces cerevisiae with specific defects in mitochondrial function. J. Bact. 122, 826-831 (1975).

2. Tzagoloff, A., Foury, F., Akai, A.: Resolution of the mitochondrial genome. In: The genetic function of mitochondria DNA (Saccone, C., Kroon, A. M., eds.) North Holland Biomedical Press, Amsterdam, 155-162 (1976).

3. Bolotin-Fukuhara, M., Faye, G., Fukuhara, H.: Temperature-sensitive respiratory-deficient mitochondrial mutations:Isolation and genetic mapping. Molec. Gen. Genet. 152, 295-305 (1977).

4. Handwerker, A., Schweyen, R. J., Wolf, K., Kaudewitz, F.: Evidence for an extrakaryotic mutation affecting the maintenance of the rho factor in yeast. J. Bact. 113, 1307-1310 (1973).

5. Lancashire, W. E.: Mitochondrial mutations conferring heat or cold sensitivity in Sacharomyces cerevisiae. In: Genetics and biogenesis of chloroplasts and mitochondria (Bucher, Th. et al, eds.). North Holland Biomedical Press, Amsterdam, 481-490 (1976).

6. Sriprakash, K.S., Molloy, P.L., Nagley, P., Lukins, H. B., Linnane, A. W.: Biogenesis of mitochondria. XLI. Physical mapping of mitochondrial genetic markers in yeast. J. Mol. Biol. 104, 485-503 (1976).

7. Morimoto, R., Lewin, A., Merten, S., Rabinowitz, M.: Restriction endonuclease mapping and analysis of grande and mutant yeast mitochondrial DNA. In: Genetics and biogenesis of chloroplasts and mitochondria (Bucher, Th. et al., eds.) North Holland Biomedical Press, Amsterdam, 519-524 (1976).

8. Schweyen, R. J., Steyrer, U., Kaudewitz, F., Dujon, B., Slonimski, P. P.: Mapping of mitochondrial genes in Sacharomyces cerivisiae:Population and pedigree analysis of retention or loss of four genetic markers in rho$^-$ cells. Molec. Gen. Genet. 146, 112-132 (1976).

9. DiFranco, A., Sanders, J.P.M., Heyting, C., Borst, P., Slonimski, P.P.: Restriction enzyme analysis and physical mapping of mitochondrial DNA from petite mutants, carrying a genetic marker for oligomycin or paromomycin resistance. In: The genetic function of mitochondrial DNA (Saccone, C.,Kroon, A.M., eds.). North Holland Biomedical Press, Amsterdam, 291-304 (1976).

10. Rabinowitz, M., Jakovcic, S., Martin, N., Hender, F., Halbreich, A., Lewin, A., Morimoto, R.: Transcription and organization of yeast mitochondrial DNA. In: The genetic function of mitochondrial DNA (Saccone, C.,Kroon, A. M., eds.) North Holland Biomedical Press, Amsterdam, 219-230 (1976).

11. Sanders, J.P.M., Heyting, C., DiFranco, A., Borst, P., Slonimski, P.P.: The organization of genes in yeast mitochondrial DNA. In: The genetic function of mitochondrial DNA (Saccone, C., Kroon, A. M., eds.) North Holland Biomedical Press, Amsterdam, 259-272 (1976).

12. Douglas, M. G., Butow, R. A.: Variant forms of mitochondrial translation products in yeast:Evidence for location of determinants on mitochondrial DNA. Proc. Nat. Acad. Sci. USA 73, 1083-1086 (1976).

13. Douglas, M. G., Strausberg, R. L., Perlman,P.S., Butow, R.A.: Genetic analysis of mitochondrial polymorphic proteins in yeast. In: Genetics and biogenesis of chloroplasts and mitochondria (Bucher, Th. et al., eds.) North Holland Biomedical Press, Amsterdam, 435-442 (1976).

14. Perlman, P.S., Douglas, M. G., Strausberg, R. L., Butow, R.A.: Localization of genes for variant forms of mitochondrial proteins on mitochondrial DNA of Saccharomyces cerevisiae. J. Mol. Biol. In press (1977).

15. Strausberg, R. L., Butow, R. A.: Expression of petite mitochondrial DNA in vivo:Zygotic gene rescue. Proc. Nat. Acad. Sci. USA. In Press (1977).

16. Cleveland, D. W., Fischer, S. G., Kirschner, M. W., Laemmli, U.K.: Peptide mapping by limited proteolysis in sodium dodecyl sulfate and analysis by gel electrophoresis. J. Biol. Chem. 252, 1102-1106 (1977).

17. Dujon, B., Slonimski, P.P., Weill, L.: Mitochondrial genetic IX. A model of recombination and segregation of mitochondrial genomes in Saccharomyces cerevisiae. Genetics 78, 415-437 (1974).

MUTATIONS AT ANY ONE OF THREE UNLINKED MITOCHONDRIAL GENETIC LOCI BOX1, BOX4, AND BOX6, MODIFY THE STRUCTURE OF CYTOCHROME B POLYPEPTIDE(S).

M.L. CLAISSE, A. SPYRIDAKIS and P.P. SLONIMSKI
Centre de Génétique Moléculaire du C.N.R.S., Gif-sur-Yvette 91190 - FRANCE

INTRODUCTION

The BOX (1,2,3,4) and the COB (5,6,7) mutations, as well as mutations which confer resistance to antimycin, diuron, mucidin, and/or funiculosin (cf. 8 for review) have been located in a unique region of mtDNA that contains information for the biosynthesis and regulation of the coenzyme QH_2-cytochrome c reductase (coQH_2-cyt c red) and cytochrome b.
22 BOX mutations have been mapped by recombination and deletion, and 6 loci, distributed in two adjacent segments, have been distinguished in this region (4).
Enzymic and physiological studies of BOX mutants have shown that in these coQH_2-cyt c red deficient mutants the spectral absorption bands of cytochrome b cannot be observed, except in two cases (2,3). In several mutants located in segment II, the formation of cytochrome oxidase appears to be particularly sensitive to catabolite repression (2,3).
We report here modifications of mitochondrial translation products in BOX mutants which suggest three distinct genetic loci as structural gene(s) for cytochrome b polypeptide(s).

MATERIAL AND METHODS

Yeast strains.

We have analyzed 22 mit^- mutants localized in the BOX region of mtDNA and derived from the parental strain 777-3A α mit^+ op1 ade1 (1,2,3,4).

Identification of mitochondrially translated polypeptides.

Cells were grown aerobically overnight in 2 ml 1 % yeast extract, 2 % bactopeptone, and 0.5 % glucose to the early stationary phase. Then, cultures were diluted 2 to 5 with 1 % yeast extract, 2 % bactopeptone, and 0.3 % glucose, and regrown for two hours. Cells were labelled in 0.3 % glucose low sulphate medium with 250 μCi/ml ^{35}S-sulphate (The Radiochemical Center Amersham, carrier free), essentially as described in (9). Cycloheximide (0.6 mg/ml) was added 10 min prior to the addition of label.

To prepare mitochondria, cells were suspended in 0.25 M mannitol, 0.02 M Tris-H_2SO_4, 0.001 M EDTA, pH 7.4 (MTE buffer), plus phenylmethansulfonylfluorid (PMSF), 0.001 M, and broken with glass beads (0.45-0.50 mm ø) in a Braun homogenizer fitted with a multisample device. Mitochondria were suspended in the same MTE buffer plus PMSF.

Samples of mitochondrial proteins (200,000 cpm each) were analyzed by electrophoresis in the presence of sodium lauryl sulphate on gel slabs containing an exponential gradient (10-15 %) of polyacrylamide (9,10). The gels were fixed for at least 1h in 45 % methanol, 10 % acetic acid, dried (11), and radioautographed by exposing 24h to Kodak Kodirex X-ray films, which were subsequently developed in Kodak liquid X-ray developer LX-24 for 7 min at 20°C.

The apparent molecular weights of the polypeptides were estimated assuming 33,000 and 22,000 as molecular weights for the subunits II and III of cytochrome oxidase (12), 30,000 for the b polypeptide (6, 13) and 12,400 for horse heart cytochrome c run in parallel on gel slabs.

RESULTS

In 19 of the BOX mutants examined, the polypeptide of approximate molecular weight 30,000, attributed to cytochrome b (6,13), is not detectable. In BOX 6-1 (Fig. 1) and BOX 4-3 (Fig. 2) mutants this polypeptide is clearly distinguishable, although the labelling of the corresponding band is in both cases lower than in the wild type. BOX6-1 and BOX4-3 mutants are precisely the two exceptions exhibiting spectral absorption bands characteristic of cytochrome b, albeit both mutants are devoid of coenzyme QH_2-cytochrome c reductase activity (3). On the other hand, the fact that the labelling of the presumed cytochrome b polypeptide is weaker in mutants G925 (BOX6-1) and W7 (BOX4-3) than in the wild type, is also in agreement with spectral observations : the intensity of

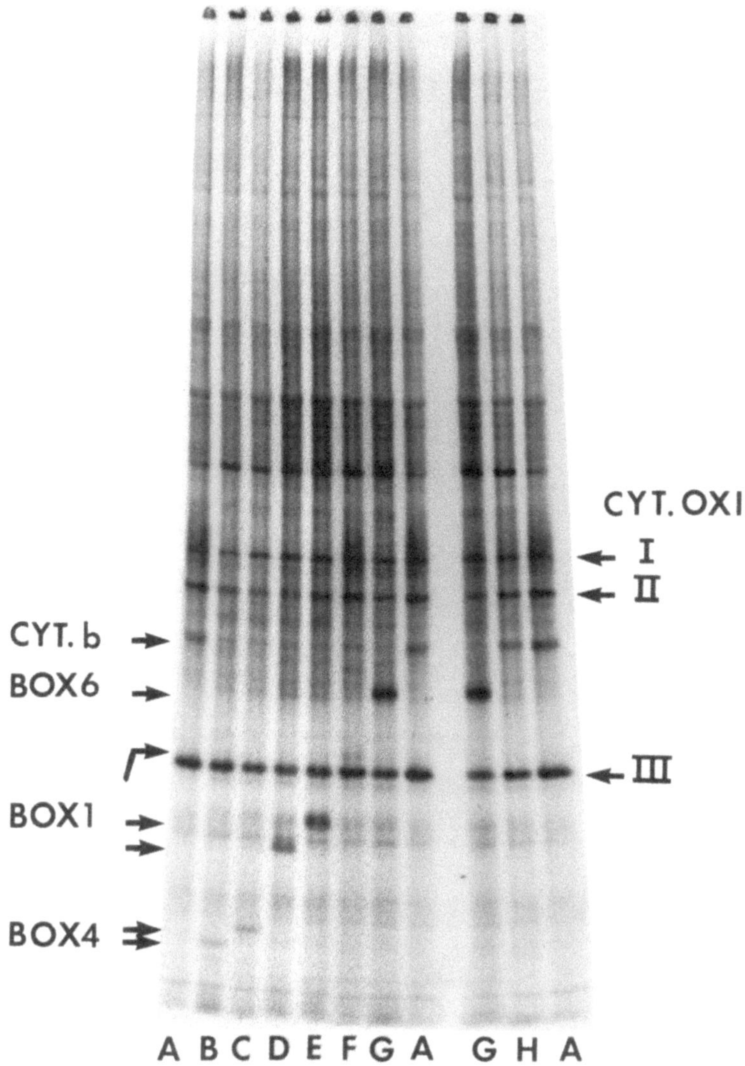

Fig. 1. Mitochondrial translation products of BOX4, BOX1, and BOX6 mutants. Cultures, labelling of cells, isolation of mitochondria, electrophoresis, and autoradiography were carried out as described in the text.
A: mit^+ strain 777-3A; B: G1334 (BOX4-1); C: G171 (BOX4-2); D: G1988 (BOX1-1); E: G2286 (BOX1-2); F: G2457 (BOX1-3); G: W267 (BOX6-2); H: G925 (BOX6-1).
The roman numerals identify cytochrome oxidase subunits I, II, and III; CYT b indicates the presumed cytochrome b polypeptide of the wild type, BOX6, BOX1, BOX4 the putative cytochrome b polypeptide fragments found in the corresponding mutants.

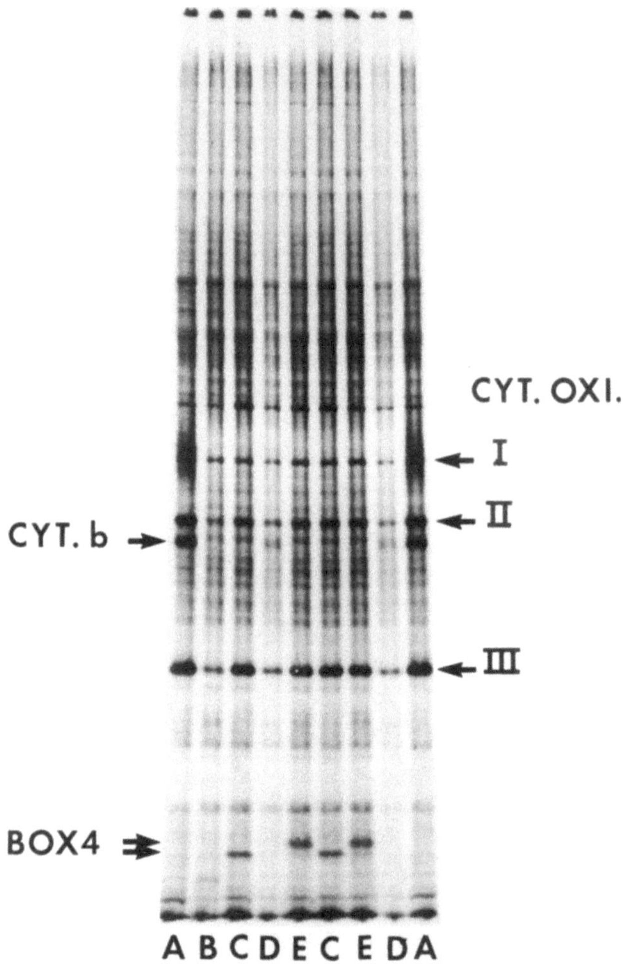

Fig. 2. Mitochondrial translation products of BOX4 mutants.
Same conditions as in Fig. 1 .
A: mit^{+} strain 777-3A; B: G839; C: G1334 (BOX4-1); D: W7 (BOX4-3); E: G171 (BOX4-2).
The meanings of other symbols are as in the legend to figure 1.

cytochrome b absorption bands is lower in these mutants than in the wild type (3).

We have clearly observed new labelled products in mutants at loci BOX4, BOX1 and BOX6.

Fig. 1 shows an autoradiography of mitochondrially synthesized polypeptides in the different BOX1 and BOX6 mutants compared with BOX4-1 and BOX4-2 mutants and with the parental isogenic wild type. New mitochondrial translation products are visible with apparent molecular weights of about 13,400 in BOX4-1, 13,700 in BOX4-2, 17,300 in BOX1-1, 18,500 in BOX1-2, 22,500 in BOX 1-3, 26,600 in BOX6-2. Furthermore, in BOX1-3 mutant, in addition to the 22,500 peptide, apparently exists a labelled product of about 27,000-28,000 molecular weight and another at the 30,000 level. However, they are poorly labelled and the resolution is not sufficient to tell whether or not the weak product at the 30,000 level migrates as the wild type cytochrome b.

We found that the intensity of label of the new polypeptides was particularly strong in mutants BOX1-1, 1-2, and 6-2, whereas it was systematically lower in mutants BOX1-3, 4-1 and 4-2.

Fig. 2 shows that although the difference of migration of the new labelled products is small between the mutated hetero-alleles BOX4-1 and BOX4-2 belonging to the same locus, it is very reproducible indeed. On the other hand, as pointed out above, BOX4-3 mutant exhibits the 30,000 polypeptide attributed to cytochrome b. G839, a mutation located at a site close to BOX4 locus, shows neither this polypeptide nor a new labelled product. According to physiological studies (3) BOX3 and BOX2 loci might control the assembly processes rather than the structure of $coQH_2$-cyt c red itself. Indeed we did not observe any major new mitochondrial translation product in BOX3 mutants, nor in BOX5 mutants. For BOX2 mutants the situation is less clear at the present time. We did not observe as important modifications as in BOX1 and BOX6 loci ; nevertheless further investigations of BOX2 locus appear interesting and are being carried on.

DISCUSSION

We have seen that the two mutations BOX6-1 and BOX4-3, located in the most distant loci in the BOX region and exhibiting cytochrome b spectral absorption bands (but no activity), show a mitochondrial translation product which migrates in the same way as the polypeptide attributed to cytochrome b. A complete identification of this polypeptide could only be performed by immunology and sequence analysis. Nevertheless, in the present state of knowledge on mitochondrial translation products, it is generally assumed that this 28,000-32,000 polypeptide, labelled in presence of cycloheximide, corresponds to cytochrome b (6,13).

We have demonstrated that in at least five mutants devoid of cytochrome b spectrum and coenzyme QH_2-cytochrome c reductase activity, the 30,000 polypeptide is absent and instead new mitochondrial translation products appear. All these new products migrate faster than the wild type polypeptide and the migration distances are specific for each mutation studied. It is therefore logical to conclude that these new polypeptides are mutated cytochrome b.

Until now, two mitochondrial mutants, in particular M9-228, were shown to produce a modified translation product presumed to be cytochrome b (5,6,7) and were allocated to the COB2 locus (6,7). After correction (4) made to the genetic map of COB mutants it appears that the mutational site M9-228 belongs to the locus BOX1 rather than to the locus COB2. It is thus satisfactory to find that two other heteroalleles at this locus, BOX1-1 and BOX1-2, synthesize a structurally modified cytochrome b. However, the main finding of the present work is the demonstration that mutations at other genetic loci, namely BOX4 and BOX6, also modify the structure of cytochrome b polypeptide(s). It is interesting to notice that the order of the genetic loci (4) is BOX4, BOX1, BOX6, while the apparent molecular weights of the putative cytochrome b fragments follow that order : 13,400 and 13,700 in BOX4 mutants, 17,300 and 18,500 in BOX1 mutants, and 26,600 in BOX6 mutants. This could be interpreted as indicating that a single structural gene for cytochrome b apoprotein covers the entire mtDNA region and that this gene is read from BOX4 to BOX6 and that BOX4-1, BOX4-2, BOX1-1, BOX1-2, and BOX6-2 mutations generate chain termination codons.

This hypothesis, although tempting, has to be first reconciled with the folowing observations :

1. There is a discrepancy between polypeptide lengths and genetic distances. For instance, one finds 0.5% of wild type recombinants between BOX1-1 and BOX1-2 sites (4) while the apparent molecular weights of polypeptides

synthesized by the mutants are respectively 17,300 and 18,500. One finds 0.02 % of wild type recombinants between BOX4-1 and BOX4-2 mutants (4) and the apparent molecular weights of labelled fragments are 13,400 and 13,700. On the contrary, between BOX4 and BOX1 loci, one finds 8 % of wild type recombinants and the two loci are unlinked by recombination (4). If the hypothesis of a single structural gene were true, this high level of recombinants would have to correspond to some 100 to 200 base pairs only (cf. molecular weights of ca 13 kd and ca 17 kd), which seems a priori improbable. However, a strong "map expansion" phenomenon (cf. 4) could possibly account for such a discrepancy.

2. There is evidence that two genetic loci (BOX2 and BOX3) situated between BOX4 and BOX6 might be regulatory genes (3), which is conceivable in the case of an operon but is not consistent with the hypothesis of a single structural gene read continuously from BOX4 to BOX6.

3. Polypeptides, presumed to be cytochrome b fragments, appear in strains mutated at loci BOX4, BOX1 and BOX6, which are precisely the three genetic loci allelic to loci that modify the drug sensitivity of the $coQH_2$-cyt c red (cf. 8) enzymatic complex in which cytochrome b is an essential component (for reviews see 13, 14). Indeed, mutants resistant to diuron, mucidin, antimycin A, and funiculosin, have been located as follows (cf. 8) : DIU2, ANA1 and MUC1 are allelic or closely linked to BOX4 ; DIU1, ANA2 and FUN1 are allelic or closely linked to BOX1 ; MUC2 is allelic or closely linked to BOX6. Genetic analysis of these respiratory competent drug resistant loci and their allelism relationships with $coQH_2$-cyt c red deficient (BOX) loci suggest the existence of three structural genes (8).

Thus our results, together with the physiological and genetic data reported in (3,4,8), suggest very strongly the existence of at least three structural genes for yeast cytochrome b, which in turn would imply at least three different cytochromes b or polypeptide chains of cytochrome b.

ACKNOWLEDGEMENTS

We gratefully acknowledge Dr. O. Groudinsky for helpful advice. This work has been supported by D.G.R.S.T. Grants n° 75-70748 and 77-70240.

REFERENCES
==========

1. KOTYLAK, Z. and SLONIMSKI, P.P. (1976) in "The Genetic Function of Mitochondrial DNA". Eds C. SACCONE and A.M. KROON. Elsevier/North Holland Biomedical Press, Amsterdam, pp. 143-154.
2. PAJOT, P., WAMBIER-KLUPPEL, M.L., KOTYLAK, Z. and SLONIMSKI, P.P. (1976) in "Genetic and Biogenesis of Chloroplasts and Mitochondria" Eds Th. BUCHER et al., Elsevier/North Holland Biomedical Press, Amsterdam, pp. 443-451.
3. PAJOT, P., WAMBIER-KLUPPEL, M.L. and SLONIMSKI, P.P. (1977). This volume.
4. KOTYLAK, Z. and SLONIMSKI, P.P. (1977). This volume.
5. SLONIMSKI, P.P. and TZAGOLOFF, A. (1976). Eur. J. Biochem., 61,27-41.
6. TZAGOLOFF, A., FOURY, F. and AKAI, A. (1976). Molec. Gen. Genet., 149, 33-42.
7. FOURY, F. and TZAGOLOFF, A. (1976). Mol. Gen. Genet., 149, 43-50.
8. COLSON, A.M. and SLONIMSKI, P.P. (1977). This volume.
9. DOUGLAS, M. and BUTOW, R.A. (1976). Proc. Natl. Acad. Sci. USA 73, 1083-1086.
10. DOUGLAS, M., FINKELSTEIN, D. and BUTOW, R. (1977). Methods in Enzymology. Eds S. FLEISCHER and L. PACKER. Academic Press, New York, in press.
11. CABRAL, F. and SCHATZ, G. (1977). Methods in Enzymology. Eds S. FLEISCHER and L. PACKER. Academic Press, New York, in press.
12. POYTON, R.O. and SCHATZ, G. (1975). J. Biol. Chem., 250, 752-761.
13. WEISS, H. (1976). Biochim. Biophys. Acta, 456, 291-313.
14. RIESKE, J.S. (1976). Biochim. Biophys. Acta, 456, 195-247.

STRUCTURAL AND REGULATORY MUTATIONS AFFECTING MITOCHONDRIAL GENE PRODUCTS

H. R. Mahler, D. Hanson, D. Miller and T. Bilinski*
Chemistry Department, Indiana University, Bloomington, Indiana 47401, USA and
D. M. Ellis, N.J. Alexander and P. S. Perlman
Department of Genetics, The Ohio State University, Columbus, Ohio 43210, USA

INTRODUCTION

The discovery of mit⁻ mutants constitutes a highly significant advance in mitochondrial genetics (1). These respiration deficient strains of *Saccharomyces cerevisiae* (and *carlsbergenesis*) with a genetic lesion in their mitochondrial (mt) DNA differ from the classical cytoplasmic *petite* or *rho*⁻ mutants in their ability to i) support mitochondrial protein synthesis (1-4); ii) retain one or more of the complexes of the phosphorylating respiratory chain (3,5-6); iii) regain the missing (or otherwise modified) wild-type function by crosses with an appropriate *rho*⁻ tester (5-8); iv) remain viable in the presence of the *op-1* mutation (9); and v) retain a set of antibiotic resistance mutations spanning the mitochondrial genome (10). These properties singly or in combination, can and have been used operationally for the selection and characterization of *mit*⁻ mutants (1-11).

Useful and revealing as previous investigations on such mutants have been, they - with some exceptions (3,12,13) - have not yet addressed three fundamental questions: a) the specific change in mtDNA responsible for a particular

* Present Address: Institute of Biochemistry and Biophysics, Warszawa, 02-532, 36 Rakowiecka Street, Poland.

mutation: b) the specific alteration in a polypeptide gene product resulting from this change; c) the specific assignment of such alterations as being of structural or regulatory significance in the biogenesis of the affected mitochondrial function. In this communication we provide some evidence concerning these questions.

RESULTS

In a previous communication (10) we outlined our objectives in studying mit⁻ mutants and presented preliminary biochemical and genetic data on a small group of such mutants. Here we provide further analysis of the original mutants and augment those results with data on additional ones. The new mutants were all detected by screening using rho^{o} and mixed rho^{-} tester stocks as outlined in ref. 10. Several mapping procedures have been described by others for mit^{-} mutants; discriminating class analysis (5) (in crosses to ρ^{+} tester strains with suitable drug markers); allelism mapping (5,8) (in crosses with a library of defined petite mutants encompassing, in their aggregate, the whole genome (5-8). We have chiefly utilized the latter method using a collection of 400 petite mutants calibrated against a set of mit^{-} strains provided by Dr. Tzagoloff. Our library of petite mutants is sufficiently large to discriminate nearby pairs of mutant loci as well as sites within several of the loci (notably oxi3 and cob2). We have used the nomenclature first applied to Tzagoloff's mutants (5,14) since we have established allelism relationships only with reference to his mutants.

Gene Products in Some mit^{-} Mutants

Polypeptide patterns. We used standard techniques (15) for the labeling, identification, analysis, and display of mitochondrially translated polypeptides. Six major products are observed on slab gel autoradiograms of the wild type (Fig. 1, slots 1 and 10); their tentative identification in order of decreasing mass and mobility are var1 (15), cytochrome oxidase subunit I (cox I (16)), cytochrome oxidase subunit II (cox II (16)), cytochrome b apoprotein (cyt b (17)), cytochrome oxidase subunit III (cox III (16)) and ATPase subunit 9 (16,18). In addition many preparations show several weaker bands, the most prominent of which is one with an assigned molecular weight of 28.5 x 10^3. In all cases shown the assignment of a mutation as affecting a subunit of either cytochrome oxidase or cytochrome b can be correlated with a specific enzymatic defect in the appropriate enzymatic activity. The observed patterns comparing wild-type and several representative mit⁻ mutants of the same strain are shown in Fig. 1 (Slots 2-9) and disclose that, in addition to the loss of single major bands (Class I), several other classes can also be discerned: no apparent alteration of the wild-type mobilities (Class II) and emergence of novel bands not present in wild-type (Class III). In principle class III can be further subdivided into sub-classes according to whether the appearance of the new bands is accompanied by a loss of a wild-type band (IIIA) or not (IIIB) and whether as revealed by fingerprint patterns (see below) there exists an obvious relationship between the new band and a wild-type band (IIIA-1 and IIIB-1) or not IIIA-2 and IIIB-2).

Polypeptide changes associated with defined mit⁻ mutations. Representatives of these various classes of altered polypeptides result from mutations in a number of defined mit loci. In our studies so far at least one mutant at each of the loci oxi2, oxi3 and cob2 belongs to class I. It is tempting from these observations to conclude that there is a direct coding relationship between the single missing protein (due to a deletion or nonsense mutation, etc.) and the particular mit locus in question. At this stage of our analysis of mit⁻ mutants tentative gene/protein assignments are as follows: i) oxi3: cox I; ii) cob2: cyt b; iii) oxi2: cox III. Our two mutants mapping near oxi1 (E37 and E65) have unusual phenotypes and will be discussed below. However, it remains possible that at least some mit loci specify a product required for the successful expression of the structural gene

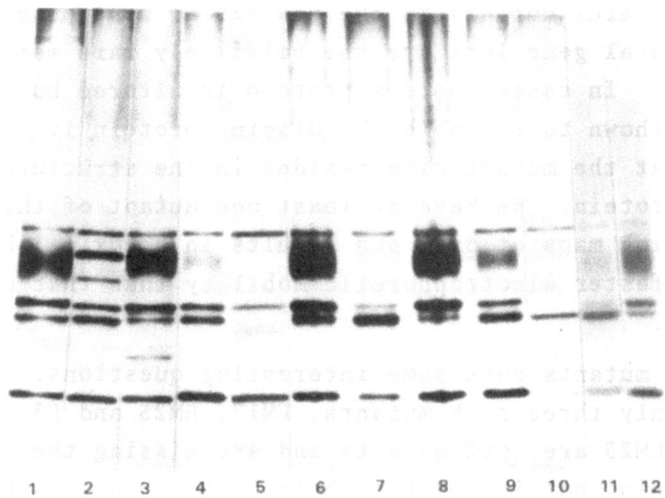

FIGURE 2: Additional mutant phenotypes. 1) 41-6/161; 2) M44; 3) M113; 4) m80; 5) M45; 6) E33-30°; 7) E33-37°; 8) E56-30°; 9) E56-37°; 10) E65; 11) 41-6/161 (5% glucose); 12) 41-6/161.

are two new major bands; in M44 both new bands are in the cox I region, but cox I is absent. In M113 one new band is near cox I (obscured by cox I on this gel but see ref. 10), the other is between cyt b and cox III and no ρ^+ bands are missing. In fact, M113 grows on glycerol and was initially detected by its unusual pigmentation (altered shade of red pigment due to the <u>ade2</u> mutation). Whether these mutations are regulatory (increase the rate of synthesis of an otherwise minor mitochondrial translational product) or structural (affect the processing or stability of a mitochondrial gene product) has not yet been assessed for these mutants. Other, perhaps less interesting, nuclear mutants have also been observed; for instance, mutant M80 (Slot 4) has reduced cytochrome oxidase activity but has no missing or altered major products of mitochondrial protein synthesis. It therefore resembles certain Class II <u>mit</u>$^-$ mutants. ii) Double mutants:

Mutant M45 (slot 5) is missing cox I and cyt b and thus resembles the box mutants described in refs. 5, 9, 19-21. In agreement with the studies of Tzagoloff et al. (21) on such mutants, we have found that M45 is a double mutant with separate lesions in cob2 and oxi3, both of which are stable. Using our collection of petite tester strains previously calibrated by means of a collection of mit⁻ testers obtained from Dr. Tzagoloff, we found that M45 is restored to wild-type only when crossed to those petites capable of restoring both cob2 and oxi3 mutations; no wild-type recombinants are obtained when cob2+ oxi3° or cob2° oxi3+ petites are used. iii) Thermosensitive mutants: We have isolated a number of heat sensitive conditional (8,22-25) mitochondrial mutants. Two of these (E33 and E56) are analyzed in Fig. 2 (slots 6-9). Each has a normal band pattern when the cells are labeled at 30°; when labeled at 37°, mitochondrial protein synthesis continues but incorporation into cox I, II and III is greatly diminished while that into all other bands appears normal. It is not yet known whether the simultaneous disappearance of the cytochrome oxidase subunits is due to a thermosensitive regulatory site or an alteration in the primary sequence of a polypeptide that renders it and associated polypeptides sensitive to proteolysis at 37°. Preliminary mapping data locate E33 near oxi3; this is consistent with the altered synthesis of cox I, but cannot explain the effect of the E33 mutation on cox II and III. Deletion mapping studies of E33 show that it is not an oxi3-oxi1 double mutant. Analysis of other conditional mutants will certainly be valuable in the elucidation of structural genes on mtDNA. iv) Regulatory mutants: In our first communication on this topic (10) we suggested that pleiotropic effects on band patterns might be indicative of regulatory effects of the mit genes responsible for them. Such is the case only if it can be determined that, unlike M45, multiple lesions are referable to a mutation at a single locus. Mutant E33 (and probably E56) are examples of a conditional mutation of this sort. The mutant E65 provides an example of such a regulatory phenotype under ordinary conditions. This mutant, when grown under standard conditions (galactose), is deficient in cytochrome oxidase and cytochrome aa_3. Its polypeptide pattern (slot 10) shows a virtually complete absence of cox I coupled with a greatly reduced incorporation into cox II. This is similar to the pattern exhibited by wild-type cells grown under conditions of glucose repression (slots 11 and 12). Mapping data suggest that E65 is located close to oxil. Mutant E37 (Fig. 1, slot 3) maps near E65 (and oxil) but has a different phenotype; it is missing cox I only. These observations suggest that at least some mit^- mutations may be concerned with the regulation of the amount of cox proteins synthesized or assembled into cytochrome oxidase and that this function is one of the primary targets of catabolite repression.

Fingerprint Analysis of Mitochondrial Proteins.

Analysis of polypeptide products by SDS-polyacrylamide gel electrophoresis as described here and in related studies can at best provide information only concerning the size of such molecules. Therefore, inferences concerning the molecular consequences of class II and III mutations require the supplemental application of techniques designed to probe alterations in amino acid sequence. Fingerprinting techniques are simple and of great and proven utility for this purpose.

To this end we have adapted the method of Cleveland et al. (26) for fingerprinting the polypeptides separated by electrophoresis in SDS-acrylamide gels (Fig. 1). We label the strain in question with $^{35}SO_4^=$ to obtain mitochondrial proteins of very high specific activity (1-2 x 10^5 cpm per gel slot), separate the bands, dry the gel, cut out the band(s) of interest and load the gel pieces containing them into the sample wells of a second slab gel. They are then rehydrated in the electrode buffer and digested with either chymotrypsin or papain for 30 min and the resultant peptides in the limit digest ($\leq$ 20 peptides with either enzyme for even the largest polypeptide) analyzed and displayed as before. Results of this kind of analysis are shown in Fig. 3 for all major wild-type polypeptides (slots 1-3, 5 and 6) and the novel bands in the mitochondrial class III mutants EM17 (slot 4), E3 and E20 (slots 7 and 8) and the nuclear mutant M113 (slot 9 for the band of high mobility).

From these and related experiments, we draw the following inferences: i) The method provides a reliable and straightforward means of identifying and distinguishing all of the major wild-type polypeptides. ii) The cox III bands of E3 and E20, which appear to have slightly different mobilities from the wild-type cox III, do indeed give distinct peptide patterns, though they are obviously closely related. iii) EM17 is a mutant in cob2, a mitochondrial gene implicated in the synthesis of cytochrome b and, in fact, the mitochondria of EM17 exhibit a specific deficiency in this band and cytochrome b-linked enzyme activities. However, the new band appears to be not at all related to cytochrome b (or any of the other major products of mitochondrial translation). EM17 is therefore assigned to class IIIA-2 and the molecular event giving rise to its new band becomes of the greatest interest. iv) The fast extra band in the nuclear mutant M113 also appears to be unrelated to any other major product. There is a band with a similar mobility which is occasionally discernible in the wild-type, but because it is very weak it has not been possible to analyze it in this way. The possibility exists, however, that the nuclear mutation in M113 leads to an amplification of a normally minor product.

Restriction Analysis of mit^- Mutants

Among the mutants in our collection there are several that are missing (or have an altered form of) a product of mitochondrial protein synthesis and are unable to revert to ρ^+. such mutations are likely to be correlated with deletions in their mitochondrial (mt)DNA and as such may be very useful in

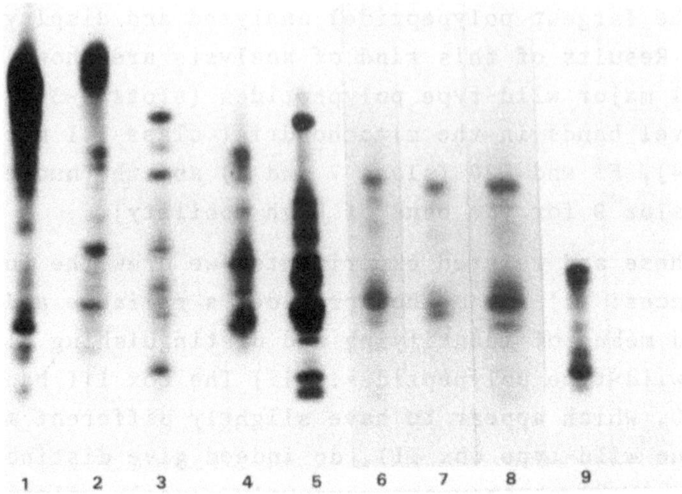

FIGURE 3: Fingerprint analysis of mitochondrial products. 1) cox I; 2) var 1; 3) cyt b; 4) EM17 extra band; 5) cox II; 6) cox III (41-6/161); 7) cox III (E3); 8) cox III (E20); 9) M113 fast extra band. Each gel slot was treated with 50-75 μg papain.

the physical mapping of the structural genes for these polypeptides. It has already been shown that some <u>mit</u>⁻ mutants represent deletions, some of which are quite large (12). Rapid progress in the physical mapping of mitochondrial genes has been made recently with the aid of petite deletion mapping (or more correctly, retention or "reverse deletion" mapping) (5-8,27-34). The precise localization of a gene by that method requires finding an extensively deleted petite, retaining <u>only</u> the gene in question. Since most mitochondrial genes are not contiguous on mtDNA, it is difficult to use genetic procedures alone to select the most suitable petites among this class, since for the reason just cited they will also retain significant (and variable amounts of DNA flanking the gene. The analysis of <u>mit</u>⁻ mutants with small deletions therefore represents an attractive alternative method for locating structural genes with considerable precision. This approach is a direct application of the type of deletion mapping done in other genetic systems. In this section we will describe our results with several mit⁻ mutants that illustrate the utility of this approach.

We first analyzed the restriction pattern of mtDNA isolated from the wild-type parent of our mit⁻ mutants, strain 41-6/161. MtDNA was isolated from stationary phase glucose-grown cells by first separating a mitochondrial fraction by the "handshake" method (35). The mitochondria were lysed with detergent and the lysate run on a preparative CsCl density gradient; the DNA was located in the gradient fractions by restriction analysis of aliquots of roughly one-third of the fractions. Those fractions with mtDNA of maximum purity were retained for further analysis. A single gradient purification is usually sufficient to obtain mtDNA of high molecular weight and low level of contamination with nucDNA. In our hands this general protocol has been suitable for use with various ρ^+ strains, mit⁻ mutants and petites, using as little as 5 gm (wet weight) of cells per preparation.

We established a six enzyme restriction map of strain 41-6/161; some of the relevant data are presented in Fig. 4 and summarized in Fig. 5. Other data (not shown) used to confirm the map assignments included: i) partial digestions with Eco R1 and Hinc II; ii) multiple enzyme digestions; iii) analysis of mtDNA from other ρ^+ strains; iv) analysis of mtDNA from petite mutants. Molecular weights were calculated using bacteriophage lambda DNA (digested with Eco R1 or Hind III) as a standard and run in the same or adjacent gel slot as the mtDNA fragments. Our map closely resembles that of Morimoto et al. (31) for strain MH41-7B for the enzyme Eco R1, Bam I, Sal and Pst. Our map lacks the Hind III site in Eco R1 band 1 but has the other Hind III sites; as mentioned elsewhere (36) we find that ome⁻ strains (such as 41-6/161) lack this Hind III site while ome⁺ strains have it.

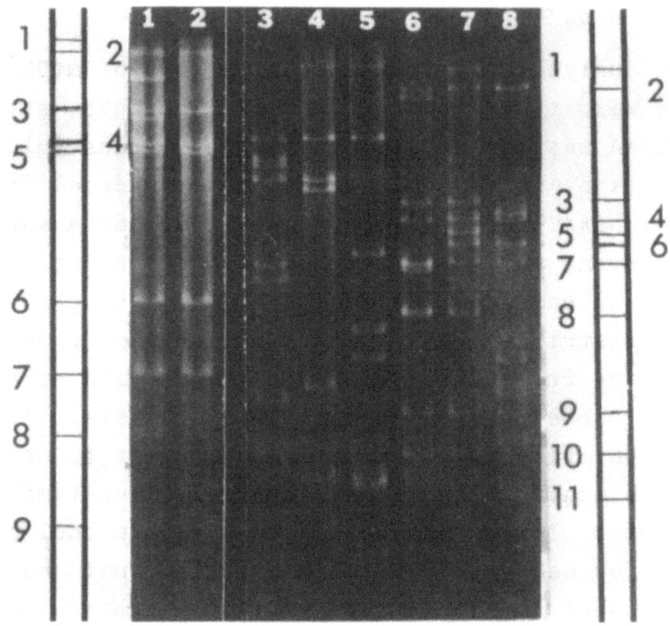

FIGURE 4: Agarose (0.7%) gel electrophoresis patterns of mtDNA from strain 41-6/161. 1) Eco R1 + Sal; 2) Eco R1 + Pst; 3) Eco R1 + Bam I; 4) Eco R1; 5) Eco R1 + Hind III; 6) Hinc II + Hind III; 7) Hinc II; 8) Hinc II + Bam I. A schematic pattern of an Eco R1 digestion is shown on the left (same scale as slots 1 and 2) and a schematic pattern of a Hinc II digestion is shown on the right (same scale as slots 3-8).

For preliminary investigation of this form of deletion mapping we selected several of our non-reverting mit$^-$ mutants representing four of the six mit loci. Figure 6 presents selected restriction patterns of mtDNA from two of these mutants. As is summarized in Fig. 5 (bottom) two of these mutants (CAD245 and E69) are deleted in the vicinity of the relevant locus estimated by petite deletion mapping. The other two mutants EM17 (cob2) and E65 (oxi1) have no restriction fragment differences when treated with Eco R1, Hinc II or Eco R1 plus Hinc II. Thus they cannot be deleted for more than 100 base pairs.

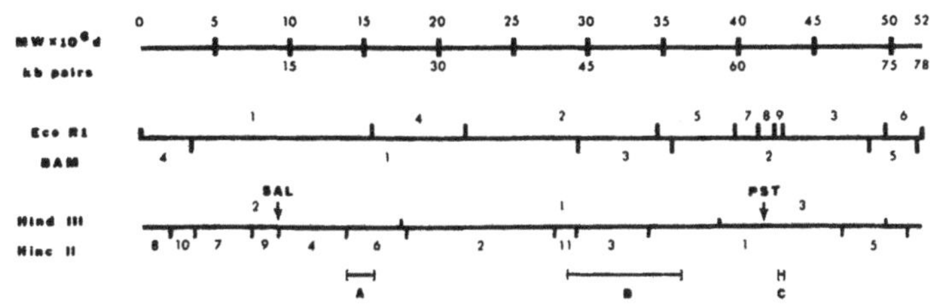

FIGURE 5: Restriction fragment map of mtDNA from strain 41-6/161. The map was prepared using standard procedures as outlined in the text. It closely resembles that of Morimoto <u>et al</u>. (31) for strain MH41-7B for the enzymes Eco R1, Bam I, Sal and Pst. Our strain has no Hind III site in Eco R1 fragment 1 (Hinc II fragment 9); as mentioned elsewhere (36) we find that ω^- strains (such as 41-6/161) lack this Hind III site while ω^+ strains have it. The Hind III site in Eco R1 fragment 4 has been shown by others (see ref. 31) to be a cluster of three sites; we have not yet analyzed the number of sites there, but our data are consistent with there being at least two sites. The last line of the figure summarizes deletion data obtained with two <u>mit</u>$^-$ mutants (E69, A; and CAD245, B, C). Relevant data are presented in other figures and are described in the text.

The class I mutant E69 maps at <u>oxi2</u> and lacks (Figure 1) cox III, the smallest protein subunit of cytochrome oxidase. It has a deletion of 0.17 x 10^6 daltons in the region between <u>cap</u> and <u>par</u> (where <u>oxi2</u> is known to map) in Hinc II band 6 (to the left of the Eco R1 site) (Fig. 5A). Since no other alterations were observed in the restriction pattern it is likely that the deletion includes at least part of the <u>oxi2</u> locus. Since the <u>oxi2</u> locus should be at least 0.36 x 10^6 daltons in length and since no cox III product is made in this mutant, the deletion probably includes the 5' phosphate

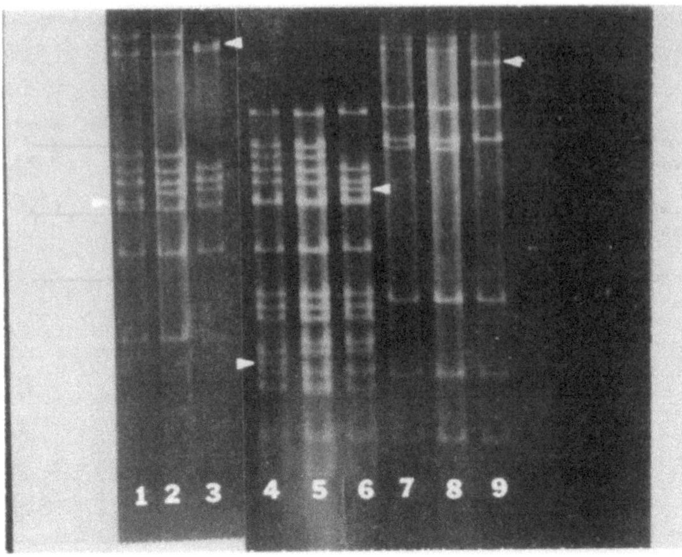

FIGURE 6: Agarose (0.7%) gel electrophoresis patterns of mtDNA from two mit^- mutants (CAD245 and E69). Slots 1, 4 and 7 are of E69; slots 2, 5 and 8 are of 41-6/161 (the ρ^+ parent of the mits); slots 3, 6 and 9 are of CAD245. Samples 1-3, 4-6, and 7-9 were digested with Hinc II, Hinc II + Eco R1, and Eco R1, respectively. Bands having altered mobility in the mutants are denoted by arrows.

end of the gene. It is not likely that this small deletion has removed the entire gene so that this mutant should be very useful in developing a fine structure map of this gene (of the kind already presented for the oxi3 locus by Slonimski and Tzagoloff (ref. 5)).

Mutant CAD245 also belongs to class I; it maps at oxi3 and lacks cox I (see Fig. 1), the largest subunit of cytochrome oxidase. Based on recombination studies with oxi3 mutants provided by Dr. Tzagoloff it is clear that CAD245 is defective in only a portion of the oxi3 locus; it yields ρ^+ recombinants when crossed to strains M18-65 (mapping near M15-98, ref. 5) and M36-35 (mapping near M16-41; ref. 5) but not with M37-62 (resembling M5-16; ref. 5). The two former strains map on the left side of the oxi3 fine structure map (as originally drawn, ref. 5) while the latter strain appears to be a large deletion of the right side of the locus (A. Tzagoloff, personal communication).

There are two deletions in the mtDNA isolated from CAD245 (Fig. 5B, C and Fig. 7): i) 7.55×10^6 daltons, mapping between par and oli2; ii) 0.15×10^6 daltons, mapping near oli2 and between oli2 and oli1. The former deletion resembles that reported for another oxi3 mutant (M5-16) by Rabinowitz's group (12), although our deletion is more than twice as large. Some of this difference may be explained by the presence of "extra" sequences in our ρ^+ strain (37). We have isolated a petite mutant capable of restoring CAD245 and have analyzed mtDNA from four restored ρ^+ diploids. In each case both deletions were replaced by the appropriate sequences from the petite mutant.

It is likely that the large deletion includes part of the oxi3 locus. We have characterized several petite mutants capable of restoring Tzagoloff's mutants M18-65 and M36-35 but not M37-62. Restriction analysis of mtDNA isolated from these petites shows that they all have in common the right side of Hinc II fragment 2; some retain the fragment 2-fragment 11 junction (and part of fragment 11) but others retain none of fragment 11. Therefore, the mutants M18-65 and M36-35 probably map in fragment 2, a region retained by CAD245. Since the petites are deleted for all sequences from Hinc II band 11 through oli2 and up to band 8 it is clear that M36-35 cannot map on the oli2 side of the large deletion of CAD245. These data appear to orient the oxi3 genetic fine structure map with the par locus to the left of mutant M18-65 and the oli2 locus to the right of mutant M37-62 (Fig. 7).

Since CAD245 has a breakpoint in fragment 11 and another far to the right of that position (in fragment 1) it is likely that fragment 11 is within the oxi3 gene (based on its proximity to the mutant sites of M18-65 and M36-35). While it has already been proposed that oxi3 is a very large locus

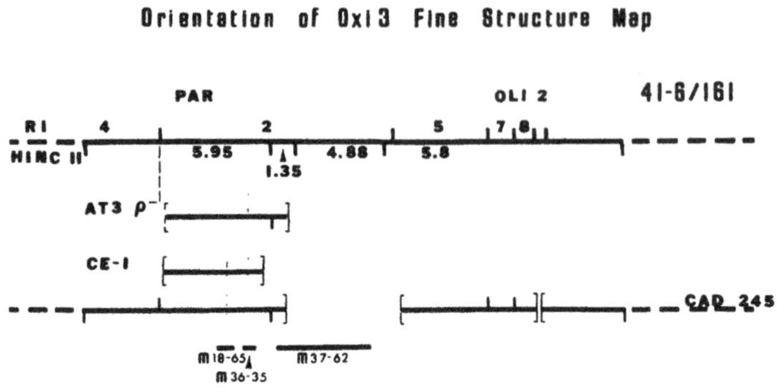

FIGURE 7: Orientation of the oxi3 fine structure map. The figure dipicts the portion of the ρ^+ physical map between par and oli2 as well as relevant portions of mtDNA from two petite mutants and the mit⁻ mutant, CAD245. The latter three strains restore Tzagoloff's oxi3⁻ mutants M18-65 and M36-35 but not M37-62. The order of the three tester mit⁻ strains is according to the published fine structure map of oxi3 (5); the sizes of the three oxi3⁻ mutants are not to scale.

(possibly containing several genes) it is likely that our deletion extends well past the right end of the structural gene. Since CAD245 is defective only in cytochrome oxidase activity we conclude that there are no other genes for major protein bands in the deleted flanking region of the genome. Clearly there are no genes in that region necessary for mitochondrial protein synthesis, transcription, or DNA replication.

The second deletion in CAD245 poses an interesting problem. It is probably not at all responsible for the cytochrome oxidase defect of this mutant since it maps well outside of the known oxi loci. Since CAD245 makes all other major products of mitochondrial protein synthesis (Fig. 1) and has normal cytochrome b-linked enzyme activities and normal oligomycin-sensitive ATPase it is possible that this missing region is not necessary for respiratory sufficiency; that is, we may have detected, for the first time by deletion analysis, a spacer region as postulated by Bernardi's group (38,39). Alternatively this region may be in a new locus whose inactivation does not affect a major product of mitochondrial protein synthesis or any enzyme activity currently thought to be associated with products of the mitochondrial genome. Since every wild-type strain analyzed in our laboratory (more than nine independent strains) has Eco R1 band 9 identical to that present in strain 41-6/161, it is not likely that this deletion merely removes a portion of one of the "extra" sequences characterized by Borst's group (32) already shown to be gratuitous to respiratory sufficiency; our strains do vary among themselves with regard to presence or absence of the large "extra" sequences.

A distinction between these hypotheses can be made by separating the two deletions by a cross. To this end we performed the cross outlined in Figure 8 between CAD245 and an <u>oli2^r</u> tester (strain K615/44: ρ^+ <u>cap^s</u> <u>ery^s</u> <u>oli1^s</u> <u>oli2^r</u> <u>par^s</u> <u>oxi3$^+$</u>) and isolated ρ^+ diploid recombinants of genotype <u>oli1^s</u> <u>oli2^s</u> <u>par^s</u> <u>oxi3$^+$</u>. At least two crossovers are required in the region of interest (between <u>oli1</u> and <u>par</u>) to yield this genotype, one in interval A and another in interval B. Considering the relative sizes of intervals B1 and B2 we expected the A-B2 double crossover to be more frequent than the A-B1 one; the A-B2 crossover should yield a ρ^+ genome with the small deletion from CAD245 (Fig. 8).

Mitochondrial DNA was isolated from four recombinants and analyzed by restriction endonuclease treatment. Surprisingly in each case, Eco R1 band 9 (see Fig. 4,5) was that of the ρ^+ tester strain (normal); thus instead of the expected double crossover (A-B2) we obtained the A-B1 double or a gene conversion event at <u>oli2</u>. Genetic analysis of this cross (complete genotype outputs) showed that this recombinant was not a rare progeny type, as might be expected if only the A-B1 exchange could yield ρ^+ progeny. We have not yet analyzed mit⁻ (respiration deficient) progeny of this cross for the A-B2 event.

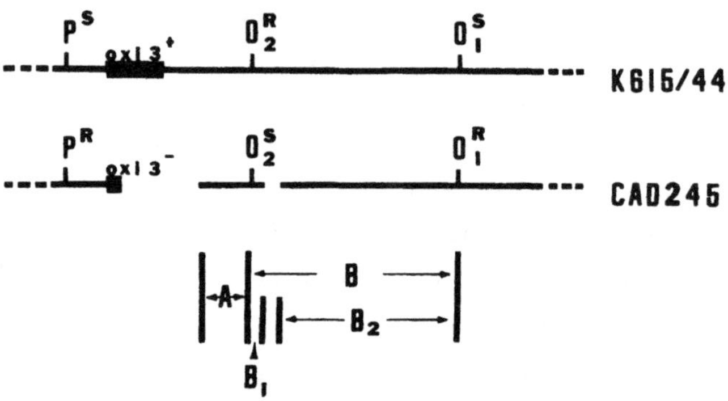

FIGURE 8: Schematic diagram of the cross between CAD245 and K615/44. This cross was performed in an attempt to obtain wild-type recombinants with a deletion in Eco R1 band 9 originating in strain CAD245.

DISCUSSION

The data presented here on class I and II mutants are consistent with the conclusion that oxi1, oxi3 and cob2 are the structural genes for cox II, cox I and cyt b, respectively. A stronger relationship can be established when a specific alteration in the primary sequence of a polypeptide is uniquely referrable to a mutant site in one particular mit locus. This test has been met using the class IIIA-1 mutant E3, which maps at oxi2 and affects cox III. We have also studied mutant E69 which maps at oxi2 and eliminates cox III from the electrophoretic profiles. E69 is a deletion mutant, and the deletion of 0.17 x 10^6 daltons has been located in a restriction fragment (Eco R1 plus Hinc II) of 1.8 x 10^6 daltons, and located between the cap and par markers.

Our physical and deletion mapping studies provide useful insights into the structure of oxi3 and cob2 loci. The fine structure genetic map of the oxi3 locus has been oriented on the physical map, and in relation to the portions of the flanking drug markers oli2 and par. Four mutant sites in the cob2 locus have been ordered and oriented relative to the flanking drug markers oli1 and oli2.

We have identified several mutants affecting the synthesis of two mitochondrial translational products but which do not appear to be doubly mutant at the two relevant mit loci. Some of these are allelic to mutations at loci believed to define structural genes based on studies with class I and II mutants; thus these novel alleles indicate regulatory, processing or assembly functions of the affected locus or its products. We have also analyzed one mutant, M45, which is a double mutant and as a result is deficient in two protein bands.

Two of our nuclear mutants may provide a novel means for the identification of mitochondrial translation products since they cause the appearance of new major polypeptide bands that are unrelated (by finger printing) to the other major products found in ρ^+ strains. These, along with the class III mutants, raise important questions about the origin and control of mitochondrial translation products. We have been able to establish that the extra band of the cob2 mutant EM17 is not related to any of the normal major products, as is also the case for the fast extra band of the nuclear mutant M113, but their precise origin is still uncertain. It should also be possible to isolate mitochondrial mutants in M44 and M113 that affect their extra bands, and thus to assess the involvement of the mitochondrial genome in specifying those proteins.

Another approach to the identification of new coding regions of the mitochondrial genome is illustrated by our analyses of

CAD245. This mutant has two deletions, one affecting the oxi3 locus and the other located elsewhere. So far our attempts to construct a ρ^+ strain having the small deletion near oli2 have been unsuccessful. This segment of 0.15 x 10^6 daltons must include either a portion of a new genetic locus or an interlocus (spacer) region. Studies are in progress aimed at making this distinction.

ACKNOWLEDGEMENTS

We are grateful to Dr. A. Tzagoloff for providing us with a collection of his mit⁻ mutants. The highly competent technical assistance of Mary Ensign, Cindy Ellis, Robbin Palmer, Jeanette Johnson and Richard Head is acknowledged with pleasure and gratitude. Supported by research grant GM12228 (to H. R. Mahler), GM19607 (to P. S. Perlman and C. W. Birky, Jr.) and GM22837 (subcontract to P. S. Perlman) all from NIH-NIGMS. H. R. Mahler is a recipient of Research Career Award K06 05060 from the same institute. N. J. Alexander is an Ohio State University Postdoctoral Fellow. D. Hanson is a recipient of a Proctor and Gamble Fellowship.

REFERENCES

1. Tzagoloff, A., Akai, A., Needleman, R. B.: Assembly of the mitochondrial membrane system: Isolation of nuclear and cytoplasmic mutants of Saccharomyces cerevisiae with specific defects in mitochondrial functions. J. Bact. 122, 826-831 (1975).
2. Tzagoloff, A., Akai, A., Needleman, R. B.: Properties of cytoplasmic mutants of Saccharomyces cerevisiae with specific lesions in cytochrome oxidase. Proc. Nat. Acad. Sci. USA 72, 2054-2057 (1975).
3. Tzagoloff, A., Akai, A., Needleman, R. B., Zulch, G.: Assembly of the mitochondrial membrane system: Cytoplasmic mutants of Saccharomyces cerevisiae with lesions in enzymes of the respiratory chain and in the mitochondrial ATPase. J. Biol. Chem. 250, 8236-8242 (1975).

4. Flury, U., Mahler, H. R., Feldman, F.: A novel respiratory-deficient mutant of Saccharomyces cerevisiae. J. Biol. Chem. 249, 6130-6137 (1974).
5. Slonimski, P. P., Tzagoloff, A.: Localization in yeast mitochondrial DNA of mutations expressed in a deficiency of cytochrome oxidase and/or coenzyme QH_2-cytochrome c reductase. Europ. J. Biochem. 61, 27-41 (1976).
6. Groot Obbink, D. J., Hall, R. M., Linnane, A. W., Lukins, H. B., Monk, B. C., Spithall, T. W., Trembath, M. K.: Mitochondrial genes involved in the determination of mitochondrial membrane proteins. in The Genetic Function of Mitochondrial DNA (Saccone, C., Kroon, A. M., eds.) North-Holland Biomedical Press:Amsterdam, 163-174 (1976).
7. Schweyen, R. J., Weiss-Brummer, B., Backhaus, B., Kaudewitz, F: Localization of seven gene loci on a circular map of the mitochondrial genome of Saccharomyces cerevisiae. in The Genetic Function of Mitochondrial DNA (Saccone, C., Kroon, A. M., eds.) North-Holland Biomedical Press:Amsterdam, 251-258 (1976).
8. Bolotin-Fukuhara, M., Faye, G., Fukuhara, H.: Temperature-sensitive respiratory deficient mitochondrial mutations: Isolation and genetic mapping. Molec. Gen. Genet. 152, 295-305 (1977).
9. Kotylak, Z., Slonimski, P. P.: Joint control of cytochromes a and b by a unique mitochondrial DNA region comprising four genetic loci. in The Genetic Function of Mitochondrial DNA (Saccone, C., Kroon, A. M., eds.) North-Holland Biomedical Press:Amsterdam, 143-154 (1976).
10. Mahler, H.R., Bilinski, T., Miller, D., Hanson, D., Perlman, P. S., Demko, C.A.: Respiration deficient mutants with intact mitochondrial genomes: Casting a wider net. in Genetics and Biogenesis of Chloroplasts and Mitochondria (Bücher, Th., Neupert, W., Sebald, W., Werner, S., eds.) North-Holland Biomedical Press:Amsterdam, 857-863 (1976).

11. Rytka, J., English, K. J., Hall, R. M., Linnane, A. W., Lukins, H. B.: The isolation and simultaneous physical mapping of mitochondrial mutations affecting respiratory complexes. in Genetics and Biogenesis of Chloroplasts and Mitochondria (Bücher, Th., Neupert, W., Sebald, W., Werner, S., eds.) North-Holland Biomedical Press:Amsterdam, 427-434 (1976).

12. Rabinowitz, M., Jakovcic, S., Martin, N., Hendler, F., Halbreich, A., Lewin, A., Morimoto, R.: Transcription and organization of yeast mitochondrial DNA. in The Genetic Function of Mitochondrial DNA (Saccone, C., Kroon, A. M., eds.) North-Holland Biomedical Press: Amsterdam, 219-230 (1976).

13. Tzagoloff, A., Foury, F., Akai, A.: Assembly of the mitochondrial membrane system. XVIII. Genetic loci on mitochondrial DNA involved in cytochrome b biosynthesis. Molec. Gen. Genet. 149, 33-42 (1976).

14. Tzagoloff, A., Foury, F., Akai, A.: Resolution of the mitochondrial genome. in The Genetic Function of Mitochondrial DNA (Saccone, C., Kroon, A. M., eds.) North-Holland Biomedical Press:Amsterdam, 155-161 (1976).

15. Douglas, M.G., Butow, R.A.: Variant forms of mitochondrial translation products in yeast: Evidence for location of determinants on mitochondrial DNA. Proc. Nat. Acad. Sci. USA 73, 1083-1086 (1976).

16. Douglas, M. G. , Kendrick, E., Boulikas, P., Perlman, P. S., Butow, R.A.: Electrophoretic variants of yeast mitochondrial translation products: Genetic and biochemical analysis. in The Genetic Function of Mitochondrial DNA (Saccone, C., Kroon, A. M., eds.) North-Holland Biomedical Press:Amsterdam, 199-207 (1976).

17. Katan, M. B., Groot, G. S. P.: Complex III of yeast: Subunit composition and biosynthesis. in Genetics and Biogenesis of Chloroplasts and Mitochondria (Bücher, Th., Neupert, W., Werner, S., eds.) North-Holland Biomedical Press:Amsterdam, 273-280 (1976).

18. Tzagoloff, A., Akai, A., Foury, F.: Assembly of the mitochondrial membrane system. XVI. Modified form of the ATPase proteolipid in oligomycin-resistant mutants of Saccharomyces cerevisiae. FEBS Lett. 65, 391-395 (1976).

19. Pajot, P., Wambier-Kluppel, M. L., Kotylak, Z., Slonimski, P. P.: Regulation of cytochrome oxidase formation by mutations in a mitochondrial gene for cytochrome b. in Genetics and Biogenesis of Chloroplasts and Mitochondria (Bücher, Th., Neupert, W., Sebald, W., Werner, S., eds.) North-Holland Biomedical Press:Amsterdam, 443-452 (1976).

20. Cobon, G. S., Groot Obbink, D. J., Hall, R. M., Maxwell, R., Murphy, M., Rytka, J., Linnane, A. W.: Mitochondrial genes determining cytochrome b (complex III) and cytochrome oxidase function. in Genetics and Biogenesis of Chloroplasts and Mitochondria (Bücher, Th., Neupert, W., Sebald, W., Werner, S., eds.) North-Holland Biomedical Press:Amsterdam, 453-460 (1976).

21. Foury, F., Tzagoloff, A.: Assembly of the mitochondrial membrane system. XIX. Genetic characterization of mit$^-$ mutants with deficiencies in cytochrome oxidase and coenzyme QH_2-cytochrome c reductase. Molec. Gen. Genet. 149, 43-50 (1976).

22. Storm, E. M., Marmur, J.: A temperature sensitive mitochondrial mutation of Saccharomyces cerevisiae. Biochem. Biophys. Res. Commun. 64, 752-759 (1975).

23. Handwerker, A., Schweyen, R. J., Wolf, K., Kaudewitz, F.: Evidence for an extrakaryotic mutation affecting the maintenance of the rho factor in yeast. J. Bact. 113, 1307-1310 (1973).

24. Trembath, M. K., Monk, B. C., Kellerman, G. M., Linnane, A. W.: Biogenesis of mitochondria. 36. The genetic and biochemical analysis of a mitochondrially determined cold sensitive oligomycin resistant mutant of Saccharomyces cerevisiae with affected mitochondrial ATPase assembly. Molec. Gen. Genet. 141, 9-22 (1975).

25. Lancashire, W. E.: Mitochondrial mutations conferring heat or cold sensitivity in Saccharomyces cerevisiae. in Genetics and Biogenesis of Chloroplasts and Mitochondria (Bücher, Th., Neupert, W., Sebald, W., Werner, S., eds.) North-Holland Biomedical Press:Amsterdam, 481-490 (1976).

26. Cleveland, D. W., Fischer, S. G., Kirschner, M. W., Laemmli, U. K.: Peptide mapping by limited proteolysis in sodium dodecyl sulfate and analysis by gel electrophoresis. J. Biol. Chem. 252, 1102-1106 (1977).

27. Molloy, P. L., Linnane, A. W., Lukins, H. B.: Biogenesis of mitochondria: Analysis of deletion of mitochondrial antibiotic resistance markers in petite mutants of Saccharomyces cerevisiae. J. Bact. 122, 7-18 (1975).

28. Sriprakash, K. S., Molloy, P. L., Nagley, P., Lukins, H. B., Linnane, A. W.: Biogenesis of mitochondria. XLI. Physical mapping of mitochondrial genetic markers in yeast. J. Mol. Biol. 104, 485-503 (1976).

29. Fukuhara, H., Bolotin-Fukuhara, M., Hsu, H.J., Rabinowitz, M.: Deletion mapping of mitochondrial transfer RNA genes in Saccharomyces cerevisiae by means of cytoplasmic petite mutants. Molec. Gen. Genet. 145, 7-19 (1976).

30. Schweyen, R. J., Steyrer, U., Kaudewitz, F., Dujon, B., Slonimski, P. P.: Mapping of mitochondrial genes in Saccharomyces cerevisiae: Population and pedigree analysis of retention or loss of four genetic markers in rho^- cells. Molec. Gen. Genet. 146, 117-132 (1976).

31. Morimoto, R., Lewin, A., Merten, S., Rabinowitz, M.: Restriction endonuclease mapping and analysis of grande and mutant yeast mitochondrial DNA. in Genetics and Biogenesis of Chloroplasts and Mitochondria (Bücher, Th., Neupert, W., Sebald, W., Werner, S., eds.) North-Holland, Biomedical Press:Amsterdam, 519-525 (1976).

32. Heyting, C., Sanders, J. P. M.: The physical mapping of some genetic markers in the 21S ribosomal region of the mitochondrial DNA in yeast. in The Genetic Function of Mitochondrial DNA (Saccone, C., Kroon, A. M., eds.) North-Holland Biomedical Press:Amsterdam, 273-280 (1976).

33. DiFranco, A., Sanders, J. P. M., Heyting, C., Borst, P., Slonimski, P. P.: Restriction enzyme analysis and physical mapping of mitochondrial DNA from petite mutants, carrying a genetic marker for oligomycin or paromomycin resistance. in The Genetic Function of Mitochondrial DNA (Saccone, C., Kroon, A. M., eds.) North-Holland Biomedical Press:Amsterdam, 291-304 (1976).

34. Lazowska, J., Jacq, C., Cebrat, S., Slonimski, P.P.: Fine structure map constructed by electron microscopy and restriction endonucleases of the mitochondrial DNA segment conferring erythromycin resistance. in The Genetic Function of Mitochondrial DNA (Saccone, C., Kroon, A. M., eds.) North-Holland Biomedical Press:Amsterdam, 325-336 (1976).

35. Lang, B., Burger, G., Doxiadis, I., Thomas, D.Y. Bandlow, W., Kaudewitz, F.: A simple method for the large-scale preparation of mitochondria from microorganisms. Anal. Biochem. 77, 110-121 (1977).

36. Butow, R. A., Vincent, R. D., Strausberg, R. L., Zanders, E., Perlman, P. S.: Genetic,physical and biochemical analysis of a mitochondrial gene. This volume (1977).

37. Sanders, J. P. M., Heyting, C., Borst, P.: The variability of the mitochondrial genome of Saccharomyces strains. in Genetics and Biogenesis of Chloroplasts and Mitochondria (Bücher, Th., Neupert, W., Sebald, W., Werner, S., eds.) North-Holland Biomedical Press: Amsterdam, 511-518 (1976).

38. Bernardi, G., Prunell, A., Fonty, G., Kopecka, H., Strauss, F.: The mitochondrial genome of yeast: Organization, evolution and the petite mutation. in The Genetic Function of Mitochondrial DNA (Saccone, C., Kroon, A. M., eds.) North-Holland Biomedical Press:Amsterdam, 185-198 (1976).
39. Prunell, A., Bernardi, G.: The mitochondrial genome of wild-type yeast cells. VI. Genome organization. J. Mol. Biol. 110, 53-74 (1977).

A STRUCTURAL ANALYSIS OF THE OXI-3 REGION ON YEAST mtDNA

L.A.GRIVELL AND A.F.M.MOORMAN

Section for Molecular Biology, Laboratory of Biochemistry, University of Amsterdam, Kruislaan 320, Amsterdam (The Netherlands)

INTRODUCTION

Recent genetic studies show that three separate regions of the yeast mitochondrial genome are involved in the specification of the mitochondrially-synthesized subunits of cytochrome c oxidase and/or their integration into the mitochondrial membrane (1,2). Two of these regions (oxi-1 and oxi-2) can be relatively sharply defined by both recombination analysis and deletion mapping as lying in the quadrant of the genome between the markers C and P (3,4). The third region (oxi-3) is much more diffuse by both criteria and apparently spans a stretch of the genome between the markers P and oli-2 (3), which greatly exceeds that minimally required to specify subunit I of cytochrome c oxidase, whose synthesis is blocked in nearly all mutants at this locus (3). However, even more puzzling is the finding of Sanders et al. (5) that this locus is bisected by a site at which large scale insertions and deletions occur in various related yeast strains, so that the region can show a variation in length of up to 5000 base pairs. This variation is without apparent effect on either the synthesis of subunit I or its size (Groot, G.S.P. and Van Harten-Loosbroek, N., unpublished).

It is obvious that an exact localization of the structural gene for subunit I within the oxi-3 region and a characterization of the sequences flanking it are essential requirements

---

Non-standard abbreviations: SSC, 0.15 M NaCl, 0.015 M Na-citrate (pH 7.0); TCA, trichloroacetic acid.

for the unravelling of the enigmatic properties of this region of mtDNA. We describe in this article results of preliminary attempts to achieve this aim by transcription-translation of restriction fragments of mtDNA _in vitro_ and identification of the products synthesized. These results are correlated with a fine-structure map of the oxi-3 region obtained by restriction enzyme analysis of mtDNA from mutants at this locus. Possible explanations for the picture which emerges are presented and discussed.

RESULTS AND DISCUSSION

Fine-structure mapping of mutants in the oxi-3 region

Several of the mutants located by Slonimski and Tzagoloff (3) in the oxi-3 region behave genetically as large deletions in that they are unable to restore the wild-type genotype to several groups of non-allelic mutants. That this is in fact so for the mutant M5-16 has been shown by Rabinowitz _et al_. (6): digests of mtDNA from this strain with the restriction endonuclease HhaI gave a pattern consistent with the notion that the mutant DNA contains a deletion some 5000 base pairs long, extending across a Hha cleavage site. Since, in principle, it should be possible to localize the site of the structural gene for subunit I as the deleted region common to several such mutants, we attempted to pin-point the lesions in this and three other similar mutants more accurately. These are the mutants M10-150, M11-125 and M12-120, originally isolated by Alex Tzagoloff (3) and kindly made available to us by him. They behave genetically as overlapping deletion or translocation mutations and their positions on the genetic map of the oxi-3 region are shown in Fig. 1. None of these mutants is capable of the synthesis of subunit I as shown by sodium dodecylsulphate-acrylamide gel electrophoresis of submitochondrial particles isolated from cells pulse-labelled with (^{35}S)methionine in the presence of cycloheximide (3; Van Harten-Loosbroek, N. and Groot, G.S.P., unpublished).

The mutants are derived from the _Saccharomyces cerevisiae_

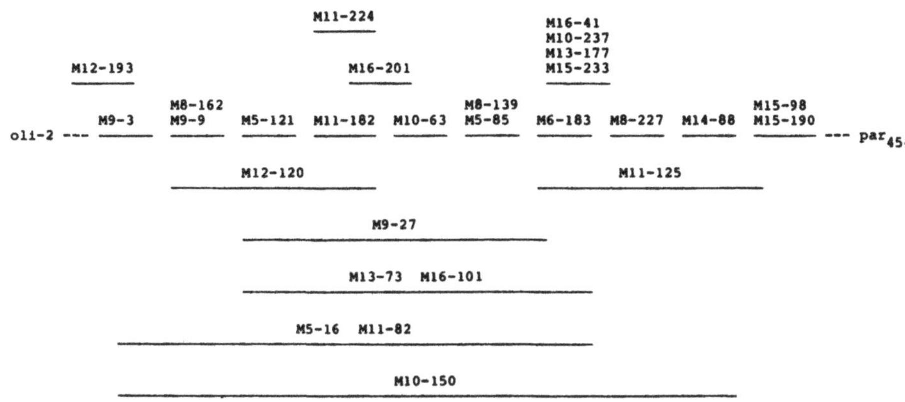

Fig. 1. A genetic map of the oxi-3 region, taken from Slonimski and Tzagoloff (3). Overlapping lines represent lack of restoration of growth on glycerol when the mt^- mutants are crossed with each other. Non-overlapping lines indicate restoration of growth.

strain D273-10B, for which - unfortunately - no detailed physical mapping data are available. In order to provide information on this point, mtDNA from this strain was digested with the enzymes EcoRI, HindII+III and HhaI either apart or in combination. Some of the results of this analysis are presented in Plate I (tracks a-d). Digestion with EcoRI and HindII+III (tracks a and b) shows that D273-10B is basically highly homologous with S. cerevisiae JS1-3D mtDNA, for which a physical map has been constructed (5). Only two differences in fragmentation pattern are demonstrable and these involve fragment 2 (TT3 in Fig. 2), which in D273-10B is about 100 base pairs longer than its counterpart in JS1-3D, and fragment 8 (RT8), which is 50 base pairs longer than the corresponding JS1-3D fragment. Differences between the two strains seem greater when Hha digestion patterns are compared (tracks c and d). These differences are more apparent than real, however, and can be explained by postulating that D273-10B incorporates several features of both Saccharomyces carlsbergensis and JS1-3D physical maps:

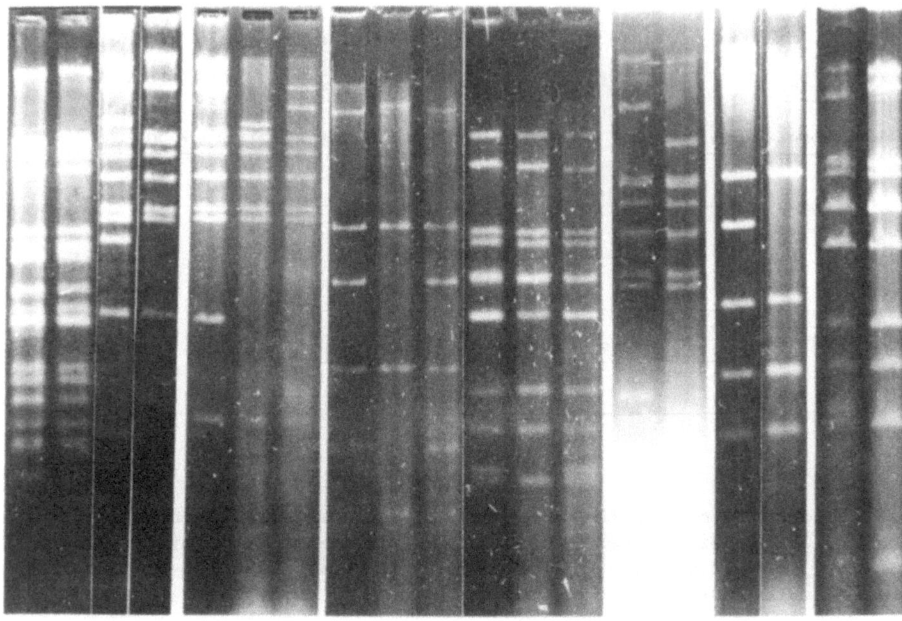

Plate I. Restriction enzyme analysis of the mtDNAs of S. cerevisiae JS1-3D, D273-10B and of the mt^- mutants M10-150, M11-125 and M5-16. Digestion was carried out with the restriction endonucleases HhaI, EcoRI and HindII+III in 7 mM $MgCl_2$, 50 mM NaCl, 5 mM Tris-HCl (pH 7.9), 1 mM dithiothreitol at 37°C for 2-4 h and the digests were analysed on 0.7% agarose slab gels (18 h; 1 v/cm). After staining in ethidium bromide (0.5 µg/ml), gels were illuminated with short wavelength UV light and photographed on Copex Pan Rapid film through a Kodak 23A filter. a) JS1-3D mtDNA x EcoRI x HindII+III; b) D273-10B mtDNA x EcoRI x HindII+III; c) JS1-3D mtDNA x HhaI; d, e and n) D273-10B mtDNA x HhaI; f, i and l) M10-150 mtDNA x HhaI, EcoRI and HindII+III, respectively; g, j and m) M11-125 mtDNA x HhaI, EcoRI and HindII+III, respectively; h and p) D273-10B mtDNA x EcoRI; k and r) D273-10B mtDNA x HindII+III; o, q and s) M5-16 mtDNA x HhaI, EcoRI and HindII+III, respectively. In some digests, the high-molecular-weight fragments are under-represented, because the preparations analysed consist of randomly broken mtDNA, rather than intact, full-length molecules.

1. D273-10B lacks fragments corresponding in migration to LL9 and LL11 of JS1-3D, but has an LL1 which is apparently correspondingly longer. It would seem that point mutations or small deletions have eliminated the cleavage sites between these fragments which are adjacent on the JS1-3D map. This situation is somewhat analogous to that in S. carlsbergensis (7).

2. D273-10B possesses an extra Hha cleavage site in the fragment corresponding to LL3 in JS1-3D. This can be seen from the presence of two "new" fragments, which migrate differently from any of the JS1-3D fragments. These are LL4 and LL10 and their combined lengths are roughly equal to that of LL3 in JS1-3D. Their probable locations, based on double digestion and DNA-DNA hybridization analysis (data not shown), are shown in Fig. 2.

No significant differences in size between D273-10B and JS1-3D are observed for fragments derived from the oxi-3

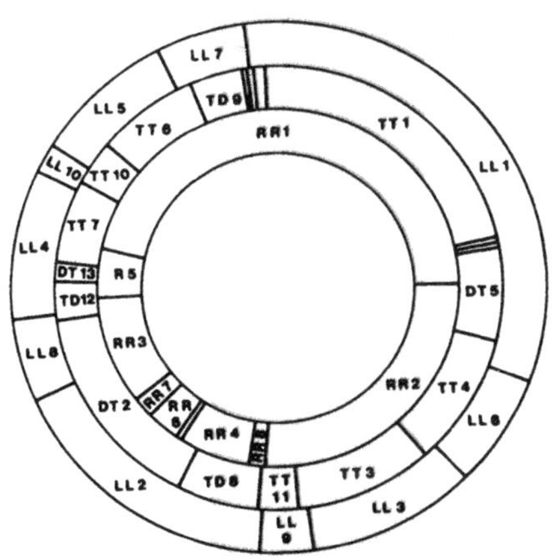

Fig. 2. A physical map of S. cerevisiae D273-10B mtDNA, showing cleavage sites for the restriction endonucleases HindII (T), HindIII (D), EcoRI (R) and HhaI (L). Nomenclature of DNA fragments is based on recommendations made by Sanders et al. (7), whereby the sites at each end of the fragment are identified by a single letter code. The order of site identification is in a clockwise direction and each fragment is given a number, which indicates, in order of decreasing length, its position in the first fragmentation pattern in which it is observed.

region. This, when taken together with the high degree of homology between the two strains in the remaining parts of the genome would seem to be sufficient justification for applying the physical map, previously constructed for JS1-3D and modified as noted above, for fine mapping of oxi-3 mutants.

M10-150

Digests of M10-150 mtDNA by EcoRI, HindII+III and HhaI are shown in Plate I. In the HhaI digestion (track f), fragments LL2 and LL9 are missing and are replaced by a new fragment of 8200 base pairs. This is consistent with the presence of a deletion of about 7000 base pairs lying across the HhaI site between LL2 and LL9. Digestion with EcoRI (track i) confirms and extends this conclusion: it shows that fragments RR4, RR6 and RR8 are missing and that RR2 is between 1000-2000 base pairs longer than in D273-10B (19 000 base pairs). For the HindII+III digest (track l), the differences are that band 6, which is double in D273-10B, is single suggesting that either TT7 or TD8 is absent and that TT11 is missing; none of the other fragments is detectably altered. The most reasonable model consistent with all these observations is shown in Fig. 3 and is that a 7000 base pair deletion is indeed present and that it stretches from just within RR6, across RR9, RR4 and RR8 and ends in RR2 just beyond the HhaI site linking LL2 with LL9. This has been further verified by performing EcoRI x HindIII and EcoRI x HindII+III digestions (data not shown).

M5-16

A similar line of reasoning can be applied to the mutant M5-16. In confirmation of the report by Rabinowitz et al. (6), fragments LL2 and LL9 are replaced by a new fragment 5400 base pairs shorter than their combined lengths (track o). In analogy with M10-150, this suggests the presence of a deletion of this length lying across the HhaI site between these two fragments. This picture is confirmed by the EcoRI and HindII+III digestion patterns (tracks q and s). In the EcoRI digest, RR4

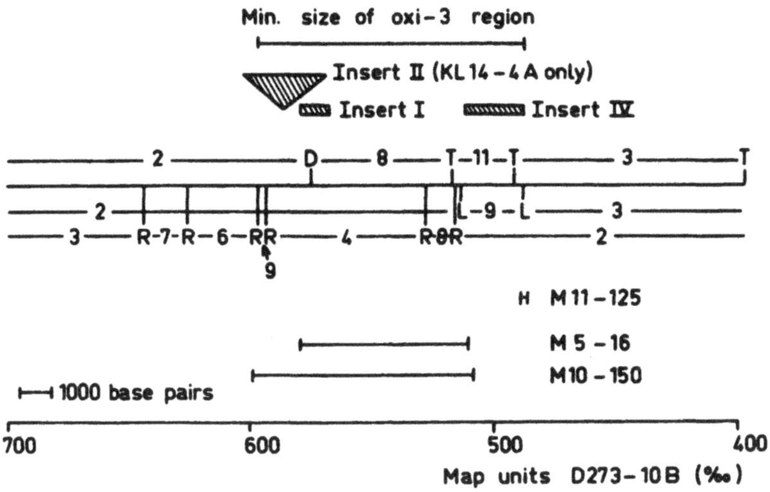

Fig. 3. A fine-structure map of the oxi-3 region in S. cerevisiae D273-10B, indicating the positions of deletions in mt⁻ mutants M5-16, M10-150 and M11-125 as determined by restriction enzyme analysis. Approximate positions of sequences present in this strain and S. cerevisiae KL14-4A, which are absent from S. carlsbergensis are also indicated. Identification of the fragments and of the restriction sites at their termini follows the principles laid down in the legend to Fig. 2. The molecular weights used are those given by Sanders et al. (7).

and RR8 are missing, while the remaining fragments are not detectably altered. In the HindII+III digest, TT2 is about 2000 base pairs longer than in D273-10B, band 6 in the digest is single as in M10-150, suggesting that TD8 is missing and TT11 is missing. The observations are satisfactorily accounted for by placing one end of the deletion just within DT2 and extending it across TD8 into the fragment which TT11 and LL9 have in common (Fig. 3).

M11-125

The only change detectable in M11-125 is in the HhaI cleavage pattern (track g). As also reported by Rabinowitz et al. (6), LL3 and LL9 are missing and are replaced by a new fragment of a length not significantly different from the sum of the two (9500 base pairs), so that the HhaI site linking these two fragments must have been removed either by a point mutation or by insertion or deletion of less than 100 base pairs, which is the smallest change readily detectable in a fragment the length of TT3 (see Fig. 3). The lack of changes in the other digestion patterns is not consistent with the finding of Rabinowitz et al. (6) that HpaI and HincII sites are also missing from this DNA, since these sites are also recognized by HindII. At this point, we have no good explanation for this discrepancy, apart from suggesting the possibility that different clones of this mutant might have changed since their separation.

M12-120

So far, no change in fragmentation pattern has been detected for the mutant M12-120 with any of the enzymes used (cf. ref. 6). It is possible that this is a point mutation, but this would seem unlikely in view of its genetic behaviour. An alternative and interesting possibility is that this mutant contains an inverted sequence either in a region lacking a restriction site, or in one symmetrically arranged around a site. We are currently investigating this latter alternative by use of finer cutting restriction enzymes.

Of course, at this stage we cannot be entirely certain that the lesions identified above are those solely responsible for the mutant phenotypes, since the mutants were isolated after applied mutagenesis (3) and mtDNA can apparently tolerate a wide variety of rearrangements (5). However, assuming - for the present - that this is indeed so, the data presented above have the following intriguing aspects:

1. The 7000 base pair deletion in M10-150 puts a new lower limit on the size of the oxi-3 locus. This is minimally some 6400 base pairs in S. carlsbergensis, 8800 base pairs in D273-10B and 11 400 base pairs in the strain KL14-4A, which contains a further insertion of 2600 base pairs in the fragment homologous with RR4 of S. carlsbergensis and S. cerevisiae

JS1-3D (7). Since there are mutants mapping outside the M10-150 deletion (3), the real size of oxi-3 is likely to be even greater.

2. The location of the lesion detected in M11-125 over the HhaI site on TT3 extends the oxi-3 region in a new sense: fragments TT11 and LL10 in JS1-3D mtDNA (LL9 in D273-10B) represent the major part of an insertion of 1500-2100 base pairs (insert IV; ref. 8), which carries information absent from S. carlsbergensis mtDNA (Meijlink, F.C.P.W. and Sanders, J.P M., unpublished). The mutation in M11-125 thus lies on the edge of this insertion, perhaps extending a short way into TT3. We are forced to the conclusion that the sites identified by the various mutations as having functions in the specification of subunit I of cytochrome c oxidase can be physically separated into at least two and probably three distinct units without noticeable effect on the synthesis of this subunit. It is also worth noting that the deletions in M5-16 and M10-150 also terminate at or near the edge of insert IV, suggesting perhaps that sequences surrounding the insert are a source of genetic variability in this region.

3. The mutation site of M11-125 possibly represents a mutational hot-spot. If our estimate of the size of the deletion is correct, it means that at least five groups of non-allelic mutants are clustered within less than 100 base pairs (see Fig. 1). We can speculate that this kind of clustering might also apply for other complementable mutations, so that the oxi-3 region could in fact consist of genetically-sensitive and genetically-silent sequences interspersed with each other.

4. Comparison of the physical with the genetic mapping data reveals a discrepancy between the two concerning the location of the lesions in the mutants M5-16 and M11-125. According to the genetic map (Fig. 1; ref. 3), the mutations overlap; according to the physical map, they are separated by minimally 1700 and maximally 2100 base pairs. One possible reason for this is that the intervening sequences are part of an insertion, likely to be absent from the S. carlsbergensis strain

used to construct mt^- mutants with different mating-type alleles for use in the original complementation tests (3). Unless both mutants to be complemented possess the inserted sequence, heteroduplex formation in this region will not be possible, so that the two mutations will be scored as overlapping.

Location of the gene for subunit I of cytochrome *c* oxidase by transcription-translation *in vitro*

We have previously shown that mtDNA is capable of directing the synthesis of antigenic determinants of cytochrome *c* oxidase in cell-free extracts of *Escherichia coli*, even after cleavage with restriction endonucleases (8). Fragments derived from cleavage of JS1-3D mtDNA with HindII+III were, therefore, separated on a preparative scale on 0.7% agarose slab gels, eluted and used to programme protein synthesis. The products made were then identified immunologically using antisera directed primarily against either subunit II alone, or against both subunits I and II. The results obtained are presented in Table I and are summarized schematically in Fig. 4. Interpretation is somewhat complicated by a combination of two factors: first, the antisera used are not entirely specific, but react weakly with subunits other than those against which they were raised (8); second, many of the bands in the HindII+III digest of JS1-3D mtDNA are double and contain fragments derived from different regions of the physical map. The picture which emerges, however, is that regions active in directing the synthesis of antigenic determinants of cytochrome *c* oxidase are located primarily on TT1, DT2, TT3 and/or TT4 and TT11 and/or TD12. In addition, the products of DT5 and/or TT7 give a lower - but probably significant - level of immunoprecipitation. The products of DT2, TT3 + TT4 and of TT11 + TD12 react to a greater extent with anti-subunits I + II than with anti-subunit II, while those of TT1 react about equally well with both antisera. We provisionally conclude, therefore, that determinants of subunit II are synthesized by TT1 and of subunit I by

TABLE I

ACTIVITY OF HindII+III FRAGMENTS OF S. cerevisiae JS1-3D mtDNA IN THE SYNTHESIS OF ANTIGENIC DETERMINANTS OF CYTOCHROME c OXIDASE

Approximately 200 µg of mtDNA was completely digested with HindII+III in 50 mM Tris-HCl (pH 7.9), 7 mM $MgCl_2$, 1 mM dithiothreitol for 2 h at 37°C and fragments were separated by electrophoresis through 0.7% agarose slab gels 40 cm in length. DNA was recovered by a procedure involving excision of bands from the gel, solution of the agarose by incubation for 30 min at room temperature in 5 M $NaClO_4$, 0.3 M NaCl and passage of the solution over a column of hydroxylapatite. The column was then washed extensively with 5 M $NaClO_4$, 0.3 M NaCl, followed by 0.05 M and 0.15 M sodium phosphate (pH 6.8) and DNA was eluted with 0.4 M sodium phosphate (pH 6.8). DNA-containing fractions were pooled, extracted twice with amyl alcohol to remove ethidium bromide and dialysed overnight against 0.1 x SSC. Where necessary, the DNA was concentrated by alcohol precipitation and 3 µg of each fraction was assayed for protein synthesis activity in E. coli extracts, as described by Moorman et al. (8). Immunoprecipitation values have been corrected for a contribution by endogenous product.

DNA fragment	TCA precipitable (cpm/5 µl)	% precipitable by		
		control serum	anti-subunits I + II	anti-subunit II
None	56 724	0.5	0.7	0.7
TT1	264 227	0.3	10.1	8.1
DT2	236 443	0.5	5.4	3.0
TT3 + TT4	347 067	0.6	3.7	2.6
DT5 + TT6	312 365	0.7	2.9	2.3
TT7 + TD8	298 805	0.4	1.6	1.1
TD9	201 682	0.4	1.7	1.1
TT10	97 841	<0.2	1.3	1.1
TT11 + TD12	149 072	0.2	7.3	3.7

the remaining fragments. Subunit III is not localised by these results, but there is some preliminary evidence (Cabral, F., personal communication) that mutants at the oxi-2 locus either lack or contain an altered form of this subunit.

From Fig. 4 can be seen that the position of TT1 on the physical map corresponds roughly with the oxi-1 and oxi-2 loci. For the oxi-3 region, the situation is complex, with the active fragments DT2 and TT3 + TT4 together with TT11 + TD12 separated by TD8, the products of which do not react with either of the

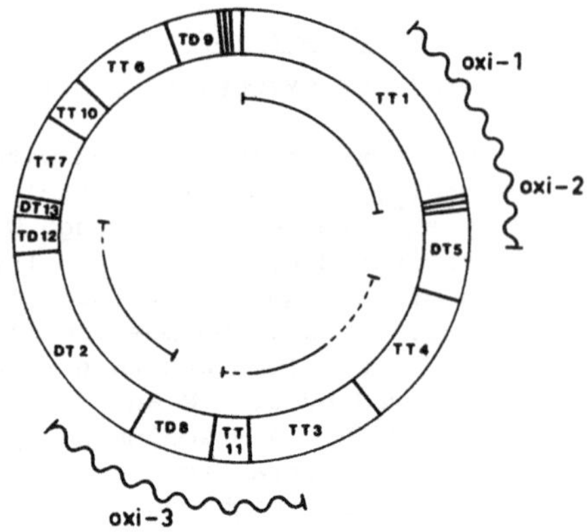

Fig. 4. A preliminary localization of the genes for cytochrome c oxidase on the physical map of JS1-3D mtDNA, based on the data presented in Table I. |-- —— --| denotes a fragment directing the synthesis of antigenic determinants of cytochrome c oxidase in cell-free extracts of E. coli; |~~~~~| denotes the approximate positions of genetically identified loci, involved in the synthesis of subunits of cytochrome c oxidase.

antisera. Unfortunately, we cannot yet unambiguously decide which of the fragments in the bands TT11 + TD12 is active. If it is TT11, activity must be confined to a region of this fragment adjoining RR8, some 400-500 base pairs long, since the remainder of the fragment carries - in the form of insert IV - information which is absent from S. carlsbergensis mtDNA. This is clearly insufficient to accomodate a complete gene for subunit I. If the active fragment is TD12, then there is either a gene lying across the HindII site between it and DT2, or there is more than one gene situated in this region of the map, with the second copy being located somewhere on DT2.

These results, combined with the physical mapping data, create a paradoxical situation. TD8, which is the one fragment missing from both M5-16 and M10-150, and thus which might reasonably be supposed to carry the structural gene for cytochrome c oxidase subunit I, is one which fails to direct the synthesis of antigenic determinants of this subunit in vitro. Although the trivial explanation that this fragment specifies parts of the subunit not recognized by our antisera cannot be completely excluded at this stage, this explanation would seem far-fetched in view of the length of the fragment. We are thus forced to the conclusion that at least two physically separated regions, active in the synthesis of antigenic determinants of subunit I of cytochrome c oxidase, are located on either side of fragment TD8. Even more intriguing, however, is the fact that TD8 is transcribed in all the strains so far studied (9; Moorman, A.F.M., Van Ommen, G.J.B. and Grivell, L.A., in preparation), so that the sequences present on it seem likely to be involved in the expression of the gene for subunit I in a role other than the specification of the primary structure of subunit I. Models which could account for these findings are discussed more fully elsewhere in this volume (10), but can be briefly summarized here. First, we can postulate that the whole of the oxi-3 region is transcribed into a giant precursor messenger RNA, which has to be processed before it can be read. Many of the oxi-3 mutations would then be clustered at sites involved in processing, while the insertions and deletions found by Sanders et al. (5) in certain strains presumably occur at less sensitive sites, or in non-conserved regions and can thus be tolerated without expression of the oxidase gene being affected. Alternatively, the active sequences defined by in vitro protein synthesis experiments could represent parts of active genes, which are brought together at the stage of transcription in a manner analogous to that recently reported for the messenger RNAs of adenovirus (11) and which might also apply to the 21S ribosomal RNA cistron on mtDNA (10).

Clearly, much of this is speculation, but it is speculation which is open to direct experimental verification and we can expect that - whatever the outcome - the results will be interesting.

ACKNOWLEDGEMENTS

We thank Professor Piet Borst, Johan Sanders and Frits Meijlink for informative discussions and for making unpublished data available to us. We are grateful to Dr J.L.Talen and Frits Meijlink for gifts of EcoRI and HhaI, respectively, to Dr R.O.Poyton for generous gifts of antisera against cytochrome c oxidase and to Mrs G.Reymer-Van Dam for technical assistance, Hans Bos for drawing the figures and Dick Lindenaar for the photography. This work was supported in part by a grant to P.Borst from The Netherlands Foundation for Chemical Research (SON) with financial aid from The Netherlands Organization for the Advancement of Pure Research (ZWO).

REFERENCES

1 Tzagoloff, A., Foury, F. and Akai, A. (1976) Resolution of the mitochondrial genome. In: The Genetic Function of Mitochondrial DNA (Saccone, C. and Kroon, A.M., Eds), North-Holland, Amsterdam, pp. 155-161.
2 Cobon, G.S., Groot Obbink, D.J., Hall, R.M., Maxwell, R., Murphy, M., Rytka, J. and Linnane, A.W. (1976) Mitochondrial genes determining cytochrome b (complex III) and cytochrome oxidase function. In: Genetics and Biogenesis of Chloroplasts and Mitochondria (Bücher, Th., Neupert, W., Sebald, W. and Werner, S., Eds), North-Holland, Amsterdam, pp. 453-460.
3 Slonimski, P.P. and Tzagoloff, A. (1976) Localization in yeast mitochondrial DNA of mutations expressed in a deficiency of cytochrome oxidase and/or coenzyme QH_2-cytochrome c reductase. Europ.J.Biochem. 61, 27-41.
4 Choo, K.-B., Nagley, P., Lukins, H.B. and Linnane, A.W. (1977) Biogenesis of mitochondria. 47. Refined physical map of the mitochondrial genome of Saccharomyces cerevisiae determined by analysis of an extended library of genetically and molecularly defined petite mutants. Mol.Gen. Gen. 153, 279-288.
5 Sanders, J.P.M., Heyting, C. and Borst, P. (1976) The variability of the mitochondrial genome of Saccharomyces strains. In: idem ref. 2, pp. 511-517.
6 Rabinowitz, M., Jakovcic, S., Martin, N., Hendler, F., Halbreich, A., Lewin, A. and Morimoto, R. (1976) Transcription and organization of yeast mitochondrial DNA. In: idem ref. 1, pp. 219-230.
7 Sanders, J.P.M., Heyting, C., Verbeet, M.P.H., Meijlink, F.C.P.W. and Borst, P. (1977) The organization of genes in yeast mitochondrial DNA. III. Comparison of the physical maps of the mitochondrial DNAs from three wild-type Saccharomyces strains. Mol.Gen.Gen., submitted.
8 Moorman, A.F.M., Grivell, L.A., Lamie, F. and Smits, H.L. (1977) Identification of mitochondrial gene products by DNA-directed protein synthesis in vitro. Biochim.Biophys.Acta, submitted.
9 Van Ommen, G.J.B. and Groot, G.S.P. (1977) A transcription map of yeast mitochondrial DNA. This volume.
10 Borst, P., Bos, J.L., Grivell, L.A., Groot, G.S.P., Heyting, C., Moorman, A.F.M., Sanders, J.P.M., Talen, J.L., Van Kreijl, C.F. and Van Ommen, G.J.B. (1977) The physical map of yeast mitochondrial DNA anno 1977. This volume.
11 Sambrook, J. (1977) Adenovirus amazes at Cold Spring Harbor. Nature 268, 101-104.

YEAST MITOCHONDRIAL MESSENGER RNA IS NOT CORRECTLY TRANSLATED IN HETEROLOGOUS CELL-FREE SYSTEMS

A.F.M.MOORMAN, F.N.VERKLEY, F.A.M.ASSELBERGS[∇] AND L.A.GRIVELL

Section for Molecular Biology, Laboratory of Biochemistry, University of Amsterdam, Kruislaan 320, Amsterdam (The Netherlands) and [∇]Department of Biochemistry, University of Nijmegen, Geert Grooteplein Noord 21, Nijmegen (The Netherlands)

INTRODUCTION

In vitro translation of mtRNA is an inescapable aspect of the characterization of mitochondrial messenger RNAs in terms of the protein they specify and is an important element in our attempts to identify and localize genes on the physical map of mtDNA (1).

Making use of the fact that mtRNA is translated in cell-free extracts of *Escherichia coli* to give antigenic determinants of cytochrome *c* oxidase (2), we have recently succeeded in isolating a messenger RNA fraction, sedimenting at 11S in sucrose gradients, which has activity in the synthesis of determinants of this complex *in vitro* (3). Unfortunately, although 25% of the products synthesized *in vitro* is precipitable by antisera against subunits of cytochrome *c* oxidase, electrophoretic analysis reveals that complete polypeptides are missing and that the products consist entirely of short polypeptides. This result is unexpected since messenger RNAs from a variety of sources have been shown to be accurately translated by *E. coli* extracts (4-8). The lack of complete polypeptides seriously hampers product identification, since the only means of identification is an immunological one and the resolution of this approach is dependent on the specificity of the available antisera (9). This article, therefore, describes our attempts to rule out trivial explanations for this phenomenon and to achieve correct translation of a mitochondrial messenger RNA in the *E. coli* and several other cell-free systems of eukaryotic origin.

RESULTS

The E. coli cell-free system

E. coli cell-free systems have been reported to be capable of the correct synthesis of eukaryotic proteins (4-8). Indeed, addition of mtRNA to an E. coli protein-synthesizing system results in a high stimulation of protein synthesis and leads to the synthesis of determinants of cytochrome c oxidase and ATPase (2,3). Using antisera against subunits of these complexes as probes for specific messenger RNA (mRNA) activities, we were able to isolate an 11S RNA fraction specifically enriched in messengers for subunit(s) of cytochrome c oxidase (Fig. 1). The genes for these RNAs have been localized on the physical map of Saccharomyces carlsbergensis and S. cerevisiae

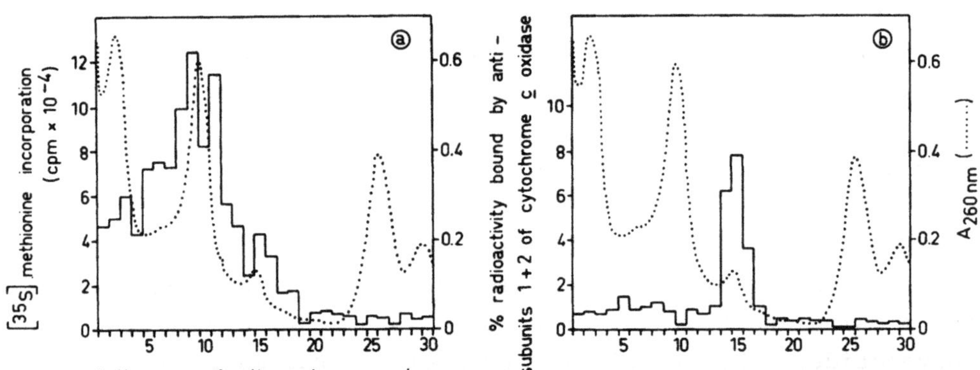

Fig. 1. Identification of mRNA for subunits of cytochrome c oxidase. See for details of RNA isolation, centrifugation, protein synthesis and immunoprecipitation refs 2 and 9. 100 µg mtRNA from Saccharomyces carlsbergensis NCYC-74 was fractionated by sedimentation through isokinetic sucrose gradients containing 3% formaldehyde. RNA of each fraction was assayed for activity in protein synthesis as described (2) except that in the translation assay phenylmethylsulphonylfluoride was omitted and the ammonium chloride concentration was raised to 100 mM. (^{35}S)Methionine was 61 Ci/mmol. 5-µl samples were assayed for hot trichloroacetic acid-precipitable radioactivity (a). In vitro product of each fraction was assayed in double precipitation tests for binding to antibodies raised against subunits 1 + 2 of cytochrome c oxidase (b). Values given have been corrected for binding to pre-immune serum (about 1%).

mtDNAs by hybridization to restriction fragments, using the techniques described by Southern. The position of these genes corresponds roughly to the OXI-1, 2 and 3 loci as located by genetic studies (3).

Nevertheless, as shown in Plate I, _in vitro_ products

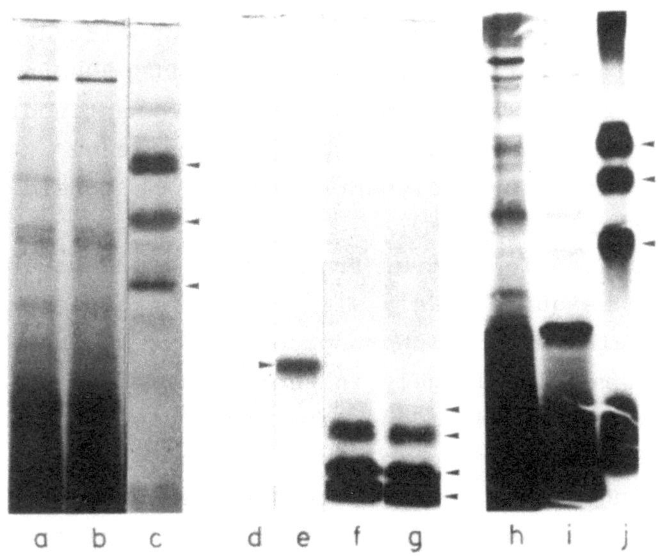

Plate I. Electrophoretic and immunological characterization of the products synthesized in an _E. coli_ cell-free system under direction of mtRNA. See for conditions of protein synthesis, immunoprecipitation and electrophoresis refs 2 and 9. (^{35}S)Methionine-labelled products were fractionated on a 13.5% polyacrylamide sodium dodecylsulphate gel. After electrophoresis gels were sliced, stained to visualize marker proteins, dried and fluorographed. a) _In vitro_ products synthesized under direction of 11S mtRNA; b) _in vitro_ products synthesized under direction of total mtRNA; c) _in vivo_ products of mitochondrial protein synthesis: submitochondrial particles were prepared from yeast cells pulse-labelled with (^{35}S)methionine in the presence of cycloheximide: arrows indicate the positions of the main products migrating at M_r 42 000, 33 000 and 23 000; d) endogenous product; e) _in vitro_ product synthesized under direction of bacteriophage MS2 RNA: the arrow indicates the position of the coat protein (M_r 14 000); f) immunoprecipitates of _in vitro_ products of 11S mtRNA (see track a) with antibodies raised against subunits 1 + 2 of cytochrome _c_ oxidase; g) immunoprecipitates of _in vitro_ products of total mtRNA (see track b) with antibodies raised against subunits 1 + 2 of cytochrome _c_ oxidase; h) _in vitro_ products synthesized under direction of total mtRNA in the presence of 50 µg extra _E. coli_ tRNA; i) _in vitro_ products synthesized under direction of 11S mtRNA in the presence of 50 µg extra mitochondrial tRNA: 4S RNA, isolated by sucrose gradient centrifugation, was used as source of mitochondrial tRNA; j) _in vivo_ products of mitochondrial protein synthesis (see track c).

specified by 11S RNA (track a), like those specified by total mtRNA (track b), consist largely of short polypeptides, absent from the in vivo products of mitochondrial protein synthesis (track c). In contrast, the homologous mRNA of bacteriophage MS2 is correctly translated, giving only coat protein (M_r 14 000) and some read-through product (track e). From both 11S RNA (track f) and total mtRNA (track g) specified product, the same short polypeptides with M_r 7400, 6400, 5000 and 4100 can be bound in a double precipitation test by antibodies raised against subunits 1 + 2 of cytochrome c oxidase.

Why are such short polypeptides made? We considered the following possibilities:

1. Is substantial out-of-frame translation occurring? This possibility is quite likely with such a heterologous mRNA, but cannot explain why only short products are synthesized. Random translation results largely in short polypeptides (M_r <2500), since three out of 64 codons is a stop codon, but does not exclude the synthesis of complete products, since one out of the three reading frames is correct. Moreover, 25% of the in vitro product can be bound by anti-subunits 1 + 2 of cytochrome c oxidase, so a significant fraction of the in vitro product must be synthesized in the correct reading frame.

2. Are ionic requirements not optimal for initiation and/or elongation on mtRNA? Variation of the concentration of magnesium in the range from 5-15 mM and of ammonium chloride or K-acetate from 25-300 mM did not result in the appearance of larger material. The highest activity of protein synthesis was observed at 10 mM magnesium-acetate and 100 mM ammonium chloride, values at which correct products of bacteriophage MS2 RNA, T_7 DNA or PM2 DNA are made (2,9). These data suggest that the ionic conditions used should permit correct translation of mtRNA. The fact that they are not, suggests that other factors could be limiting or that other barriers to correct translation exist.

3. Are the short in vitro products due to either ribonuclease or protease activity present in the E. coli system?

This possibility cannot be completely excluded but is rather unlikely.

i) The degradation must be specific for mtRNA and/or mitochondrial protein.

ii) Translation time (3-30 min) has no influence on the size of the product.

iii) Addition of the protease inhibitor phenylmethylsulphonylfluoride gives no difference (see ref. 2).

iv) Addition of calcium ions, which are a constituent of the more complex Zubay system (11) and which have been reported to inhibit ribonuclease activity in an E. coli cell-free system (12), results in a 6-fold extra stimulation of protein synthesis; however, less immunoprecipitation is found by antibodies raised against subunits 1 + 2 of cytochrome c oxidase (Table I) and much more small material can be observed on gels. It is possible that products are made other than cytochrome c oxidase. We prefer the interpretation that more aspecific initiation is occurring - a well-known effect of divalent cations (contrast ref. 13).

4. Does synthesis of mitochondrial proteins require vectorial translation by membrane-bound ribosomes? The mitochondrial translation products are strongly hydrophobic and mitochondrial ribosomes stick fairly firmly to the membrane: high salt and detergent are necessary to isolate them in reasonable yield and in pure state (14). As shown in Table I, mtRNA-directed protein synthesis is indeed stimulated by Triton X-100 and artificial membranes like liposomes, while bacteriophage MS2 RNA-directed protein synthesis is not stimulated (not shown). The liposomes were made according to Eytan et al. (15) who reported the incorporation of cytochrome c oxidase into these liposomes. The percentage immunoprecipitation remains the same, which means that more senseful products are made; electrophoretic analysis, however, reveals the same pattern of incomplete polypeptides. Ideally, a requirement for vectorial translation can only be rigorously demonstrated or excluded by checking the effect of adding mitochondrial membranes. This has not been

TABLE I

TRANSLATION OF mtRNA IN AN E. coli CELL-FREE SYSTEM UNDER VARIOUS CONDITIONS

Protein synthesis and immunoprecipitations were performed as described previously (2,9). The complete system contained all components necessary for protein synthesis (2) and 0.5 mg/ml mtRNA. 100% activity in protein synthesis, as determined by (^{35}S)methionine incorporation in hot trichloroacetic acid-insoluble material, equals 77 014 cpm/5 µl. Liposomes were prepared according to Eytan et al. (15). The phospholipid composition was phosphatidylethanolamine-phosphatidylcholine-phosphatidylserine (3:2:1). Immunoprecipitations were carried out with an input of in vitro product of about 50 000 cpm. Values given have been corrected for reaction with pre-immune serum (about 0.7%). N.D., not determined.

Additions/omissions	Activity in protein synthesis (% of complete system)	% radioactivity bound by anti-subunits 1 + 2 of cytochrome c oxidase (% of input)
Complete system	100	10.7
- mtRNA	3.1	0.1
+ 5 mM CaCl	607.7	2.8
+ 0.6% Triton X-100	169.5	9.7
+ Liposomes (5 mg/ml)	145.2	10.1
+ E. coli tRNA (0.7 mg/ml)	100.8	N.D.
+ Mitochondrial tRNA (0.7 mg/ml)	7.9	N.D.

done for various reasons:

i) Mitochondrial membrane preparations contain ribonuclease and protease activity.

ii) Electrophoretic analysis is unreliable, since mitochondrial proteins aggregate specifically with in vitro products of mtRNA-directed protein synthesis (9,10).

iii) Immunoprecipitation values will be too high since in vitro products will be enclosed in membrane vesicles which consist of proteins which are antigenic themselves.

In conclusion, it is still open whether or not the short polypeptides result from the lack of a suitable matrix to incorporate the newly synthesized mitochondrial proteins.

5. Does translation of mtRNA require specific tRNAs, which

are not present in the E. coli extract? The unusual base composition of mtDNA (16) makes it likely that a different range of tRNAs will be necessary, because of a different use of synonymous codons. Table I shows that addition of mitochondrial tRNA to the protein-synthesizing system reduced activity by 92%, while extra E. coli tRNA has no effect. A tRNA-dependent system consisting of purified ribosomes and supernatant purified over DEAE-cellulose (17) gave no stimulation at all with mitochondrial tRNA (data not shown). This low activity is unexpected since the same batch of mitochondrial tRNA can be loaded by E. coli synthetases and for methionine we have found equivalent loading rates for E. coli and mitochondrial tRNA (Moorman, A.F.M., unpublished). Electrophoretic analysis revealed that considerably fewer short products were made. Moreover, a relatively pronounced product at M_r 10 000 is present (Table I; track i). Apparently less initiation is occurring, but still the product does not resemble the in vivo product of mitochondrial protein synthesis. Clearly more experiments on the identification of this product are necessary. As yet we cannot decide whether there is a specific tRNA requirement or not.

The mitochondrial translation system

Although it was initially assumed that mitochondrial ribosomes shared great similarities with bacterial ribosomes, it has gradually become apparent that there are numerous important differences (18). Furthermore, considerable translational barriers have been reported to exist between different species of bacteria (19). Specificity of mRNA recognition apparently resides in the 16S ribosomal RNA (rRNA) and proteins of the small ribosomal subunit (20). It might be expected that E. coli and mitochondrial ribosomes are sufficiently different for recognition of a heterologous mRNA to constitute a real problem. We have thus tried to translate mitochondrial mRNAs in a system in which mitochondrial ribosomes replace those of E. coli.

As can be seen in Table II, a slight stimulation of

TABLE II

TRANSLATION OF mtRNA ON MITOCHONDRIAL RIBOSOMES

"High salt" mitochondrial ribosomes were prepared exactly as described previously (14). "Low salt" mitochondrial ribosomes were prepared with two modifications: puromycin incubation of the mitochondria, just before lysis, was omitted and ribosomes were purified over a sucrose cushion, containing 50 mM ammonium chloride instead of 500 mM. E. coli ribosomes and supernatant were prepared as described (17), omitting passage of the supernatant over DEAE-cellulose. Conditions of protein synthesis were essentially as described (2). Incubations were for 15 min at 30°C. The concentration of ammonium chloride was 50 mM, of ribosomes 5 A_{260}/ml, of supernatant 5 A_{280}/ml and of mtRNA, if added, 0.5 mg/ml. (^{35}S)Methionine (140 Ci/mmol) was 2.5 µM. Activity of protein synthesis was measured as (^{35}S)methionine incorporation into hot trichloroacetic acid-precipitable material. Values have been corrected for a zero time control of 3199 cpm. Immunoprecipitation has been corrected for 2.1% binding to pre-immune serum. N.D., not determined.

Additions	Activity in protein synthesis (cpm/10 µl)	% radioactivity bound by anti-subunits 1 + 2 of cytochrome c oxidase
"Low salt" mit ribosomes	10 009	21.1
"Low salt" mit ribosomes + mtRNA	11 230	19.7
"High salt" mit ribosomes	288	N.D.
"High salt" mit ribosomes + mtRNA	610	N.D.
E. coli ribosomes	2 593	0.3
E. coli ribosomes + mtRNA	51 927	7.8

activity in protein synthesis can be observed with those ribosomes on which still mRNA probably is attached, since the mitochondria were not incubated with puromycin and the high salt purification of the ribosomes was omitted. Addition of mtRNA has no effect, demonstrating that no re-initiation is occuring. Immunoprecipitation is rather high, but this result is difficult to interpret since it can easily be explained by coprecipitation with membrane vesicles present in the ribosome preparation. Electrophoretic analysis again reveals absence of complete products. The low activity of the mitochondrial ribosomes is understandable in the light of the finding that RNA isolated from purified mitochondrial ribosomes is largely degraded (Groot, G.S.P. and Van Ommen, G.J.B., unpublished).

Finally, we have attempted to set up a completely homologous mitochondrial translation system, using mitochondrial extracts. Such extracts have been reported by Küntzel and co-workers (21) to work in Neurospora crassa, where they synthesize highly hydrophobic polypeptides with a M_r around 10 000. Unfortunately, our attempts to do this were unsuccessful. Mitochondrial S-30 extracts even inhibit active E. coli extracts. Apparently the system contains ribonuclease or protease activity (cf. ref. 22).

Eukaryotic cell-free systems

In view of the lack of success with the E. coli cell-free system, translation of 11S mtRNA in other cell-free systems was attempted, since they are also able to accept mRNAs from a wide variety of sources.

The wheat germ cell-free system

As shown in Fig. 2, mtRNA gives a reasonable stimulation of protein synthesis in wheat germ extracts at ionic conditions normally used for this system (23,24). Alfalfa mosaic virus (AMV) RNA, used as control, is correctly translated giving mainly synthesis of coat protein (M_r 25 000) (cf. ref. 25). Products synthesized at all different concentrations

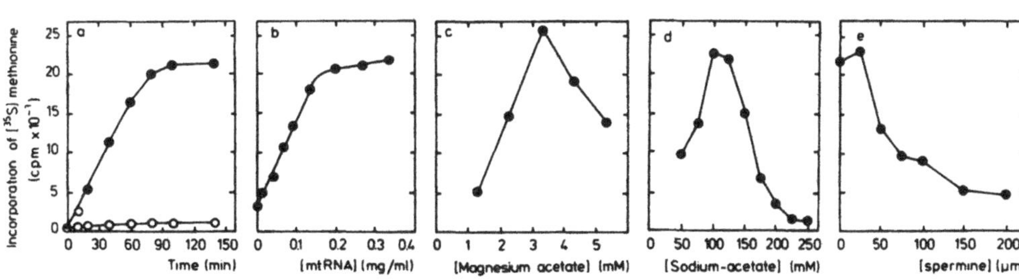

Fig. 2. Characterization of the mtRNA-directed protein synthesis in a wheat germ cell-free system. Extracts were prepared according to Marcu et al. (24). Protein synthesis was for 90 min at 25°C. Stimulation of protein synthesis, as determined by (^{35}S)methionine incorporation per 5 μl, was measured as a function of time (a), RNA concentration (b), magnesium concentration (c), sodium-acetate concentration (d) and spermine concentration (e). ●—●, Activity due to mtRNA; O—O, endogenous activity.

tested (see Fig. 2) were analysed on gels and gave no large qualitative differences (not shown). Plate II, track b, shows the in vitro products of 11S mtRNA. It has no bands in common with the in vivo products of mitochondrial protein synthesis

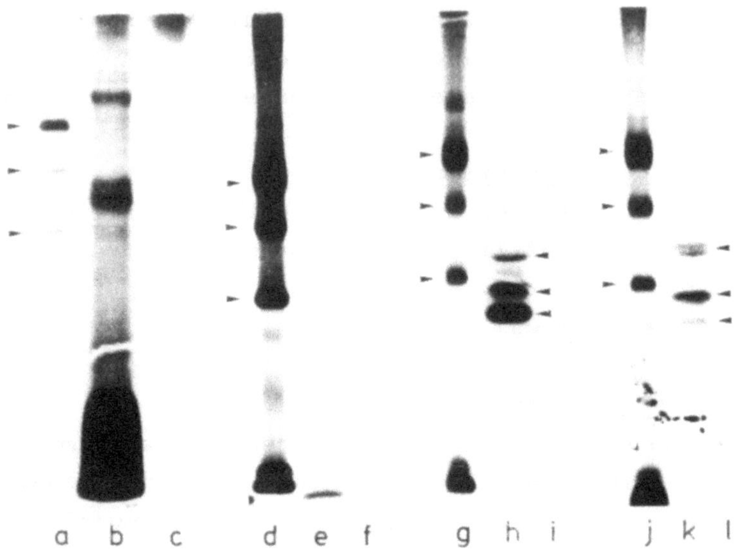

Plate II. Electrophoretic and immunological characterization of mtRNA-specified products, synthesized in different eukaryotic translation systems. See for details legend to Plate I. a) In vivo products of mitochondrial protein synthesis: submitochondrial particles were prepared from yeast cells pulse-labelled with (^{35}S)methionine in the presence of cycloheximide: arrows indicate the positions of the main products migrating at M_r 42 000, 33 000 and 23 000; b) in vitro products synthesized under direction of 11S mtRNA in a wheat germ cell-free system; c) endogenous product; d) in vivo products of mitochondrial protein synthesis (over exposed); e) immunoprecipitate of in vitro products synthesized in the wheat germ system (track b) with antibodies raised against subunits 1 + 2 of cytochrome c oxidase; f) immunoprecipitate of in vitro products synthesized in the wheat germ system (track b) with pre-immune serum; g) in vivo products of mitochondrial protein synthesis; h) in vitro products specified by crystallin mRNA in an oocyte system (26), precipitated by antibodies directed against α- and β-crystallin: arrows indicate the positions of β-crystallin (M_r 26 000) and two α-crystallin proteins (M_r 22 000 and 20 000) (cf. ref. 28); i) in vitro products specified by 11S mtRNA in an oocyte system (26), precipitated by antibodies raised against subunits 1 + 2 of cytochrome c oxidase; j) in vivo products of mitochondrial protein synthesis; k) in vitro products specified by crystallin mRNA in a reticulocyte system (29), precipitated by antibodies directed against α- and β-crystallin; l) in vitro products specified by 11S mtRNA in a reticulocyte system (29), precipitated by antibodies raised against subunits 1 + 2 of cytochrome c oxidase.

(track a). Remarkable is a polypeptide migrating at M_r 26 000. A polypeptide with similar molecular weight is also present among the products synthesized by E. coli cell-free extracts. Only 0.5-1.0% of the incorporated radioactivity is precipitable by antisera directed against subunits of cytochrome c oxidase. This is not significantly higher than that found for pre-immune sera. Despite this, the immunoprecipitates were analysed on gels (Plate II, tracks d-f). It is clear from these that the wheat germ system, even under direction of purified 11S RNA, does not synthesize products which share antigenic determinants with subunits of cytochrome c oxidase.

The Xenopus laevis oocyte system

The oocyte system translates with high fidelity a variety of mRNAs (26). Moreover, the synthesis of mitochondrial proteins has been previously reported (27). Indeed, we find correct translation with 10S calf-lens mRNA for crystallin , as shown in the electrophoretic analysis of the immunoprecipitate with an antiserum directed against α- and β-crystallin (Plate II, track h). However, no product was bound by antiserum raised against subunits 1 + 2 of cytochrome c oxidase (track i). 11S mtRNA did not inhibit the oocyte translation system. Oocytes injected with H_2O, 10S calf-lens mRNA for crystallin or 11S mitochondrial mRNA for subunits of cytochrome c oxidase incorporate about the same amount of radioactive amino acids. Our results contrast with those of Eggit and Scragg (27). However, in their experiments proof is lacking that complete proteins are made in vitro, identical to those synthesized in vivo.

The reticulocyte system

If treated with Micrococcus nuclease this system has an extremely low endogenous activity, which is an advantage for the further characterization of the product (29). The system responds very efficiently to 10S calf-lens mRNA for crystallin (Table III) and with high fidelity as shown in the electrophoretic analysis of the immunoprecipitates (Plate II, track k).

TABLE III

TRANSLATION OF mtRNA IN A RETICULOCYTE CELL-FREE SYSTEM

Protein synthesis was performed exactly as described in ref. 29.

Micrococcus nuclease treatment	Additions	Incorporation of (^{35}S)methionine into trichloroacetic acid-insoluble material (cpm/2 µl)
-	None	17 048
-	11S mtRNA (20 µg/ml)	9 874
+	None	602
+	11S mtRNA (20 µg/ml)	895
+	10S crystallin RNA (8 µg/ml)	15 922

However, the response to 11S mtRNA is marginal and electrophoretic analysis of the immunoprecipitates with antibodies raised against subunits 1 + 2 of cytochrome c oxidase shows that virtually no product is bound by this anti-oxidase serum.

DISCUSSION

We have found that various cell-free translation systems, known to translate a variety of messengers with high fidelity, are incapable of the complete synthesis of mitochondrial proteins under the conditions we have used. If - as would seem likely from the results presented - this is not due to trivial explanations, it could mean that a strong translational barrier exists between the mitochondrial translational machinery and that of both bacteria and of the eukaryotic cell sap. This is remarkable because eukaryotic cell-free systems are capable of translating bacterial messengers correctly, albeit with very low efficiency (23,30,31).

An alternative possibility is that the short polypeptides, synthesized in the *E. coli* cell-free system, are the true primary translation products, which *in vivo* in the mitochondrion are linked to each other to give full-length proteins. Such an unprecedented mechanism has been proposed by Küntzel and coworkers, who have observed the synthesis of short mitochondrial proteins both in vivo and in vitro (21,32). Moreover, they have reported that the largest subunit (M_r 42 000) of cytochrome c oxidase from yeast consists of a complex of small polypeptides (33; contrast ref. 34). However, proteolytic degradation cannot completely be ruled out and short pulse-label experiments either in intact cells or in isolated mitochondria have never revealed short products (35).

The incompatibility of the mitochondrial translation system with other systems implies that reconstruction of a completely homologous mitochondrial system will be of crucial importance in mRNA identification and in the study of the regulation of mitochondrial gene expression.

ACKNOWLEDGEMENTS

We thank Professor P.Borst and Dr G.S.P.Groot for stimulating discussions; Dr R.O.Poyton for generous gifts of antisera; Mr E.J.Stuik for advice on the development of the wheat germ system; Mr K.J.Hellingwerf for the preparation of liposomes; Mrs N.Van Harten-Loosbroek for a gift of submitochondrial particles and Mrs G.Reymer-Van Dam for valuable technical assistance. This work was supported in part by a grant to P.Borst from The Netherlands Foundation for Chemical Research (SON) with financial aid from The Netherlands Organization for the Advancement of Pure Research (ZWO).

REFERENCES

1 Sanders, J.P.M., Heyting, C., DiFranco, A., Borst, P. and Slonimski, P.P. (1976) The organization of genes in yeast mitochondrial DNA. In: The Genetic Function of Mitochondrial DNA (Saccone, C. and Kroon, A.M., Eds), North-Holland, Amsterdam, pp. 259-272.

2 Moorman, A.F.M., Grivell, L.A., Katan, M.B., Van Ommen, G.J.B. and Meiland, C.M. (1976) Translation of mitochondrial messenger RNAs in vitro as an approach to the identification of gene products of yeast mitochondrial DNA. In: Genetics, Biogenesis and Bioenergetics of Mitochondria (Bandlow, W., Schweyen, R.J., Thomas, D.Y., Wolf, K. and Kaudewitz, F, Eds), De Gruyter, Berlin, pp. 241-258.

3 Moorman, A.F.M., Van Ommen, G.J.B. and Grivell, L.A. (1977) Transcription in yeast mitochondria: Isolation and physical mapping of messenger RNA for subunits of cytochrome c oxidase and ATPase. In preparation.

4 Rekosh, M.R., Lodish, H.F. and Baltimore, D. (1970) Protein synthesis in Escherichia coli extracts programmed by poliovirus RNA. J.Mol.Biol. 54, 327-340.

5 Gielkens, A.L.J., Salden, M.H.L., Bloemendal, H. and Konings, R.N.H. (1972) Translation of oncogenic viral RNA and eukaryotic messenger RNA in the E. coli cell-free system. FEBS Letters 28, 348-352.

6 Hartley, M.R., Wheeler, A. and Ellis, R.J. (1975) Protein synthesis in Chloroplasts. V. Translation of messenger RNA for the large subunit of

fraction I protein in a heterologous cell-free system. J.Mol.Biol. 91, 67-77.
7 Struhl, K., Cameron, J.R. and Davis, R.W. (1976) Functional genetic expression of eukaryotic DNA in Escherichia coli. Proc.Natl.Acad.Sci.U.S. 73, 1471-1475.
8 Greenblatt, J.F., Allet, B. and Weil, R. (1976) Synthesis of the tumour antigen and the major capsid protein of Simian Virus 40 in a cell-free system derived from Escherichia coli. J.Mol.Biol. 108, 361-379.
9 Moorman, A.F.M., Grivell, L.A., Lamie, F. and Smits, H.L. (1977) Identification of mitochondrial gene products by DNA-directed protein synthesis in vitro. Biochim.Biophys.Acta, submitted.
10 Moorman, A.F.M. and Grivell, L.A. (1976) Coupled transcription-translation of yeast mitochondrial DNA in vitro as a means of gene identification and mapping. In: idem ref. 1, pp. 281-289.
11 Zubay, G., Chambers, D.A. and Cheong, L.C. (1970) Cell-free studies on the regulation of the lac operon. In: The Lactose Operon (Beckwith, J.R. and Zipser, D., Eds), Cold Spring Harbor Laboratory, New York, pp. 375-391.
12 Cremer, K. and Schlessenger, D. (1974) Ca^{2+} ions inhibit messenger ribonucleic acid degradation, but permit messenger ribonucleic acid transcription and translation in deoxyribonucleic acid-coupled systems from Escherichia coli. J.Biol.Chem. 249, 4730-4736.
13 Hendler, F., Halbreich, A., Jakovcic, S., Patzer, J., Merten, S. and Rabinowitz, M. (1976) Characterization and translation of yeast mitochondrial DNA. In: Genetics and Biogenesis of Chloroplasts and Mitochondria (Bücher, Th., Neupert, W., Sebald, W. and Werner, S., Eds), North-Holland, Amsterdam, pp. 679-684.
14 Grivell, L.A., Reijnders, L. and Borst, P. (1971) Isolation of yeast mitochondrial ribosomes highly active in protein synthesis. Biochim. Biophys.Acta 247, 91-103.
15 Eytan, G.D., Matheson, M.J. and Racker, E. (1976) Incorporation of mitochondrial membrane proteins into liposomes containing acidic phospholipids. J.Biol.Chem. 251, 6831-6837.
16 Bernardi, G., Piperno, G. and Fonty, G. (1972) The mitochondrial genome of wild-type yeast cells. I. Preparation and heterogeneity of mitochondrial DNA. J.Mol.Biol. 65, 173-189.
17 Gold, L.M. and Schweiger, M. (1971) Synthesis of bacteriophage-specific enzymes directed by DNA in vitro. In: Methods in Enzymology (Moldave, K. and Grossman, L., Eds), Vol. 20, Academic Press, New York, pp. 537-542.
18 Borst, P. (1972) Mitochondrial nucleic acids. Ann.Rev.Biochem. 41, 333-376.
19 Lodish, H.F. (1976) Translational control of protein synthesis. Ann. Rev.Biochem. 45, 39-72.
20 Goldberg, M.L. and Steitz, J.A. (1974) Cistron specificity of 30S ribosomes heterologously reconstituted with components from Escherichia coli and Bacillus stearothermophilus. Biochemistry 13, 2123-2129.
21 Küntzel, H. and Blossey, H.-Ch. (1974) Translation products in vitro of mitochondrial messenger RNA from Neurospora crassa. Europ.J.Biochem. 74, 165-171.
22 Richter, D., Herrlich, P. and Schweiger, M. (1972) Phage DNA directed enzyme synthesis in in vitro system from yeast mitochondria. Nature New Biol. 230, 74-76.
23 Davies, J.W. and Kaesberg, P. (1973) Translation of virus mRNA:

Synthesis of bacteriophage Qβ proteins in a cell-free extract from wheat embryo. J.Virol. 12, 1434-1441.

24 Marcu, K. and Dudock, B. (1974) Characterization of a highly efficient protein synthesizing system derived from commercial wheat germ. Nucl. Acids Res. 1, 1385-1397.

25 Van Vloten-Doting, L., Rutgers, T., Neeleman, L. and Bosch, L. (1975) In vitro translation of the RNAs of Alfalfa Mosaic Virus. Inserm 74, 233-242.

26 Gurdon, J.B., Lane, C.D., Woodland, H.R. and Marbaix, G. (1971) Use of frog eggs and oocytes for the study of messenger RNA and its translation in living cells. Nature 233, 177-182.

27 Eggit, M.J. and Scragg, H.J. (1975) Messenger activity of ribonucleic acid from yeast mitochondria. Biochem.J. 149, 507-512.

28 Berns, T.J.M., Schreurs, V.V.A.M., Van Kraaikamp, M.W.G. and Bloemendal, H. (1973) Synthesis of lens protein in vitro: Translation of calf-lens messengers in heterologous systems. Europ.J.Biochem. 33, 551-557.

29 Pelham, H.R.B. and Jackson, R.J. (1976) An efficient mRNA-dependent translation system from reticulocyte lysates. Europ.J.Biochem. 67, 247-256.

30 Beck, D.E. and Gassen, H.G. (1977) Translation of bacteriophage T_4 mRNA into active lysozyme by a wheat germ cell-free system. Biochem.Biophys. Res.Commun. 74, 16-24.

31 Sagher, D., Grosfeld, H. and Edelman, H. (1976) Large subunit ribulosebisphosphate carboxylase messenger RNA from Euglena chloroplasts. Proc. Natl.Acad.Sci.U.S. 73, 722-726.

32 Küntzel, H., Pieniażek, N.J., Pieniażek, D. and Leister, D.E. (1975) Lipophilic proteins encoded by mitochondrial and nuclear genes in Neurospora crassa. Europ.J.Biochem. 54, 567-575.

33 Küntzel, H., Marahiel, M.A., Leister, D.E. and Nelson, P. (1976) Polypeptides encoded by mitochondrial genes in ascomycetes. In: idem ref. 13, pp. 159-166.

34 Poyton, R.O. and Schatz, G. (1975) Cytochrome c oxidase from bakers' yeast. III. Physical characterization of isolated subunits and chemical evidence for two different classes of polypeptides. J.Biol.Chem. 250, 752-761.

35 Groot, G.S.P. and Van Harten-Loosbroek, N. (1976) The use of isolated mitochondria in the study of the mitochondrial biogenesis in yeast. In: idem ref. 2, pp. 269-277.

EFFECT OF MITOCHONDRIAL MUTATIONS ON CYTOCHROME c OXIDASE IN YEAST

F. Cabral, M. Solioz, D. Deters, Y. Rudin, G. Schatz
Biochemistry Department, Biozentrum der Universität Basel, CH-4056 Basel, Switzerland
and
L. Clavilier, O. Groudinsky, P. Slonimski
Centre de Génétique Moléculaire du C.N.R.S., F-91190 Gif-sur-Yvette, France

INTRODUCTION

Mitochondria synthesize a limited number of hydrophobic mitochondrial polypeptides, including three subunits of cytochrome c oxidase (1,2,3). Recently strong evidence has accumulated that at least some of these polypeptides are coded for by mitochondrial DNA (2,4,5). However, it has not been possible to identify a mitochondrial gene which codes for a functionally defined polypeptide.

The discovery of mit^- mutations (6,7) has now opened up the possibility of exploring the effects of these mutations on mitochondrially made proteins and on their assembly into functional complexes. Mit^- mutants are unable to grow on a nonfermentable carbon source but are still capable of mitochondrial protein synthesis. Here we report how mit^- mutations in the OXI 1 and OXI 3 loci affect cytochrome c oxidase. Our results show that the OXI 1 locus codes for subunit II of cytochrome c oxidase whereas the OXI 3 locus appears to code for subunit I of cytochrome c oxidase. Moreover, a revertant of an OXI 3 mutant has been found which exhibits an altered cytochrome c oxidase absorption spectrum. This is the first example of a mutationally altered, yet active cytochrome c oxidase with an altered absorption spectrum. This altered enzyme may provide new insights into the catalytic mechanism of cytochrome c oxidase.

METHODS

Yeast cells were grown to late log phase in 1% yeast extract, 2% bactopeptone and 0.5% glucose and labelled in the presence of cycloheximide with ^{35}S-sulfate as described by Douglas *et al*. (8). Mitochondrial translation products were separated on high resolution SDS polyacrylamide gels and located on the dried gels by radioautography (8). The wild-type strain is D273 - 10B. The mutants examined have been previously described (6, 7).

Spectra were recorded on an Aminco DW2 split beam/dual wavelength spectrophotometer either at room temperature or at liquid nitrogen temperature. Whole cell spectra were recorded at liquid nitrogen temperature using the dual wavelength mode with the reference wavelength set at 615 nm.

RESULTS

Mutations in OXI 1:

Figure 1 shows radioautographs of SDS polyacrylamide gels of whole mitochondria from wild-type and several OXI 1 mutants which had been labeled with ^{35}S-sulfate in the presence of cycloheximide. The major products seen under these conditions include cytochrome *c* oxidase subunits I, II and III (arrows) and another band, probably apocytochrome *b*, between subunits II and III. In some OXI 1 mutants no change is detectable in the mitochondrial translation products. However, in many OXI 1 mutants, subunit II is missing and in others it is replaced by a faster-moving band. The position of the other mitochondrial translation products is unaffected by these mutations.

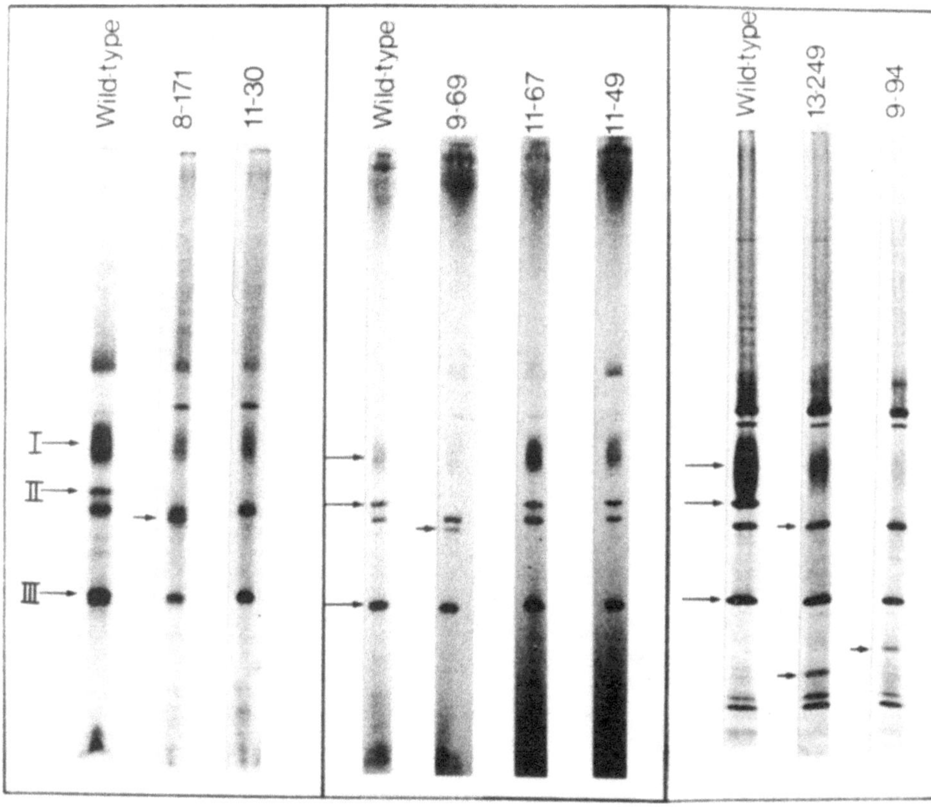

Fig. 1. Mitochondrial translation products of OXI 1 mutants. Wild-type and mutant yeast cells were labeled as described in the text. Mitochondria were isolated, dissociated in SDS and subjected to polyacrylamide gel electrophoresis. Gels were then dried and radioautographed. Roman numerals I, II and III refer to the mitochondrially made subunits of cytochrome c oxidase. The shorter arrows indicate new bands seen in the OXI 1 mutants.

From several mutants exhibiting a new band in place of subunit II, spontaneous or induced revertants were obtained. These revertants have regained cytochrome c oxidase spectra and activity as well as the ability to grow on glycerol. Examination of these revertants by gel electrophoresis shows that the new band seen in the parent mutant is replaced by another band which migrates either faster or slower than the band in the

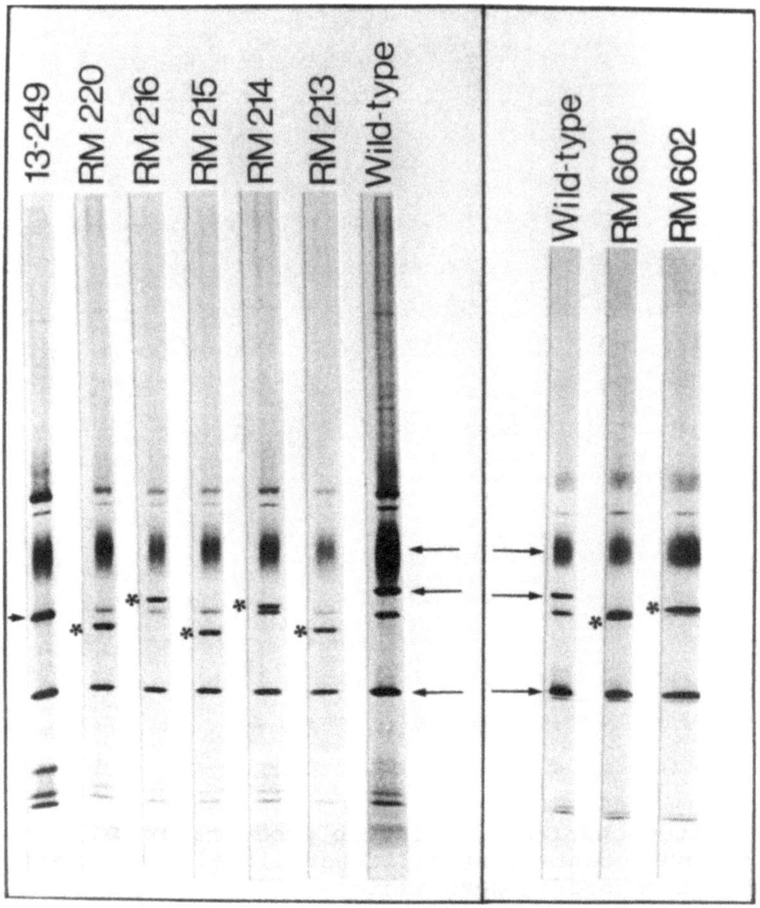

Fig. 2. Mitochondrial translation products of revertants from OXI 1 mutants. Labeling and subsequent treatment of mitochondria are as described in the text and the legend to Figure 1. Large arrow indicates position of new band in mutant 13-249. (This band nearly co-migrates with the band which is probably the apoprotein of cytochrome b.) The asterisks denote the position of new bands seen in the revertants. RM 213 to RM 220 are revertants of mutant 13-249. RM 601 and RM 602 are revertants of mutant 9-69 shown in Figure 1.

parent mutant, but always faster than wild-type subunit II (Figure 2). The other mitochondrially-made polypeptides are not affected by the reversions.

The behavior of the OXI 1 mutants and revertants immediately suggested that the mutation had occurred in the gene coding for subunit II. To be certain, however, it was necessary to show that the new bands were related to subunit II. This we have done in two ways: (a) immunochemically and (b) by one-dimensional peptide mapping.

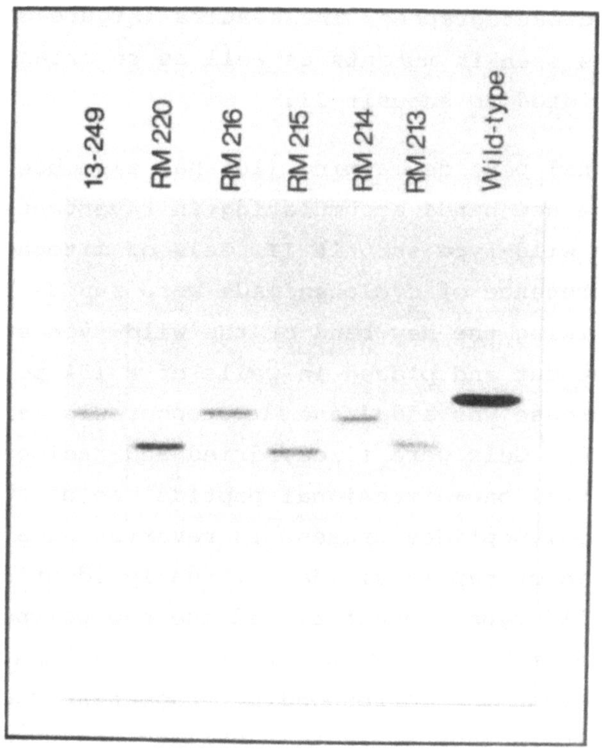

Fig. 3. Immunochemical analysis of an OXI 1 mutant and of several revertants derived from it. Labeling conditions and subsequent treatment are described in the text. The number 13-249 identifies the mutant. RM 213 to RM 220 are revertants of this mutant.

For immunochemical analysis, mitochondria are labeled in vivo with ^{35}S-sulfate in the presence of cycloheximide and electrophoresed in an SDS-polyacrylamide gel slab as described above. The gel slab is then layered onto a 1% agarose gel that contains subunit II-specific antiserum. Upon incubation for 16 h, proteins diffuse from the polyacrylamide gel into the agarose gel where subunit II and antigenically related material is precipitated by the antiserum (9). The agarose gel is then washed, dried and radioautographed. The results (Figure 3) show that the new bands seen in mutants as well as revertants are immunologically related to subunit II.

One dimensional peptide mapping (10) has also been used to show that the new bands accumulating in revertants share sequences with wild-type subunit II. Gels of mitochondria labeled in the presence of cycloheximide were run as before; gel pieces containing the new band or the wild-type subunit II were then cut out and placed in wells of a 15% polyacrylamide SDS gel. Protease was added and electrophoresis carried out as described (10). Gels were fixed, dried and radioautographed to reveal a partial one-dimensional peptide map of subunit II and the shorter polypeptides present in revertants. S.aureus protease, nagarse or papain yielded virtually identical fingerprints for wild-type subunit II and the new polypeptides except that one or more fragments from the revertant polypeptide moved faster than the corresponding fragments from wild-type subunit II. It is thus clear that the new bands seen in the OXI 1 revertants share sequences with subunit II. By determining free electrophoretic mobilities as well as mobilities in the presence of a cationic detergent (not shown) we could substantiate our view that the faster mobility of the revertant subunit II in SDS gels reflects indeed a smaller molecular weight and not merely abnormal binding of SDS.

Cytochrome c oxidase was isolated from one OXI 1 revertant. The enzyme contained a shortened subunit II, yet was still able to catalyze the oxidation of ferrocytochrome c in vitro. This result opens up new approaches for studying the structure of cytochrome c oxidase.

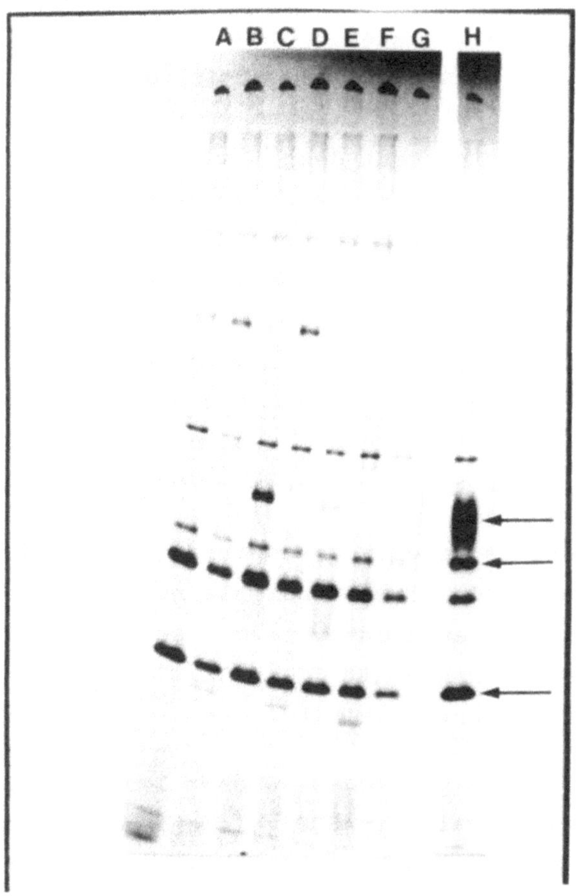

Fig. 4. Mitochondrial translation products of OXI 3 mutants. Labeling and gel procedures are described in the text and in the legend to Fig. 1. The positions of cytochrome c oxidase subunits I, II and III are indicated in the wild-type by arrows. Tracks A to G are different OXI 3 mutants; track H is the wild-type.

Mutations in OXI 3:

Mutations in two other mitochondrial loci (OXI 2 and OXI 3) affect cytochrome c oxidase (6,7). We have examined a number of mutants in OXI 3. As found with OXI 1 mutants, only one mitochondrial translation product is affected but now it is cytochrome c oxidase subunit I. Figure 4 shows several of these OXI 3 mutants. Subunit I is often missing and instead one or more new bands appear in the gel. These new bands do not cross-react with antiserum against wild-type subunit I in the immune-replica test described above even though a strong reaction is observed with wild-type subunit I. The "new" bands are thus either unrelated to subunit I or represent subunit I fragments which have lost the antigenic sites of the wild-type polypeptide. Some OXI 3 mutants (data not shown) exhibit an apparently unaltered subunit I, at least as judged by one-dimensional electrophoresis. However, in no case is there a detectable alteration of any mitochondrial translation product other than cytochrome c oxidase subunit I.

Many OXI 3 mutants revert to respiring cells which can again grow on glycerol. On electrophoretic analysis, most of these revertants show an apparently normal subunit I. Unlike the results obtained with subunit II in OXI 1 revertants, none of these OXI 3 revertants exhibits a bona fide subunit I fragment.

Although cytochrome c oxidase subunit I of some OXI 3 revertants migrates normally in SDS gels and reacts immunologically with antiserum versus subunit I, the structure of the holoenzyme is apparently changed, since some revertants exhibit an altered cytochrome c oxidase spectrum. If the absorption spectrum of such revertant cells is examined at liquid nitrogen temperature, the α absorption band of the reduced oxidase is shifted about 3 nm towards shorter wavelengths (Fig. 5). From one such revertant cytochrome c oxidase was isolated. The in *vitro* activity of the isolated revertant cytochrome *c* oxidase (expressed as nmoles of cytochrome *c* oxidized/min/nanomole heme *a*) is approximately five times lower than the corresponding activity of the isolated wild-type enzyme. As expected, the reduced purified enzyme still exhibits the shift in the absorption maximum of the α band (Fig. 6). The Soret peak is also blue-shifted. Changes in absorption maxima are also seen in the oxidized enzyme and in the carbon monoxide complex of the reduced enzyme (not shown).

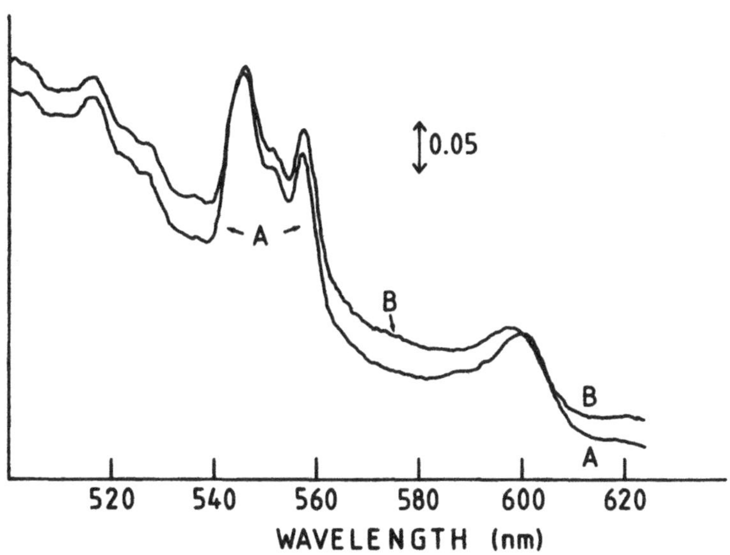

Fig. 5. Low temperature absorption spectrum of wild-type yeast cells and cells of an OXI 3 revertant. Wild-type or revertant cells were suspended in 25% glycerol, a few grains of dithionite were added and after 2 to 3 minutes the suspension was frozen in liquid nitrogen. Spectra were recorded as described in the text. Cytochrome *c*, c_1 and *b* and the α-band of cytochrome *c* oxidase can be seen. A = wild-type, λ_{max} = 602.5 nm; B = revertant, λ_{max} = 599.5 nm.

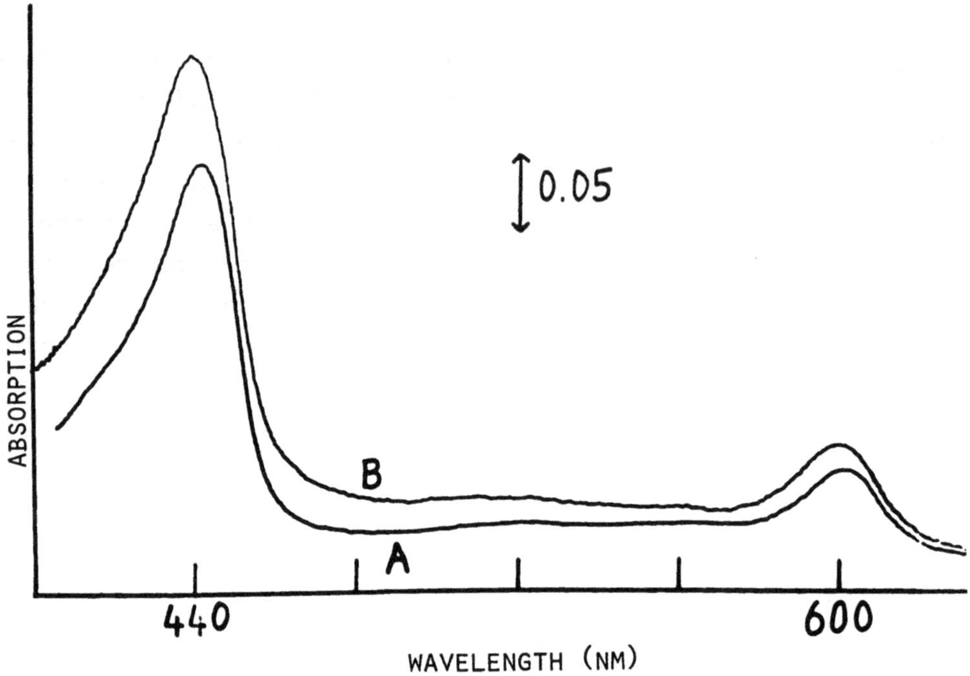

Fig. 6. Room temperature absorption spectrum of purified cytochrome c oxidase reduced with dithionite. Both α and Soret peaks are changed in the revertant. A = wild-type, λ_{max} = 604 nm and 443.5 nm; B = revertant, λ_{max} = 602 nm and 441.5 nm.

DISCUSSION

It is clear that mutations in OXI 1 lead to changes in subunit II of cytochrome c oxidase. Moreover there can be no doubt that, at least in some cases, the new bands present in these mutants are chemically related to subunit II. They cross-react with anti-II antiserum and, when analyzed by one-dimensional peptide mapping, yield a fragment pattern which is almost identical to that of wild-type subunit II. We believe these observations are most simply explained by a direct gene-product relationship between the OXI 1 locus and cytochrome c oxidase subunit II. Other interpretations, for example that OXI 1 codes for a regulatory protein or a "modifier" protein speci-

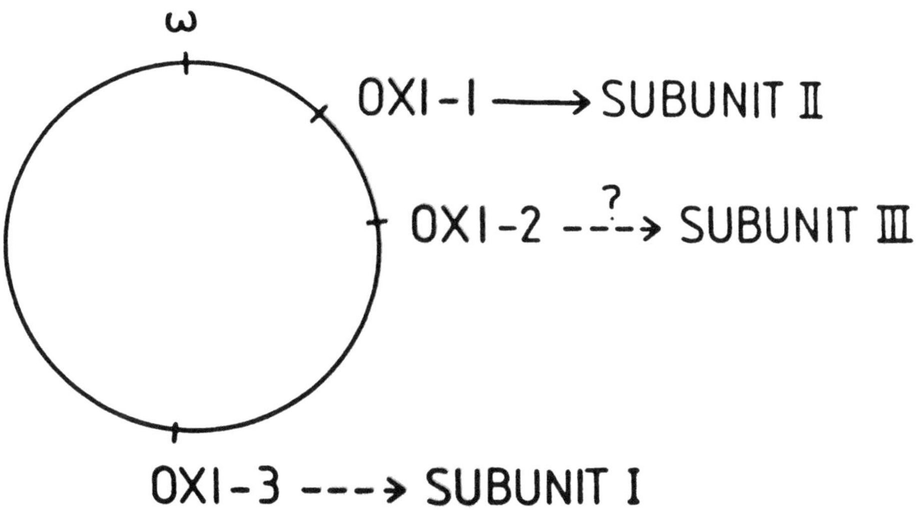

Fig. 7. Schematic illustration of mitochondrial DNA with its three OXI loci. OXI 1 codes for subunit II of cytochrome c oxidase. OXI 3 most probably codes for subunit I. ω denotes the locus described by Dujon and Michel (11).

fically affecting subunit II, are difficult to reconcile with our data. In such cases it would be hard to explain why subunit II species from different mutants and revertants show such widely different mobilities on gels. Finally, extensive control experiments appear to exclude the possibility that the altered electrophoretic mobilities are simply caused by abnormal binding of SDS. We conclude that the OXI 1 locus contains the structural gene for cytochrome c oxidase subunit II.

We interpret our observations in the following way. A small deletion or insertion in the OXI 1 locus causes a shift in the reading frame and produces a new stop codon at some distance from the mutation site. Since the mutation site as well as the size of the deletion or insertion can vary, different mutants would contain subunit II polypeptides of different sizes. Respiratory competent revertants may arise by a second deletion or insertion which restores the original reading frame. The result would be a chemically related but shortened subunit. Of course, such a hypothesis will remain speculative until detailed peptide mapping and sequence data are available.

In the case of the OXI 3 mutants, the results are not yet unambiguous since a chemical relationship between wild-type subunit I and the bands found in mutants has not yet been established. However, it is clear that mutants in OXI 3 are affected mainly or exclusively in subunit I of cytochrome c oxidase.

The discovery of a revertant with a modified cytochrome c oxidase spectrum may offer a promising new approach to the study of the structure and function of the enzyme.

The change in the absorption spectrum could be caused by a change in the structure of the heme a or by a change in the heme environment. The first possibility is unlikely since the absorption spectra of the pyridine hemochromogen complexes prepared from the wild-type and mutant enzymes are identical (data not shown). We thus suspect that the environment of the heme a is altered in the revertant enzyme.

In summary, we conclude that OXI 1 codes for subunit II of cytochrome c oxidase (Figure 7). Most probably OXI 3 codes for subunit I.

ACKNOWLEDGEMENT

O.G. gratefully acknowledges receipt of a short-term fellowship from EMBO, supported in part by Grant No. 77-70240 from D.G.R.S.T.

REFERENCES

1. Tzagoloff, A., Rubin, M.S. & Sierra, M.F., Biochim. Biophys. Acta 301, 71-104 (1973).

2. Schatz, G. & Mason, T.L., Annu. Rev. Biochem. 43, 51-87 (1974).

3. Schatz, G. in Genetics, Biogenesis and Bioenergetics of Mitochondria, eds. Bandlow, W., Schweyen, R.J., Thomas, D.Y., Wolf, K. & Kaudewitz, F. (de Gruyter, Berlin) pp. 163-177 (1976).

4. Moorman, A.F.M. & Grivell, L.A. in The Genetic Function of Mitochondrial DNA, eds. Saccone, C. & Kroon, A.M. (North Holland, Amsterdam) pp. 281-289 (1976).

5. Rabinowitz, M., Jakovcic, S., Martin, N., Hendler, F., Halbreich, A., Lewin, A. & Morimoto, R. in The Genetic Function of Mitochondrial DNA, eds. Saccone, C. & Kroon, A.M. (North Holland, Amsterdam) pp. 219-230 (1976).

6. Tzagoloff, A., Akai, A., Needleman, R.B. & Zulch, G., J. Biol. Chem. 250, 8236-8242 (1975).

7. Slonimski, P.P. & Tzagoloff, A., Eur. J. Biochem. 61, 27-41 (1976).

8. Douglas, M. & Butow, R.A., Proc. Natl. Acad. Sci. USA 73, 1083-1086 (1976).

9. Showe, M., Isobe, E. and Onorato, L., J. Mol. Biol. 107, 55-69 (1976).

10. Cleveland, D.W., Fischer, S.G., Kirschner, M.W. and Laemmli, U.K., J. Biol. Chem. 252, 1102-1106 (1977).

11. Dujon, B. & Michel, F. in The Genetic Function of Mitochondrial DNA, eds. Saccone, C. & Kroon, A.M. (North Holland, Amsterdam) pp. 175-184 (1976).

TRANSCRIPTION MAP OF YEAST MITOCHONDRIAL DNA

G.J.B.VAN OMMEN AND G.S.P.GROOT

Department of Molecular Biology, Laboratory of Biochemistry, University of Amsterdam, Kruislaan 320, Amsterdam (The Netherlands)

INTRODUCTION

In the last two years, our knowledge of the organization of the genes on mtDNA of *Saccharomyces* has greatly increased. Sanders and co-workers (1) have constructed a physical map of the mtDNA of several yeast strains with a number of restriction enzymes. The genes for ribosomal RNAs (rRNAs) (2) and 4S RNAs (3) have been localized on this map by means of hybridization of the labelled RNAs to restriction fragments. Furthermore, several mitochondrial genetic markers could be positioned on this map by hybridization of complementary RNA (cRNA) transcribed from the mtDNA of deletion mutants retaining only one of these markers (1,4-6). This work, together with the genetic mapping studies of Tzagoloff, Slonimski and Linnane, has resulted in the picture shown in Fig. 1 (see refs 7-9).

Recently, Moorman and co-workers (10) have provided direct biochemical evidence for the presence of structural genes on mtDNA for some of the mitochondrial proteins and they have tentatively identified genes for subunits of cytochrome *c* oxidase and of the ATPase complex. In this paper we report the localization of genes for several minor constituents of mtRNA likely to be messenger RNAs (mRNAs). By fractionating labelled mtRNA according to size and hybridizing the fractions with restriction fragments of *Saccharomyces carlsbergensis* mtDNA, we identified at least 13 discrete components apart from rRNAs and

---

Abbreviation: SSC, 0.15 M NaCl, 0.015 M sodium citrate (pH 7.0).

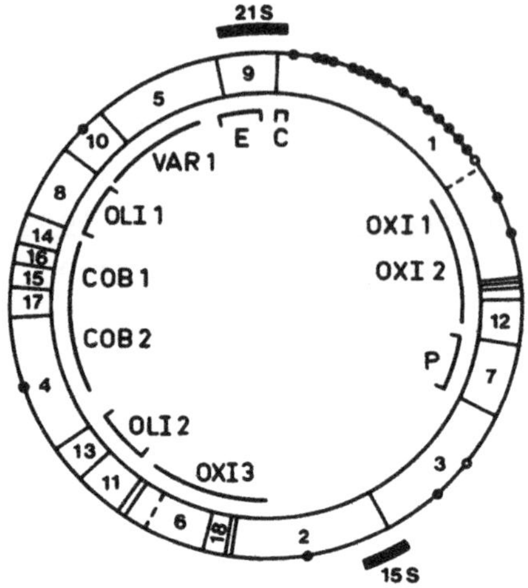

Fig. 1. Physical map of S. carlsbergensis mtDNA. The fragments obtained after digestion with the restriction endonucleases HindII+III + EcoRI are numbered according to decreasing size. Inside the fragment map the position of the genetic loci involved in deficiency of complex III activity (COB), cytochrome c oxidase (OXI) or involved in the coding of the polymorphic ribosome-associated protein (VAR1) are indicated (——). Loci involved in resistance against chloramphenicol (C), erythromycin (E), paromomycin (P) or oligomycin (OLI), for which also physical mapping data are available, are indicated (⊢—⊣). The positions of the 21S and 15S rRNAs are indicated (▬). ●, 4S RNA; O, (F)met tRNA.

tRNAs. The coding regions for these transcripts correspond closely to the positions of the known genetic markers and in combination with mitochondrial rRNAs and tRNAs these account for about 45% of the coding capacity of the mtDNA.

RESULTS

The electropherogram of the ^{32}P-labelled mtRNA preparation is shown in Fig. 2. In addition to the 21S and 15S mitochondrial rRNAs and the 4S RNA, several discrete minor components can be detected. We have calculated their S_E values relative to 21S and 15S rRNAs. The visible, minor species migrate as 24, 22, 19, 18, 12 and 11 S_E, respectively. This number obviously is a minimum estimate, as several components may co-migrate or be obscured by the rRNAs. The region from the top of the gel to about 10 S_E was cut into 1-mm slices and counted (Fig. 3).

Fig. 2. Electropherogram of S. carlsbergensis mtRNA. S. carlsbergensis cells were grown in the presence of 0.1 mM phosphate and 20 mCi/l ^{32}P-labelled phosphate. Cells were harvested at 1.0 A_{550nm} and mtRNA was prepared as described (11). 20-µl samples containing 8 µg RNA in 0.1 x SSC were heated to 100°C for 1 min, quickly chilled to 0°C and electrophoresed on 2.3% polyacrylamide-0.5% agarose as described (3). The gels were fixed and stained with methylene blue.

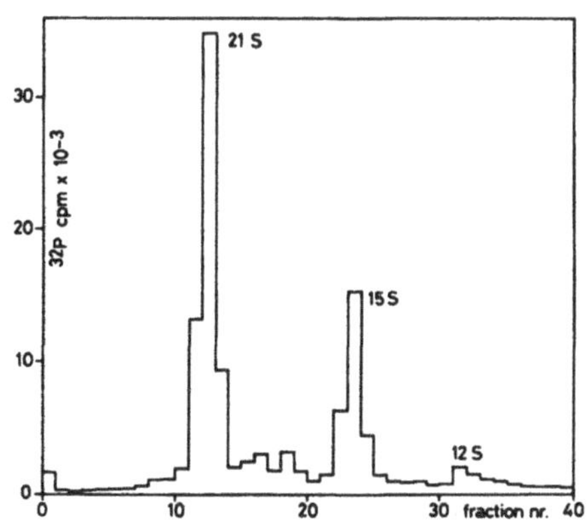

Fig. 3. Distribution of radioactivity in the gel shown in Fig. 2. The first 4 cm of the gel shown in Fig. 2 (to approximately 10 S_E) was sliced in 1-mm parts, which were counted (Cerenkov).

The slices were eluted and the eluate of each slice was hybridized to a nitrocellulose strip containing the HindII+III + EcoRI restriction fragments of S. carlsbergensis mtDNA (1). After removal of the non-hybridized RNA, the strips were mounted in the fraction order and fluorographed as described earlier (3). The result is shown in Fig. 4.

The rather even hybridization to all the fragments that is noted in the first fraction is considered to be due to unremoved DNA. This point and the very faint hybridization around

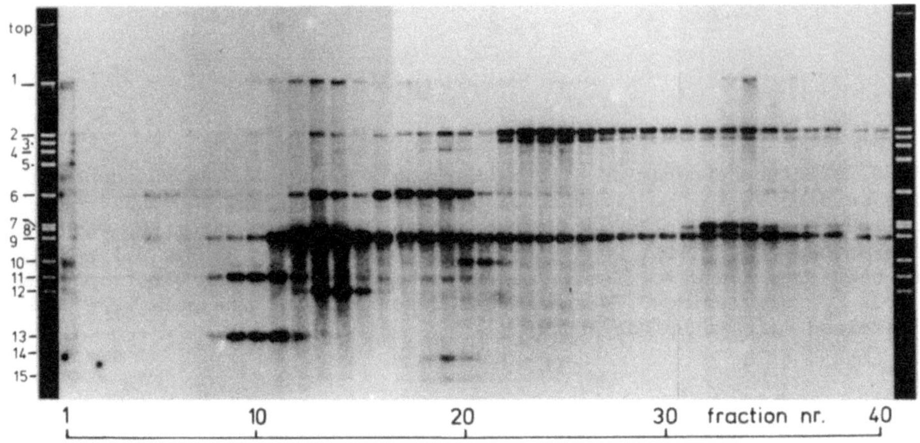

Fig. 4. Hybridization of fractionated S. carlsbergensis mtRNA with restriction fragments of mtDNA. 0.5-μg portions of fragmented S. carlsbergensis mtDNA were separated in parallel on a 0.7% agarose slab gel. The fragments were transferred to a nitrocellulose sheet according to Southern (15) with the exception that whole gels were handled to ensure an even transfer of all fragments. The sheet was cut into strips which were allowed to hybridize to the eluted RNA fractions, washed, dried and fluorographed (3). Extreme right and left: HindII+III + EcoRI digests of mtDNA after staining with EthBr. Fraction 1-40: fluorogram of the hybridized strips.

fractions 5 and 6 are discussed below. The first hybridization that we regard as significant is to fragments 11 and 13. It follows the radioactivity profile from fractions 8 to 11 and coincides with the 24 S_E and 22 S_E RNA species present in fractions 9 and 11, respectively. No other hybridization is seen with RNA from this region of the gel except for the rising hybridization of the 21S rRNA to fragment 9 and a faint hybridization of partial digestion products of mtDNA that contain either fragment 11 or both fragments 11 and 13. As shown in Fig. 5, we conclude that both the 24 S_E and 22 S_E RNAs are transcripts of the same region overlapping at least one EcoRI site and possibly extending into a 200 base pair EcoRI fragment to the right of fragment 11 that has run off the gel. The trivial explanation that the 24 S_E and 22 S_E RNAs are conformational isomers is not very likely since both components have been detected under many different electrophoresis conditions.

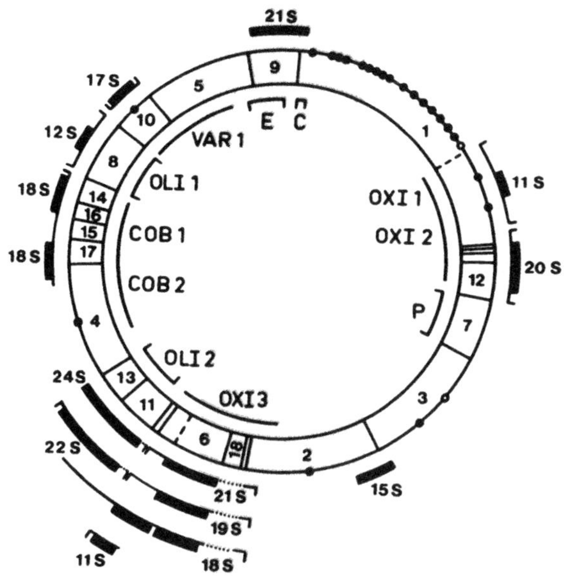

Fig. 5. Physical map of S. carlsbergensis mtDNA. The map presented in Fig.1 has been completed with the transcripts found. For details, see text.

Hence, we consider this as the first direct evidence for RNA processing in yeast mitochondria.

The next hybridization to show up is a dominant hybridization to fragment 9, due to the 21S rRNA. A marked feature of this hybridization is the very long tailing as opposed to the steep appearance (within about three fractions). This is in apparent contradiction with the evidence for the intactness of the RNA preparation shown in Fig. 2. Apparently, the method is extremely sensitive so that even trace amounts of broken molecules are detected. Fraction 15 e.g. contains only about 80 ng of RNA (so at most this amount of rRNA) and it still hybridizes to around half saturation value with the rDNA. The lower limit of detection obviously lies in the order of one to several ng, probably even less. Another 21 S_E RNA hybridizes to fragment 6. Overlap onto fragment 2 is not excluded as the linking fragment 17 has run off the gel and fragment 2 does hybridize faintly. Immediately next to the 21 S_E region a species of about 20 S_E hybridizes to fragment 12. The estimated length of this RNA is some 3600 bases, whereas fragment 12 is only 1900 base pairs long. The adjacent fragment 7 shows no hybridization so the gene does not extend into this direction. On the other hand, fragment 1 behaves similarly as fragment 12 in hybridizing equally well with fractions 13 and 14. It is, therefore, likely that the gene for the 20 S_E RNA extends into fragment 1, traversing the small fragments between fragments 1 and 12.

A strong hybridization to fragment 6 associated with a faint one to fragment 2 again shows up at 19 S_E and this occurs once more at 18 S_E. We believe that this is no coincidence and that the possibility mentioned for the 21 S_E RNA, that is a gene extending over fragment 17 into fragment 2, also holds for the 19 S_E and 18 S_E RNAs. Obviously, this would mean a common gene. The need for these three species to share at least part of their sequence follows independently from the fact that their added length would greatly exceed that of fragment 6 (3700 + 3000 + 2600 as compared to 4400 nucleotides). We shall return to this point in the Discussion.

Hybridization in the 18 S_E region is also found to fragments 4, 11, 14 and 15. Since fragment 14 is too short to accomodate a whole 18 S_E gene and fragment 8 does not hybridize, this gene must extend into fragment 16 (run off the gel). It is not likely that this gene extends as far as fragment 4. The RNA then has to contain at least 2900 nucleotides (the added lengths of fragments 15, 16 and 18) apart from sharing enough sequence with fragment 14 to show an appreciable hybridization. For this reason we have drawn two 18 S_E RNA genes in this region, the minimum model to account for the observed hybridization to fragments 14, 15 and 4. More detailed analysis is needed to clarify this.

The 18 S_E RNA that hybridizes to fragment 11 could have sequences in common with the 24 S_E RNA. The other possibility is that its gene is immediately adjacent to the 24 S_E RNA gene extending into fragment 6. We favour the latter interpretation because of results obtained with other strains. These suggest the presence of two 18 S_E species originating from both sides of fragment 6. Next to the 18S region there is a 17 S_E species that hybridizes to fragment 10. Somewhat further down the gel we see the 15S rRNA hybridization to fragments 2 + 3. 12 S_E RNA hybridizes to fragment 8, a slower 11 S_E component (not discernable in Fig. 2) to fragment 6. More detailed studies have indicated that the gene for the slow 11 S_E RNA lies in the 5500 base pair region of fragment 1 closest to fragment 12. Competition hybridization experiments and data from other strains (not shown) have indicated that the faster 11 S_E RNA is contained within the sequence of the 21 S_E non-ribosomal RNA and located at the left side of fragment 6. Finally, we see a 12 S_E peak in the hybridization with fragment 9, which carries the gene for 21 S rRNA. This may indicate an in vivo degradation product of the 21 S_E RNA.

DISCUSSION

The data presented here show the existence of at least 13 discrete RNA transcripts of yeast mtDNA, distinct from rRNAs and tRNAs. Their localization on the physical map is shown in Fig. 5. The agreement of this transcription map with the genetic map is striking. We consider it likely that the genetic loci indeed represent structural genes and that the transcripts we identified are mRNAs of these genes. That this in fact is so for the 12 S_E and 11 S_E RNAs will be shown by Moorman et al. (16). Support for the mRNA function of the larger minor species is provided by the finding that RNA sedimenting between 24 and 15S is highly active in directing in vitro protein synthesis (11). The only locus for which no corresponding transcript can be found is P. The implications of this will be discussed by Borst et al. (12), elsewhere in this volume.

A second observation is that some transcripts apparently undergo processing of some sort. The first and relatively simple example is the 24 S_E RNA, originating from the vicinity of OLI-2, that seems to be converted into a 22 S_E. An even more interesting phenomenon is found in the adjacent region. Here it looks as if the primary transcript(s) undergo(es) extensive processing. This may be correlated to the curious behaviour of the OXI-3 locus in genetic and physical mapping analyses (13). The conclusion that these transcripts do get processed hinges on the absence of aggregation during the fractionation of the RNA. Evidence for this can be derived from the following observations:

1. The RNA species detectable in Fig. 2 show up reproducibly under various electrophoresis conditions (not shown).

2. Whenever aggregation occurs, the 21S rRNA is always the first species to remain on the top of the gel (Van Ommen, G.J. B., unpublished observations). That this is not the case in our experiment can be readily inferred from the absence of any specific hybridization to fragment 9 in the top fraction.

3. In nearly all the cases, comparison of hybridization patterns taken together with length estimations exclude the possibility that larger species consist of combinations of smaller molecules, whether identical or different. One possible exception might be the region of the gel at about 30 S_E (around fractions 5 and 6). We cannot decide from the present data whether this 30 S_E RNA is a large transcript from the whole fragment 2-17-6()-11-13 region or that the hybridization here is caused by an aggregation product. A minor dimer aggregation product of the 21S rRNA that migrates slightly slower may also be present since the ribosomal fragment hybridizes without concomitant hybridization to the adjacent fragments.

Finally, the sum of the calculated molecular weights of the non-overlapping RNAs, including the rRNAs and the 4S RNAs, is approximately 32 000 nucleotides. This means that at least 47% of the S. carlsbergensis mtDNA is transcribed into structural RNAs and mRNA-like species. This value is in good agreement with the observation of Rabinowitz and co-workers (14) who found with an entirely different method that about 60% of the genome is transcribed asymmetrically. Although the method we used is extremely sensitive, the possibility exists that we missed transcripts present only in minute amounts. Since only 10% of the mitochondrial genome is required to code for the structural genes of the known mitochondrial translation products and about 50% is transcribed into species apparently corresponding to the genes for these proteins, the possibility exists that these transcripts are actually polycistronic mRNAs also coding for minor mitochondrial translation products. It is more likely, however, that a much greater part of the mtDNA is involved in the correct expression of the genes for subunits of the mitochondrial complexes than can be inferred from the molecular weights of the polypeptides.

ACKNOWLEDGEMENTS

We like to thank Dr L.A.Grivell, Mr A.F.M.Moorman and Prof. P.Borst for their continuing advice and very helpful discussions. We are grateful for the kind cooperation of Mr D.Lindenaar, Mr H.Bos and Ms M.A.Tilanus in making this manuscript ready on time in the proper form. This work was supported in part by a grant to P.Borst from The Netherlands Foundation for Chemical Research (SON) with financial aid from The Netherlands Organization for the Advancement of Pure Research (ZWO).

REFERENCES

1 Sanders, J.P.M., Heyting, C., DiFranco, A., Borst, P. and Slonimski, P.P. (1976) The organization of genes in yeast mitochondrial DNA. In: The Genetic Function of Mitochondrial DNA (Saccone, C. and Kroon, A.M., Eds), North-Holland, Amsterdam, pp. 259-272.
2 Sanders, J.P.M., Heyting, C. and Borst, P. (1975) The organization of genes in yeast mitochondrial DNA. I. The genes for large and small ribosomal RNA are far apart. Biochem.Biophys.Res.Commun. 65, 699-707.
3 Van Ommen, G.J.B., Groot, G.S.P. and Borst, P. (1977) Fine-structure physical mapping of 4S RNA genes on mitochondrial DNA of Saccharomyces cerevisiae. Mol.Gen.Gen., in press.
4 Heyting, C. and Sanders, J.P.M. (1976) The physical mapping of some genetic markers in the 21S ribosomal region of the mitochondrial DNA of yeast. In: idem ref. 1, pp. 273-280.
5 DiFranco, A., Sanders, J.P.M., Heyting, C., Borst, P. and Slonimski, P.P. (1976) Restriction enzyme analysis and physical mapping of mitochondrial DNA from petite mutants, carrying a genetic marker for oligomycin or paromomycin resistance. In: idem ref. 1, pp. 291-304.
6 Morimoto, R., Lewin, A., Merten, S. and Rabinowitz, M. (1976) Restriction endonuclease mapping and analysis of grande and mutant yeast mitochondrial DNA. In: Genetics and Biogenesis of Chloroplasts and Mitochondria (Bücher, Th., Neupert, W., Sebald, W. and Werner, S., Eds), North-Holland, Amsterdam, pp. 519-524.
7 Saccone, C. and Kroon, A.M. (Eds) (1976) The Genetic Function of Mitochondrial DNA, North-Holland, Amsterdam.
8 Bandlow, W., Schweyen, R.J., Thomas, D.Y., Wolf, K. and Kaudewitz, F. (Eds) (1976) Genetics, Biogenesis and Bioenergetics of Mitochondria,

De Gruyter, Berlin.
9 Bücher, Th., Neupert, W., Sebald, W. and Werner, S. (Eds) (1976) Genetics and Biogenesis of Chloroplasts and Mitochondria, North-Holland, Amsterdam.
10 Moorman, A.F.M., Grivell, L.A., Lamie, F. and Smits, H.L. (1977) Identification of mitochondrial gene products by DNA-directed protein synthesis in vitro. Biochim.Biophys.Acta, submitted.
11 Moorman, A.F.M., Grivell, L.A., Katan, M.B., Van Ommen, G.J.B. and Meiland, C.M. (1976) Translation of mitochondrial messenger RNAs in vitro as an approach to the identification of gene products of yeast mitochondrial DNA. In: idem ref. 8, pp. 241-258.
12 Borst, P., Bos, J.L., Grivell, L.A., Groot, G.S.P., Heyting, C., Moorman, A.F.M., Sanders, J.P.M., Talen, J.L., Van Kreijl, C.F. and Van Ommen, G.J.B. (1977) The physical map of yeast mitochondrial DNA anno 1977. This volume.
13 Grivell, L.A. and Moorman, A.F.M. (1977) A structural analysis of the oxi-3 region on yeast mtDNA. This volume.
14 Hendler, F., Halbreich, A., Jakovcic, S., Patzer, J., Merten, S. and Rabinowitz, M. (1976) Characterization and translation of yeast mitochondrial RNA. In: idem ref. 6, pp. 679-684.
15 Southern, E.M. (1975) Transfer of DNA fragments to Millipore filters after restriction endonuclease cleavage followed by agarose gel electrophoresis. J.Mol.Biol. 98, 503-517.
16 Moorman, A.F.M., Verkley, F.N., Asselbergs, F.A.M. and Grivell, L.A. (1977) Yeast mitochondrial messenger RNA is not correctly translated in heterologous cell-free systems. This volume.

A ONE-STEP METHOD FOR RHO$^+$cRNA/RHO$^-$DNA HYBRIDIZATION IN MAPPING OF THE MITOCHONDRIAL GENOME OF SACCHAROMYCES CEREVISIAE

G. Grosch, I. Doxiadis, B. Lang, R.J. Schweyen and F. Kaudewitz

Genetisches Institut der Universität München, Maria-Ward-Str.1a
D-8000 München, Germany

INTRODUCTION

In the last few years, several approaches have been made in mapping of the mitochondrial genome of Saccharomyces cerevisiae: "genetic mapping" by analysis of co-retention of markers in rho$^-$ clones (1); "molecular mapping" by hybridization of rho$^-$ DNA to rho$^-$ DNA (2); "physical mapping" by restriction enzyme analysis of rho$^+$ and rho$^-$ mit DNA* (3,4). This report presents a simplified approach towards the molecular mapping of the mitochondrial genome. It is based on a selection of certain classes of genetically defined rho$^-$ clones and makes use of a quick electrophoretic system for the separation of mit and nuc DNA.

RESULTS AND DISCUSSION

Slab gel electrophoresis has been used for the separation of different species DNA molecules in work with bacteria (5). This attempt was adopted for the separation of mit and nuc DNA of yeast (Fig.1). After electrophoresis of nucleic acids from lysed spheroplasts or mitochondria a defined band of NaOH resistant nucleic acids could be detected.

* Abbreviations: mit DNA (mitochondrial DNA)
  nuc DNA (nuclear DNA)
  cRNA (complementary RNA)

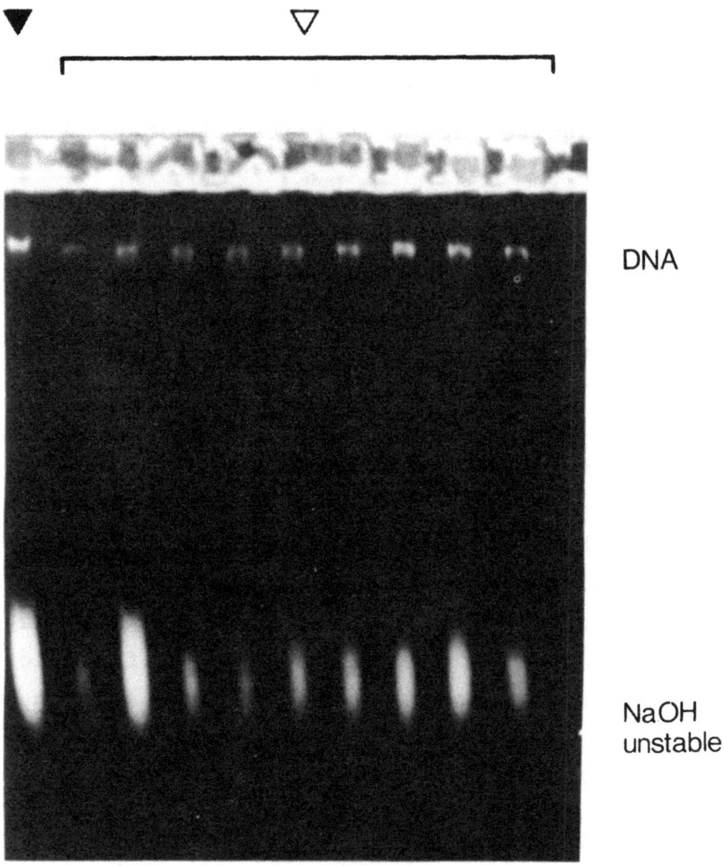

Fig. 1 Separation of nucleic acids from lysed mitochondria by gel electrophoresis

A 0.6 agarose slab gel (Tris-acetate, pH 7.6; 1µg Ethidium-bromide) was loaded with mitochondria lysed in the presence of 2% Sarcosyl NL30, at 60°C, for 10 min. Electrophoresis was performed at 4°C, for 14 hrs (10 V/cm). The gel was photographed using a red/orange filter combination.

Mitochondria from rho^+ (▶), from different rho^- (▷) strains.

In order to show that this band consists of <u>mit</u> DNA, a ^{32}P DNA from such a band was co-centrifuged with total yeast DNA (^{3}H) (Fig. 2)

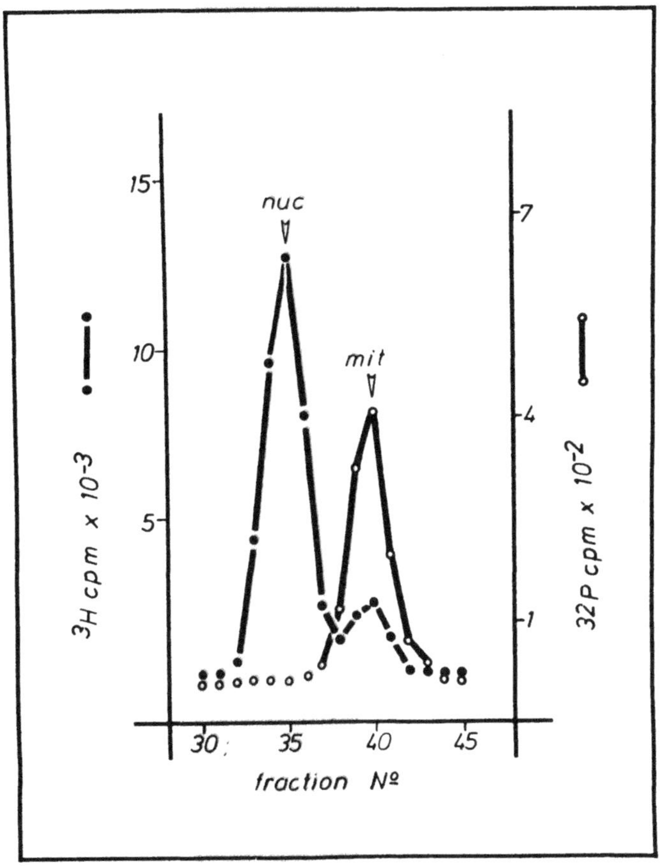

Fig. 2 Sodium iodide density gradient
A ^{32}P labelled band was sliced out of the gel, solubilized in NaJ, and mixed with ^{3}H labelled total DNA. The refractometer index was adjusted to 1.4295, 1µg Ethidiumbromide was added, and the gradient run for 63 hrs, at 40.000 rpm, 21°C.

It is evident from the results that i) the density of the ^{32}P labelled band corresponds to mit DNA; ii) no contamination with nuc DNA can be detected. For the estimation of the molecular weight of the mit DNA its mobility was compared to known molecular weight markers in different agarose gel concentrations (Fig.3).

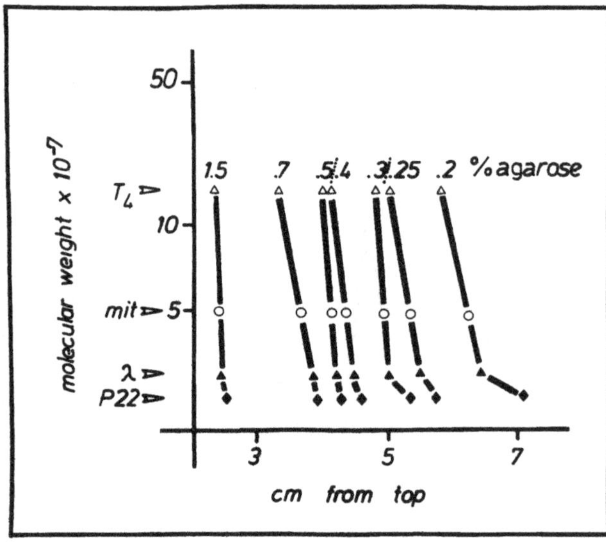

Fig. 3 Mobility of mit DNA

Co-electrophoresis of T4, lambda, P_{22}, and mit DNA in different agarose gel concentrations (1.5; 0.7; 0.3; 0.25 and 0.2%) was performed. In order to calculate the molecular weight of the mit DNA the logarithm of the molecular weight of the known markers was plotted against the respective distance of migration from the top of the gel.

The molecular weight of the mit DNA (see Fig.3) is higher than lambda DNA, lower than T_4 DNA. It is estimated to be about 50×10^6 Daltons.

The mit DNA bands from rho^+ and different rho^- strains were sliced out of the gel -following the method of Southern (6)-, transferred onto nitrate cellulose filters, and hybridized with rho^+cRNA (7). The hybridization level of rho^+ mit DNA bound to the filters with rho$^+$cRNA was taken as 100%. The percentage of hybridization of rho$^+$cRNA to rho$^-$ mit DNA is regarded as representing the length of the genome segment retained in the respective rho$^-$ Clone (Table 1).

rho$^-$ clone studied	marginal markers [a)] retained on rho$^-$ genome segment		length of genome segment in	
			% hybridization rho$^+$cRNA/rho$^-$DNA	units on the genetic map[b)]
G 41	cob-2282	oxi3-2501	25	25
G 81	cap	cob-1282	18	20 - 35
G 71	cap	cob-1282	24	20 - 35
H 21	cob-2011	par	46	45 - 60
H 31	tsm8	cap	8	8 - 25
G 31	oli1	cob-3881	11	5 - 40
G 91	cob-1431	cob-1282	13	2 - 45
F 6/2	par	oxi2-2532	17	15 - 40
F 6/3	oli2	par	26	29 - 57
rho$^+$	cap	cap	100	100

Table 1 Determination of the length of mit DNA segments retained in rho$^-$ clones

a) Markers are given in clock-wise orientration according to the genetic map as presented by Schweyen et al., this volume.

b) Minimal and maximal length as concluded from the genetic analysis.

The rationale of the mapping approach is shown in Fig. 4. The length of the mit DNA sequences retained in three rho$^-$ clones has been determined, which contain part of the cob1 (cf. Schweyen et al., this volume) markers and either only the oli1 marker or the oli1 and cap markers. Oli1 is concluded to be located within a segment of 11% (map units) distant from cob1, cap in a segment between 11 and 24%. Its location is further defined by the analysis of a fourth rho$^-$ mit DNA of 8 units length, covering cap and tsm8. The combined analysis of the third and fourth rho$^-$ clone reveals that the location of cap is in a distance of 16-18 and tsm8 in a distance of 24-26 map units from cob1.

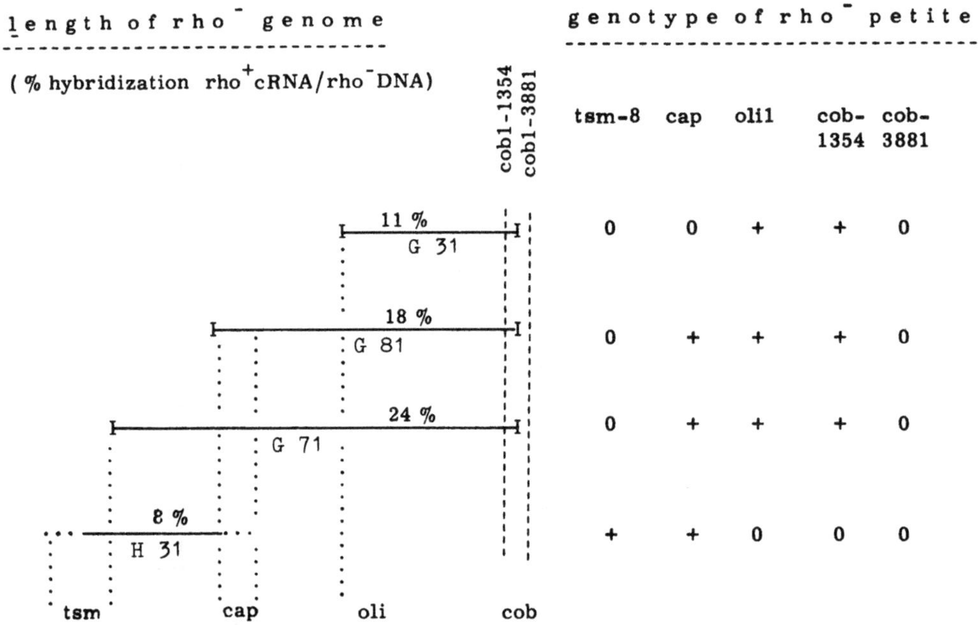

Fig. 4 Location of the markers cap and tsm

The other markers involved in the analysis shown in Table 1, were localized in a similar way. From this results a circular map of mit DNA as shown in Fig.5. The location of some of the loci is rather precise and in good agreement with their location of genetic and physical maps shown in this volume (e.g. the loci oxi3, cap, tsm, oxi1). In order to obtain a complete molecular map of the mitochondrial genome, another set of rho$^-$ clones has to be included in this study.

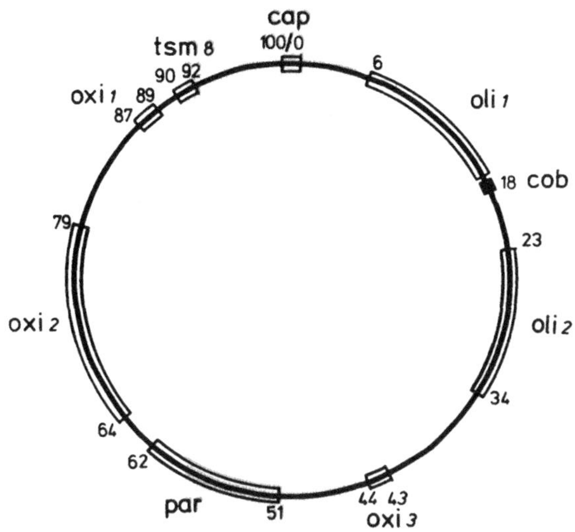

Fig. 5 Molecular map of the mitochondrial genome
Data shown in Table 1 and a few additional results not shown were used for the construction of this map.

CONCLUSIONS

The rationale of the method described is to start from a low number of cells, to concentrate mit DNA in a defined band, and to hybridize against cRNA. The method is quick, simple, and allows the analysis of many rho⁻ clones simultaneously. The use of rho⁻ clones in which the deletion in mit DNA starts or ends between two defined mutational sites of the same cluster, renders the location of genetic markers on the mitochondrial genome both quickly and precisely.

ACKNOWLEDGEJENT

The authors wish to thank Dr. D.Y. Thomas for encouragement of this work and for many helpful suggestions. This work was supported by the Deutsche Forschungsgemeinschaft

REFERENCES

1. R.J. Schweyen, B. Weiss-Brummer, B. Backhaus, F. Kaudewitz: Localization of seven gene loci on a circular map of the mitochondrial genome of Saccharomyces cerevisiae (1976) in: "The Genetic Fundtion of Mitochondrial DNA" (Saccone, C. and A.M. Kroon, eds.) pp. 251-258
2. J.P.M. Sanders, C. Heyting, A. Di Franco, P. Borst and P.P. Slonimski: The organization of genes in yeast mitochondrial DNA (1976) in: "The Genetic Function of Mitochondrial DNA" (Saccone, C. and A.M. Kroon, eds.) pp. 259-271
3. M. Rabinowitz, S. Jankovcic, N. Martin, F. Hendler, A. Halbreich, A. Lewin and R. Morimoto: Transcription and organization of yeast mitochondrial DNA (1976) in: "The Genetic Function of Mitochondrial DNA" (Saccone, C. and A.M. Kroon, eds.) pp. 219-230
4. K.S. Sriprakash, P.L. Molloy, P. Nagley, H.B. Lukins and A. W. Linnane: Biogenesis of mitochondria: XLI. Physical mapping of mitochondrial genetic markers in yeast. J.Mol. Biol. 104 (1976, 485-503
5. M. Takahashi, T. Ogino, K. Baba: Estimation of relative molecular length of DNA by electrophoresis in agarose gels. BBA 174 (1969), 183-187
6. M. Southern: Detection of specific sequences among DNA fragments separated by gel electrophoresis. J.Mol.Biol. 98 (1975), 503-517
7. J.O. Bishop: Molecular hybridization of ribonucleic acid with a large excess of deoxyribonucleic acid. Bioch.J. 126 (1972), 171-185

ALTERED AMINO ACID SEQUENCE OF THE DCCD-BINDING PROTEIN OF THE NUCLEAR OLIGOMYCIN-RESISTANT MUTANT AP-2 FROM NEUROSPORA CRASSA

Sebald, W., Sebald-Althaus, M. and Wachter, E.
Institut für Physiologische Chemie, Physikalische Biochemie und Zellbiologie der Universität München

1. INTRODUCTION

A major constituent of the mitochondrial ATPase complex from *Neurospora crassa* is a polypeptide of molecular weight 8 300 which reacts selectively with the ATPase inhibitor dicyclohexyl carbodiimide[1-4]. The DCCD-binding protein is soluble in neutral chloroform/methanol as the DCCD-binding protein from beef heart [5,6]. The protein has been purified to homogeneity and its amino acid sequence has been determined[7,8].

Previous studies have shown that the DCCD-binding protein of *Neurospora crassa* is translated on extramitochondrial ribosomes[2,3,9]. The present communication presents data which indicate that the structural gene is located in the nucleus. In an accompaning paper it is shown that the homologous DCCD-binding protein from *Saccharomyces cerevisiae* is coded for on mitochondrial DNA.

2. ISOLATION AND GENETIC ANALYSIS OF OLIGOMYCIN-RESISTANT MUTANT AP-2

2.1. Isolation and purification. Conidia germinated for 2 hours were submitted to mutagenesis with 2-aminopurine[10]. The mutagenized culture was plated on glycerol complete medium containing 0.5 µg oligomycin per ml. After 7 days incubation, 2 to 5 resistant cells were obtained per 10^9 conidia. They were purified by growth through a race tube of 20 cm length containing Vogel´s minimal medium plus 0.5 µg oligomycin per ml.

Reciprocal crosses were performed with the mutant AP-2 and an adenine requiring strain (ad-1). Tetrad analysis revealed a 4 : 4 segregation of both the drug resistance and the auxotrophic marker. Thus, the mutation conferring oligomycin-resistance in strain AP-2 is located in the nucleus.

One oligomycin resistant prototrophic spore obtained from the cross was used for further genetical and biochemical analysis.

2.2. Determination of linkage group. Mutant AP-2 was crossed with tester strains for linkage groups I to VI (alcoy[11]) and VII (nic-3; me-7). Linkage only was observed with the markers for chromosome VII. The result of a random spore analysis is shown in table 1. 30% recombinants between oli and nic-3 and 9% recombinants between oli and me-7 were obtained. This indicates that in mutant AP-2 the oli allele is located on linkage group VII near me-7. The number of the analysed spores does not allow a determination of the gene order.

TABLE 1

Mapping of Nuclear Oligomycin-resistant Mutant AP-2

Mutant AP-2 was crossed with a strain carrying two auxotrophic markers for linkage group VII (nic-3; me-7). The conidia from 100 random spores were analysed for oligomycin-sensitivity and for nicotinamide and methionine requirement.

Cross: A nic-3 me-7 OLI X a NIC-3 ME-7 oli

	Random spores	
Phenotype	*Number*	*Percentage*
nic me OLI	38	66
NIC ME oli	28	
nic ME oli	9	25
NIC me OLI	16	
NIC ME OLI	3	5
nic me oli	2	
NIC me oli	3	4
nic ME OLI	1	

3. PROPERTIES OF MUTANT AP-2

The growth of mutant AP-2 proceeds at oligomycin concentrations up to 3 µg per ml (Fig. 1). The wild type ceases to grow at inhibitor concentrations above 0.1 µg per ml. Surprisingly, the mutant grows poorly in the absence of the drug. Among 20 oligomycin-resistant mutants analysed up to now, only strain AP-2 showed this oligomycin-dependence. Cold-sensitivity is another special feature of this strain. It is unable to grow at 15 °C.

Figure 1. *Oligomycin-resistant and -dependant Growth of Nuclear Mutant AP-2.*

Mycelial elongation was determined in "race tubes" on Vogel's minimal medium plus 2% glycerol in the presence of the indicated amounts of oligomycin.

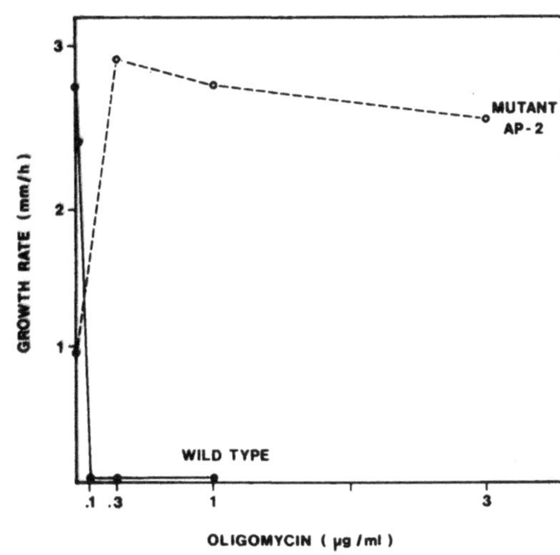

The oligomycin-resistance of mutant AP-2 is retained in vitro as demonstrated in Fig. 2. The ATPase activity of submitochondrial particles isolated from the mutant is inhibited half-maximally at 4 µg oligomycin per mg protein. This value is 10 to 20 times higher than in the wild type. The sensitivity to venturicidin is unaltered in the mutant, whereas the sensitivity to DCCD appears to be increased.

The resistance to oligomycin in vitro indicates that a constituent of the ATPase complex has been modified in mutant AP-2. It was a reasonable working hypothesis that this modification resides in one of the hydrophobic polypeptides of the ATPase membrane factor which is the target of oligomycin.

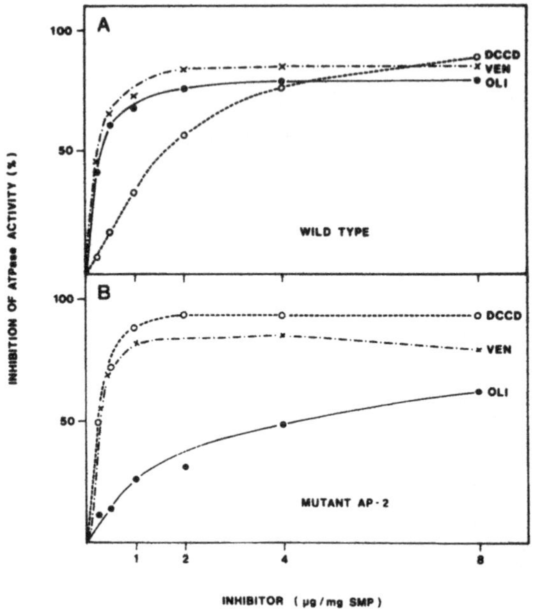

Figure 2. *Inhibition by Oligomycin (●-●), Venturicidin (x-x) and Dicyclohexylcarbodiimide (DCCD) (o--o) of the ATPase Activity of Submitochondrial Particles from Neurospora Wild Type and Mutant AP-2.*

Submitochondrial particles suspended (5 mg/ml) in 10 mM Tris-acetate, pH 7.4, were incubated at 0 °C with the indicated amounts of inhibitor for 5 min (OLI, VEN) and 4 h (DCCD). Aliquots were withdrawn for determination of ATPase activity.

4. ALTERED AMINO ACID SEQUENCE OF THE DCCD-BINDING PROTEIN IN MUTANT AP-2

In Neurospora, two hydrophobic polypeptides of the ATPase complex are mitochondrial translation products[9]. The extremely hydrophobic DCCD-binding protein, however, is synthesized outside the mitochondria. Thus, it was a distinct possibility that this cytoplasmically synthesized polypeptide had been altered in the nuclear mutant AP-2. Therefore the primary structure of the mutant polypeptide was studied.

The isolated DCCD-binding protein was cleaved with cyanogen bromide, the fragments were purified and their amino acid compositions were determined. The fragments are sufficiently small and pure to detect an alteration in the mutant in comparison with the corresponding fragments from the wild type (table 2). The only significant difference in the amino acid composition is observed in BrCN fragment III. Here one phenylalanine less and one additional serine is present. The same amino acid ex-

TABLE 2

Amino Acid Composition of Cyanogen-bromide and N-bromosuccinimide Fragments of the DCCD-binding Protein from Neurospora crassa Wild Type and Mutant AP-2

The cleavage with BrCN and NBS as well as the isolation of the fragments are described in Ref. 8. The fragments were hydrolysed with 5.7 N HCl at 105 °C for 24 hours.

	BrCN fragments								NBS fragment II of BrCN fragment III	
	I/V		II		III		IV			
	WT	AP-2	WT	AP-2	WT	AP-2	WT	AP-2	WT	AP-2
Aspartic acid			0.98	1.04	2.93	3.09			1.11	1.10
Threonine	1.04	1.02			1.07	1.10				
Serine	1.73	1.59	0.95	0.98	1.80	2.97				0.99
Glutamic acid	2.45	2.08	1.19	1.05	1.91	2.13			1.16	1.20
Proline					1.28	1.17				
Glycine			1.10	1.19	10.3	9.80			2.09	2.37
Alanine	2.97	2.73			9.55	9.95	1.05	1.06	3.03	3.21
Valine			2.03	2.00	2.93	3.11	0.96	0.98	1.00	1.04
Isoleucine	1.03	1.01			4.92	4.76			1.83	1.98
Leucine			1.13	1.13	8.73	8.80	1.00	0.98	2.90	2.85
Tyrosine	0.92	0.85			1.02	0.88				
Phenylalanine	1.12	1.07			5.54	4.03			2.99	2.03
Lysine	0.92	0.97	0.74	0.99						
Arginine					1.48	2.16				
Homoserine	+	+	+	+	+	+	+	+	+	+

change occures in a 17 residue peptide obtained by cleavage of BrCN fragment III with N-bromosuccinimide (NBS II).

The amino acid sequence of this peptide (NBS II) is shown in Fig. 3. The phenylalanine in position 61 of the wild type sequence is substituted by a serine residue in mutant AP-2. This data indicate that the structural gene of the DCCD-binding protein has been mutated in the oligomycin-resistant strain AP-2. Thus, in Neurospora this gene is located in the nucleus.

Wild type:

Ala-Ile-Leu-Gly-Phe[61]-Ala-Phe-Val-Glu[65]-Ala-Ile-Gly-Leu-Phe--Asp-Leu-Met

Mutant AP-2

Ala-Ile-Leu-Gly-<u>Ser</u>-Ala-Phe-Val-Glu-Ala-Ile-Gly-Leu-Phe--Asp-Leu-Met

Figure 3. *Altered Amino Acid Sequence of the DCCD-binding Protein in the Nuclear Oligomycin-resistant Mutant AP-2 of Neurospora crassa.*

The amino acid sequence of NBS peptide II from BrCN fragment III (see table 2) was determined by automated solid-phase Edman degradation (Ref. 8). The amino acid phenylthiohydantoins were identified by gas chromatography and mass spectroscopy.

---

5. CONCLUSIONS

The genetical and biochemical studies with *Neurospora crassa* described in this paper resulted in a clear correlation of a nuclear mutation conferring oligomycin-resistance in vivo and in vitro with an amino acid exchange in the DCCD-binding protein of the mitochondrial ATPase complex. The nuclear location of the structural gene identified by this classical approach fits with the extramitochondrial translation of this polypeptide. The presented data indicate also that strong hydrophobicity is not a pecularity of mitochondrial gene products. The DCCD-binding protein from *Neurospora crassa* is the most hydrophobic polypeptide found up to now in the mitochondrial membrane of this organism.

In *Saccharomyces cerevisiae* probably three genes have been identified on mitochondrial DNA which confer oligomycin-resistance in vivo and in vitro[12,13]. In *Aspergillus nidulans* nuclear and cytoplasmic mutants have been isolated which contain an oligomycin-resistant ATPase complex[14]. In *Neurospora crassa* it remains to be established whether several genes are present which determine the oligomycin-sensitivity of the mitochondrial ATPase complex. 20 mutants from this organism analysed up to now are all nuclear and probably carry a mutation in the same gene as described in this paper for mutant AP-2

6. REFERENCES

1. Sebald, W. and Jackl, G.: Biogenesis of mitochondrial ATPase in Neurospora crassa. In Electron Transfer Chains and Oxidative Phosphorylation (E. Quagliariello et al. eds.) North-Holland, Amsterdam, p. 193 (1975).
2. Sebald, W., Graf, Th. and Wild, G.: Cytoplasmic synthesis of the DCCD-binding protein in Neurospora crassa. In Genetics and Biogenesis of Chloroplasts and Mitochondria (Th. Bücher et al. eds.), North-Holland, Amsterdam, p. 167 (1976).
3. Sebald, W.: Biogenesis of mitochondrial ATPase. Biochim. Biophys. Acta, 463, 1 - 27 (1977).
4. Sebald, W.: Identification of the DCCD-binding protein of the mitochondrial ATPase complex from Neurospora crassa and Saccharomyces cerevisiae. Manuscript in preparation.
5. Cattell, K.J., Lindop, C.R., Knight, I.G. and Beechey, R.B.: The identification of the site of action of DCCD as a proteolipid in mitochondrial membranes. Biochem. J. 125, 169 (1971).
6. Stekhoven, F.S., Waitkus, R.F. and van Moerkerk, H.Th.B.: Identification of the DCCD-binding protein in the oligomycin sensitive ATPase from bovine heart mitochondria. Biochemistry, 11,1144 (1972).
7. Sebald, W., Sebald-Althaus, M., Lukins, H.B. and Wachter, E.: Biogenesis of the mitochondrial ATPase complex. The dual origin of subunits and genes. In Membrane ATPases (H. Peeters et al. eds.), European Press, Gent, in press.
8. Wachter, E. and Sebald, W.: The amino acid sequences of the DCCD-binding protein of the mitochondrial ATPase complex from Neurospora crassa and Saccharomyces cerevisiae. Manuscript in preparation.

9. Jackl, G. and Sebald, W.: Identification of two products of mitochondrial protein synthesis associated with mitochondrial ATPase from Neurospora crassa. Eur. J. Biochem. 54, 97 (1975).

10. Shannon, C., Enns, R., Wheelis, L., Burchiel, K. and Criddle, R.S.: Alterations in mitochondrial ATPase activity resulting from mutation of mitochondrial DNA. J. Biol. Chem. 248, 3004 (1973).

11. Perkins, D.D., Newmayer, D., Taylor, C.W. and Bennet, D.C.: New markers and map sequences in Neurospora crassa, with a description of mapping by duplication coverage, and of multiple translocation stocks for testing linkage. Genetics 40, 247 (1969).

12. Lancashire, W.E. and Griffiths, D.E.: Studies on energy-linked reactions; Genetic analysis of venturicidin-resistant mutants. Eur. J. Biochem. 51, 403 (1975).

13. Groot Obbink, D.J., Hall, R.M., Linnane, A.W., Lukins, H. B., Monk, B.C., Spithill, T.W. and Trembath, M.K.: Mitochondrial genes involved in the determination of mitochondrial membrane proteins. In The Genetic Function of Mitochondrial DNA (C. Saccone and A.M. Kroon eds.) North-Holland, Amsterdam, p. 163 (1976).

14. Rowlands, R.T. and Turner, G.: Physiological and biochemical studies of nuclear and extranuclear oligomycin-resistant mutants of Aspergillus nidulans. Molec. gen. Genet. 132, 73 (1974).

This work was supported by the Deutsche Forschungsgemeinschaft, Sonderforschungsbereich 51, Medizinische Molekularbiologie und Biochemie.

ALTERED AMINO ACID SEQUENCE OF THE DCCD-BINDING PROTEIN IN THE OLI 1 RESISTANT MUTANT D273-10B/A21 OF *SACCHAROMYCES CEREVISIAE*

E. Wachter and W. Sebald
Institut für Physiologische Chemie, Physikalische Biochemie und Zellbiologie der Universität München
and
A. Tzagoloff
The Public Health Research Institute of the City of New York

1. INTRODUCTION

Studies on mitochondrial translation products in *Saccharomyces cerevisiae* and *Neurospora crassa* during the last six years yielded corresponding results in both organisms as far as subunits of cytochrome oxidase and cytochrome *b* are concerned [1-5]. A puzzling difference between the two organisms, however, had to be noticed in the case of the mitochondrial contribution to the ATPase complex. In the yeast ATPase large amounts of a mitochondrial translation product are present (subunit 9), which exhibits the properties of a proteolipid. This product can be recovered in a high molecular weight form of 45 000 and a low molecular weight form of 8 000 daltons[6,7]. In *Neurospora* such a mitochondrially synthesized polypeptide can not be detected neither when the ATPase complex nor when whole mitochondria are analysed[8].

This observation initiated a series of experiments which showed the following results: (1) The ATPase proteolipid from yeast reacts selectively with the ATPase inhibitor dicyclohexylcarbodiimide (DCCD)[9] similar to the DCCD-binding proteins already identified in beef heart[10] and *Neurospora*[11] mitochondria. (2) In contrast to yeast, the DCCD-binding protein from *Neurospora* is translated outside the mitochondria[11,12]. (3) The DCCD-binding proteins from both microorganisms show extensive homology in their amino acid sequences despite the different translational origin. The homology is shown diagrammatically

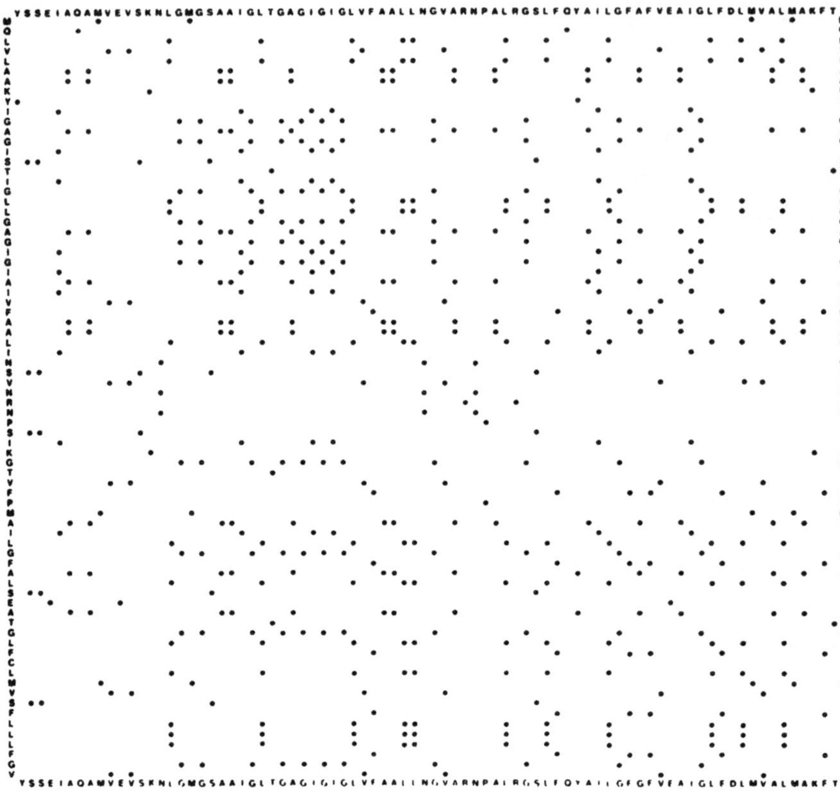

FIGURE 1. *Homology of the Amino Acid Sequences of the DCCD-binding Protein from Neurospora crassa (abscissa) and Saccharomyces cerevisiae (ordinate).*

The amino acid sequences represented in the one-letter code were determined by automated solid-phase Edman degradation of the whole protein and of peptides obtained after cleavage with BrCN, N-bromosuccinimide, Pepsin and Trypsin (Ref. 13). The positions of residues N 41, G 42, R 45, N 46, R 50, G 51, S 52, Q 55 (Neurospora) and N 35, S 36, N 38, R 39, N 40, S 42 (Saccharomyces) are not yet finally settled.

Identical amino acids are represented in the system of coordinates by a point. Homology between two sequences is indicated when a significant number of points forms a diagonal of 45 degrees from the upper left side to the lower right side.

in Fig. 1 by the straight line of points running from the upper left hand corner to the lower right hand corner. The DCCD is bound in both proteins to a glutamyl residue in position 65 (*Neurospora*) and 59 (*Saccharomyces*), respectively[9,13].

The knowledge of the amino acid sequence of the DCCD-binding protein from yeast made it feasable to approach the identification of its structural gene. Evidence has already been provided that the aggregation behaviour of the ATPase proteolipid is modified in OLI 1 resistant mutants[14,15]. Therefore the amino acid sequence in the oligomycin-resistant mutant D273-10B/A21 mapping at locus OLI 1 was studied.

2. MAPPING OF MUTANT D273-10B/A21

The spontaneous oligomycin-resistant mutant D273-10B/A21 was selected on glycerol medium containing 0.5 µg per ml rutamycin as already described[14]. The mutant was mated to strain D22 A16 known to contain the OLI 1 marker[16]. No sensitive recombinants were found amongst 7 265 diploid segregants obtained from the cross. This indicates that mutant D273-10B/A21 contains the OLI 1 resistant locus.

3. IN VITRO RESISTANCE OF THE MUTANT ATPase

The ATPase activity of submitochondrial particles from the mutant D273-10B/A21 shows a 4 to 6 fold increased resistance to oligomycin compared to the wild type (Fig. 2). The in vitro sensitivity to venturicidin is unchanged in the mutant.

These properties correspond to results obtained with other OLI 1 resistant strains described in the literature[17].

4. ALTERED AMINO ACID SEQUENCE OF THE DCCD-BINDING PROTEIN IN MUTANT D 273-10B/A21

The DCCD-binding protein was isolated from submitochondrial particles prepared from the wild type (D273-10B/A) and from the mutant. The protein was treated with cyanogen bromide and the resulting four fragments were purified[13]. Their amino acid compositions were determined. They are compiled in table 1. No significant differences between the wild type and the mutant

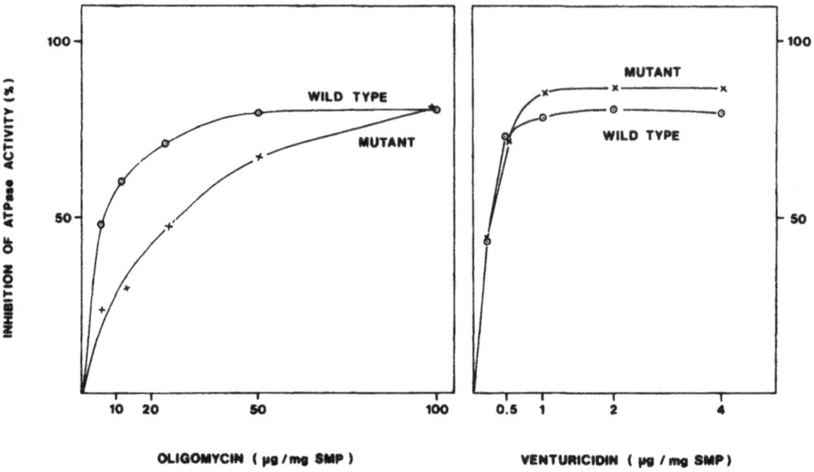

Figure 2. *Inhibition by Oligomycin (o-o) and Venturicidin (x-x) of the ATPase Activity of Submitochondrial Particles from Saccharomyces Wild Type (D273-10B/A) and Mutant D273-10B/A21.*

Submitochondrial particles suspended (5mg per ml) in 10 mM Tris-acetate, pH 7.4, were incubated with the indicated amounts of inhibitor at $0^{o}C$ for 5 min. Aliquots were withdrawn for determination of ATPase activity

are found in the amino acid composition of BrCN fragment I, II and IV. Differences between the two strains, however, are apparent in the case of BrCN fragment III (17 residues). The mutant peptide contains one leucine less and one additional phenylalanine.

This amino acid exchange could be confirmed and localized by sequencing BrCN fragment III. As shown in Fig. 3, the leucine residue in position 53 of the wild type sequence is substituted by a phenylalanine in mutant D273-10B/A21. This indicates that the structural gene of the DCCD-binding protein is affected in the OLI 1 resistant mutant.

TABLE 1

Amino Acid Composition of the Cyanogen-bromide Fragments of the DCCD-binding Protein from Saccharomyces cerevisiae Wild Type and Oligomycin-resistant Mutant D273-10B/A21

The cleavage with BrCN and the isolation of the fragments are described in Ref. 4. The peptides were hydrolysed with 5.7 N HCl at 105 °C for 24 hours.

	BrCN fragments							
	I		II		III		IV	
	WT	A 21	WT	A 21	WT	A 21	WT	A 21
Cysteic acid					0.65	0.45		
Aspartic acid			2.99	3.08				
Threonine			1.94	2.07	0.95	0.96		
Serine			2.73	2.86	1.14	1.11	0.83	0.88
Glutamic acid			1.30	1.17	0.98	0.99		
Proline			2.05	2.25				
Glycine			6.52	7.15	2.04	1.98	1.00	1.05
Alanine			5.84	6.33	2.76	2.77		
Valine			3.27	3.33			1.89	1.96
Isoleucine			5.74	6.22	1.16	0.98		
Leucine			4.71	5.09	3.84	3.14	2.75	2.89
Tyrosine			0.93	1.04				
Phenylalanine			1.57	1.79	2.04	3.02	2.06	2.22
Lysine			2.09	1.97				
Arginine			1.02	1.11				
Homoserine	+	+	+	+	+	+		

Wild type:

Ala-Ile-Leu[53]*-Gly-Phe-Ala-Leu-Ser-Glu*[59]*-Ala-Thr-Gly-Leu-Phe-*
-Cys-Leu-Met

Mutant D273-10B/A21:

Ala-Ile-Phe-Gly-Phe-Ala-Leu-Ser-Glu-Ala-Thr-Gly-Leu-Phe-
-Cys-Leu-Met

Figure 3. *Altered Amino Acid Sequence of the DCCD-binding Protein in the OLI 1 Resistant Mutant D273-10B/A21 of Saccharomyces cerevisiae.*

The amino acid sequence of BrCN fragment III was determined by automated solid-phase Edman degradation (Ref. 13). The amino acid phenylthiohydantoins were identified by gas chromatography and mass spectroscopy.

5. DISCUSSION

The present results obtained with *Saccharomyces cerevisiae* point to a clear correlation between a mutation at the OLI 1 locus and an alteration of the primary structure of the DCCD-binding protein. This indicates that the OLI 1 locus is located in the structural gene of this protein.

It is not yet understood how the amino acid exchange in mutant D273-10B/A21 causes the different aggregation behaviour of the DCCD-binding protein[14,15] as well as the increased oligomycin-resistance of the ATPase. The isopolar amino acid substitution (leu - phe) in position 53 of the mutant sequence occurs near to the glutamyl residue (position 59) which becomes selectively labelled by (^{14}C)DCCD[9]. It remains to be established whether other amino acid substitutions leading to oligomycin-resistance are also in this segment of the polypeptide chain.

In contrast to yeast, the structural gene of the DCCD-binding protein from *Neurospora* is found in the nucleus on chromosome VII[9,20]. Accordingly, a homologous protein can be coded for at different sites of the cell in different organsims. The genetic location correlates with the mitochondrial translation in yeast[6,7] and the extramitochondrial translation of this protein in *Neurospora*[9,11,13]. These findings raise the intriguing possibility of an evolutionary transition in which the site of synthesis and of genetic information has been transferred from the mitochondrion to the nucleo-cytoplasmic system or vice-versa. Up to now, however, no unambiguous evidence from other organisms is available which could possibly indicate that either *Neurospora* or *Saccharomyces* represents a unique case. In *Xenopus laevis* oocytes, the smallest mitochondrially made subunit of the ATPase complex exhibits a molecular weight of 17 500 daltons[18]. From the absence of a mitochondrial translation product of low molecular weight in these cells it may be concluded that the DCCD-binding protein - if present at all - is translated outside the mitochondria as in *Neurospora*. In *Aspergillus nidulans* a "intramitochondrially synthesized proteolipid associated with the mitochondrial ATPase complex has been identified as the product of a mitochondrial gene determining oligomycin resistance"[19]. It appears to be uncertain, however, whether this proteolipid corresponds to the DCCD-binding protein from yeast and *Neurospora*. The amino acid composition shows a much higher percentage of polar residues (38%), the protein is rather small (67 residues) and it is unknown whether it binds DCCD. Furthermore, the mutant analysed (OR11) does not show a uniparental inheritance as the extranuclear mutant O^R6 described by Rowlands and Turner[21]. Thus it is unclear whether mutant OR11 is mitochondrial or even cytoplasmic.

6. REFERENCES

1. Mason, T.L. and Schatz, G.: Cytochrome oxidase from baker's yeast. II. Site of translation of protein components. J. Biol. Chem. 248, 1355-1360 (1973)

2. Rubin, M.S. and Tzagoloff, A.: Assembly of the mitochondrial membrane system. X. Mitochondrial synthesis of three of the subunit proteins of yeast cytochrome oxidase. J. Biol. Chem. 248, 4275-4279 (1973)

3. Sebald, W., Machleidt, W. and Otto. J.: Products of mitochondrial protein synthesis in Neurospora crassa. Determination of equimolar amounts of three products in cytochrome

oxidase on the basis of amino acid analysis. Eur. J. Biochem. 38, 311-324 (1973)

4. Weiss, H. and Ziganke, B.: Cytochrome b in Neurospora crassa mitochondria. Site of translation of the heme protein. Eur. J. Biochem. 41, 63-71 (1973)

5. Katan, M.B., Van Harten-Loosbroek, N. and Groot, G.S.P.: The cytochrome bc_1 complex of yeast mitochondria. Site of translation of the polypeptides in vivo. Eur. J. Biochem. 70, 409-417 (1976)

6. Tzagoloff, A. and Meagher, P.: Assembly of the mitochondrial membrane system. V. Mitochondrial synthesis of subunit proteins of the rutamycin-senstive ATPase. J. Biol. Chem. 247, 594-603 (1972)

7. Sierra, M.F. and Tzagoloff, A.: Assembly of the mitochondrial membrane system. Purification of a mitochondrial product of the ATPase. Proc. Nat. Acad. Sci. U.S.A. 70, 3155-3159 (1973)

8. Jackl, G. and Sebald, W.: Identification of two products of mitochondrial protein synthesis associated with mitochondrail ATPase from Neurospora crassa. Eur. J. Biochem. 54, 97-106 (1975)

9. Sebald, W., Sebald-Althaus, M., Lukins, H.B. and Wachter, E.: Biogenesis of the mitochondrial ATPase complex. The dual origin of subunits and genes. In Membrane ATPase's (H. Peeters et al. eds.), European Press, Gent. In press.

10. Cattell, K.J., Lindop, C.R., Knight, I.G. and Beechey, R.B.: The identification of the site of action of DCCD as a proteolipid in mitochondrial membranes. Biochem. J. 125, 169-177 (1971)

11. Sebald, W.: Biogenesis of mitochondrial ATPase. Biochim. Biophys. Acta, 463, 1-27 (1977)

12. Sebald, W., Graf, Th. and Wild, G.: Cytoplasmic synthesis of the DCCD-binding protein in Neurospora crassa. In Genetics and Biogenesis of Chloroplasts and Mitochondria (Th. Bücher et al. eds.), North-Holland, Amsterdam, p. 167-174 (1976)

13. Wachter, E. and Sebald, W.: The amino acid sequences of the DCCD-binding protein from the ATPase complex of Neurospora crassa and Saccharomyces cerevisiae. Manuscript in preparation

14. Tzagoloff, A., Akai, A. and Foury, F.: Assembly of the mitochondrial membrane system. XVI. Modified form of the ATPase proteolipid in oligomycin-resistant mutants of Saccharomyces cerevisiae. FEBS Lett. 65, 391-396 (1976)

15. Groot Obbink, D.J., Hall, R.M., Linnane, A.W., Lukins, H.B., Monk, B.C., Spithill, T.W. and Trembath, M.K.: Mitochondrial genes involved in the determination of mitochondrial membrane proteins. In The Genetic Function of Mitochondrial

DNA (C. Saccone and A.M. Kroon eds.) North-Holland, Amsterdam, p. 163-173 (1976)

16. Avner, P.R. and Griffiths, D.E.: Studies on energy-linked reactions. Isolation and characterisation of oligomycin-resistant mutants of Saccharomyces cerevisiae. Eur. J. Biochem. 32, 301-311 (1973)
17. Griffiths, D.E. and Houghton, R.L.: Studies on energy-linked reactions. Modified mitochondrial ATPase of oligomycin-resistant mutants of Saccharomyces cerevisiae. Eur. J. Biochem. 46, 157-167 (1974)
18. Koch, G.: Synthesis of the mitochondrial inner membrane in cultures Xenopus laevis oocytes. J. Biol. Chem. 251, 6097-6107 (1976)
19. Marahiel, M.A., Imam, G., Nelson, P., Pieniażek, N.J., Stepień, P.P. and Küntzel, H.: Identification of an intramitochondrially synthesized proteolipid associated with the mitochondrial ATPase complex as the product of a mitochondrial gene determining oligomycin resistance in Aspergillus nidulans. Eur. J. Biochem. 76, 345-354 (1977)
20. Sebald, W., Sebald-Althaus, M. and Wachter, E.: Altered amino acid sequence of the DCCD-binding protein of the nuclear oligomycin-resistant mutant AP-2 from Neurospora crassa. These Proceedings.
21. Rowlands, R.T. and Turner, G.: Nuclear and extranuclear inheritance of oligomycin resistance in Aspergillus nidulans. Molec. gen. Genet. 126, 201-216 (1973)

This work was supported by a NATO Research Grant, No. 1290, and by the Deutsche Forschungsgemeinschaft, Sonderforschungsbereich 51, Medizinische Molekularbiologie und Biochemie.

STRUCTURE OF THE MITOCHONDRIAL AND PLASMA MEMBRANE ATPases OF THE YEAST *SCHIZOSACCHAROMYCES POMBE*

A. Goffeau, M. Boutry and J.P. Dufour
Laboratoire d'Enzymologie, Université de Louvain, 1348 Louvain-la-Neuve, Belgium

This paper compares some properties of subunits 3 and 9 of the mitochondrial ATPase of *Saccharomyces cerevisiae* and *Schizosaccharomyces pombe*. It also compares the subunit composition of the mitochondrial ATPase of *S. pombe* to that of the plasma membrane ATPase of the same yeast. The nomenclature used by Tzagoloff and Meagher (1) for numbering from 1 to 9 the ATPase subunits of *S. cerevisiae* is followed. As pointed out previously in our different papers concerning the mitochondrial ATPase of *S. pombe* (2 to 8), the haploid fission yeast *S. pombe* differs from the diploid budding yeast *S. cerevisiae*, in that it does not yield viable mitochondrial rho$^-$ mutants (9). In addition, the mitochondrial DNA of *S. pombe* has been reported to be shorter (about 8 µm) than that of *S. cerevisiae* (about 25 µm) (10,11).

SUBUNIT 3 OF THE YEAST MITOCHONDRIAL ATPases

According to Tzagoloff and Meagher (1), the subunit 3 of the *S. cerevisiae* oligomycin-sensitive ATPase is a component of 38 500 molecular weight which is observed in purified F1 (the water soluble part of oligomycin-sensitive ATPase bearing the catalytic site(s) for ATP hydrolysis). We have reproduced this data which is illustrated in Fig. 1, using a procedure very similar to that of Tzagoloff and Meagher (1), including : extraction by sonication, ammonium sulphate precipitation, DEAE chromatography and centrifugation through glycerol gradient(s) (Table 1). However, such component of about 40 000 molecular weight was not observed in F1 purified from *S. pombe*. Indeed, the small component located between subunits 2 and 4 of *S. pombe* F1 is of much higher molecular weight (47 000).

Table 1. Purification procedures for the F1-ATPase of glycerol-grown *Schizosaccharomyces pombe*

PROCEDURE A (Sonication)		PROCEDURE B (Chloroform)			
Purification steps	Specific (2) activities	Purification steps	Proteins (mg)	Total (1) activities	Specific (2) activities
Homogenate	0.8	Submitochondrial particles	780	4836	6.1
Submitochondrial particles	5.8	Chloroform extract	58	4320	75
Supernatant pH 5.6 of sonicated extract	14	DE 52 cellulose	17	2546	151
40% $(NH_4)_2SO_4$ supernatant	17	Hydroxyapatite	8	1683	204
80% $(NH_4)_2SO_4$ precipitate	33	$(NH_4)_2SO_4$	6.8	1164	170
Sephadex	29	Ultrogel	5.1	983	192
DE 52	54	$(NH_4)_2SO_4$	3.6	661	183
Glycerol gradient	102				
2nd glycerol gradient	155				

(1) in µmol phosphate x min^{-1}
(2) in µmol phosphate x min^{-1} x mg $proteins^{-1}$

The enzyme was either extracted by sonication at pH 9.0 (procedure A) and purified as previously described by ammonium sulfate fractionation followed by DEAE chromatography with the following modifications (3). The DE52 (Whatman) column was eluted with a linear gradient of solvent 2 to solvent 3. The most active fractions were precipitated with 80% ammonium sulfate and centrifuged through a 10 to 20% glycerol gradient. The most active fractions were pooled, precipitated with 80% ammonium sulfate, solubilized with 5% glycerol containing 1 mM ATP, 2 mM EDTA and 35 mM Tris-acetate, pH 7.5, and centrifuged again through a second 10 to 20% glycerol gradient.
Or the enzyme was extracted with chloroform (procedure B) using the following procedure. Eighty ml of submitochondrial particles isolated from glycerol-grown cells were suspended at 5 mg/ml in 20 mM Tris-

acetate, pH 7.5, 1 mM EDTA, 0.6 M sorbitol and mixed for 15 s with 40 ml chloroform at room temperature. The aqueous phase obtained by centrifugation at 100 000 x g for 30 min. The supernatant was fixed on top of a Whatman DE52 column (7 x 1.6 cm). After washing with 40 ml of 12 mM $(NH_4)_2SO_4$, the ATPase was eluted by 150 ml of a linear gradient of 12 mM to 120 mM $(NH_4)_2SO_4$. The active fractions eluted by approximately 40 mM $(NH_4)_2SO_4$ were pooled and put on top of a Bio-gel HTP, Bio-Rad (Hydroxyapatite) column (4 x 1,6 cm). After washing with 30 ml of 120 mM ammonium sulfate, the ATPase was eluted by 50 ml of a linear gradient of 120 mM to 1 M $(NH_4)_2SO_4$. The active fractions eluted by approximately 600 mM $(NH_4)_2SO_4$ were precipitated by $(NH_4)_2SO_4$ at 80% saturation. The pellet was suspended in 1 ml of 20 mM Tris acetate, pH 7.5, 1 mM EDTA, and 2 mM ATP (TEA) and put on top of an Ultrogel ACA 34 LKB column (2 x 120 cm). The ATPase was eluted by 120 to 130 ml of TEA and precipitated by $(NH_4)_2SO_4$ (80% saturation). The pellet was suspended in 1 ml of TEA and stored in liquid nitrogen after addition of 30% (w/v) glycerol.

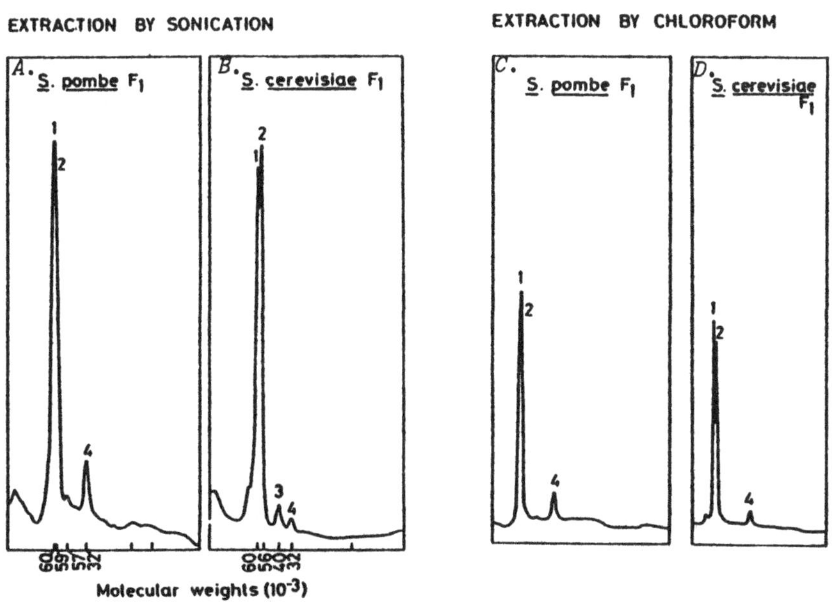

Fig. 1. Sodium dodecyl sulfate polyacrylamide gel electrophoresis of F_1 extracted by sonication or with chloroform and further purified from Schizosaccharomyces pombe and Saccharomyces cerevisiae.
The enzymes were extracted by sonication at pH 9.0 (A,B) or with chloroform (C,D) from glycerol-grown S. pombe (A,C) and S. cerevisiae (B,D) and purified under Table 1. After electrophoresis in SDS-polyacrylamide gel (according to Weber and Osborn) the F_1 components were stained with 0.25% Coomassie blue.

In order to verify whether the *S. cerevisiae* subunit 3 was a contaminant or a true component of F_1, another procedure for purification was carried out in both *S. pombe* and *S. cerevisiae*. This procedure, involving extraction by chloroform and purification by DE 52, hydroxyapatite and ultrogel chromatographies, yielded higher ATPase specific activities both in *S. pombe* and *S. cerevisiae*. Under these circumstances, the subunit 3 of about 40 000 molecular weight of *S. cerevisiae* was no longer detected whereas subunits 1, 2 and 4 were not modified. We conclude that subunit 3 is not a component of F_1 in *S. cerevisiae* as in *S. pombe*. Therefore in contrast to what has been mentioned in several reviews (12, 13, 14, 15) the γ subunit of yeast ATPase is of about 31 000 to 32 000 molecular weight (actually subunit 4 of the F_1).

Two other points should be mentioned. 1. The low detectability of the small subunits (δ and ε) of yeast F_1 is not related to the absence of subunit 3 since the latter is present under conditions where the two small subunits can be clearly detected (see ref. 3). 2. The *S. pombe* subunit 2 is of significant lower mobility than the *S. cerevisiae* subunit 2 (about 59 000 compared to 56 000 molecular weights respectively).

SUBUNIT 9 OF THE MITOCHONDRIAL ATPase OF *SCHIZOSACCHAROMYCES POMBE*

Four products of the mitochondrial protein synthesis were reported to be part of the *S. cerevisiae* oligomycin-sensitive ATPase (1), while only two mitochondrial products were recognized in the *Neurospora crassa* oligomycin-sensitive ATPase (16). The clearest difference between these two species lies in the fact that subunit 9, a protein component of low molecular weight (about 7 000) soluble in chloroform/methanol (17, 18) is the major mitochondrial product of the *S. cerevisiae* oligomycin-sensitive ATPase while no mitochondrial product of molecular weight lower than 10 000 was detected in *Neurospora* ATPase (16). A peptide with properties similar to those of the *S. cerevisiae* subunit 9 of ATPase, is however present in the *Neurospora* enzyme, but is synthesized on the cytoplasmic ribosomes (19, 20) and is of nuclear heredity (21). In *Neurospora* (16), and in derepressed *S. cerevisiae* (22), the mitochondrial products of ATPase could be immunoprecipitated with an antibody dressed against purified F_1, while in repressed *S. cerevisiae* (1), an antibody against purified oligomycin-sensitive

ATPase had to be used. Absence of precipitation by anti F_1 serum under the latter conditions indicates that the mitochondrial products were not assembled to F_1 (23).

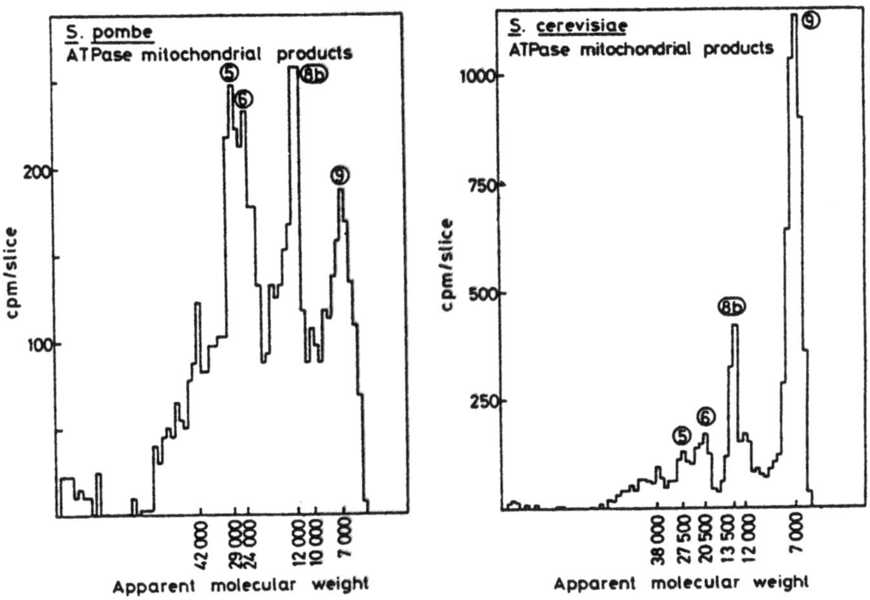

Fig. 2. Mitochondrial products of ATPase in Schizosaccharomyces pombe and Saccharomyces cerevisiae.
Two mg of submitochondrial particles labelled with L-[^3H]-leucine in the presence of cycloheximide as described under Fig. 3, were washed in 35 mM Tris-acetate pH 8.5, 1 mM EDTA, 2 mM ATP and 0.5 mM PMSF (TEA) and incubated for 1 min in 1.0 ml of TEA supplemented with 0.1% Triton-X 100. After adjustment of the pH to 7.0 with acetic acid, the suspension was centrifuged for 1 h at 200 000 x g. To the supernatant, 0.2 ml of potassium phosphate 10 mM pH 7.5 and 0.2 ml of a partially purified rabbit serum dressed against purified S. pombe F_1 were added. After 30 min of incubation at room temperature, the suspension was put at 0°C overnight. The immuno-precipitate was collected in an Eppendorf centrifuge and washed twice with 1 ml Tris-acetate 10 mM pH 7.5. About 20 000 cpm were used for gel electrophoresis using the conditions (Laemli-Blatter) described in Fig. 3.

In our hands, at least four products of mitochondrial protein synthesis were precipitated in *S. cerevisiae* by a serum dressed against F_1 purified from *S. pombe*. Fig. 2 shows that a major component of about 7 500 molecular weight is precipitated along with at least three less important components of about 13 000 (subunit 8 b ?), 20 500 (subunit 6 ?) and 27 500 (subunit 5 ?). The present data confirm essentially those of Tzagoloff and Meagher (1, 17).

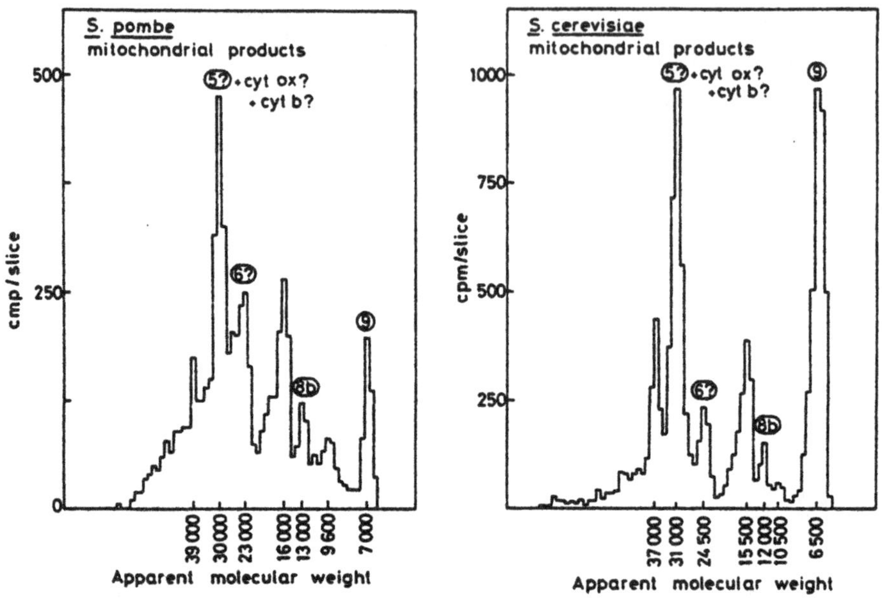

Fig. 3. Total mitochondrial products of Schizosaccharomyces pombe and Saccharomyces cerevisiae.
After inoculation of 1 x 10^6 cells/ml from an actively growing preculture, the cells were grown for 18 h in 1% glucose, 2% yeast extract (Difco) brought to pH 4.5 with HCl. One g wet weight of cells washed at 4° with 25 mM potassium phosphate pH 6.5, was incubated during 3.5 h at 30° in 100 ml of 0.3% yeast extract, 1% glucose, 2 mg/ml chloramphenicol and 25 mM potassium phosphate pH 6.5. The cells were washed with 50 ml of 25 mM potassium phosphate pH 6.5 containing 1 mg/ml cycloheximide and further incubated at 30° in 100 ml of 0.25% yeast nitrogen base (Difco), 1% glucose, cycloheximide 1 mg/ml and 25 mM potassium phosphate pH 6.5. After 10 min, 0.5 mCi L-4,5-$[^3H]$-leucine (Amersham, 58 Ci/mmol, 430 mCi/mg) was added. The incubation was continued for 2 h at 30° and the cells were washed twice with 50 ml 1 mg/ml cycloheximide and 25 mM potassium phosphate pH 6.5.
The cells were homogenized with 2 g of glass beads of 0.50 mm in 2 ml of 0.6 M sorbitol, 50 mM Tris-acetate pH 7.5, 1 mM EDTA, 1 mg/ml cycloheximide and 0.5 mM phenylmethansulfonylfluorid freshly prepared (PMSF), 2 times 45 s using the refrigerated Braun cell homogenizer at full speed. The supernatant of a 1000 x g x 5 min centrifugation was further centrifuged at 25 000 x g x 20 min. This pellet was washed with 0.6 M sorbitol, 50 mM Tris-acetate pH 7.5, 0.5 mM PMSF and sonicated for 1 min in the same medium. The sonicated particles were centrifuged and washed once at 100 000 x g for 1 h. The final submitochondrial particles exhibited 350 000 cpm/mg

for S. pombe and 210 000 cpm/mg for S. cerevisiae. About 30 000 cpm (S. cerevisiae) or 10 000 cpm (S. pombe) were used for electrophoresis on 15% polyacrylamide gel in the presence of 0.5% sodium dodecyl sulfate using the procedure of Laemli-Blatter. Before electrophoresis the samples (about 30 μg) were incubated for 20 min at room temperature in 10 μl NaOH 0.2 N and further digested overnight at room temperature in 0.1N sodium phosphate pH 7.0, 5% mercaptoethanol, 2% sodium dodecyl sulfate.

The exact same procedure applied to *S. pombe*, precipitates peptides of about 29 000, 24 000, 12 000 and 7 500 molecular weight which are possibly similar to subunits 5, 6, 8 and 9 of *S. cerevisiae*. However the low molecular weight component contains only 1/5 of the total radioactivity of the mitochondrial products precipitated in *S. pombe* by the anti F1 serum, while subunit 9 of *S. cerevisiae* contains more than 1/2 of the radioactivity of the ATPase mitochondrial products.

Compared to *S. cerevisiae*, the lower relative amount of the 7 500 molecular weight component among the mitochondrial products of *S. pombe* ATPase, could conceivably be due to the Triton treatment preceeding the immunoprecipitation. In *S. pombe* (but not in *S. cerevisiae*), Triton-extraction of the ATPase reduces the oligomycin-sensitivity to 50% of the total ATPase activity (8). However the Triton treatment is not reducing the amount of the 7 500 component in *S. pombe*. ATPase since Fig. 3 shows that direct solubilization by sodium dodecyl sulfate of the total radioactive products of mitochondrial proteins in membranes exhibiting more than 85% sensitivity to oligomycin, yields in *S. pombe* a 7 500 molecular weight component which contains only 7% of the total radioactivity compared to more than 30% in *S. cerevisiae*.

A decreased amount of low molecular weight component in the *S. pombe* ATPase compared to the *S. cerevisiae* enzyme, is also demonstrated in total ATPase and using different experimental conditions. As shown in Fig. 4, obtained in another gel electrophoresis system (see legends), the Triton-extract obtained from membranes uniformally labelled (in the absence of cycloheximide) treated with a serum dressed against the purified rutamycin-sensitive ATPase of *S. cerevisiae*, yields a total ATPase. The *S. pombe* ATPase exhibits much lower amounts of low molecular weight components than the *S. cercvisiae* enzyme.

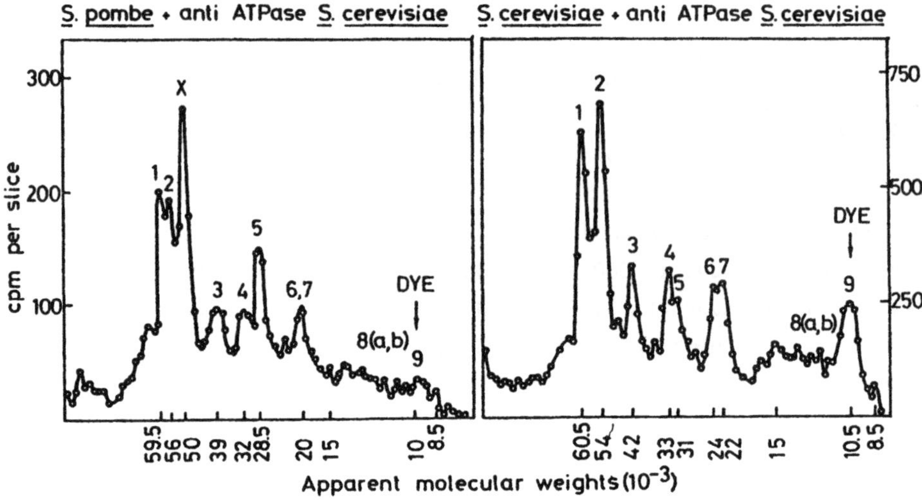

Fig. 4. Sodium dodecyl sulfate polyacrylamide gel electrophoresis of yeast Triton-extracted ATPases immunoprecipitated by an antibody dressed against the rutamycin-sensitive ATPase of Saccharomyces cerevisiae.
Submitochondrial particles of glycerol-grown S. pombe (A) or S. cerevisiae (B) were labelled with L-[^{3}H]-leucine as described under Fig. 3. 2 mg of proteins containing either 421 000 cpm/mg (S. pombe) or 691 000 cpm/mg (S. cerevisiae) were incubated at 4°C with 1 ml of 1 mM ATP, 2 mM EDTA, 35 mM Tris-acetate pH 8.5 and 3 mg Triton X-100. After 1 min, the pH were adjusted to pH 7.5 with 1 N acetic acid and the suspensions were centrifuged for 30 min at 150 000 x g. The supernatants which contained 40% of the cpm were concentrated to 100 µl by centrifugation through Amicon 50 cones and supplemented with 180 µl of rabbit serum dressed against purified rutamycin-sensitive ATPase of S. cerevisiae kindly provided by A. Tzagoloff. The immunoprecipitates were dissolved with 20 µl of 0.2 N NaOH and incubated for 15 min at 25°C before being supplemented with 80 µl of 1% sodium dodecyl sulfate, 1% β-mercaptoethanol and 164 mM sodium phosphate buffer, pH 7.5. After 5 min incubation at 100°C, samples containing 10 800 cpm (S. pombe) or 26 200 cpm (S. cerevisiae) were submitted to SDS-polyacrylamide electrophoresis according to Weber and Osborne.

It must also be mentioned that the immunoprecipitates of Figs. 2, 3 and 4 were treated with NaOH in order to depolymerize possible high molecular weight forms of *S. cerevisiae* subunit 9 (17, 18) and that in Fig. 2 and 3, the protease inhibitor phenylmethanesulfoxyl fluoride was added during homogenization and isolation of the mitochondrial particles in order to avoid possible degradation of mitochondrial products (24).

We therefore conclude that in contrast to *Neurospora*, a 7 500 molecular weight component of the ATPase is a product of the mitochondrial protein synthesis in *S. pombe*. However its relative amount among the ATPase mitochondrial products is about 1/3 to 1/5 of subunit 9 in *S. cerevisiae*. We are presently studying whether the 7 500 molecular weight component of *S. pombe* ATPase exhibit properties similar to those of the *S. cerevisiae* subunit 9 (*e.g.* solubility in chloroform/methanol and binding of the ATPase inhibitor : dicyclohexylcarbodiimide).

COMPARISON OF PLASMA MEMBRANE AND MITOCHONDRIAL ATPases IN *SCHIZOSACCHAROMYCES POMBE*

We have recently shown (25) that the plasma membrane ATPase of *S. pombe* is insensitive to oligomycin but is sensitive to dicyclohexylcarbodiimide (DCCD). This raises the interesting possibility of a similar site of action of DCCD in the mitochondrial and plasma membrane ATPases. Some genetical data obtained with *S. cerevisiae* could be in agreement with such a possibility. Several years ago, Colson *et al.* (26) demonstrated that a mitochondrial mutation of oligomycin resistance, possibly controlling the subunit 9 of mitochondrial ATPase, as recently shown for OLI1 strains (27), modifies an extra-nuclear character of multiple resistances to venturicidin and synthalin, probably related to the permeability of the plasma membrane. Similar observations of modification of the properties of plasma membrane by mitochondrial and/or episomal modifications have been reported by Rank *et al.* (28 to 31), by Wilkie and coll. (32, 33), by Griffith and coll. (34, 35), by Guerineau *et al.* (36). Rank has proposed the existence of a common control of plasma membrane and mitochondrial ATPases. Such hypothesis would be strongly supported if there was a structural similarity in the two ATPases similar to that existing between bacterial plasma membrane and mitochondrial ATPases. In this case, one could imagine a unique genetic control, possibly of episomal nature, of the biosynthesis of one or several subunits common to the two ATPases. This hypothesis seems however remote since the purified plasma membrane ATPase of *S. pombe* exhibits a subunit composition very different from that of the mitochondrial enzyme. Only one peptide of 105 000 molecular weight stains with Coomassie

Blue. In this respect, the yeast plasma membrane is rather similar to the Na^+/K^+ ATPase of mammalian plasma membrane (38) eventhough no glycoproteic component of 60 000 molecular weight is observed in yeast. However, in the low molecular weight range, an opaque band stained by periodic-Shiff (but not by Coomassie Blue) is observed in the purified plasma membrane ATPase. Since *S. cerevisiae* subunit 9 of mitochondrial ATPase has similar properties, and binds DCCD it cannot be excluded yet that the site of DCCD inhibition of the plasma membrane is located in this low molecular weight component. Further work is in progress to verify this possibility.

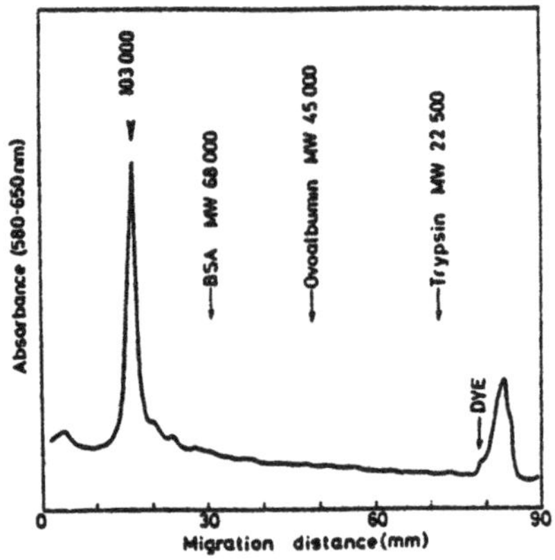

Fig. 5. Subunit composition of the plasma membrane ATPase of Schizosaccharomyces pombe.
Plasma membranes were isolated by selective pH precipitation of a crude particular fraction of glucose grown S. pombe followed by isopycnic centrifugation in a discontinuous sucrose gradient (37)*.The ATPase activity was solubilized by lysolecithin. The solubilized enzyme concentrated by centrifugation through an Amicon CF25 cone was deposited on a sucrose gradient. SDS gel electrophoresis was carried out as described by Neville et al. (39) by use of multiphasic buffer system n° J4179 of Jovin et al. (40). Proteins were stained with Coomassie Blue. The low molecular weight component is opaque but not stained.*

REFERENCES

1. Tzagoloff, A. and Meagher, P., J. Biol. Chem. *246*, 7328-7336 (1971).
2. Goffeau, A., Colson, A.M., Landry, Y. and Foury, F., Biochem. Biophys. Res. Commun. *48*, 1448-1454 (1972).
3. Goffeau, A., Landry, Y., Foury, F., Briquet, M. and Colson, A.M.,

J. Biol. Chem. *248*, 7097-7105 (1975).

4. Landry, Y. and Goffeau, A., Biochim. Biophys. Acta *376* ,470-484 (1974).
5. Goffeau, A., Briquet, M., Colson, A.M., Delhez, J., Foury, F., Labaille, F., Landry, Y., Mohar, O. and Mrena, E., in Membrane Biogenesis, Mitochondria, Chloroplasts and Bacteria, A. Tzagoloff (ed.), Plenum Press, New-York, pp. 27-61 (1974).
6. Goffeau, A., Colson, A.M., Landry, Y., Foury, F. and Briquet, M., in Biomembranes, Architecture, Biogenesis, Bioenergetics and Differentiation, L. Packer (ed.) Ac. Press, New-York, pp. 88-101 (1974).
7. Goffeau, A. and Landry, Y. in Genetics, Biogenesis and Bioenergetics of Mitochondria, W. Bandlow *et al.* (eds.) Walter de Gruyter, Berlin, pp. 303-308 (1976).
8. Landry, Y., Boutry, M., Dufour, J.P., Delhez, J. and Goffeau, A. in Membrane ATPase, Peeters (ed.) European Press, Gent (in press) (1977).
9. Heslot, H., Louis, C. and Goffeau, A. J. Bacteriol. *140*, 482-491 (1970).
10. O'Connor, R. M., McArthur, C. R. and Clark-Walker, G. D. Eur. J. Biochem. *53*, 137-144 (1975).
11. Tabak, H. F. and Weijers, P. J., FEBS Letters, *69*, 211-215 (1976).
12. Tzagoloff, A., Rubin, M. S. and Sierra, M. F., Biochim. Biophys. Acta *301*, 71-104 (1973).
13. Senior, A. E., Biochim. Biophys. Acta *301*, 249-277 (1973).
14. Pedersen, P., J. Bioenergetics *6*, 243-275 (1973).
15. Kozlov, I. A. and Skulachev, V. P., Biochim. Biophys. Acta *463*, 29-89 (1977).
16. Jackl, G. and Sebald, W., Eur. J. Biochem. *54*, 97-106 (1975).
17. Tzagoloff, A. and Akai, A., J. Biol. Chem. *247*, 6517-6523 (1972).
18. Sierra, M. F. and Tzagoloff, A., Proc. Natl. Acad. Sci. U.S. *70*, 3155-3159 (1973).
19. Sebald, W. and Jackl, G., in Electrotransfer chains and oxidative phosphorylation, E. Quagliariello *et al.* (ed.) Norht Holland, 193-198 (1975).
20. Sebald, W., Biochim. Biophys. Acta *463*, 1-27 (1977).
21. Sebald, W. (This book).
22. Ebner, E. and Schatz, G., J. Biol. Chem. *248*, 5379-5384 (1973).
23. Tzagoloff, A. and Meagher, P., J. Biol. Chem. *247*, 594-603 (1972).
24. Michel, R. Liebl, A., Hartmann, A. and Neupert, W., Hoppe Seyler's

Z. Physiol. Chem. *357*, 415-426 (1976).

25. Delhez, J., Dufour, J.P., Thines, D. and Goffeau, A., Eur. J. Biochem. (in press) (1977)
26. Colson, A.M., Goffeau, A., Briquet, M., Weigel, P., Mattoon, J. R., Molec. Gen. Genet. *135*, 309-326 (1974).
27. Tzagoloff, A., Akai, A. and Foury, F., FEBS Lett. *65*, 391-395 (1976).
28. Rank, G. H., Gerlach, J. H., Robertson, A. J., Molec. Gen. Genet. *144*, 281-288 (1976).
29. Rank, G. H., Robertson, A.J., Phillips, K., Genetics *80*, 483-493 (1975).
30. Rank, G. H., Robertson, A. J., Phillips, K., J. Bacteriol. *122*, 359-366 (1975).
31. Rank, G. H., Robertson, A. J., Gerlach, S. H., Molec. Gen. Genet. *152*, 13-18 (1977).
32. Evans, I. H., Egilsson, V., Wilkie, D. in Genetics, Biogenesis and Bioenergetics of Mitochondria, W. Bandlow *et al.* (eds.) Walter de Gruyter, Berlin, pp. 125-136 (1975).
33. Linstead, D., Evans, I., Wilkie, D., in The Biogenesis of Mitochondria, A.M. Kroon and Saccone, C. (eds.), Acad. Press, New York, pp. 179-192 (1974).
34. Griffiths, D. E., Lancashire, W. E. and Zanders, E. D., FEBS Letters *53*, 126-130 (1975).
35. Carignani, G., Lancashire, W.E. and Griffiths, D. E., Molec. Gen. Genet. *151*, 49-56 (1977).
36. Guerineau, M., Slonimski, P. P. and Avner, P.R., Biochem. Biophys. Res. Commun. *61*, 462-469 (1974).
37. Dufour, J.P. and Goffeau, A., submitted for publication.
38. Guidotti, G., T.I.B.S., *1*, 11-13 (1976).
39. Neville, D.M. and Glossmann, H., Methods Enzymol. *32*, 92-102 (1974).
40. Jovin, T. K., Dante, M. L. and Chrambach, A. in Multiphasic buffer systems output, Federal Scientific and Technical Information. United States Department of Commerce, Springfield, Virginia (1971).

PARTIAL IDENTIFICATION, STOICHEIOMETRY AND SITE OF TRANSLATION OF THE SUBUNITS OF UBIQUINONE: CYTOCHROME C OXIDOREDUCTASE

Hanns Weiss and Barbara Ziganke
European Molecular Biology Laboratory
Postfach 10 22 09, 6900 Heidelberg, W.Germany

In previous papers we have shown that solubilisation of the mitochondrial membranes from Neurospora crassa with deoxycholate at high ionic strength and subsequent chromatographic separation of the resolved membrane protein leads to the isolation of a dimeric cytochrome b. This preparation has a molecular weight of 55 000 and consists of two subunits of equal size, each carrying one protohaem. By means of protein-labelling experiments, namely in vivo incorporation of radioactive leucine in the presence of inhibitors of either cytoplasmic protein synthesis or mitochondrial protein synthesis, the two cytochrome b subunits were shown to be products of the mitochondrial translation system (for a review see reference 1).

In the present paper we describe some recent experiments which were performed to show that this dimeric cytochrome b is a constituent of ubiquinone cytochrome c oxidoreductase, a multienzyme complex which catalyses the electron transfer from reduced ubiquinone to cytochrome c.

PURIFICATION AND SOME PROPERTIES OF UBIQUINONE: CYTOCHROME C OXIDOREDUCTASE

The purification procedure is described elsewhere (2). It involves the following steps: 1. Depletion of mitochondrial membranes from cytochrome c by sonication at high ionic strength. 2. Solubilisation of the mitochondrial membranes with Triton X-100 at low ionic strength. 3. Specific binding of the ubiquinone: cytochrome c oxidoreductase to ferricytochrome c from horse heart which has been coupled to Sepharose 4B. 4.Specific release of the multienzyme complex by converting the coupled ferricytochrome c to ferrocytochrome c, and 5. Concentration of the eluted multienzyme complex by ultrafiltration and removal of the excess Triton X-100 by centrifugation on a sucrose density gradient.

In the presence of Triton X-100 the isolated multienzyme complex is soluble in a monodisperse state and binds approximately 0.3g Triton X-100 per g protein. By analytical ultracentrifugation a sedimentation coefficient $S_{20,w}$ of 7.5 was determined for the enzyme-detergent complex. The preparation contains approximately 7 µmole/g protein cytochrome b and 3,5 µmole/g protein cytochrome c_1.

At room temperature the air-oxidized form of the multienzyme complex shows light absorption bands at 560 nm, 530 nm, and 415 nm, the dithionite-reduced form at 560 nm with a shoulder at 554 nm, at 530 nm, 524 nm and 430 nm with a shoulder at 415 nm. At liquid nitrogen temperature the absorption band of the reduced cytochromes b splits into two bands with maxima at 561.5 nm and 555.5 nm.

RESOLUTION AND PARTIAL IDENTIFICATION OF THE SUBUNITS OF UBIQUINONE: CYTOCHROME C OXIDOREDUCTASE

Dodecyl sulfate gel electrophoresis of the isolated multienzyme complex reveals 6 protein bands of the apparent molecular weight 8 000, 11 000, 14 000, 30 000, 45 000 and 50 000 (Fig.1). The 30 000 molecular weight protein band is further resolved by hydroxylapatite chromatography in the presence of dodecyl sulfate (Fig.2). Three protein peaks of about equal amount are obtained.

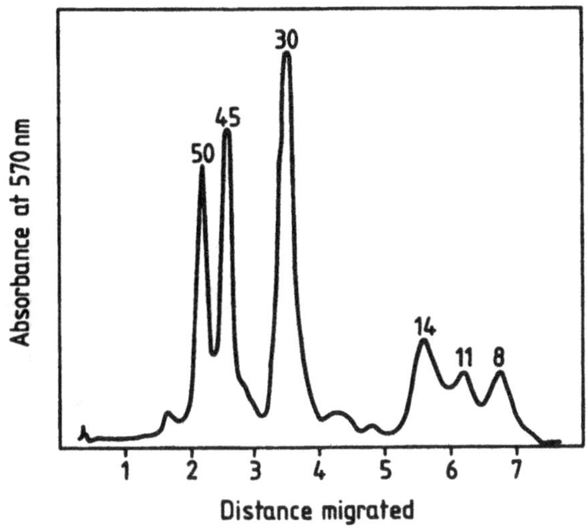

Fig.1. Densitometric tracing of Coomassie-blue-stained protein-bands obtained after dodecyl sulfate gel electrophoresis of ubiquinone cytochrome c oxidoreductase.

The first of these three peaks appears radioactively labelled when the complex is isolated from cells which have incorporated 5-amino-^{3}H- levulinic acid, a precurser of haem biosynthesis. This peak is therefore identified as the haem bearing polypeptide of cytochrome c_1, since only the c type cytochromes contain covalently bound haem. In contrast, the haem groups of cytochrome b, which being non-covalently bound to the apoprotein, are removed upon treatment with dodecylsulfate and mercaptoethanol.

The second and the third protein peaks are radioactively labelled, when the preparation is isolated from cells which have incorporated ^{14}C-leucine in the presence of cycloheximide, a specific inhibitor of cytoplasmic protein synthesis. These peaks, therefore, are identified as the two cytochrome b polypeptides, because these polypeptides are the only ones of the multienzyme complex which are translated in mitochondria (see below). When the same hydroxylapatite chromatography is performed with dimeric cytochrome b, only the latter two peaks are obtained (1).

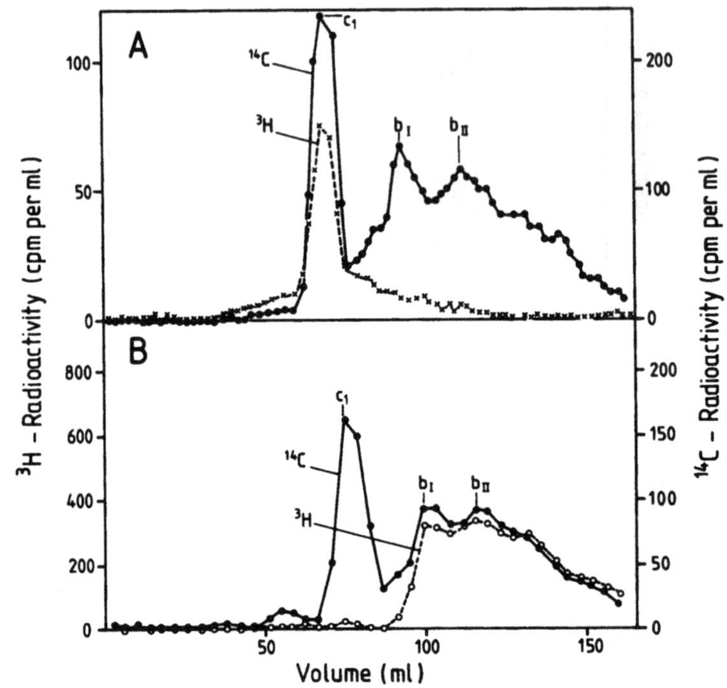

Fig.2. Dodecylsulfate hydroxylapatite chromatography of the 30 000 molecular weight polypeptides of ubiquinone: cytochrome c oxidoreductase. The complex was isolated from cells which had (A) simultaneously incorporated ^{14}C-leucine and 5-amino-^{3}H- levulinic acid and (B) incorporated ^{14}C-leucine in the absence of inhibitor and ^{3}H-leucine in the presence of cycloheximide.

ARE THE TWO CYTOCHROME B POLYPEPTIDES IDENTICAL POLYPEPTIDE CHAINS OR NOT?

The finding that the cytochrome b constituent of the multienzyme complex can be separated into two protein peaks of about equal amount suggested non-identity of the two cytochrome b polypeptides, however, no differences could be detected when comparing their protein chemical properties.

The amino acid composition of the two cytochrome b subunits is very similar. Differences are found essentially only in the lysine contents. Whether this difference, however, really exists or whether it is caused by lysine rich contaminations still present in the preparation cannot yet be decided (1).

Attempts to determine the N-terminal amino acids of the cytochrome b subunits by means of the dansyl chloride method have not given any result, nor could the protein be attacked by the solid-phase Edman degradation. This suggests that the N-terminus is blocked (1).

The C-terminal amino acids of the two cytochrome b subunits were determined by means of enzymic hydrolysis. No amino acid is released by the action of carboxypeptidase A alone. Carboxypeptidase B releases lysine from each of the two subunits. The combined action of carboxypeptidases A and B produces a release of lysine, tyrosine and leucine. These data suggest that the two subunits have the C-terminal sequence lys-tyr-leu.

Cytochrome b subunit I from ^{3}H-leucine labelled Neurospora crassa and subunit II from ^{14}C-leucine labelled Neurospora crassa were combined and cleaved with cyanogenbromide. After gelfiltration on Bio-Gel P 30 in the presence of 85% formic acid all the eluted peptides showed identical $^{3}H/^{14}C$ ratios.

Also after peptic digestion of the CNBr-peptides and subsequenting "fingerprinting" by two dimensional thin layer chromatography no significant differences in the $^{3}H/^{14}C$ ratio of the various peptide spots are found.

From these results it is suggested that the two polypeptide chains of cytochrome b have a very similar or identical primary structure.

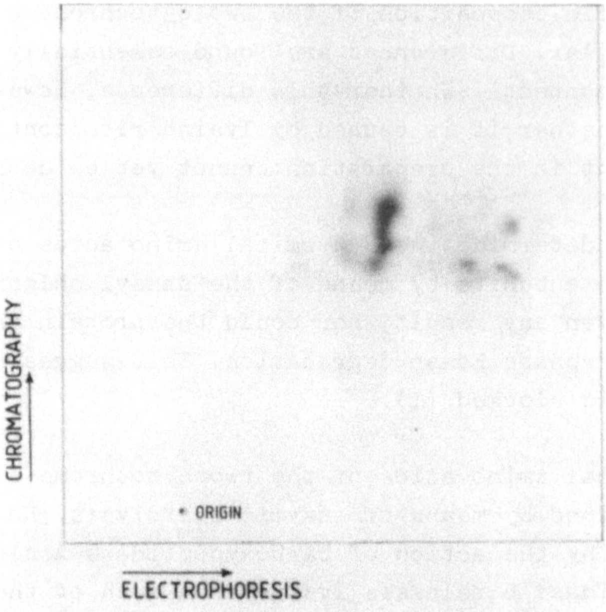

Fig.3. Peptide map of the cytochrome b subunits obtained after CNBr-cleavage and peptic digestion. (Photograph after ninhydrine staining). 120 µg cytochrome b in 100 µl 95% formic acid were incubated with 3 mg BrCN for 27 h at 25°C, lyophilised,dissolved in 100 µl 5% formic acid and with 2 µg pepsin incubated for 16h at 37°C. Thin layer chromatography was performed on microcristalline cellulose plates and developed in the first dimension with butanol/acetic acid/pyridine/water (15/3/10/12) (v/v/v/v) and in the second dimension with electrophoresis at pH 4.5.

STOICHEIOMETRY OF THE SUBUNITS OF UBIQUINONE: CYTOCHROME C OXIDOREDUCTASE

The relative amount of leucine present in each subunit, compared to the total leucine content of the multienzyme complex is obtained as follows: The leucyl-residues of Neurospora crassa protein are uniformly labelled by incubating exponentially growing hyphae in the presence of ^{3}H-leucine for 3h (3). After harvesting the cells the multienzyme complex is isolated, disaggregated with dodecyl sulfate and the subunits separated by either polyacrylamide gel electrophoresis or hydroxylapatite chromatography (2). The relative areas of the ^{3}H-radioactivity of the protein peaks obtained are assumed to be proportional to the leucine contents of the subunits (Fig.4, Table 1).

The relative amount of protein of each subunit to the amount of protein of the whole multienzyme complex is calculated by the formula:

$$\begin{matrix}\%\text{protein} \\ \text{per subunit}\end{matrix} = \frac{\begin{matrix}\%^{3}\text{H radioactivity} \\ \text{per subunit}\end{matrix} \times \begin{matrix}\mu\text{mole leucine} \\ \text{per g complex}\end{matrix}}{\mu\text{mole leucine per g subunit}}$$

The molar ratios of the subunits are determined by dividing the relative amount of protein of each subunit by its apparent molecular weight (Table 1). From the data obtained it appears that the ubiquinone: cytochrome c oxidoreductase contains one cytochrome c_1 and one each of the 8 000, 11 000, 14 000 and 50 000 molecular weight subunits, but two cytochromes b and two 45 000 molecular weight subunits.

By summing up the apparent molecular weights of the subunits a molecular weight of approximately 260 000 is obtained for the whole multienzyme complex. This value is in good agreement with the minimum molecular weight derived from the cytochrome c_1 content of the preparation. It is also consistent with the sedimentation coefficient $S_{20,W}= 7,5$ measured by analytical ultracentrifugation of the whole complex. Unfortunately molecular weight determination by sedimentation equilibrium measurement has not yet been possible, because under the experimental conditions namely 15°C for about 15h centrifugation time, the multienzyme complex dissociates.

Table 1: Molecular weights and molar ratios of the subunits of ubiquinone: cytochrome c oxidoreductase

Subunit band	Molecular weight	Relative amount of leucine	Leucine content	Relative amount of protein	Molar ratio	
		%of total	mmol/g	%of total		
1	50 000	17	0,70	19	0,38	0,9
2	45 000	35	0,67	41	0,92	2,1
3 (cytochrome c_1)	31 000	12	0,68	14	0,44(≡)	1,0
4 (cytochrome b)	27 000	24	1,01	19	0,72	1,7
5	14 000	7	0,92	6	0,43	1,0
6	11 000	4	0,70	4,5	0,45	1,0
7	8 000	2	0,66	3	0,38	0,9

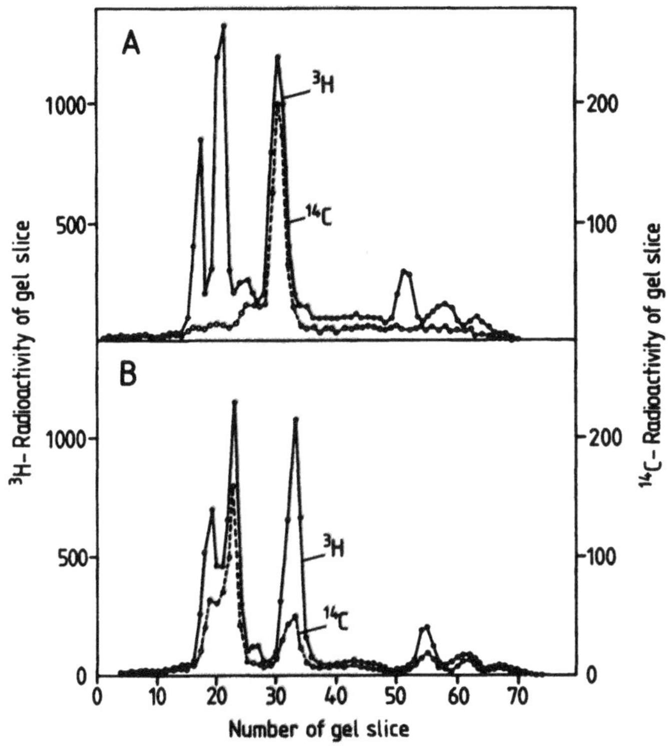

Fig.4. Dodecyl sulfate gel electrophoresis of ubiquinone: cytochrome c oxidoreductase from cells which have incorporated ^{3}H-leucine in the absence of inhibitors and ^{14}C-leucine in the presence of (A) cycloheximide and (B) chloramphenicol. For experimental detail see (1) and (2).

SITE OF TRANSLATION OF THE SUBUNITS OF UBIQUINONE: CYTOCHROME C OXIDOREDUCTASE

In order to define the translation site of the subunits of the ubiquinone cytochrome c oxidoreductase, the multienzyme complex was isolated from cells which had been labelled with ^{3}H-leucine in the absence of inhibitors, followed by labelling with ^{14}C-leucine, in the presence of either cycloheximide or chloramphenicol (1,2). The multienzyme complex was subsequently separated into subunits by gel electrophoresis or hydroxylapatite chromatography both in the presence of dodecyl sulfate (Fig. 2,4).

In cells treated with cycloheximide only the cytochrome b subunits contain significant ^{14}C-radioactivity (Fig.4A and 2B), whereas in cells treated with chloramphenicol a pronounced ^{14}C-labelling of all subunits, except of the cytochrome b subunits, takes place (Fig. 4B; hydroxylapatite chromatography of the 30 000 molecular weight band is not shown). This complementary effect of cycloheximide and chloramphenicol on the labelling of the single subunits suggests that the cytochrome b subunits are the only ones of the multienzyme complex, which are translated on mitochondrial ribosomes, whilst the other subunits are synthesized in the cytoplasm.

REFERENCES

1. Weiss, H.: Subunit Composition and Biogenesis of Mitochondrial Cytochrome c. Biochim.Biophys.Acta 456, 291-313 (1976)

2. Weiss, H., Juchs, B. and Ziganke, B.: Cytochrome bc_1-complex from Mitochondria of Neurospora crassa. Methods in Enzymology; Biomembranes, Editors:S.Fleischer and L.Packer (C.VI.A.4) in press

3. Sebald, W., Machleidt, W. and Otto, J.: Products of Mitochondrial Protein Synthesis in Neurospora crassa. Eur.J.Biochem. 38, 311-324 (1973)

Isolation and Hydrodynamic Properties of the bc_1 Complex from Beef Heart prepared by Hydroxyapatite Chromatography in Triton X-100

G. von Jagow and H.J. Kolb[+]
Institut für Physiologische Chemie und Physikalische Biochemie, Goethestr. 33, 8 München 2
[+]Institut für Diabetesforschung, Kölner Platz 1, 8 München 40

SUMMARY

The bc_1 complex from beef heart prepared by hydroxyapatite chromatography in Triton X-100 consists of six polypeptide subunits: core protein-1, core protein-2, cytochrome c_1, the two b-type cytochromes and the antimycin binding protein.

Hydroxyapatite chromatography in Triton X-100 separates not only the bilayer phospholipids, but also the boundary phospholipids from the complex. Ubiquinone is carried with the phospholipid fraction.

The two b-type cytochromes and cytochrome c_1 seem to be present in the isolated complex in their native state. They exhibit characteristic absorbance spectra, and have midpoint potentials of -10 mV, +80 mV and +250 mV, respectively.

The isolated complex has no $QH_2 \rightarrow$cytochrome c reductase activity.

The molecular weight of the complex isolated in Triton X-100 as determined by sedimentation equilibrium runs is 400,000 - 440,000 after correction for the bound detergent. The minimal molecular weight as calculated from the heme contents is about 220,000. It is therefore concluded that the complex is isolated as a dimer. The dimer carries 4 heme b and 2 heme c (cytochrome c_1) centers.

The bc_1 dimer binds 0.2 g Triton X-100/g protein. This corresponds to one Triton micelle, or about 140 molecules of Triton X-100, per bc_1 dimer.

INTRODUCTION

The cytochrome bc_1 segment of the respiratory chain includes the second phosphorylation site. Its isolation in the form of complex III, also called the "bc_1 complex", is therefore of special interest for the study of the mechanism of oxidative phosphorylation.

So far, the complex has only been isolated by salt fractionation in the ionic detergents cholate, desoxycholate according to the procedure of Hatefi and Rieske (1). The method described here consists of the isolation of the bc_1 complex from beef heart mitochondria by hydroxyapatite chromatography in the nonionic detergent Triton X-100 (2,3,4).

The pure and monodisperse preparation obtained will be used for the study of the quaternary structure of the bc_1 complex, for the isolation of the polypeptide subunits from the complex, for the characterization of the antimycin binding site, and finally for reactivation, reconstitution studies.

RESULTS

Isolation procedure and polypeptide composition of the complex

Fig.1 gives a flow chart for the isolation of the complex. The procedure consists of three steps which are given on the left of the scheme. A prelabelling of the mitochondria with carboxyatractylate was performed to facilitate the separation of the ADP,ATP carrier which under these conditions is not absorbed to the hydroxyapatite (5). In order to stabilize the native conformation of the b-type cytochromes the complex was loaded with antimycin (6).

On the right of the scheme the stepwise separation of the contaminating proteins following the application of the various buffers is indicated.

The preextraction of the mitochondria by a low Triton and salt concentration enables a separation of part of the matrix proteins. In the second step the mitochondrial membrane proteins are almost quantitatively solubilized by application of a relatively high Triton and salt concentration. The first

FLOW CHART FOR THE ISOLATION OF THE bc_1 COMPLEX FROM BEEF HEART BY HYDROXYAPATITE CHROMATOGRAPHY IN TRITON X-100

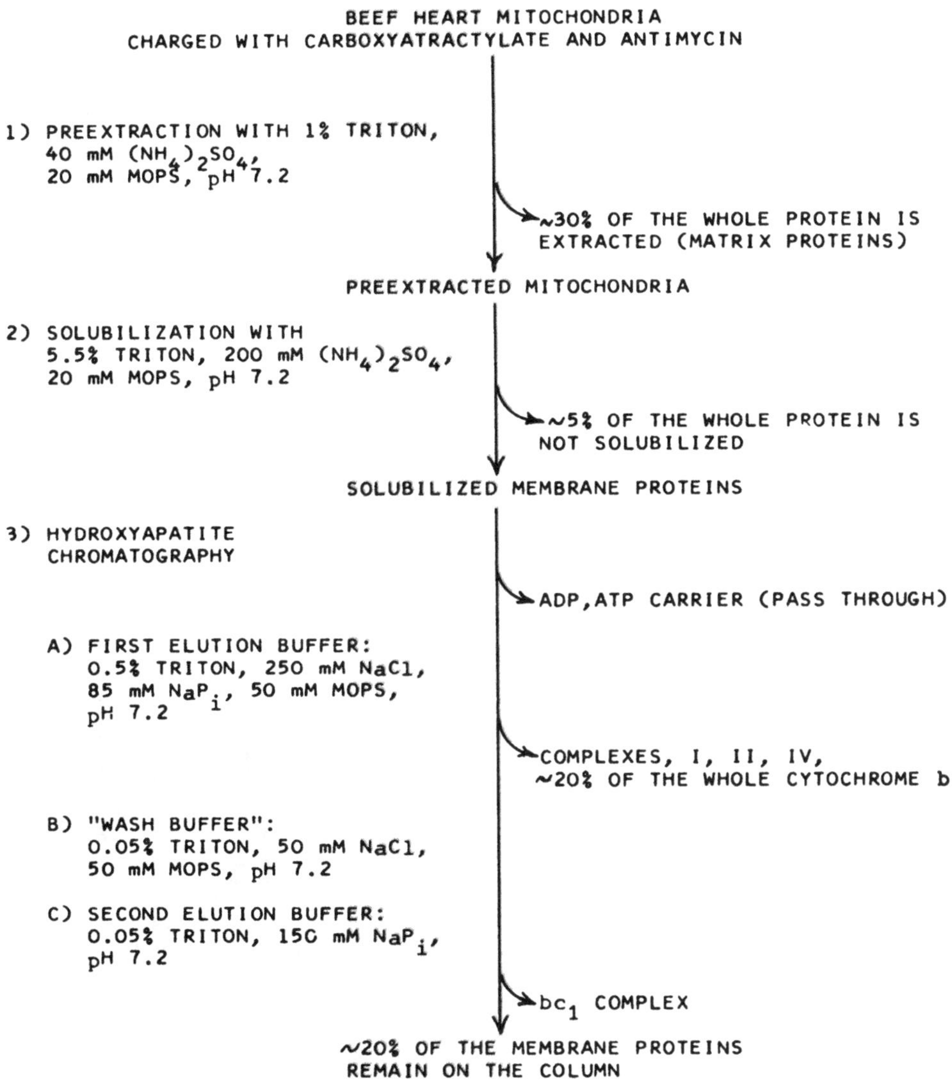

Fig. 1

and second step include a high speed centrifugation (cf.3).

Then, the solubilized proteins are applied on a hydroxyapatite column. It seems to be remarkable that it is possible to obtain a pure bc_1 complex after a simple stepwise elution with three buffers. The ADP,ATP carrier comes from the column already in the pass through. The first elution buffer elutes about 70% of the applied protein including complexes I, II, IV and about 20% of the whole cytochrome b. Subsequently the so called "wash" buffer decreases the detergent and salt concentrations. Finally, the complex can be eluted by a buffer containing either 150 mM inorganic phosphate or 12 mM sodium citrate.

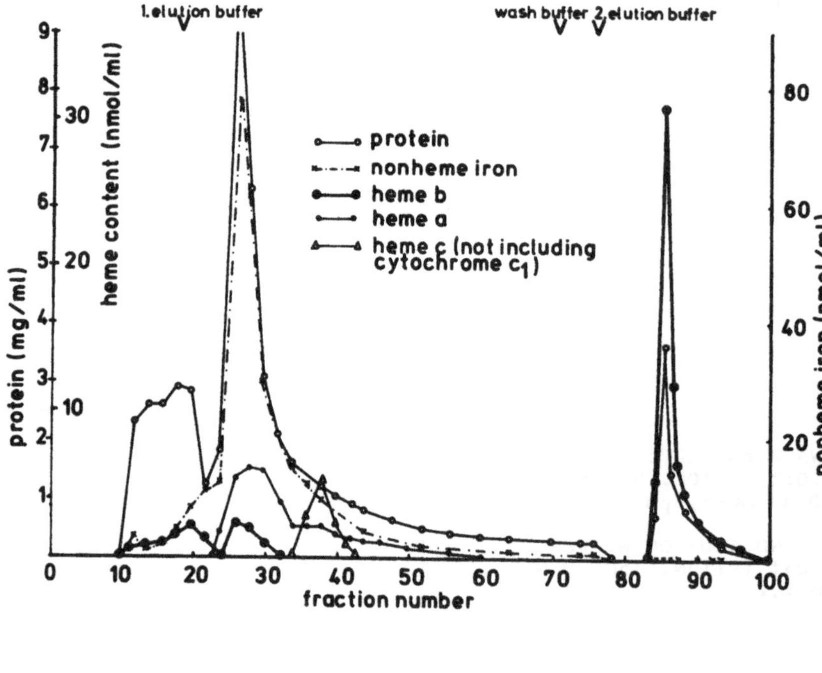

Fig. 2 Eluate profile of mitochondrial membrane proteins solubilized by Triton X-100 after hydroxyapatite chromatography. The arrows indicate the buffer changes.

In the eluate pattern (fig. 2) the heme and nonheme contents of the various fractions are given as nmol per ml, and the protein contents as mg protein per ml. In the pass through whose position is indicated by the first protein peak are also the phospholipids and the ubiquinone. Nearly the whole nonheme iron is found within the second protein peak. This may be explained by the high iron sulfur protein contents of the NADH and succinate dehydrogenase which mainly contribute to this protein fraction. The cytochrome b in this peak may be the b-type cytochrome of the succinate dehydrogenase (7). The nonheme iron curve shows clearly that the isolated complex has completely lost its Rieske iron sulfur protein (8).

Table 1 gives the enrichment and yield of the complex during the various preparation steps as indicated by the heme b and protein contents. The described preparation was done on a large scale with a hydroxyapatite column of 300 ml content. The first two steps gave only an enrichment of the specific heme b content by a factor of 1.4. But the third step brought the complex to complete homogeneity. The method has the advantage that it is relatively simple and that it gives a high yield.

Fig. 3 demonstrates the polypeptide composition of the isolated complex. It consists of six subunits: the two core proteins, cytochrome c_1, the two b-type cytochromes and the 11,000 dalton polypeptide which was characterized by Rieske (9) as the protein binding the inhibitor antimycin.

The 31,000 dalton band should be larger as it represents three subunits. Since no protein remains on the origin of the gel the reason for the low yield may be a low staining index for cytochromes b.

Table 1 ENRICHMENT AND YIELD OF THE bc_1 COMPLEX DURING PREPARATION

FRACTION	TOTAL PROTEIN (mg)	TOTAL HEME b (nmol)	HEME b PER PROTEIN (nmol/mg)	PURIFICATION (FOLD)	YIELD (%)
MITOCHONDRIA	5000	2500	0.5	1	100
PREEXTRACTED MITOCHONDRIAL MEMBRANES	3500	2400	0.68	1.4	96
SOLUBILIZED MEMBRANE PROTEINS	3300	2240	0.68	1.4	90
bc_1 COMPLEX	160	1375	8.6	17	55

THE LARGE SCALE PREPARATION WAS PERFORMED ON A COLUMN PACKED WITH 350 ML HYDROXYAPATITE.

EXTRACTION AND CHROMATOGRAPHY WERE PERFORMED AS DESCRIBED ON THE FLOW CHART AND ON THE ELUATE PROFILE.

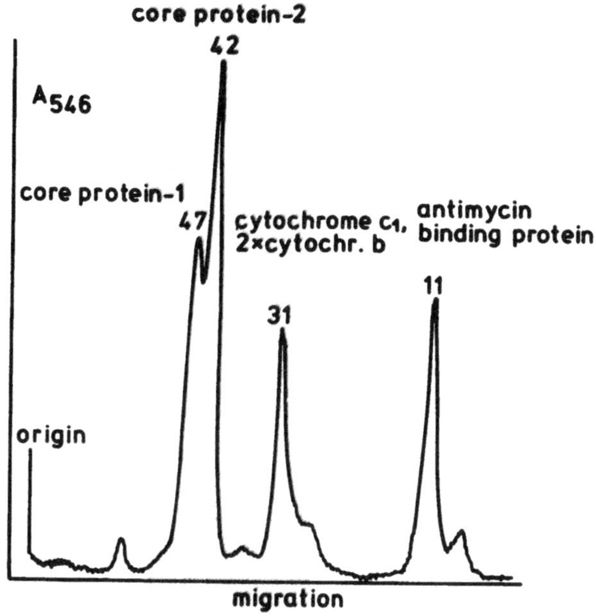

Fig. 3 Polypeptide pattern of the bc_1 complex. Dodecyl sulffate polyacrylamide gel electrophoresis and densitometry were performed as described in (3). The tentative assignment of the subunits to the various polypeptide bands of the densitogram is indicated. The numbers show the molecular weight in k dalton.

State of the isolated complex

The native state of the complex cannot be studied by its enzymatic activity, since it has lost its ubiquinol cytochrome c reductase. This may be due to the loss of the nonheme iron protein, or the depletion of phospholipids. Therefore it is only possible to test the state of the complex by studying the heme groups of the respective subunits. Native cytochromes b and c_1 subunits should exhibit characteristic absorbance spectra and midpoint potentials.

Fig. 4 shows in A) a room temperature spectrum in which the two b-type cytochromes and cytochrome c_1 of the complex are reduced simultaneously. Spectrum B) reflects the same redox state, but now recorded at -196°C. In these spectra

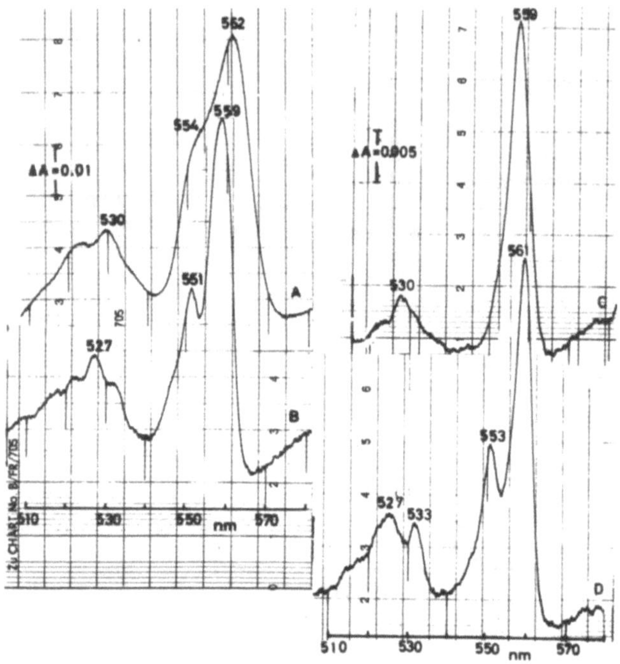

Fig. 4 Difference absorbance spectra of the bc_1 complex and of cytochromes b_K and b_T of the isolated complex

the b-type cytochromes have at room temperature a maximum at 562 nm and cytochrome c_1 a shoulder at 554 nm. At -196°C the absorbance bands become sharper, the maxima shift each by 3 nm to the blue, but the bands still overlap.

By adjustment of appropriate redox potentials it is possible to reduce the b-type cytochromes separately (cf. 3). In fig. 4C) the spectrum of cytochrome b_K and in fig. 4D) that of cytochrome b_T are depicted at low temperature. The first one has a typical symmetrical absorbance band with a maximum at 559 nm, the latter one a double banded spectrum with maxima at 561 and 553 nm.

The midpoint potentials of the b-type cytochromes were measured by aid of a redox titration according to Wilson and Dutton (10) (fig. 5). The titration was performed at 562-575

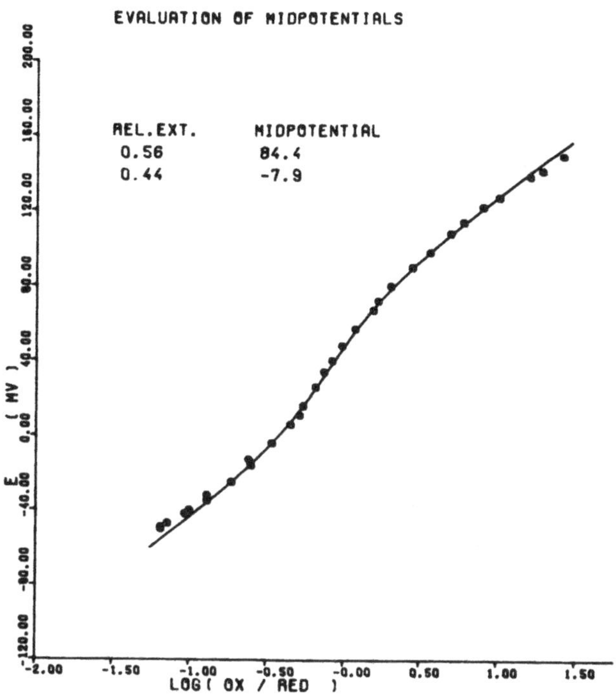

Fig. 5 The redox titration curve of the b-type cytochromes of the bc_1 complex. The redox titration was performed and analyzed as described in (3)

nm. The obtained curve was analyzed with a computer program. The analysis shows that both b-type cytochromes are present in about the same amount. The midpoint potentials for cytochromes b_T and b_K are -10 mV and +80 mV, respectively. The measured redox potentials deviate only by about 30 mV in positive direction from the native b-type cytochromes embedded in the mitochondrial membrane.

In summary, it may be stated that the typical absorbance spectra and midpoint potentials indicate that at least the heme carrying subunits of the isolated complex are present in a native state.

Amino acid composition and Triton binding of the complex

The amino acid analysis of the isolated complex was necessary for the calculation of the partial specific volume of the protein. Moreover, the polarity and the frequency of the most abundant and rarest amino acids were of interest. Table 2 gives this analysis in which also the number of residues is approximated assuming a molecular weight of 200,000 for the monomer. The rarest amino acid is cysteine. 12 of the 13 cysteinyl residues must be accounted for the two b-type cytochromes and cytochrome c_1 (cf. 4), so that one might expect that the two core proteins and the antimycin binding protein are relatively devoid of this amino acid.

The polarity of the whole complex is relatively high for a membrane protein. It amounts to 45%. The partial specific volume evaluated according to Cohn and Edsall (11) is 0.737 ml/g.

The Triton binding of the complex was studied in a preparation in which the complex was solubilized and kept dispersed by tritiated Triton X-100.

Fig. 6 shows on the left the eluate profile of such a low scale preparation, on the right, just the pattern of the complex in an enlarged scale. A column with 4 ml hydroxyapatite was used. The eluate profile can be compared with that of the large scale preparation (cf. fig. 2). Additionally in fig. 6 the drastically varying concentration of Triton can be followed by the curve of the (^{3}H)-Triton radioactivity. It reflects exactly the different detergent concentrations used in the various buffers. The solubilization buffer contained 5.0% Triton, the first elution buffer 0.5% and finally the "wash" and second elution buffers only 0.05% Triton. Different to the large scale preparation, where the complex was eluted by 150 mM sodium phosphate, in the low scale preparation the complex was eluted by 12 mM sodium citrate.

The pattern of the eluted complex on the right of fig. 6 in which are only the protein and detergent concentrations

Table 2 Amino acid composition of the bc_1 complex of beef heart prepared in Triton X-100 by hydroxyapatite chromatography devoid of the iron sulfur protein

AMINO ACID	MOL %	RESIDUES	
nonpolar			
ALANINE	9.50	171	(171.0)
VALINE	5.32	96	(95.8)
LEUCINE	9.93	179	(178.7)
ISOLEUCINE	3.65	66	(65.7)
PROLINE	8.52	153	(153.4)
METHIONINE	1.47	27	(26.5)
PHENYLALANINE	4.10	74	(73.8)
TRYPTOPHAN	1.00	18	(18.0)
GLYCINE	6.90	124	(124.2)
CYSTEINE	0.71	13	(12.8)
TYROSINE	3.77	68	(67.9)
polar			
ASPARTATE	8.40	151	(151.2)
GLUTAMATE	10.23	184	(184.1)
LYSINE	6.85	123	(123.3)
ARGININE	5.17	93	(93.0)
HISTIDINE	2.87	52	(51.7)
SERINE	6.26	113	(112.7)
THREONINE	5.36	97	(96.5)
polarity	45.14	-	-
TOTAL RESIDUES		1802	-

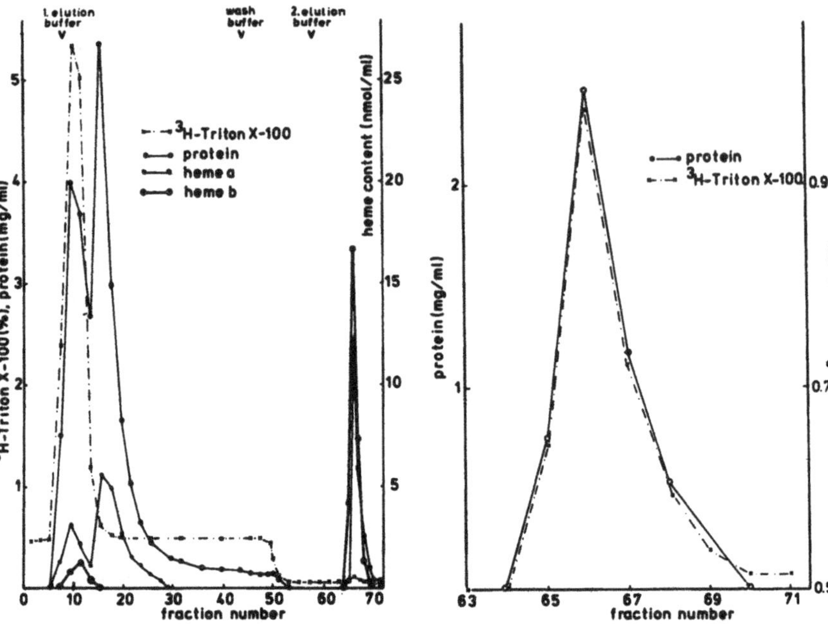

Fig. 6 Eluate profile of a low scale preparation of the bc_1 complex in (^{3}H)-Triton X-100. Triton X-100 concentration was determined by aid of the specific radioactivity of the (^{3}H)-Triton X-100. The arrows indicate the buffer changes.

are plotted, shows a constant ratio of bound Triton per bc_1 complex. A Triton binding of 0.2g/g protein was calculated from 5 chromatographic runs.

Molecular weight and state of aggregation of the complex

Fig. 7 shows a plot of a sedimentation velocity run, recorded on the left at 415 nm and on the right at 280 nm. In the former case, the complex was dispersed in 0.05% Triton, in the second, in 0.025% Triton, in order to lower the Triton absorbance. Both runs give straight lines in the plot of log r vs. time. From both curves a value of $S_{20,W}$ = 13.0-13.1 is calculated. Although sedimentation velocity runs are not as convincing as sedimentation equilibrium runs, the linearity of these curves indicates that the preparation is essentially homogenous.

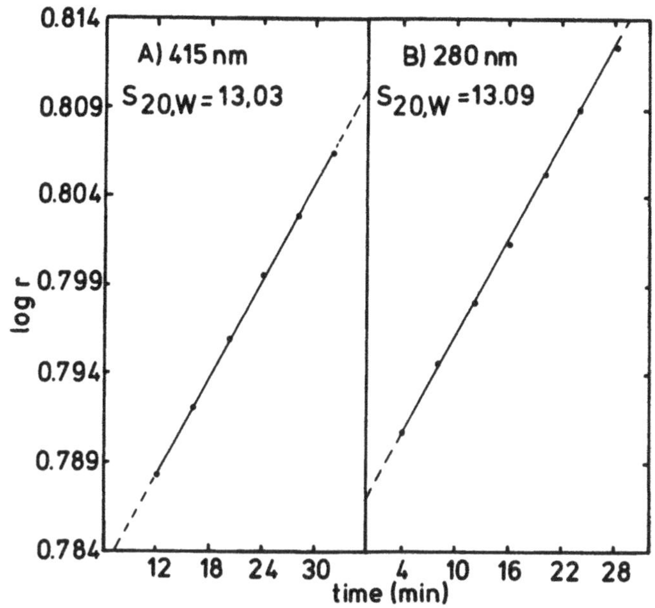

Fig. 7 Sedimentation velocity runs of the bc_1 complex in Triton X-100. The logarithm of the distance of the gradient from the center of rotation is plotted versus the time. Rotorspeed: 48,000 rpm, protein concentration 0.5 mg/ml in 20 mM MOPS, pH 7.5, temperature 21.7 °C.

Fig. 8 gives curves log c vs. r^2 of sedimentation equilibrium runs at two wavelength, respectively. The runs were performed with detergent concentrations as used in the sedimentation velocity runs. The analyses of the curves obtained at 415 nm after correction for bound Triton give a molecular weight of 400,000. The measuring points of each run could be fitted with only minor deviations by a straight line.

The measurements performed at 280 nm do not give a straight line over the whole range. In the region near the bottom of the cell the points deviate slightly from the ideal values. The relatively low detergent concentration (0.025% Triton X-100), used in these runs for optical reasons, seems to be at the lower limit of detergent concentration required for the dispersion of the complex. In this case the system appears not to be in a completely homogenous state because of a small

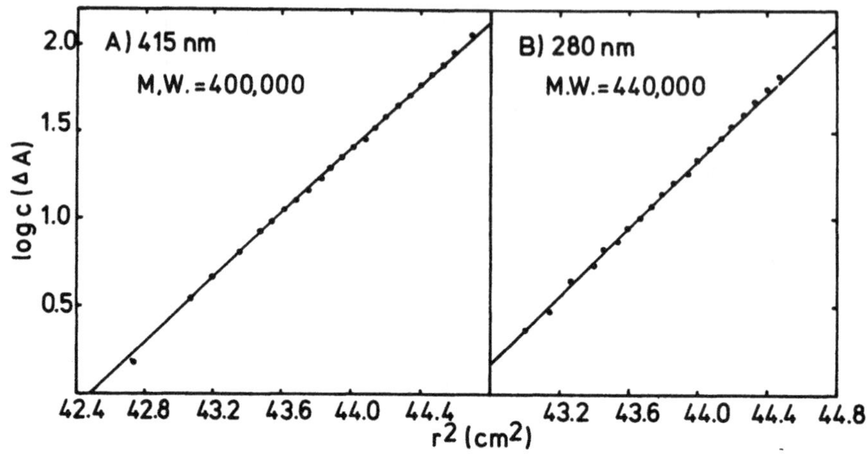

Fig. 8 Sedimentation equilibrium runs of the bc_1 complex in Triton X-100. The logarithm of recorder deflection in mm (proportional to the absorption) is plotted versus the square of the distance to the center of rotation, r^2 (cm^2). The rotor speed was 9000 rpm, the protein conc. about 0.2 mg/ml. After 20 hrs of centrifugation the equilibrium was reached.

amount of aggregation. It may be for this reason that the molecular weight determined at 280 nm of 440,000 differs slightly from that calculated from the 415 nm runs. None the less, the straight lines that were obtained in the sedimentation equilibrium runs again indicate the high homogeneity of the preparation.

OUTLOOK

To date, three protein complexes of the mitochondrial membrane have been isolated in Triton X-100 in a dimeric aggregation state: i) the ADP,ATP carrier by Klingenberg et al. (3), ii) the cytochrome oxidase (which is complex IV) by Stotz et al. (11b) and Capaldi et al. (12), and iii) the bc_1 complex (which is complex III) by v. Jagow et al. (2,3,4).

In the first case there is an indication that the dimer may be the native aggregation state of this membrane protein. It was shown by Klingenberg et al. (5) that the ADP,ATP carrier possesses only one substrate or inhibitor binding site per two subunits. Further studies are necessary to determine whether complexes III and IV are also embedded in the mitochondrial membrane as dimers.

ACKNOWLEDGEMENT

The authors are indebted to Prof. M. Klingenberg for generous support and stimulating discussion of this work. This work was supported by a grant from the Deutsche Forschungsgemeinschaft.

1. Hatefi, Y., Haavik, A.G. and Griffith, D.E. (1962) J. Biol. Chem. 237, 1681-1685

2. v.Jagow, G., Engel, W.D., Riccio, P. and Schägger, H. (1976) in Genetics and Biogenesis of Chloroplasts and Mitochondria, Th. Bücher et al., eds. Elsevier, Amsterdam, pp 267-272

3. Riccio, P., Schägger, H., Engel, W.D. and v.Jagow, G. (1977) Biochim. Biophys. Acta 459, 250-262

4. v.Jagow, G., Schägger, H., Riccio, P., Kolb, H. and Klingenberg, M. (1977) Biochim. Biophys. Acta in press

5. Riccio, P., Aquila, H. and Klingenberg, M. (1975) FEBS lett. 56, 133-138

6. Rieske, J.S. (1976) Biochim. Biophys. Acta 456, 195-246

7. Davis, K.A., Hatefi, Y., Poff, K.L. and Butler, W.L. (1973) Biochim. Biophys. Acta 325, 341-356

8. Rieske, J.S., Mac Lennan, D.H. and Coleman, R. (1964) Biochem. Biophys. Res. Commun. 15, 338-344

9. Das Gupta, U. and Rieske, J.S. (1973) Biochem. Biophys. Res. Commun. 54, 1247-1254

1o. Wilson, D.F. and Dutton, P.L. (1970) Biochem. Biophys. Res. Commun. 39, 59-64

11. Cohn, E.J. and Edsall, J.T. (1943) in Proteins, Amino Acids and Peptides, pp 370-375, Reinhold, New York

11b Love, B., Chan, S.H.P. and Stotz, E. (1970) J. Biol. Chem. 245, 6664-6668

12) Robinson, N.C. and Capaldi, R.A. (1977) Biochem. 16, 375-386

A CELL FREE SYSTEM TO STUDY SYNTHESIS AND TRANSPORT OF CYTOPLASMICALLY TRANSLATED MITOCHONDRIAL PROTEINS

R.ZIMMERMANN, H.KORB AND W. NEUPERT
Institut für Physiologische Chemie, Physikalische Biochemie und Zellbiologie der Universität München, Germany

INTRODUCTION

Previous papers have described synthesis of cytoplasmically synthesized mitochondrial proteins and their transport into the mitochondria of whole cells of *Neurospora crassa* (1,2). These *in vivo* studies have suggested the existence of extra-mitochondrial pools of mitochondrial proteins. However, such studies with intact cells can supply only limited information on the mechanism of the transport of proteins into functional locations. Therefore, an *in vitro* system has been developed to separate the processes of synthesis and transport (3,4). This report deals with further experiments to discriminate between the mechanisms currently proposed for the transport of proteins into mitochondria (2,4-6). The experiments support the earlier view that precursors of mitochondrial proteins are located in the cytosolic fraction. In addition, an attempt is made to characterize these precursors.

RESULTS AND DISCUSSION

1) Synthesis and Transport of Mitochondrial Proteins in the Cell Free System

Neurospora hyphae (wild type 74 *A*) were grown in the presence of ^{35}S-sulphate. After 12 hr of growth a chase of unlabelled sulphate was added for one hr. The cells were then washed with ice-cold distilled water and collected by filtration. The hyphae were broken by grinding with sand in an incubation mixture described previously (7) but supplemented with an ATP regenerating system and chloramphenicol (0.5 mg/ml). The homogenate was centrifuged twice at 3000 x g for 5 min. The cell free supernatant thus obtained was used for *in vitro* protein synthesis as described in a previous paper (4).

Electronmicroscopic analysis of this homogenate did not show a significant number of cytoplasmic ribosomes bound to the outer mitochondrial membrane.

Incorporation of ^{3}H-leucine into the ^{35}S-prelabelled homogenate reached a plateau within 10-15 min. The individual cellular fractions showed the labelling kinetics given in Fig.1.

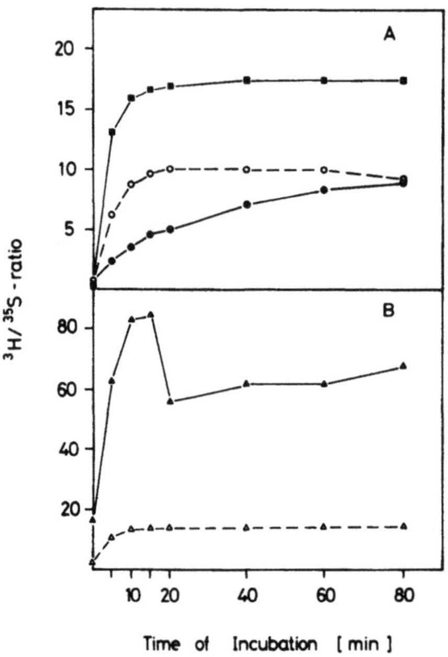

Fig.1 Incorporation of ^{3}H-leucine in a cell free system : specific labelling of the proteins of various cell fractions

A cell free homogenate of ^{35}S-labelled *Neurospora* hyphae was incubated with ^{3}H-leucine in the presence of chloramphenicol. Cell fractions were isolated by differential centrifugation. Specific ^{3}H-radioactivities are expressed by the $^{3}H/^{35}S$-ratio and plotted vs. time of incubation.

A ■———■ Homogenate
○— —○ Cytosol
●———● Mitochondria
B ▲———▲ Microsomes
△— —△ Ribosomes

The microsomal fraction showed very rapid labelling, reaching exceptionally high $^{3}H/^{35}S$-ratios at 1o-15 min, which exceeded the final $^{3}H/^{35}S$-ratio of the total homogenate by a factor of 2-6. The free ribosomes showed a lower increase of the $^{3}H/^{35}S$-ratio. Therefore, it appears that in this cell free homogenate membrane bound ribosomes are more active in the synthesis of proteins than are free ribosomes. The mitochondrial fraction also shows an increase in the $^{3}H/^{35}S$-ratio but this increase is distinctly slower than in the cytosolic fraction. However, the ratio in mitochondria does eventually reach a similar $^{3}H/^{35}S$-ratio. This incorporation into mitochondrial proteins cannot be attributed to mitochondrial protein synthesis since chloramphenicol was present in the reaction medium. Therefore, it appears that proteins synthesized on cytoplasmic ribosomes are transferred into the mitochondrial fraction.

In order to analyze the transferred proteins, gel electrophoretic separation in the presence of SDS was performed on the total mitochondrial protein. Mitochondria were isolated after 10 min and 80 min of incubation. The electrophoretic patterns are presented in Fig.2.

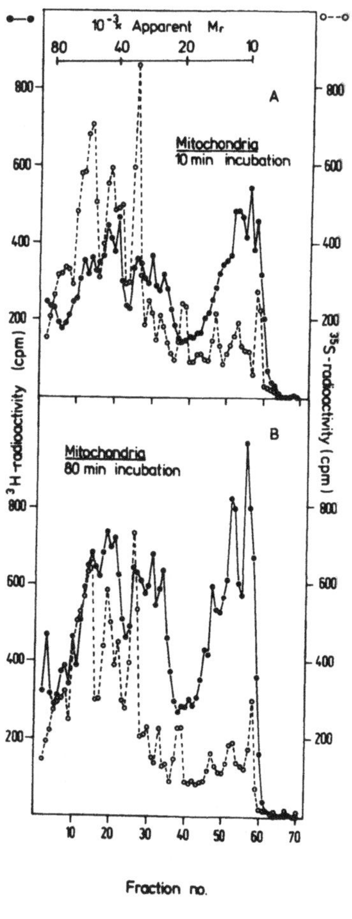

Fig.2 Gel electrophoretic analysis in the presence of SDS of ^{3}H-labelled proteins in the mitochondrial fraction

A cell free homogenate was prepared from ^{35}S-labelled hyphae and incubated with ^{3}H-leucine in the presence of chloramphenicol. Mitochondria were isolated after 10 min (A) and 80 min (B) of incubation, dissolved in SDS-containing buffer, dialysed and subjected to gel electrophoresis.

The ^3H- and ^{35}S-radioactivity patterns are not identical, although there is considerable similarity at least after 80 min of incubation. Harmey *et al.* have demonstrated that the *in vitro* synthesized proteins which are associated with the mitochondria do not represent unspecifically attached cytosolic proteins (4). Furthermore, the transferred proteins were characterized as specific mitochondrial proteins by immunoprecipitations from isolated mitochondria using antibodies directed against certain mitochondrial proteins : 1) CAT-binding protein (Fig.3) ++

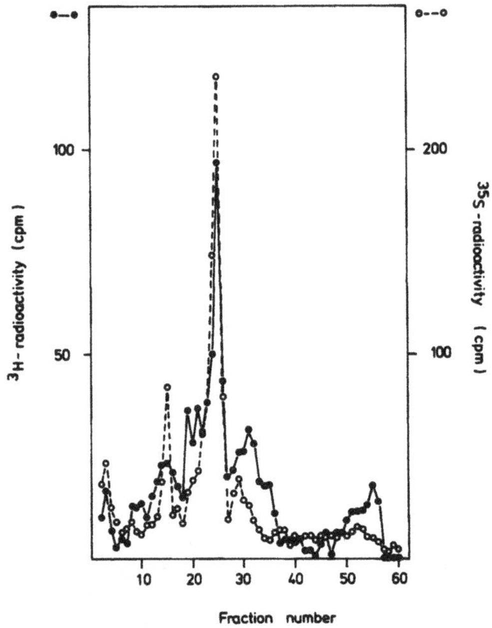

Fig.3 SDS gel electrophoresis of CAT-binding protein immunoprecipitated from the mitochondrial fraction of a cell free homogenate

A ^{35}S-prelabelled homogenate was incubated with ^3H-leucine in the presence of chloramphenicol for 20 min.

++*The electrophoretogram shows a major peak with an apparent molecular weight of 32,000. A distinct peak with an apparent molecular weight of 60,000 probably represents the dimeric form. Furthermore, in the molecular weight range of 10,000-30,000 a considerable amount of radioactivity is present. The amount of radioactivity in this region increases with the time of the in vitro incubation period. For this reason, and because these lower molecular weight proteins are not observed when the immuniprecipitation is carried out on mitochondria that have not been incubated in vitro, it seems likely that the radioactivity in this region represents proteolytic degredation products of the CAT-binding protein. Exceptional lability of the CAT-binding protein has also been observed with the beaf-heart protein (8).*

2) cytochrome *c* (see Fig.6) and 3) cytochrome *c* oxidase (Fig.4) In the latter case chloramphenicol was omitted during the incubation to obtain synthesis and labelling of mitochondrially synthesized subunits. In all the above cases the appearance of defined polypeptides in the mitochondria is apparent.

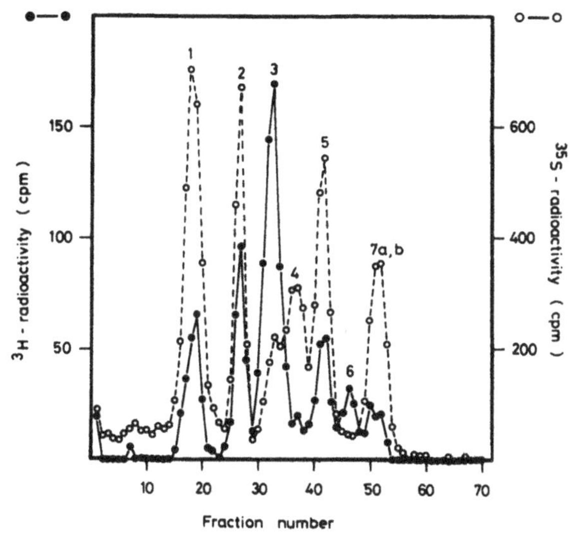

Fig.4 SDS gel electrophoresis of cytochrome *c* oxidase immunoprecipitated from the mitochondrial fraction of a cell free homogenate

A ^{35}S-prelabelled homogenate was incubated with ^{3}H-leucine. In this experiment no chloramphenicol was added. The immunoprecipitation was carried out with antibody directed against holo-cytochrome *c* oxidase.

2) Effect of Cycloheximide and Puromycin on Transfer of Proteins into the Mitochondria

A ^{35}S-prelabelled homogenate was incubated with ^{3}H-leucine for 10 min in the presence of chloramphenicol as described above. Cycloheximide and puromycin were then added to aliquots of the homogenate (final concentrations: 0.36 mM and 0.5 mM)

and incubation was continued.

The $^3H/^{35}S$-ratios of the homogenate and the various cell fractions were determined after the intervals shown in Fig.5.

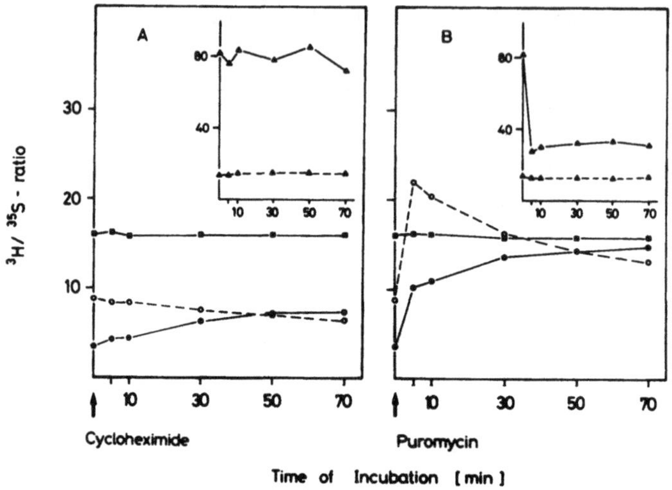

Fig.5 The effect of cycloheximide and puromycin on the transport of mitochondrial proteins into the mitochondria of a cell free homogenate

^3H-leucine was incorporated for 10 min in a cell free homogenate prelabelled with ^{35}S. The homogenate was divided into two equal portions and cycloheximide and puromycin were added (zero time) and incubation was continued. The $^3H/^{35}S$-ratios of various fractions are plotted vs.time after addition of cycloheximide (A) and puromycin (B)

■——■ Homogenate
○— —○ Cytosol ●——● Mitochondria
▲——▲ Microsomes △— —△ Ribosomes

The inhibitory effect of cycloheximide and puromycin on cytoplasmic protein synthesis is complete since the $^3H/^{35}S$-ratios of the total homogenates show no increase. When cycloheximide was present during the incubation (Fig.5A) the microsomal and ribosomal fractions showed no significant change in the $^3H/^{35}S$

ratio, whereas that of the cytosolic fraction declined slowly. When puromycin was added to an incubation mixture (Fig.5B) the $^{3}H/^{35}S$-ratio in the microsomal fraction dropped sharply but the ratio in free ribosomes was not affected. This rapid release of ^{3}H-label from the microsomes was accompanied by a corresponding increase of the $^{3}H/^{35}S$-ratio in the cytosolic fraction followed by a slow decline of this ratio.

These observations suggest that the release of polypeptides into the cytosolic fraction occurs mainly from the microsomal ribosomes. In the case of both, cycloheximide and puromycin inhibition, the mitochondrial fraction showed a significant increase in the $^{3}H/^{35}S$-ratio suggesting that transport of cytoplasmically synthesized mitochondrial proteins takes place after blocking cytoplasmic protein synthesis. The decline of the $^{3}H/^{35}S$-ratio of the cytosolic fraction indicates that the proteins are transferred into the mitochondria from that fraction. This conclusion is dependent upon the observation that the $^{3}H/^{35}S$-ratio of the total homogenate did not decrease.

In order to demonstrate that completed mitochondrial proteins are transferred after cessation of cytoplasmic protein synthesis the following experiments were carried out :
A ^{35}S-prelabelled homogenate was incubated with ^{3}H-leucine for 10 min, cycloheximide or puromycin added (zero time), and aliquots were withdrawn immediately. Incubation was continued for 70 min following which aliquots were again withdrawn. Mitochondria were isolated and matrix proteins, CAT-binding protein, and cytochrome *c* were immunoprecipitated. The immunoprecipitates of CAT-binding protein and of cytochrome *c* were analyzed by SDS gel electrophoresis and the $^{3}H/^{35}S$-ratios determined (electrophoretic analysis of immunoprecipitated cytochrome *c* is shown in Fig.6). In the case of the immunoprecipitated matrix proteins the ratio in the total immunoprecipitates were determined. The $^{3}H/^{35}S$-ratios obtained after 70 min compared to the ratios at zero time are presented in Table 1 for the various proteins. The ratio for total

3) Experiments to Demonstrate the Presence of Precursors in the cytosolic Fraction

In previous papers it was reported that mitochondrial matrix proteins show kinetics characteristic of precursors outside the mitochondria. In experiments with intact cells as well as in *in vitro* experiments putative precursors were detected in the cytosolic fraction (1,3). This work extends these observations with experiments to detect precursors for the CAT-binding protein and for cytochrome *c*.

A ^{35}S-prelabelled homogenate was incubated with ^{3}H-leucine as described above for 10 min. The cytosolic and mitochondrial fraction were then isolated and immunoprecipitated were performed with antibodies against CAT-binding protein.

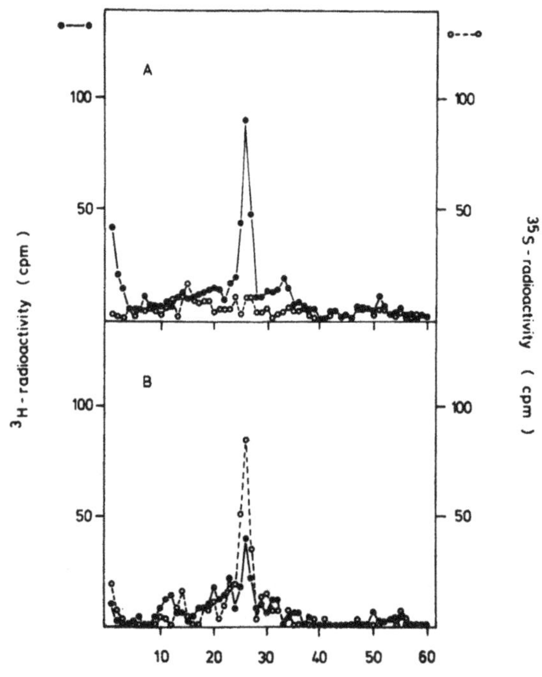

Fig.7 SDS gel electrophoresis of CAT-binding protein immunoprecipitated from the cytosolic fraction (A) and the mitochondrial fraction (B) of a cell free homogenate

A ^{35}S-prelabelled homogenate was incubated with ^{3}H-leucine for 10 min

After 1o min incubation only a small amount of ^{3}H-label co-migrates with the ^{35}S-prelabelled CAT-binding protein from mitochondria (Fig.7B). No ^{35}S-radioactivity could be

mitochondrial proteins is also given in the Table. For comparison the $^{3}H/^{35}S$-ratios determined following incubation with no inhibitor present (control) is included in the Table. The results given in the Table lead to the conclusion that specific and completed polypeptides are transported into the mitochondria when cytoplasmic translation is blocked. This demonstrates the existence of extramitochondrial pools which are exhausted after blocking protein synthesis.

Table 1

Incorporation of ^{3}H-leucine in total mitochondrial protein, matrix proteins, cytochrome *c* and CAT-binding protein after inhibition of cytoplasmic protein synthesis with cycloheximide or puromycin (values were determined as described in the text)

$\frac{^{3}H/^{35}S\ (70\ min)}{^{3}H/^{35}S\ (10\ min)}$	*Mitochondria*	*Matrix Proteins*	*Cytochrome c*	*CAT Protein*
Control	2.2	6.5	2.3	3.2
Cycloheximide	1.6	1.7	1.8	2.5
Puromycin	4.1	4.5	1.8	---

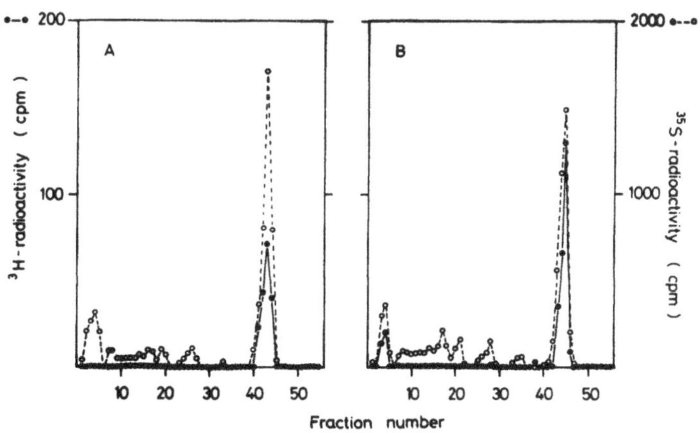

Fig.6 Gel electrophoretic analysis in the presence of SDS of cytochrome *c* immunoprecipitated from mitochondria

A cell free homogenate prelabelled with ^{35}S was incubated with ^{3}H-leucine for A) 10 min B) 10 min, then in the presence of cycloheximide for a further 70 min.

precipitated from the cytosolic fraction demonstrating the specific localisation of the CAT-binding protein in the mitochondria. However, a distinct peak of ^{3}H-radioactivity with the same electrophoretic mobility as the CAT-binding protein was apparent in the cytosol (Fig.7A). We consider this ^{3}H-labelled material to be a putative extramitochondrial precursor of the CAT-binding protein. However, precursor-product kinetic studies and chemical comparison of the putative precursor with the mitochondrial membrane component must be carried out before definite conclusions can be drawn.

In a comparable experiment, in which cells were not prelabelled with ^{35}S, the cytosolic fraction was immunoprecipitated with a specific antibody against apo-cytochrome *c* isolated from holo-cytochrome *c* according to Fisher *et al.*(9).

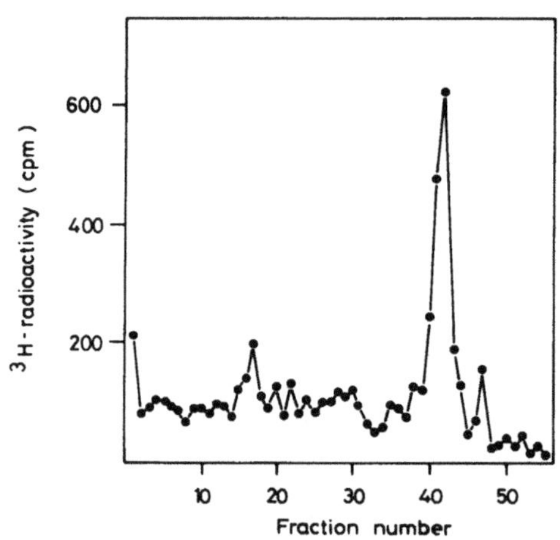

Fig.8 SDS gel electrophoresis of apo-cytochrome *c* immunoprecipitated from the cytosolic fraction of a cell free homogenate

A ^{35}S-prelabelled cell free homogenate was incubated with ^{3}H-leucine for 10 min in the presence of chloramphenicol.

The electrophoretic pattern of the immunoprecipitation displays a single band comigrating with cytochrome *c* (Fig.8). In separate experiments it was shown that this antibody did not cross react with holo-cytochrome *c*. Immunoprecipitates obtained with anti-apo-cytochrome *c* using mitochondria isolated from ^{35}S-prelabelled homogenates incubated *in vitro* with ^{3}H-leucine contained neither ^{35}S- nor ^{3}H-radioactivity. Preliminary experiments, in which apo-cytochrome *c* was precipitated from the cytosol and holo-cytochrome *c* from the mitochondria showed that following cycloheximide inhibition the ^{3}H-radioactivity disappeared from the cytosolic fraction and increased in the mitochondrial fraction. These data suggest that apo-cytochrome *c* is transported into the mitochondria *in vitro* . However, as in the case of matrix proteins the putative cytosolic precursor has not been chemically shown to be identical with the apo-protein of integrated cytochrome *c* .

FINAL REMARKS

The data we have presented here and in previous papers (1-4) lead us to propose a working hypothesis to describe synthesis and transport of cytoplasmically translated mitochondrial proteins : Mitochondrial proteins are synthesized as precursors on cytoplasmic ribosomes (free or bound) with subsequent release into the cytosol. These precursors differ in their structure and/or conformation from the integrated functional proteins in the mitochondria. Only precursors are transported into the mitochondria (see also ref.4). The outer mitochondrial membrane may not present a barrier for these precursors. Precursors of proteins of the inner mitochondrial membrane may be trapped by structural or conformational changes (a) in the course of assembly reactions which lead to phospholipid containing supermolecular structures and (b) by attachement of prosthetic groups to the apo-proteins. Precursors for proteins of the mitochondrial matrix proteins may be bound to acceptor sites on the inner membrane and subsequently translocated.

The molecular mechanisms underlying these proposed reactions remain to be determined. Comparison with other cellular reactions in which protein translocation across membranes takes place may be helpful to put questions which can be experimentally tested. Such translocation reactions are (a) the transfer of nascent chains of secretory proteins across the microsomal membrane which appears to involve the clipping of a " signal sequence " (10,11) and (b) the translocation of fragment *A* of diphtheria toxin across the cellular membrane, in which a second polypeptide chain (fragment *B*) seems to play an essential role in binding and transfer of the toxic fragment *A* (12).

ACKNOWLEDGEMENTS

This work was supported by the Deutsche Forschungsgemeinschaft, Schwerpunkt "Biochemie der Morphogenese". We wish to thank Heide Rothe and Dietlinde Rose for skillful technical assistance. We are grateful to F. Nargang for help in preparing the manuscript

REFERENCES

1) Hallermayer, G. and Neupert, W. : Studies on the Synthesis of Mitochondrial Proteins in the Cytoplasm and on their Transport into the Mitochondrion, in "Genetics and Biogenesis of Chloroplasts and Mitochondria" (Bücher, Th., *et al.*,eds.), North Holland Publishing Company, Amsterdam pp 807-812 (1976)
2) Hallermayer, G., Zimmermann, R. and Neupert, W. : Kinetic Studies on the Transport of Cytoplasmically Synthesized Proteins into the Mitochondria in Intact Cells of *Neurospora crassa*, submitted for publication
3) Harmey, M.A., Hallermayer, G. and Neupert, W. : *In vitro* Synthesis and Transport into Mitochondria of Cytoplasmically Synthesized Proteins, in "Genetics and Biogenesis of Chloroplasts and Mitochondria" (Bücher, Th., *et al.*, eds.), North Holland Publishing Company, Amsterdam pp 813-818 (1976)
4) Harmey, M.A., Hallermayer, G., Korb, H. and Neupert, W. : Transport of Cytoplasmically Synthesized Proteins into the Mitochondria in a Cell Free System from *Neurospora crassa* submitted for publication
5) Sabatini, D.D. and Kreibich, G.: Functional Specialization of Membrane-Bound Ribosomes in Eucaryotic Cells, in "The Enzymes of Biological Membranes" (Martonosi,A.,ed.)

John Wiley & Sons, London, New York, Vol.2 pp531-579 (1976)

6) Butow,R.A., Bennet, W.F., Finkelstein, D.B. and Kellems, R.E. Nuclear-Cytoplasmic Interactions in the Biogenesis of Mitochondria in Yeast, in "Membrane Biogenesis" (Tzagaloff,A., ed.) Plenum Publishing Cooperation, New York pp 155-199 (1975)

7) Rücker, A.v., Werner, S. and Neupert, W. :Synthesis of Cytochrome Oxidase Components in Isolated Mitochondria of *Neurospora crassa* , FEBS Letters, 47 29o-294 (1974)

8) Klingenberg, M., Aquila, H., Riccio, P., Buchanan, B.B., Eiermann, W. and Hackenberg, H.: Characterization of the Isolated ADP/ATP Carrier in "Electron Transfer Chains and Oxidative Phosphorylation" (Quagliariello, E., *et al.*,ed.) North Holland Publishing Company, Amsterdam pp 431-438 (1975)

9) Fisher, W.R., Taninchi, H. and Anfisen, C.B. : On the Role of Heme in the Formation of the Structure of Cytochrome *c*, J. Biol. Chem., 248, 3188-3195 (1973)

10) Blobel, G. and Dobberstein, B.: Transfer of Proteins across Membranes (I), J. Cell Biol., 67, 835-851 (1975)

11) Blobel, G. and Dobberstein, B. : Transfer of Proteins across Membranes (II), J. Cell Biol., 67, 852-862 (1975)

12) Boquet, D. and Pappenheimer, A.M., Jr. : Interaction of Diphtheria Toxin with Mammalian Cell Membranes, J. Biol. Chem., 251, 5770-5778 (1976)

DNA-Binding Properties of Mitochondrially Synthesized Proteins

G.Rödel and W.Bandlow
Genetisches Institut, Maria-Ward-Str. 1a, D-8000 München 19, Germany

Introduction

Mitochondrial circular DNA (mitDNA) of *Saccharomyces cerevisiae* has a contour length of about 25 µm corresponding to 75,000 base pairs (1). About 16 % of the total genome is likely to code for known polypeptides and structural RNA's: 20 tRNA's (2), 2 rRNA's (3), 3 subunits of cytochrome *c* oxidase (4), 1 or 2 subunits of the bc_1-complex (5-8), 3 subunits of the oligomycin sensitive ATPase complex (9, 10) and one ribosomal protein (11, 12). Even if one considers that AT - rich "spacers" of mitDNA (about 50 % of the mitochondrial genome) may not be transcribed (13), the residual DNA sequences have a coding capacity of at least 20 more proteins.The function of these is not known at present. Evidence is accumulating that some of these unknown proteins may be involved in the regulation of mitochondrial and cytoplasmic gene expression (14-16).
More refined analytical techniques like SDS polyacrylamide gradient slab gel electrophoresis or two-dimensional electrophoresis reveal the presence of about 20-25 mitochondrially synthesized components. Most of these are present in much smaller concentrations than those which have already been identified and which account for approximately 80 % of the total mitochondrially synthesized proteins.
In order to characterize one of the unknown proteins and to identify its function the inclusion of an enrichment step into the analytical procedure is necessary. One possibility to selectively enrich one or few components is to exploit particular properties of the protein,*e.g.* its capability to bind to DNA. For our purposes here it does not matter whether the DNA-binding is specific and has a defined role in the function of the protein (as in the case of enzymes involved in replication,transcription, recombination,repair or control of gene expression) or whether it is unspecific (as known for cytoplasmically synthesized cytochrome *c*).
In this paper we present a study of the DNA-binding properties of mito-

chondrially synthesized proteins of Saccharomyces cerevisiae wild type and of a mitochondrial mutant,tsm8, (retarded in oxidative adaption and conditionally disturbed in the control of mitochondrial gene expression, 17-19), grown under various conditions.

Results

In order to look for mitochondrially synthesized DNA-binding proteins two approaches have been taken:

a)Mitochondrially synthesized proteins labelled with $^{35}SO_4^=$ in the presence of cycloheximide (CHX) were solubilized in 1 % Triton X-100 and incubated with calf thymus single stranded DNA immobilized on nitrocellulose filters (20). After 90 minutes incubation the amount of radioactive protein bound to the filters was determined.

b)Mitochondria labelled as described in "a" were lysed with 2 % Sarkosyl NL35, centrifuged on a NaI/ethidiumbromide gradient and mitochondrial and contaminating nuclear DNA (nucDNA) isolated separately (21).Both DNA's were found to be contaminated with bound proteins.After DNase digestion the labelled proteins of the DNA fractions were analysed on SDS polyacrylamide slab gels.

Using the DNA-filter technique it could be demonstrated that a small percentage of mitochondrially synthesized proteins binds to DNA.Interestingly the amount of proteins bound to DNA-filters depends on the growth conditions: about 0.2 % of the proteins labelled in the presence of CHX (i. e. 0.01 % of total mitochondrial protein) were bound when the cells were grown in 5 % glycerol,whereas more than 0.9 % were bound after growth in 10 % glucose (Fig.1).The origin of the DNA on the filters was of minor importance for the protein binding (Fig.2). Fig.3 shows that the binding of mitochondrially synthesized proteins to DNA was influenced by the divalent metal-ions Mn^{++} (A) and Mg^{++} (B) in the form of an optimum curve. The maximally stimulating effect of Mn^{++} (2-3 times) was greater than that of Mg^{++} (1.5-2 times). The DNA-binding of mitochondrially synthesized components of derepressed cells cannot be stimulated by addition of cAMP, cGMP or phosphodiesterase,whereas dithionite diminishes the binding by about 25 % (data not shown).

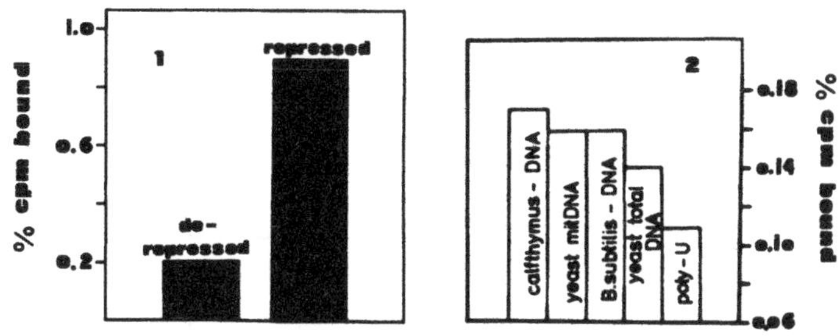

Fig.1 and 2: Nitrocellulose filters (ø 7 mm) were loaded with equal quantities of DNA as described by Gillespie and Spiegelman (20) and added to mitochondrial proteins labelled with $^{35}SO_4^{=}$ in the presence of CHX and solubilized in a 1 % Triton X-100 containing buffer solution. For each incubation 140000 cpm of labelled protein were applied. After 90 minutes incubation the filters were extensively washed and the amount of radioactive protein bound to the filters was determined.

Fig.1: Comparison of the DNA-bound radioactive material isolated from derepressed and repressed cells, respectively.

Fig.2: Percentage of labelled protein bound to DNA from various sources and to poly U.

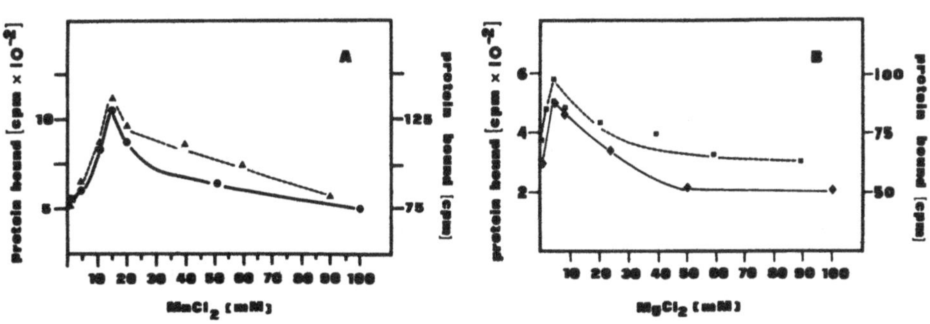

Fig.3: Influence of the concentration of Mn^{++} (A) and Mg^{++} (B) on the extend of protein binding to DNA (calf thymus DNA)

(- - -) mitochondrial proteins labelled with $^{35}SO_4^{=}$ in the presence of CHX from derepressed cells (right ordinate)

(———) mitochondrial proteins labelled with $^{35}SO_4^{=}$ in the presence of CHX from repressed cells (left ordinate)

Analysis of the mitochondrial translation products associated with isolated mitDNA and with contaminating nucDNA reveals that in addition to the quantitative difference between glucose and glycerol grown cells a qualitative difference is found (Fig.4).

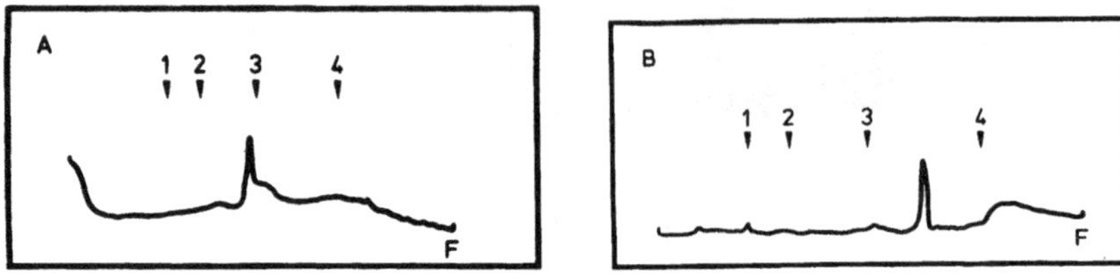

Fig.4: Densitogram of an autoradiogram obtained after SDS polyacrylamide gel electrophoresis of mitochondrial proteins (labelled with $^{35}SO_4^{=}$ in the presence of CHX) associated with mitDNA isolated from derepressed (A) and repressed (B) cells. Arrows indicate positions of marker proteins: 1 = bovine serum albumin (67,000); 2 = ovalbumin (45,000); 3 = carboanhydrase (29,000); 4 = myoglobin from horse (17,800). The molecular weight of the protein bands were calculated to 33,000 (A) and 24,000 (B). F = front

In derepressed cells one major mitochondrially synthesized component of a molecular weight of about 33,000 was found in the mitDNA fraction, whereas in the nucDNA fraction no defined band was detectable. In repressed cells one mitochondrial translation product of 24,000 dalton was found associated with mitDNA. The same polypeptide can also be detected by analyzing the nucDNA fraction. The 33,000 dalton component is absent on mitDNA. Identical results were obtained when the mitochondria were lysed in the presence of cAMP, phosphodiesterase or dithionite prior to the NaI/ethidiumbromide gradient centrifugation (data not shown). Parallel analysis of the mutant SM201 tsm8 either grown at 23 °C or 36 °C yielded comparable results. Therefore, it seems to be clear that the temperature - sensitive step does not influence the binding capability to DNA of one of the mitochondrially synthesized polypeptides. None of the isolated proteins showed an immunoprecipitation line in an Ouchterlony double diffusion test with antisera raised either against F_1 - ATPase or against total cytochrome c oxidase (data not shown).

Discussion

The data presented clearly demonstrate the existence of DNA-binding components made by the mitochondrial translation system. So far it is not possible to decide whether this binding reflects an *in vivo* situation or an *in vitro* artefact (e.g. a basic protein normally integrated into the membrane which could come in contact with mitDNA and/or contaminating nucDNA during lysis of mitochondria). The fact that the DNA-binding proteins of repressed cells differ from those of derepressed cells appears to be caused by a regulatory process. However, the difference in DNA-binding of mitochondrially synthesized proteins may either result from a regulated process (*e.g.* a protein which is not made or not integrated into the membrane under the other growth condition binds to DNA during lysis) or reflects the regulatory action of a repressor or gene activator as a response to the altered growth conditions. The finding that the origin of the DNA bound to the filters is of minor importance for the binding of proteins is not in conflict with a specific binding, since any of the DNA's used is large enough to contain sequences required for DNA-protein complex formation (22).

The important question whether the proteins associated with mitDNA are known mitochondrial translation products cannot be answered at present. The negative result of the double diffusion test is rather considered as a slight indication that the proteins found may not be identical with the known subunits II or III of cytochrome *c* oxidase or with subunit VI of the ATPase complex (these polypeptides are mitochondrially made and have similar molecular weights as the DNA-binding proteins found, 4,23,9). Because of the known difficulties to precipitate subunits with antisera raised against holo enzymes and because of the low antigenicity of some polypeptides, *e.g.* subunit III of cytochrome *c* oxidase an unambiguous answer on the non-identity with one of the known mitochondrially synthesized polypeptides cannot be given.In the case that one of these components really binds to DNA we have to ask for the reasons why the binding capacity is influenced by divalent metal ions and limited to special growth conditions. Although unequivocal proofs are lacking,it may well be that mitochondrially synthesized proteins play a role in replication [24], transcription (mitochondrial cofactor), repair and recombination or - most likely - control of gene expression (14, 16) of the mitochondrial genome.

References

1) Borst, P., Sanders, J. P. M. and Heyting, C. (1976) In "Genetics, Biogenesis and Bioenergetics of Mitochondria" (Bandlow, W., Schweyen, R. J., Thomas, D. Y., Wolf, K. and Kaudewitz, F., eds) pp. 85 - 97 De Gruyter, Berlin
2) Reijnders, L. and Borst, P. (1972) Biochem. Biophys. Res. Commun. 47, 126 - 133
3) Reijnders, L., Kleisen, C. M., Grivell, L. A. and Borst, P. (1972) Biochem. Biophys. Acta 272, 396 - 407
4) Mason, T. L. and Schatz, G. (1973) J. Biol. Chem. 248, 1355 - 1360
5) Katan, M. B. and Groot, G. S. P. (1975) In "Electron Transfer Chains and Oxidative Phosphorylation" (Quagliariello, E., Papa, S., Palmieri, F., Slater, E. C. and Siliprandi, N., eds) pp. 127 - 132 North - Holland, Amsterdam
6) Weiss, H. (1972) Europ. J. Biochem. 30, 469 -478
7) Weiss, H. and Ziganke, B. (1974) Europ. J. Biochem. 41, 63 - 71
8) Weiss, H. and Ziganke, B. (1977) this volume
9) Jackl, G. and Sebald, W. (1975) Europ. J. Biochem. 54, 97 - 106
10) Goffeau, A., Boutry, M. and Dufour, J. P. (1977) this volume
11) Groot, G. S. P. (1974) In "The Biogenesis of Mitochondria" (Kroon, A. M. and Saccone, C., eds) pp. 443 - 452 Academic Press, New York
12) Douglas, M.C., Strausberg, R.L., Perlman, P.S. and Butow, R.A. (1976) In "Genetics and Biogenesis of Chloroplasts and Mitochondria" (Bücher, Th., Neupert, W., Sebald, W. and Werner, S., eds) pp. 435 - 442 North - Holland, Amsterdam
13) Bernardi, G., Prunell, A., Fonty, G., Kopecka, H. and Strauss, F. (1976) In "The Genetic Function of Mitochondrial DNA" (Saccone, C. and Kroon, A. M., eds) pp. 185 - 198 North Holland, Amsterdam
14) Barath, Z. and Küntzel, H. (1972) Proc. Nat. Acad. Sci. U.S. 69, 1371 - 1374

15) Beauchamp, P. M. and Gross, S. R. (1976) Nature 261, 338 - 340
16) Rouslin, W. (1976) Biochem. Biophys. Res. Commun. 73, 364 - 369
17) Handwerker, A., Schweyen, R. J., Wolf, K. and Kaudewitz, F. (1973) J. Bacteriol. 113, 1307 - 1310
18) Bechmann, H., Krüger, M., Böker, E., Bandlow, W., Schweyen, R. J. and Kaudewitz, F. (1977) Molec. Gen. Genet. (in the press)
19) Bandlow, W., Metzke, R., Klein, A., Kotzias, K., Doxiadis, I., Bechmann, H., Schweyen, R. J. and Kaudewitz, F. (1977) Europ. J. Biochem. 76, 373 - 382
20) Gillespie, D. and Spiegelman, S. (1965) J. Mol. Biol. 12, 829 - 842
21) Lang, B., Burger, G., Doxiadis, I., Thomas, D. Y., Bandlow, W. and Kaudewitz, F. (1977) Analyt. Biochem. 77, 110 - 121
22) Mitra, S. and Zubay, G. (1975) Biochem. Biophys. Res. Commun. 67, 857 - 863
23) Tzagoloff, A. and Meagher, P. (1971) J. Biol. Chem. 246, 7328 -7336
24) Williamson, D. H. (1970) In "Control of Organelle Development" (Miller, P. L., ed) pp.247 - 276 Cambridge University Press, London

SOME PRELIMINARY OBSERVATIONS ON THE EFFECTS OF CERULENIN ON THE FORMATION OF MITOCHONDRIAL AND CYTOPLASMIC PRECURSORS OF ELECTRON TRANSFER COMPLEXES III AND IV IN ADAPTING, LIPID-LIMITED *SACCHAROMYCES CEREVISIAE*.

W. Rouslin
Department of Pharmacology and Cell Biophysics
University of Cincinnati College of Medicine, Cincinnati, Ohio 45267, USA

The probable importance of mitochondrial lipids for respiratory enzyme synthesis and function has been suspected for some time (1, 2). During the aeration of bakers' yeast cells grown anaerobically on lipid-limited media, total fatty acids and sterols increase in parallel with increasing respiratory capacity (3). Observed changes in mitochondrial lipid composition which accompany respiratory adaptation include marked increases in the levels of long chain unsaturated fatty acids, ergosterol and ubiquinone (4, 5). While these changes in lipid composition suggest an integral role for lipids in the adaptation process, important questions concerning possible regulatory roles for membrane lipids in the emergence of functional respiratory enzymes remain largely unanswered. For example, to what extent is the appearance of particular lipids during adaptation coupled to the synthesis and/or the functional integration of known mitochondrially and cytoplasmically translated peptidal subunits of inner membrane respiratory enzyme complexes? Does the apparent oxygen dependence of the appearance of certain mitochondrial and cytoplasmic precursors of respiratory enzyme complexes (6, 7) merely reflect the oxygen requirements of specific steps in pathways for the synthesis of unsaturated fatty acids, ergosterol and ubiquinone (8, 9, 10), or are the oxygen inductions of the syntheses of lipids and protein components mutually independent events, at least in their initial stages?

In the present preliminary study, the fatty acid and sterol synthesis inhibitor, cerulenin (11), was used to examine the extent to which fatty acid and sterol availability might be coupled to the synthesis and/or membrane attachment of mitochondrial inner membrane proteins in general, and of mitochondrial and cytoplasmic precursors of electron transfer complexes III and IV in particular, during respiratory adaptation. While these studies are at an early stage at this writing, the results, thus far, are sufficiently interesting to warrent their inclusion here.

METHODS AND MATERIALS

The wild-type *Saccharomyces cerevisiae* strain D273-10B(αPET[ρ^+]) was used. The cells were grown anaerobically in a 0.5% galactose, yeast extract medium as described previously (7) except that Tween 80 and ergosterol were absent. The cells were harvested in early stationary phase, washed once in adaptation buffer and then resuspended to a cell concentration of 25 mg wet wt/ml and shaken at 28°C in the dark. For the labeling experiments (Table I) the cells were incubated for 15 min before the addition of [^{14}C] leucine. Labeling was then carried out for 60 min after which the samples were processed for counting as described earlier (12). For the two-phase adaptation experiments (Fig. 2) the cells were chilled to 0°C while still anaerobic in the carboy. Where appropriate, the cells were poisoned while still in the carboy with cycloheximide (CHI) at a final concentration of 50 µg/ml, or immediately upon harvesting with chloramphenicol (CAP) at 4 mg/ml. The cells were incubated under phase I conditions for 3 hrs, washed 4 times with cold phase II buffers and then incubated for an additional 6 hrs in phase II. In the two-phase adaptation experiments and in all other experiments in which enzyme activities were assayed, samples were taken at the times indicated (or always after 6 hrs of aeration in the experiments depicted in Fig. 1 and in Table II) and assayed for reduced coenzyme Q cytochrome *c* reductase and for cytochrome *c* oxidase as described earlier (7). Where added, cerulenin (CRL) was present at a final concentration of 2 µg/ml (or as indicated in Fig. 1), Tween 80 at 0.2% (V/V) and ergosterol at 0.001% (W/V).

The [U-^{14}C]-leucine used was obtained from New England Nuclear. The coenzyme Q_2 was a generous gift supplied by F. Hoffmann-La Roche & Co., Basel, Switzerland. The cytochrome *c* was Sigma Chemical Co. Type III horse heart material. The cerulenin used in initial experiments (not reported here) was a generous gift supplied by Dr. Satoshi Omura of Kitasato University, Tokyo, Japan. The cerulenin used in later experiments (including those reported here) was obtained from Makor Chemicals, Ltd., Jerusalem, Israel. All other chemicals used were of reagent grade.

RESULTS

The labeling studies reported in Table I show that, in the absence of both CHI and CAP, CRL affected a relatively minor decrease in the total counts

TABLE I

In Vivo ^{14}C-Leucine Incorporation into a Washed, Crude Mitochondrial Membrane Fraction.[a]

Additions	Lipid-Limited[b] cpm/mg protein	% of control	Lipid-Supplemented[c] cpm/mg protein	% of control
NONE	156,970	100.0 (100.0)	51,393	100.0
NONE + CRL	131,325	83.6	-	-
CHI	15,812	100.0 (10.0)	3,906	7.6
CHI + CRL	3,084	19.5	-	-
CAP	17,822	100.0 (11.3)	46,177	89.8
CAP + CRL	16,283	91.3	-	-
CHI + CAP	1,047	-	467	-
CHI + CAP + CRL	882	-	-	-

[a]The cells were aerated in 50 ml batches in 250 ml flasks at 28° in the dark. 10 µCi [^{14}C]leucine was added to each flask 15 min after the beginning of the incubation and labeling was carried out for 60 min.

[b]Tween 80 and ergosterol were absent from the growth medium.

[c]Tween 80 and ergosterol were present in the growth medium at final concentrations of 0.2% (V/V) and 0.001% (W/V), respectively.

incorporated into the membrane fraction studied. Exposure to CRL in the presence of CHI, however, resulted in a substantial inhibition of CHI-insensitive synthesis suggesting that much of the relatively small CRL-mediated effect seen in the absence of both CHI and CAP was very likely due to a rather specific curtailing of either the synthesis or the membrane attachment of mitochondrially translated peptides. That this might be the case is again suggested by the rather small diminution of incorporation observed when CRL was added in the presence of CAP. This latter result suggests that either the synthesis or the membrane attachment of cytoplasmically translated membrane proteins was not significantly decreased by the addition of CRL per se in the absence of concomitant mitochondrial protein synthesis. Thus, one general conclusion which may be drawn from these incorporation data is that the prevention of aerobic fatty acid and sterol synthesis by the addition of CRL rather specifically prevents the buildup of mitochondrially translated proteins.

When lipid-supplemented cells are labeled with radioactive leucine in the presence of either CHI or CAP, one typically observes results such as those shown in the right half of Table I. CHI usually abolishes roughly 90% of the incorporation into mitochondrial membranes while CAP blocks only about 10% of the total. Table I shows that the behavior of lipid-limited cells was at variance with this pattern. In the presence of CHI alone, the cells did exhibit the expected 90% inhibition of incorporation. Surprisingly, exposure to CAP alone resulted in a similar high degree of inhibition. This latter result may be interpreted in different ways. One trivial explanation is that CAP, itself, prevents aerobic fatty acid and/or sterol biosynthesis as has, indeed, been suggested to be the case by Gordon, et al. (13) and thus causes almost as much inhibition of incorporation as does CAP in combination with CRL. This explanation is thought to be largely incorrect for the experiments reported here in that the addition of CRL in the absence of CHI and CAP resulted in only a relatively minor inhibition of the appearance in the membrane of newly synthesized proteins. A far more intriguing explanation of this observation which additionally is more consistent with the data reported in Table I is that lipid deprivation in concert with the inhibition of mitochondrial protein synthesis by CAP prevents either the synthesis or the membrane attachment of roughly 90% of the cytoplasmically translated mitochondrial membrane proteins.

Figure 1 shows the effect of varying CRL levels upon the activities of reduced coenzyme Q cytochrome *c* reductase and of cytochrome *c* oxidase observed after 6 hrs of aeration. The induction of cytochrome *c* oxidase activity was shown to be somewhat more sensitive to CRL inhibition in that its appearance was diminished proportionately by somewhat lower CRL concentrations, and its activity was reduced to a lower residual value by CRL levels greater than 1.0 µg/ml. The higher residual activity observed for reduced coenzyme Q cytochrome *c* reductase is seen even at very high CRL

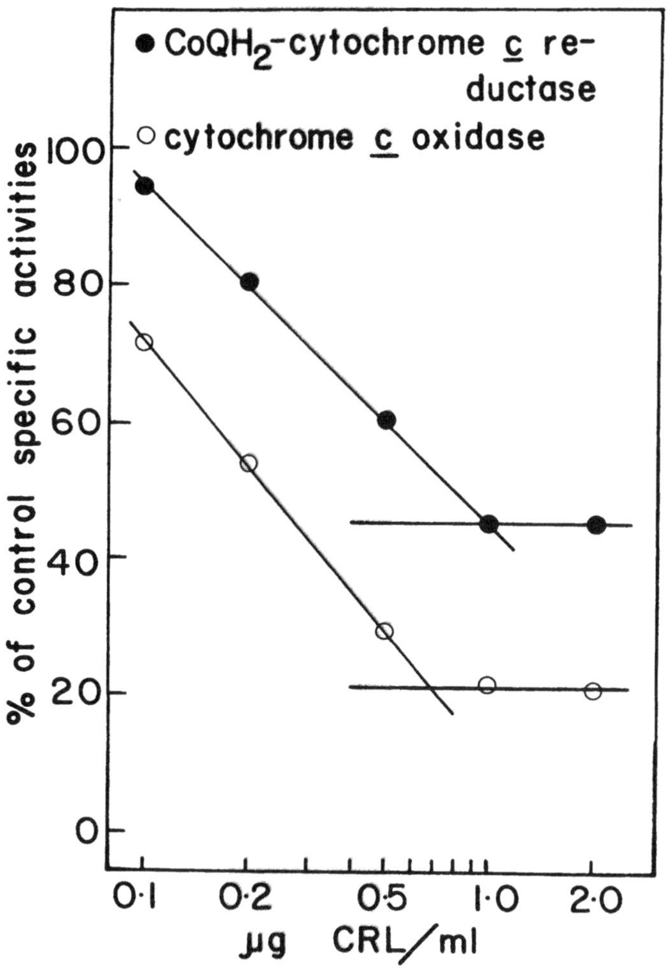

Figure 1. The effect of varying CRL concentrations upon the activities of reduced coenzyme Q cytochrome *c* reductase and of cytochrome *c* oxidase after 6 hrs of aeration.

concentrations and is believed to be due to the expression of an anaerobically made bc_1 complex precursor which has been discussed earlier (12).

The specific activity data reported in Table II again show the degrees to which the inductions of reduced coenzyme Q cytochrome *c* reductase and

TABLE II

Specific Activities of Reduced Coenzyme Q Cytochrome *c* Reductase (QCR) and of Cytochrome *c* Oxidase (CO) after 6 hrs of Aeration - Expressed as μmoles of Cytochrome *c* Reduced or Oxidized per min per mg Protein.

Additions	QCR	% of control	CO	% of control
NONE	0.324	100.0	0.499	100.0
CRL	0.142	43.8	0.110	22.2
CRL + Tween 80 & Ergosterol[a]	0.226	69.7	0.250	50.1

[a]Tween 80 and Ergosterol were present in the adaptation buffer at final concentrations of 0.2% (V/V) and 0.001% (W/V), respectively.

cytochrome *c* oxidase were diminished by the addition of CRL to a final concentration of 2.0 μg/ml. As in Fig. 1, the residual level of reduced coenzyme Q cytochrome *c* reductase activity was approximately double that of cytochrome *c* oxidase. Of particular interest was the partial reversal of the CRL-mediated inhibition of enzyme induction observed upon the addition of Tween 80 and ergosterol to the adaptation buffer.

Normally, CRL binding and its inhibition of fatty acid synthesis are irreversible in yeast (14). The ability to partially reverse its effects by the introduction of an exogenous source of an unsaturated fatty acid and a sterol was itself of interest, but it also made possible the performance of two-phase enzyme induction experiments analogous in design to others which have been described earlier (7, 12). These two-phase experiments were designed to temporally separate mitochondrial from cytoplasmic translational activity, thereby allowing the experimental manipulation of each genetic system somewhat apart from the other for a brief period of time. In the experiments depicted in Fig. 2, the cells were aerated first in CHI and then in CAP, *or* first in CAP and then in CHI. During phase I of the experiments in which CHI was added first, mitochondrially made precursors were allowed to build up first. Then, upon the removal of CHI and its replacement by CAP, cytoplasmically made precursors were incorporated into the membrane resulting in the formation of functional enzyme complexes. The levels of enzyme activity obtained during the second phase of these

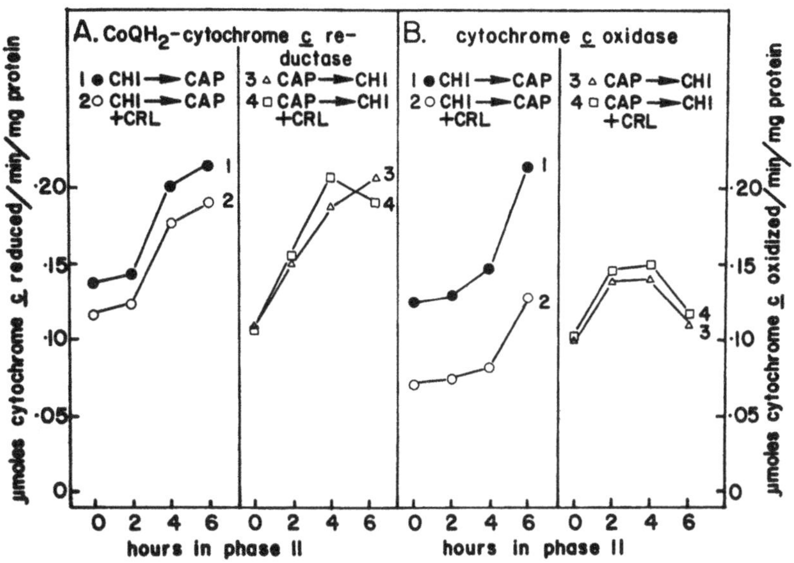

Figure 2. The two-phase inductions of (A) reduced coenzyme Q cytochrome *c* reductase and (B) cytochrome *c* oxidase. The experimental protocol is described in Methods and Materials. Tween 80 and ergosterol were present in all phase II incubations at final concentrations of 0.2% (V/V) and 0.001% (W/V), respectively.

experiments are considered to be a measure of the precursor levels built up during the first half of these experiments (12). In those two-phase enzyme inductions in which CAP was added first, the order of participation of the two translational systems was reversed, thereby allowing the prior buildup of cytoplasmically made precursors followed by the formation of mitochondrially made components in phase II.

In these two-phase induction studies, CRL was either absent or present during phase I. In those experiments in which CHI was present during phase I (traces 1 and 2), the imposition of lipid deprivation was coincident with mitochondrial translational activity. In those experiments in which CAP was present during phase I (traces 3 and 4) the lipid limiting condition was concomitant with cytoplasmic translation. While the kinetics of the phase II enzyme inductions shown in Fig. 2 are atypical, particularly for cytochrome *c* oxidase, the major CRL-mediated effects were observed in those experiments in which CRL was present during mitochondrial translational activity (traces 1 and 2). The effects are particularly marked for cytochrome *c* oxidase. In agreement with the labeling data presented in Table I, these two-phase enzyme induction experiments again suggest that it is mitochondrial translational activity which is especially sensitive to lipid deprivation during respiratory adaptation.

DISCUSSION

The final cell concentrations observed in yeast cultures grown anaerobically under lipid-limiting conditions were only 10 to 20% those of anaerobic, lipid-supplemented cultures (data not shown). Moreover, the viability of a low proportion of these cells was questionable (cf. ref. (4)). Nevertheless, and notwithstanding earlier suggestions that such unsaturated fatty acid-depleted cells may have an impaired mitochondrial protein synthesizing system (15), the data reported in Table I show that the cells used in the present study were capable of proportionately normal levels of CHI-insensitive protein synthetic activity. The specific radioactivities of the membrane fraction samples from lipid-limited cells, indeed, were reproducibly roughly 3-fold higher than those from lipid-supplemented cells, an observation for which we lack an immediate explanation.

The data reported in Table I indicate that a primary effect of CRL in lipid depleted cells is a rather specific curtailing of the formation of mitochondrially translated inner membrane components. Perhaps the most unexpected and interesting new relationship which emerges from the data reported in Table I is that concerning the comparative effects of CAP on lipid-limited and on lipid-supplemented cells. While CAP blocked roughly 10% of the total counts incorporated into the mitochondrial membranes in lipid-supplemented cells, it abolished nearly 90% of all new membrane protein synthesis and/or attachment in the lipid-limited cells. This large inhibitory effect of CAP is believed to be due to the combined effects of lipid deprivation and the inhibition of mitochondrial translational activity by CAP. If this explanation is correct, then it would appear that the

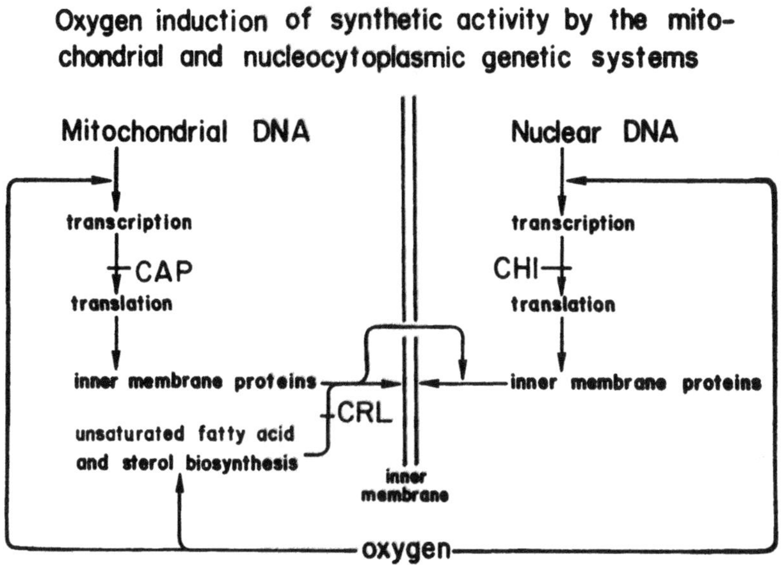

Figure 3. A schematic representation of proposed interactions between unsaturated fatty acids and sterols and the synthesis and/or membrane attachment of mitochondrially and cytoplasmically translated mitochondrial inner membrane proteins.

synthesis and/or the membrane attachment of a large proportion of the cytoplasmically translated mitochondrial inner membrane proteins is dependent upon a cooperative or synergistic action of mitochondrially translated proteins together with inner membrane lipids. Thus, the relatively more hydrophobic mitochondrial translation products together with membrane lipids appear to have a strong effect on the appearance of the relatively more hydrophilic cytoplasmically translated peptides within the mitochondrial inner membrane. These relationships are summarized schematically in Fig. 3.

Fig. 1 shows that the induction of cytochrome *c* oxidase was inhibited proportionately to a greater extent at any given concentration of CRL than was that of reduced coenzyme Q cytochrome *c* reductase. This observation suggests that the synthesis and/or the membrane attachment of different mitochondrially synthesized components may be inhibited to different degrees at any given concentration of CRL. If this is the case, then one might propose that the different mitochondrially translated peptides present within electron transfer complexes III and IV are differentially dependent upon lipid availability for their appearance within the membrane; that those peptides which are extremely hydrophobic in nature might have a more stringent requirement for fatty acid and sterol availability than those which are somewhat less hydrophobic.

ACKNOWLEDGEMENT

This work was supported by U.S. Energy Research and Development Administration Contract Number: EY-76-S-02-3516.*000.

REFERENCES

1. Klein, H.P.: Synthesis of lipids in resting cells of *Saccharomyces cerevisiae*. J. Bact. 69, 620-627 (1955).

2. Parks, L.W. and Starr, P.R.: A relationship between ergosterol and respiratory competency in yeast. J. Cell. Comp. Physiol. 61, 61-65 (1963).

3. Kovac, L., Subik, J., Russ, G. and Kollar, K.: On the relationship between respiratory activity and lipid composition of the yeast cells. Biochim. Biophys. Acta 144, 94-101 (1967).

4. Paltauf, F. and Schatz, G.: Promitochondria of anaerobically grown yeast. II. Lipid composition. Biochemistry 8, 335-339 (1969).

5. Gordon, P.A. and Stewart, P.R.: The effect of antibiotics on lipid synthesis during respiratory development in *Saccharomyces cerevisiae*. Microbios 4, 115-132 (1971).

6. Mason, T.L. and Schatz, G.: Cytochrome *c* oxidase from bakers' yeast. II. Site of translation of the protein components. J. Biol. Chem. 248, 1355-1360 (1973).

7. Rouslin, W.: Oxygen dependence of promitochondrial and cytoplasmic protein synthesis in the formation of electron transfer complexes III and IV in adapting bakers' yeast. Arch. Biochem. Biophys. 168, 685-692 (1975).

8. Bloomfield, D.K. and Bloch, K.: The formation of Δ^9-unsaturated fatty acids. J. Biol. Chem. 235, 337-345 (1960).

9. Tchen, T.T. and Bloch, K.: On the mechanism of cyclization of squalene. J. Amer. Chem. Soc. 78, 1516-1517 (1956).

10. Rudney, H. and Raman, T.S.: Biosynthesis of ubiquinones and vitamin K in microorganisms. In: "Vitamins and Hormones," Vol. 24. (R.S. Harris, I.G. Wool and J.A. Loraine, Eds.) New York: Academic Press, pp.531-549 (1966).

11. Omura, S.: The antibiotic cerulenin, a novel tool for biochemistry as an inhibitor of fatty acid synthesis. Bact. Revs. 40, 681-697 (1976).

12. Rouslin, W.: Partial independence of synthesis and membrane attachment of mitochondrial and cytoplasmic precursors of electron transfer complexes III and IV in adapting bakers' yeast. Arch. Biochem. Biophys. 178, 408-418 (1977).

13. Gordon, P.A., Lowden, M.J. and Stewart, P.R.: Effects of chloramphenicol isomers and erythromycin on enzyme and lipid synthesis induced by oxygen in wild-type and petite yeast. J. Bact. 110, 504-510 (1972).

14. D'Agnolo, G., Rosenfeld, I.S., Awaya, J., Omura, S. and Vagelos, P.R. Inhibition of fatty acid synthesis by the antibiotic cerulenin. Specific inactivation of β-ketoacyl-acyl carrier protein synthetase. Biochim. Biophys. Acta 326, 155-166 (1973).

15. Linnane, A.W. and Crowfoot, P.D.: Biogenesis of the yeast mitochondrial membranes. In: "Membrane Biogenesis: Mitochondria, Chloroplasts and Bacteria." (A. Tzagoloff, Ed.) New York: Plenum Press, pp. 99-124 (1975).

AN INDUCTOR-REPRESSOR MODEL COORDINATING THE EXPRESSION OF MITOCHONDRIAL AND NUCLEAR GENES THAT SPECIFY MITOCHONDRIAL PROTEINS IN YEAST (*CANDIDA UTILIS*)

Ezzatollah Keyhani and Jacqueline Keyhani
St. Jude Children's Research Hospital
Department of Biochemistry
Memphis, Tennessee 38101

Mitochondrial biogenesis requires the coordination of two distinct genetic systems, one located in the cytoplasm and the other in the mitochondria. Approximately 90% of mitochondrial proteins are coded by nuclear DNA, are synthesized on the cytoplasmic ribosomes and are subsequently transported to mitochondria. Mitochondrial protein synthesis, as so far characterized, involves three subunits of cytochrome *c* oxidase (1,2), four subunits of oligomycin-sensitive ATPase (2) and cytochrome *b* (3). The mitochondrial protein synthesis system resembles the bacterial system more closely than it does the cytoplasmic system. How these two distinct genetic systems are coordinated remains unresolved.

Recently, Barath and Küntzel (4), based on their observation that ethidium bromide or chloramphenicol induces the synthesis of mitochondrial enzymes, proposed a simple model for coordination of nuclear and mitochondrial protein synthesis. In their view, a mitochondrial repressor protein(s) functions at both the mitochondrial and nuclear DNA levels and therefore regulates the synthesis of mitochondrial proteins. This repressor scheme predicts two phases of mitochondrial biogenesis: an initial phase of cytoplasmic protein synthesis followed by the production of mitochondrial proteins, among which is a repressor protein that acts on the nuclear genome. Variations of this model have been presented by Lloyd *et al.* (5) and Williamson (6).

In this study, we report that low concentrations of adriamycin, an anthracycline compound with demonstrated cytotoxic activity (7), stimulate cytochrome biosynthesis and increase the incorporation of radioactive leucine in all classes of mitochondrial polypeptides. The stimulation of cytochrome biosynthesis is accompanied by stimulation of respiration in intact yeast cells and substrate oxidation in isolated mitochondria.

STIMULATION OF CYTOCHROME BIOSYNTHESIS AND RESPIRATION

Table I shows the effect of an increasing concentration of adriamycin on the cytochrome content and respiration of yeast cells grown for 16 hr. Compared to control values, the amounts of cytochromes *a*, *b*, and *c* all increased as drug concentration was increased from 0 to about 25 μg/ml in the culture medium; thereafter, cytochrome content decreased, reaching zero (for cytochrome *a*) at a drug concentration of 1 mg/ml (data not shown). The oxygen uptake of cells, with ethanol used as substrate, increased 2-fold as the adriamycin concentration was increased from 0 to 25 μg/ml, and then decreased, reaching nearly zero at 1 mg of adriamycin per ml. The stimulation of respiration by adriamycin in intact cells was corroborated by measurements of substrate oxidation in isolated mitochondria. With NADH, succinate and TMPD-ascorbate as substrates, oxygen consumption was consistently higher in adriamycin-treated cells than in controls (Table II).

Table III shows that the stimulation of cytochrome biosynthesis by adriamycin follows roughly the same pattern as the control experiment. *Candida utilis* cells were grown for 16 hr in the presence of the drug concentration, 25 μg/ml, that produced the maximum biosynthetic response. At regular intervals, 150-ml aliquots of cells were removed, centrifuged, and washed; the total cytochrome content of each pellet was calculated from difference spectra. As shown in the table, the concentration of cytochromes *a*, *b* and *c* in both drug-treated and control cells reached a maximum value at about 4 hr of cell growth and then declined at progressively slower rates through 16 hr. Since the drug-induced stimulation and subsequent decline of cytochrome biosynthesis was proportional to normal patterns, adriamycin appears to act on the same mechanism that regulates cytochrome biosynthesis in untreated cells.

STIMULATION OF MITOCHONDRIAL PROTEIN BIOSYNTHESIS

The stimulation of cytochrome biosynthesis by adriamycin was confirmed by determining the amount of radioactive leucine incorporated into mitochon-

TABLE I

CYTOCHROME CONTENT AND RESPIRATION OF YEAST CELLS AFTER 16 HR OF CULTURE IN THE PRESENCE OF VARIOUS CONCENTRATIONS OF ADRIAMYCIN

Adriamycin (μg/ml)	Cytochrome Content (% of control) cyt. *a*	cyt. *b*	cyt. *c*	O_2 Uptake (ng-atoms x min^{-1} x 10^{-8} cells)
0	100	100	100	14.87
10	124	182	112	21.08
25	155	217	188	31.86
50	126	176	147	27.65
100	96	143	118	23.77
500	---	---	---	2.96
1000	---	---	---	0.10

The amounts of cytochromes were calculated from dithionite-reduced minus oxidized difference spectra according to published methods (8). The interference of adriamycin with the α band absorption of cytochromes *b* and *c* was corrected according to ref. (9).

TABLE II

SUBSTRATE OXIDATION BY MITOCHONDRIA ISOLATED FROM YEAST GROWN IN THE PRESENCE OF ADRIAMYCIN, 25 μg/ml

	O_2 Uptake (ng-atoms x min^{-1} x mg^{-1} prot.)* NADH (38 μM)	Succinate (7.5 mM)	TMPD (50 μM) + Ascorbate (7.5 mM) + AA (6 μg)
Normal	11.33	25.8	27.2
Adriamycin	14.33	34.4	45.38
% Increase by Adriamycin	27	33	66.8

* In the presence of 25 μM pentachlorophenol as uncoupler
TMPD: N,N'-tetramethyl-*p*-phenylenediamine; AA: Antimycin A

drial proteins of yeast cells grown in the presence or absence of the drug.

Figure 1A shows that at a drug concentration of 25 μg/ml, the mitochondria of treated cells pulse labeled for 1 hr incorporated nearly one and a half times as much [^{3}H]Leu as did the untreated controls. On the other hand, the amount of [^{3}H]Leu incorporated into the cytoplasmic protein fraction

TABLE III

CYTOCHROME CONTENT AT DIFFERENT TIMES OF CULTURE
IN THE ABSENCE OR PRESENCE OF ADRIAMYCIN (AD, 25 μg/ml)

Growth (hr)	Cyt. *a*		Cyt. *b*		Cyt. *c*	
	Control	+AD	Control	+AD	Control	+AD
	(nmoles x 10^{-10} cells)					
0	0.62	0.62	0.111	0.111	0.194	0.194
2	0.66	0.84	0.132	0.174	0.242	0.288
4	0.67	0.98	0.169	0.281	0.285	0.355
6	0.49	0.84	0.139	0.187	0.222	0.283
8	0.52	0.92	0.122	0.196	0.185	0.245
12	0.48	0.77	0.112	0.188	0.150	0.239
16	0.45	0.76	0.085	0.155	0.124	0.215

Cytochrome amounts were determined as described in Table I.

was slightly higher than the control value. At drug concentrations of 25 to 50 μg/ml, the incorporation of [^{3}H]Leu into mitochondrial proteins decreased rapidly and then began a more gradual decline to approximately 80% of the control at 1 mg/ml.

The distribution of radioactivity among mitochondrial polypeptides in the presence or absence of adriamycin, as determined by $NaDodSO_4$ polyacrylamide gel electrophoresis, is illustrated in Figs. 1, B, C and D. At a drug concentration of 25 μg/ml (Fig. 1B) the incorporation of [^{3}H]Leu by some polypeptides was two to three times higher than the corresponding control values. The ratio of incorporation of [^{3}H]Leu into mitochondrial proteins in the presence *versus* the absence of adriamycin shows a general enhancement of protein synthesis. At high concentrations of adriamycin, 1 mg/ml, the distribution of radioactivity among polypeptides indicated, within the experimental error, that all classes of polypeptides were present in decreased amounts in adriamycin-treated cells (Fig. 1D). The effects of an intermediate drug concentration, 200 mg/ml, is shown in Fig. 1C. Under these conditions, some polypeptides were synthesized in amounts higher than the control, while some were lower.

When [^{3}H]Leu incorporation into a single protein, cytochrome *c* oxidase, was determined by immunoprecipitation after 1 hr of pulse labeling

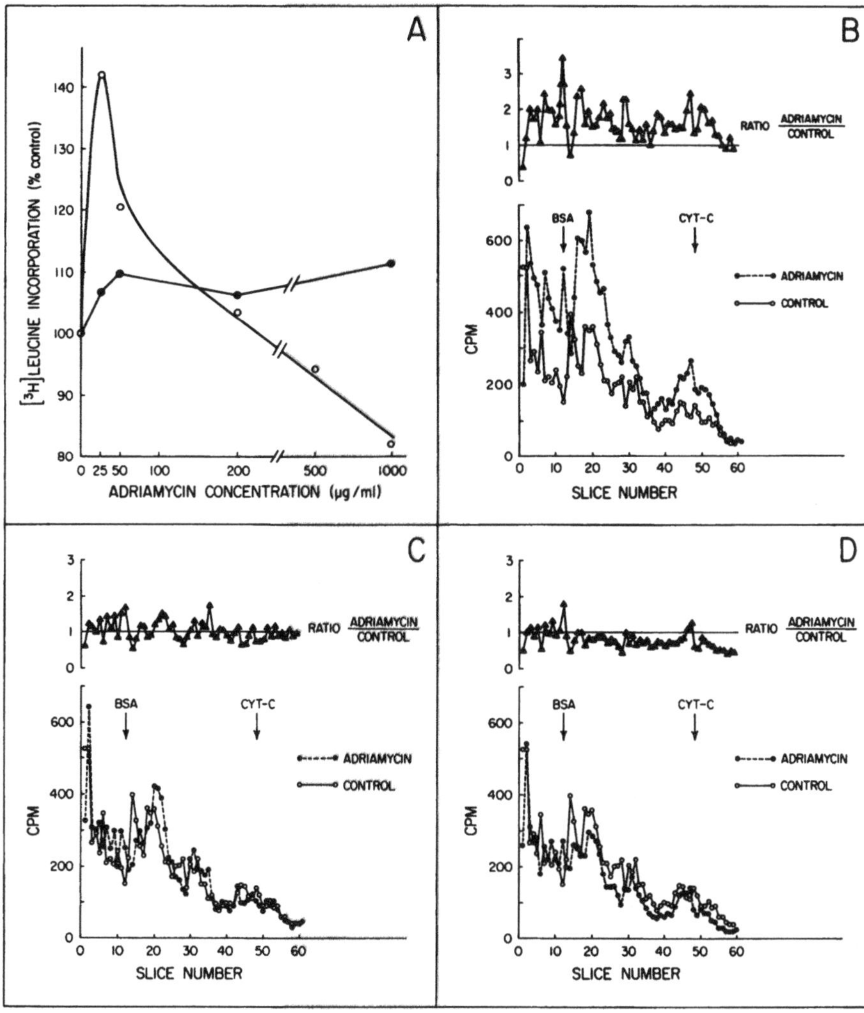

Figure 1 - Incorporation of [^{3}H] leucine into mitochondrial proteins with increasing concentration of adriamycin. Cells were pulse labeled for 1 hr between the 2nd and 3rd hr of growth. (A) Overall [^{3}H]Leu incorporation. O—O, mitochondrial proteins; ●—●, postmitochondrial supernatant. (B-D) Distribution of radioactivity in mitochondrial proteins of control (O) and of cells incubated with 25, 200 or 1000 µg adriamycin/ml, respectively. The proteins in the mitochondrial pellet were analyzed by $NaDodSO_4$-polyacrylamide gel electrophoresis in cylindrical gels (12 x 0.6 cm) at 3 mA/gel. The gels contained 7.5% acrylamide and 0.375% bisacrylamide. Mitochondrial samples contained 50 µg protein each. Electrophoresis was stopped when the tracking dye was about 2 cm from the bottom of the gel. After extraction with NCS Tissue Solubilizer, the radioactivity in gel slices was counted in a Packard Tri-Carb liquid scintillation spectrometer. Upper panels of B-D show ratios of incorporation of [^{3}H]Leu into mitochondrial proteins in the presence *versus* the absence of adriamycin.

(between the second and third hours of growth) in the presence of 25 µg of drug per ml, the specific radioactivity in the immunoprecipitate was 135% of the control value. At concentrations of 25-50 µg/ml, the amount of specific radioactivity decreased rapidly and was only about 50% of the control value when the concentration of drug was increased to 1 mg/ml (Fig. 2).

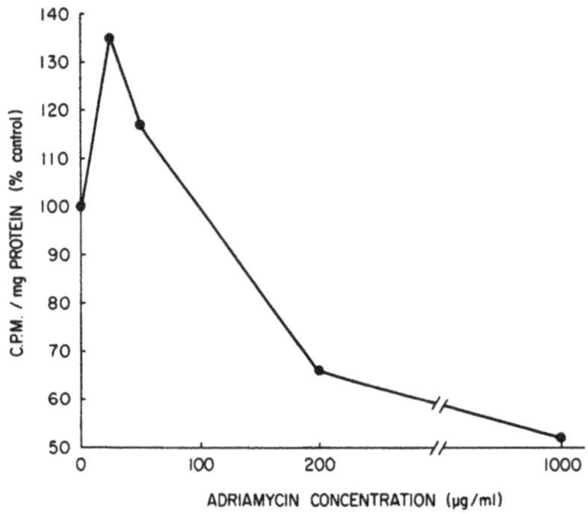

Figure 2 - Radioactivity in immunoprecipitated cytochrome *c* oxidase from yeast mitochondria as a function of increasing adriamycin concentration. Immunoprecipitates were obtained by a previously described method (10).

Thus, adriamycin stimulated protein synthesis in *Candida utilis*, which consequently led to an increase in the net synthesis of all enzymes involved in the electron transport chain and to an increase in cellular respiration and substrate oxidation by isolated mitochondria. Adriamycin appears to be a "superinducer" of protein synthesis (11), but whether this effect is directly attributable to the drug or is mediated through an inductor protein is not clear. Nevertheless, if the effect is protein-mediated, protein synthesis should be super-repressed at the end of the induction, due to an increase in the amount of sequentially activated repressor protein(s). Accordingly, when yeast cells were pulse labeled for 1 hr between the second and third hours of growth (a period corresponding to an increase in the amount of cytochromes in yeast cells, Table III) and the fifth and sixth hours of growth (a period corresponding to a decrease in the amount of cytochromes), in either the presence or absence of adriamycin, specific radioactivity was decreased at the second interval for both the drug-treated and control cells (data from ref. 9). The repression indices (percent decrease in amount of specific radioactivity determined during the second *versus* the first labeling period) showed a twofold higher repression for adriamycin-treated cells, consistent with the concept of a super-repression following superinduction (also see ref. 9).

DISCUSSION

Reconciling the stimulation of protein synthesis induced by adriamycin in *Candida utilis* with the known biological activity of the drug in cell cultures is difficult. In a wide range of experimental systems, adriamycin was shown to inhibit RNA and DNA polymerase activities (12-14) and to block mitotic activity (7, 15). Several alternative explanations can be considered. One possibility is post-transcriptional intervention, in which adriamycin, by inhibiting RNA synthesis directly and protein synthesis indirectly, might retard cytochrome decay or enhance cytochrome biosynthesis. Actinomycin D, for example, specifically enhances the synthesis of tyrosine aminotransferase (11, 16) by inhibiting inactivation of the mRNA that codes for this enzyme. With this drug, however, the enhancement of protein synthesis was limited to one or two proteins. Since, in our study, adriamycin produced a general stimulation of mitochondrial protein synthesis and a net increase in the amount of cytochrome formation, it is unlikely that it affected a post-transcriptional control. Further, the possibility that protein decay was prevented by adriamycin seems remote, since there was a net increase in the specific radioactivity of proteins in drug-treated cells (Fig. 1).

Adriamycin in low concentrations has been shown to intercalate between adjacent base pairs on the DNA strand, with a preference for certain nucleotide sequences (12). Such a mode of attachment provides a molecular basis for transcriptional intervention, in which the drug could either inactivate, repress or stimulate synthesis of an inductor protein. Moreover, adriamycin (at low concentration) should be more specific for mitochondrial DNA, which is small and circular, than for large supercoiled nuclear DNA.

If adriamycin-induced stimulation of protein synthesis in *Candida utilis* is controlled by a repressor protein, one could postulate that the drug inactivates the repressor, leading to stimulation of transcription and translation and hence to an increase in protein synthesis. Under such conditions, the pattern of stimulation of protein synthesis should be significantly different from the control experiment (see Table III). On the other hand, there is yet no indication of inactivation of any protein by adriamycin. Purified cytochrome *c*, for example, was not degraded when incubated with the drug up to 10 hr. Moreover, Goodman *et al.* (12) have shown that adriamycin, although inhibiting RNA and DNA polymerases, does not interact with them. On the other hand, if adriamycin stimulates transcription of mitochondrial DNA, either directly or by stimulating mitochondrial replication, there should be an increase in the synthesis of mitochondrial proteins, including the repressor, so that further transcription in the mitochondria and in the nucleus, as postulated by the repressor model, would be blocked. To the contrary, we observed a stimulation of cytochrome biosynthesis and incorporation of radioactive leucine into mitochondria.

Fig. 3 illustrates the biochemical model most consistent with our results. In this scheme, adriamycin, by stimulating mitochondrial transcription, promotes the synthesis of an inductor protein that acts on both the mitochondrial and nuclear genomes to stimulate and maintain the production of mRNA in the mitochondria and nucleus. A repressor subsequently produced in the mitochondria inhibits further transcription in both the mitochondrial and nuclear genomes. This model is also applicable to our control experiments in which yeast cells transferred to fresh medium showed proportionately the same pattern of cytochrome synthesis as observed for drug-treated cells. Furthermore, our data indicate that the inductor, and possibly the repressor, acts only on the nuclear DNA region responsible for the mitochondrial proteins. Stimulation of mitochondrial protein synthesis by adriamycin was not accompanied, under our experimental conditions, by a significant increase in the synthesis of the cytoplasmic proteins not destined for the mitochondria. This inductor-repressor system takes into account a nuclear origin for the DNA and RNA polymerases that catalyze the replication and transcription of mitochondrial DNA; it assumes that the genes coordinating mitochondrial and nuclear protein synthesis are located in the mitochondria. The coordination of mitochondrial and nuclear genes coding for mitochondrial proteins is crucial to present concepts of mitochondrial biogenesis; the inductor-repressor model postulated in this paper provides a rational basis for such control.

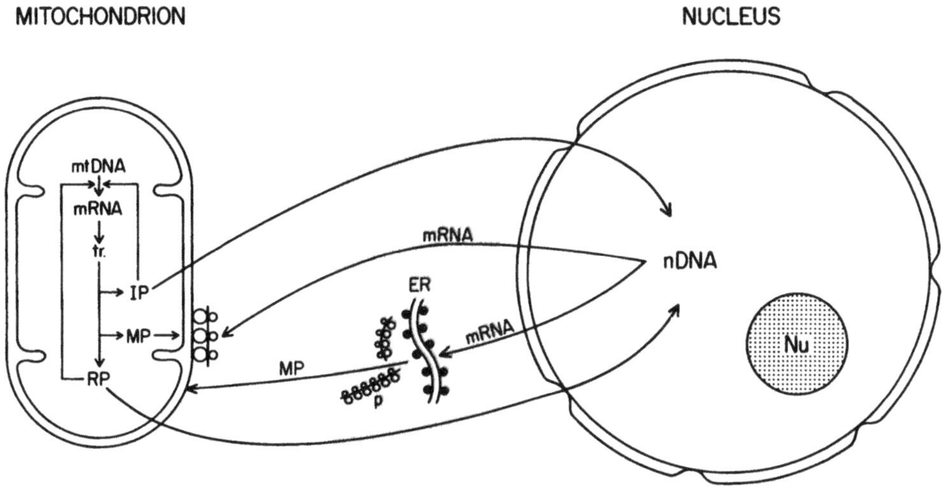

Fig. 3 - Proposed model for interplay between nuclear and mitochondrial DNA coding for mitochondrial proteins. According to this scheme, mitochondrial DNA gives rise to an inductor protein(s) (IP) that acts on both mitochondrial and nuclear DNA to induce transcription of genes coding for mitochondrial proteins (MP). The mRNA coded by nuclear DNA could be translated either on free ribosomes or on ribosomes bound to endoplasmic reticulum (ER) or the outer mitochondrial membrane. Subsequently, a repressor protein(s) (RP) coded by mitochondrial DNA is synthesized in mitochondria and inhibits further transcription of mitochondrial and nuclear genes coding for mitochondrial proteins. Nu = nucleolus; p = polysomes.

ACKNOWLEDGEMENTS

We thank Ms. Eva Bland for technical assistance, Mr. John Gilbert for editorial assistance and Ms. Pat Nicholas for secretarial assistance. This research is supported by ALSAC and a postdoctoral fellowship grant from NCI (CA 05176).

REFERENCES

1. Mason, T.L. and Schatz, G.: Cytochrome *c* oxidase from baker's yeast. II. Site of translation of the protein components. J. Biol. Chem., 248, 1355-1360 (1973).
2. Tzagoloff, A., Rubin, M.S. and Sierra, M.F.: Biosynthesis of mitochondrial enzymes. Biochem. Biophys. Acta, 301, 71-104 (1973).
3. Weiss, H. and Ziganke, B.: Cytochrome *b* in *Neurospora crassa* mitochondria. Site of translation of the heme protein. Eur. J. Biochem., 41, 63-71 (1974).
4. Barath, Z. and Küntzel, H.: Cooperation of mitochondrial and nuclear genes specifying the mitochondrial genetic apparatus in *Neurospora crassa.* Proc. Natl. Acad. Sci. U.S.A., 69, 1371-1374 (1972).
5. Lloyd, D., Turner, G., Poole, R.K., Nicholl, W.G. and Roach, G.I.: A hypothesis of nuclear-mitochondrial interactions for the control of mitochondrial biogenesis based on experiments with *Tetrahymena pyriformis.* Sub. Cell Biochem., 1, 91-95 (1971).
6. Williamson, D.H.: The effect of environmental and genetic factors on the replication of mitochondrial DNA in yeast. Symp. Soc. Exp. Biol., 23, 247-276 (1970).
7. DiMarco, A., Gaetani, M. and Scarpinato, B.: Adriamycin (NSC-123, 127): a new antibiotic with antitumor activity. Cancer Chemother. Rep., 53, 33-37 (1969).
8. Keyhani, E. and Chance, B.: Cytochrome biosynthesis under copper-limited conditions in *Candida utilis.* FEBS Lett., 17, 127-132 (1971).
9. Keyhani J. and Keyhani, E.: Adriamycin-induced stimulation of protein synthesis in *Candida utilis*: implication for control of mitochondrial biogenesis. Submitted for publication.
10. Keyhani, J. and Keyhani, E.: Porphyrin *a* as a precursor of heme *a* in *Candida utilis.* FEBS Lett., 70, 118-122 (1976).
11. Thompson, E.B., Granner, D.K. and Tomkins, G.M.: Superinduction of tyrosine aminotransferase by actinomycin D in rat hepatoma (HTC) cells. J. Mol. Biol., 54, 159-175 (1970).
12. Goodman, M.F., Bessman, M.J. and Bachur, N.R.: Adriamycin and daunorubicin inhibition of mutant T4DNA polymerases. Proc. Natl. Acad. Sci. U.S.A., 71, 1193-1196 (1974).
13. Meriwether, W.D. and Bachur, N.R.: Inhibition of DNA and RNA metabolism by daunorubicin and adriamycin in L1210 mouse leukemia. Cancer Res., 32, 1137-1142 (1972).
14. Chandra, P., Zunino, F., Götz, A., Gericke, D. and Thorbeck, R.: Specific inhibition of DNA-polymerases from RNA tumor viruses by some new daunomycin derivatives. FEBS Lett., 21, 264-268 (1972).
15. Silvestrini, R., Lenaz, L., DiFronzo, G. and Sanfilippo, O.: Correlations between cytotoxicity, biochemical effects, drug levels and therapeutic effectiveness of daunomycin and adriamycin on sarcoma 180 ascites in mice. Cancer Res., 33, 2954-2958 (1973).
16. Tomkins, G.M., Levinson, B.B., Baxter, J.D. and Dethlefsen, L.: Further evidence for post-transcriptional control of inducible tyrosine aminotransferase synthesis in cultured hepatoma cells. Nature New Biology, 239, 9-14 (1972).

Derepression of Mitochondrial Enzymes in Yeast Wild Type and a Mitochondrial Conditional Mutant, tsm8, Retarded in Oxidative Adaptation

Wolfhard Bandlow
Genetisches Institut, Maria-Ward-Str. 1a, D-8000 München 19, Germany

Introduction

Repression by glucose of a variety of extra- and intramitochondrial enzymes is a long known and extensively described phenomenon in bacteria (1,2) and yeast (3-6). Enzymes involved in the catabolism of carbon compounds other than glucose (e.g. those of the citric and glyoxylic acid cycles, of gluconeogenesis, of disaccharide cleavage, and of respiratory and energy conserving function) are drastically reduced in the presence of glucose. The regulatory process occurring upon the addition of glucose enables the cell to adapt its metabolism to the most economical use of carbon sources with respect to growth rate. Although a great body of knowledge has been accumulated about the effects on cellular metabolism of glucose added to non-fermentatively growing yeast cells, almost nothing is known about regulatory processes causing these effects.

One promising way of elucidating regulatory processes is to compare the rate and the extent of changes in enzyme activity after alteration of growth conditions in wild type and mutant cells which show pleiotropic effects after alteration of growth conditions. A mitochondrial conditional mutant was isolated in this laboratory (7) and shown to be retarded in oxidative adaptation even at permissive temperature (8) and to be disturbed in the control of mitochondrial gene expression at non-permissive temperature (9). This mutant was used to study the effect of the lesion on derepression kinetics and on expression at non-permissive temperature of several mitochondrial enzymes some of which contain mitochondrially synthetized components.

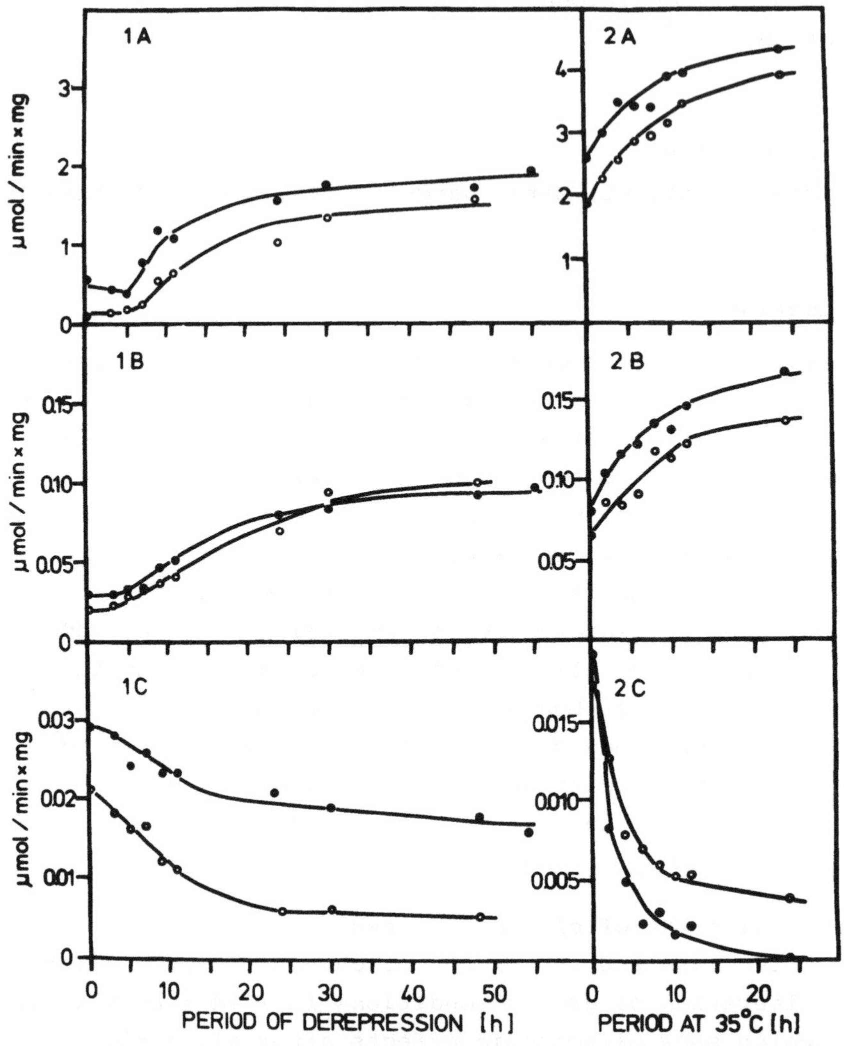

Fig.1. Responses to a transfer of cells to non-fermentative conditions after pre-growth in glucose of mitochondrial malate (A), isocitrate ($NADP^+$) (B) and glutamate ($NADP^+$) dehydrogenases (c). Cells were pre-grown in 10% glucose, transferred to 5% glycerol containing medium and aliquots harvested, broken with glass beads, and mitochondria prepared as will be detailed elsewhere. The enzymes were assayed at 25°C as described in ref.4. o——o, SM 202, wild type; •——•, SM 201 tsm8, mutant.

Fig.2. Responses to a shift of cells to non-permissive temperature (35°C), after pre-growth in glycerol medium at 23°C of malate (A), isocitrate ($NADP^+$) (B), and glutamate ($NADP^+$) dehydrogenases (C). All enzymes were assayed at 25°C. o——o, SM 202, wild type; •——•, SM 201 tsm8, mutant.

Results

Two basic experiments were performed: i) cells of the wild type SM 202 and of isonuclear mutant strain SM 201 tsm8, pre-grown in high glucose (10% glucose, 23°C, $2x10^8$ cells per ml) were transferred to glycerol medium (23°C); ii) cells of the two strains, pre-grown at 23°C in glycerol medium ($2x10^8$ cells per ml) were shifted to non-permissive temperature (35°C, glycerol). In both cases, aliquots were harvested at various periods after the change, cells broken, mitochondria prepared and assayed for enzymatic activities, as indicated in the legends.

Figs. 1A and B reveal that mitochondrial citric acid cycle enzymes like malate and isocitrate ($NADP^+$) dehydrogenases are derepressed with about equal velocities in the two strains. Glutamate dehydrogenase ($NADP^+$) - by contrast - decreases after the transfer to non-fermentative conditions (cf. ref.4) as shown in Fig.1C. But again the kinetics of decrease are comparable in the two strains.

Figs. 2A, B and C illustrate the effect of the temperature shift on malate, isocitrate ($NADP^+$) and glutamate ($NADP^+$) dehydrogenases. The specific activities of all three enzymes increase equally with the period at 35°C in the two strains.

Fig. 3A-D shows the derepression behaviour of enzymes whose function is linked to respiration. The increase in activities of cellular respiration, cytochrome c oxidase, NADH : cytochrome c oxidoreductase and δ-aminolevulinic acid synthetase is severely delayed in the mutant strain when compared to the wild type. This result suggests that mutation tsm8 has some retarding effect on the rate of synthesis of both, respiratory protein components (cf. also 13,14) and heme. The temperature shift experiment ((Fig. 4A-D) demonstrates that the four activities are diminished under non-permissive conditions in the mutant whereas they increase - at least transiently - in the wild type.

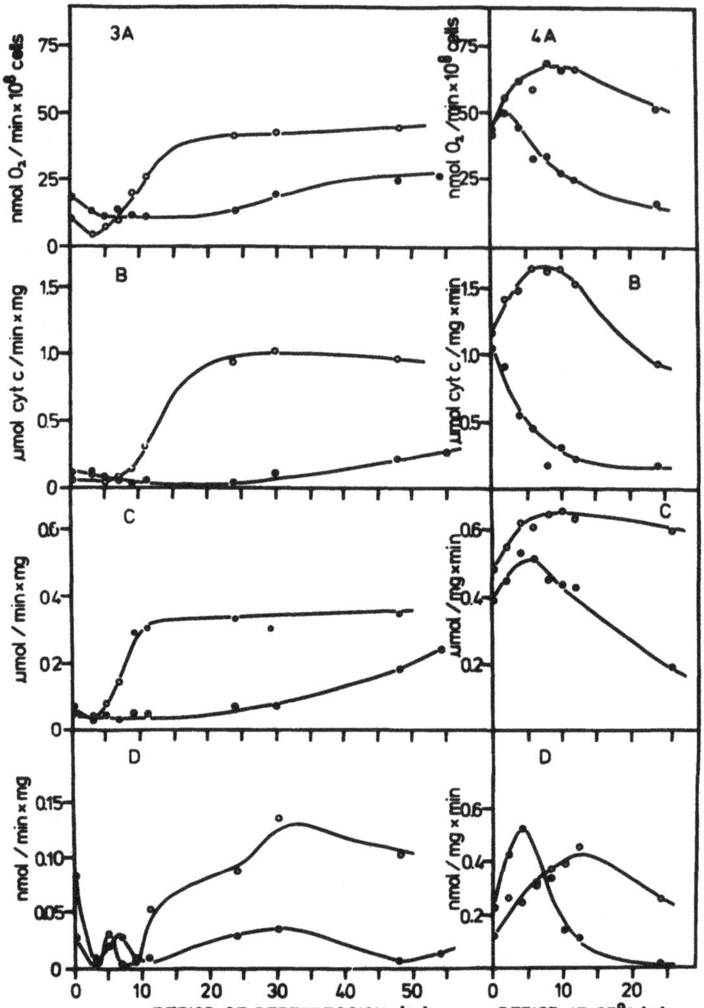

Fig.3. Kinetics of derepression of whole cell respiration (A), cytochrome c oxidase (B), NADH : cytochrome c oxidoreductase (C) and δ-aminolevulinic acid synthetase (D). Cell respiration and NADH : cytochrome c oxidoreductase as described (10). Cytochrome c oxidase activity was measured spectrophotometrically with dithionite reduced cytochrome c as substrate and the activity calculated from the first order rate constant (11). δ-Aminolevulinic acid synthetase was assayed with ^{14}C succinate as substrate (12). o——o, SM 202, wild type; ●——●, SM 201 tsm8, mutant.

Fig.4. Responses to temperature shift (as in Fig.2) of whole cell respiration (A), cytochrome c oxidase (B), NADH : cytochrome c oxidoreductase (C), and δ-aminolevulinic acid synthetase (D). Assays as under Fig.3. o——o, SM 202, wild type; ●——●, SM 201 tsm8, mutant.

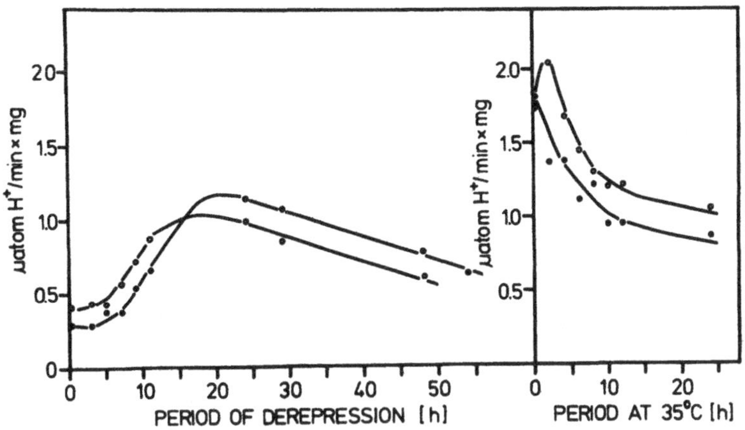

Fig. 5. Kinetics of derepression of oligomycin-sensitive ATPase. The enzyme was assayed at pH 8.8 with a pH-stat combined with a sensitive pH-meter (9). o——o, SM 202, wild type; •——•, SM 201 tsm8, mutant.

Fig. 6. Response to a temperature shift (as in Fig.2) of oligomycin-sensitive ATPase. o——o, SM 202, wild type; •——•, SM 201 tsm8, mutant.

From Fig. 5 it may be seen that oligomycin-sensitive ATPase does not follow the behaviour of the respiratory enzymes tested, but is derepressed with about the same velocity in the two strains. Thus it appears not to be influenced by mutation tsm8. The same conclusion may be derived from Fig. 6 which shows that oligomycin-sensitive ATPase of the mutant strain is only slightly more sensitive to growth at the elevated temperature than the wild type enzyme.

Discussion

It was previously shown (9) that mitochondrial RNA synthesis is the function which responds earliest to a shift of mutant cells to non-permissive temperature. A decrease of transcription rate to a residual level of 10-15% was found both in glycerol and in glucose grown cells. The decrease was found to be reversed by returning to permissive conditions. This result indicates that the temperature effect acts in addition to glucose repression, and is not identical with the glucose effect.

Mitochondrial DNA, protein and ATP syntheses and respiration responded with a significant lag to the decrease in transcription rate. In addition it was found (13, and Bandlow, Kotzias, Klein, unpublished) that the SDS polyacrylamide gel electrophoretic pattern of mitochondrially synthesized proteins is significantly altered when cells of the mutant strain, tsm8, are shifted to 35°C, whereas that of mitochondria from the wild type is not. Bands I and II, with molecular weights of 42 000 and about 30 000 daltons (which presumably contain the mitochondrially synthesized subunits I and II of cytochrome oxidase and cytochrome b) are not made any more under these conditions in the mutant, whereas subunit III of cytochrome oxidase is unaffected for at least 15 h (13). It was concluded from these data that at non-permissive temperature, mutant tsm8 is disturbed in the control of expression of part of the mitochondrial genes, and that strain SM 201 tsm8 is likely to be a mitochondrial regulatory mutant. Since it was found in addition that even at permissive temperature, mutant tsm8 is retarded in oxidative adaptation after anaerobic or aerobic growth on glucose (8), it was tempting to study the kinetics of derepression of glucose repressible enzymes in order to see which of these are affected by mutation tsm8 and may, thus, be controlled by the mitochondrial tsm8 gene product. This paper shows that citric acid cycle enzymes are derepressed with comparable velocity in the wild type and in mutant tsm8. This indicates that these matrix enzymes are not controlled by the tsm8 gene product.

On the other hand, as was also expected from spectra taken during derepression and at various times after the temperature shift (14) and from the alterations of the gel electrophoretic band pattern of respiratory components, respiratory activities which involve either Complex III or IV, are significantly delayed in derepression and are diminished after the temperature shift. Both in the wild type and in the mutant, NADH : cytochrome c oxidoreductase is derepressed faster than cytochrome c oxidase, a result which tends to agree with earlier findings of

Kim and Beattie (15). Since δ-aminolevulinic acid synthetase behaves in a similar fashion, mitochondrial respiration appears to be under twofold control of the tsm8 gene product. Firstly, the mitochondrial synthesis of subunits I and II of cytochrome oxidase and cytochrome b is controlled; and, secondly, heme biosynthesis. Surprisingly, δ-aminolevulinic acid synthetase is retarded although it is coded for by the nucleus. An interrelationship between heme biogenesis and the biogenesis of mitochondrial respiratory components was also suggested from earlier experiments with rat liver mitochondria (16). It follows that mutation of the tsm8 gene affects the expression of both mitochondrial and nuclear genes. Mitochondrial transcription rate decreases in mutant tsm8 without measurable lag immediately after shifting the cells to non-permissive conditions (9). Respiratory functions and δ-aminolevulinic acid synthetase decrease only after a short initial increase in activity. This lag indicates that none of the enzymes is temperature-sensitive per se (a conclusion which is also supported by measurements of these enzymes at 35°C in isolated mitochondria, 9). The initial increase suggests that initially a pool - most likely the pool of messengers - is exhausted and not refilled after the temperature shift. Only then does the activity decrease. The length of the lag appears to be indicative of the turnover and the size of the pool.

All the enzymes tested are, no doubt, under control of glucose repression. Some of these contain mitochondrially synthesized subunits, and others are imported from the cytoplasm. Only part of the enzymes of both groups is additionally controlled by the tsm8 gene product. The comprisal of mitochondrially synthesized constituents is not an absolute requirement for the control of gene expression by the tsm8 gene product, as shown in the case of δ-aminolevulinic acid synthetase. On the other hand, the expression of only some mitochondrially synthesized subunits containing enzyme complexes is under the control of the tsm8 gene product. Oligomycin-sensitive ATPase (and energy conservation, 9), although containing mitochondrially synthesized components, is not controlled by the tsm8 gene product.

Acknowledgements

This work was greatly supported by the skillful technical assistance of Waltraud Hilscher and Gerlinde Dörr. I thank Prof. F. Kaudewitz for his encouragement and continuous interest in the work, and the Deutsche Forschungsgemeinschaft for financial support.

References

1. Epps, H.M.R., Gale, E.F.: Biochem. J. 36, 619-623 (1942)
2. Neidhardt, F.C., Magasanik, B.: Nature 178, 801-802 (1956)
3. Spiegelman, S., Reiner, J.M.: J. gen. Physiol. 31, 175-193, (1947)
4. Polakis, E.S., Bartley, W.: Biochem. J. 97, 284-297 (1965)
5. Polakis, E.S., Bartley, W., Meek, G.A.: Biochem. J. 97, 298-302 (1965)
6. Perlman, P.S., Mahler, H.R.: Arch. Biochem. Biophys. 162, 248-271 (1974)
7. Handwerker, A., Schweyen, R.J., Wolf, K., Kaudewitz, F.: J. Bacteriol. 113, 1307-1310 (1973)
8. Bechmann, H., Krüger, M., Böker, E., Bandlow, W., Schweyen, R.J., Kaudewitz, F.: Molec. gen. Genet. (1977) (in press)
9. Bandlow, W., Metzke, R., Klein, A., Kotzias, K., Doxiadis, I., Bechmann, H., Schweyen, R.J., Kaudewitz, F.: Europ. J. Biochem. 76, 373-382 (1977)
10. Lang, B., Burger, G., Bandlow, W.: Biochim.Biophys.Acta 368, 71-85 (1974)
11. Smith, L., Conrad, H.E.: Arch. Biochem. Biophys. 63, 403-413 (1956)
12. Biggs, D.W., Condie, L.W., Sedman, R.M., Tephly, T.R.: J. Biol. Chem. 251, 4996-4501 (1976)
13. Bandlow, W., Schweyen, R.J. Biochem. Biophys. Res. Commun. 67, 1078-1085 (1975)
14. Bandlow, W.: In "Genetics, Biogenesis and Bioenergetics of Mitochondria" (Bandlow, W., Schweyen, R.J., Thomas, D.Y. Wolf, K., Kaudewitz, F. eds.) pp 179-202, De Gruyter, Berlin (1976)

15. Kim, J.C., Beattie, D.S.: Europ. J. Biochem. 36, 509-518 (1973)
16. Beattie, D.S.: Arch. Biochem. Biophys. 147, 136-142 (1971)

Isolation and Preliminary Characterization of Yeast Mutants Affected in Carbon Catabolite Repression

M. Ciriacy
Institut für Mikrobiologie der Technischen Hochschule
Darmstadt, D-6100 Darmstadt, GFR

Carbon catabolite repression is a well known but still unsolved regulatory phenomenon in the yeast Saccharomyces cerevisiae. It manifests itself as a block by glucose (or other readily fermentable sugars) of the development of mitochondria and of the cell's respiratory capacity as well as on the synthesis of numerous enzymes involved in oxidative metabolism (5, 8, 9, 10). Besides this, glucose causes repression of maltase and invertase synthesis (14). The problem of carbon catabolite repression is further complicated by the fact that glucose does not only repress enzyme syntheses but also induces ready proteolytic degradation of some gluconeogenic enzymes (catabolite inactivation, 6). Several attempts have been made to elucidate the genetic basis of this regulatory mechanism. The isolation of constitutive mutants, i. e. mutants in which glucose does no longer repress enzyme synthesis, is crucial following this approach. Such mutants are available (3, 7, 11, 12, 14, 15). Up to date however, none of these attempts has yielded mutants with a general lack of carbon catabolite repression. In all cases, constitutivity was only found for those functions which are related to the mutant selection procedure. Furthermore, no selective system has been known for the isolation of mutants affected in glucose repression of mitochondrial functions.

A relatively strong selective system for the screening of the latter mutant type is now available using yeast strains completely deficient in alcohol dehydrogenase activity (ADH^-).

The ADH^- phenotype is the result of mutations in the structural genes coding for three different ADH-forms in yeast (for details see: 2, 3). ADH^- cells hardly can grow on 8%glucose medium (doubling time: 8-10h). No growth was observed under anaerobiosis. Moreover, ADH^- were rho^o-lethal (4). Obviously, ADH^- cells are entirely dependent on some mitochondrial functions being intact. The glycolytic energy supply is restricted because of the lack of cytoplasmic NAD-regeneration normally provided by the ADH reaction. Some experiments recently carried out by Rytka et al. (personal communication) do suggest that functioning mitochondria in ADH^- cells are rather necessary for NAD/NADH balance than for ATP production in oxidative phosphorylation. However, mitochondrial functions are repressed when cells grow on glucose. Indeed, this could be shown in ADH^- cells for two mitochondrial enzymes, succinate dehydrogenase and malate dehydrogenase.

Using such mutants we hoped to isolate glucose-insensitive strains as revertants on YEP8%glucose plates. For this purpose, 10^7cells/plate were irradiated with UV. After seven days good growing colonies were picked up and tested for ADH activity to rule out ADH^+ revertants. Out of 64 "revertants" (still ADH^-) eight were selected because of their derepressed levels of succinate dehydrogenase and of malate dehydrogenase when grown on YEP8% glucose liquid medium. To test whether ADH^+ was epistatic over the glucose-resistant phenotype, the newly isolated mutants were crossed to a glucose-sensitive, ADH^+ wild type. A consistent 2:2 segregation of glucose resistance versus glucose sensitivity was observed indicating nuclear gene mutation(s). Recombination between glucose resistance (gene designation: $CCR80^r$) and ADH^+ occurred without significant effect on the repressed enzyme levels as compared to the original mutants. All further experiments were performed with ADH^+, glucose-resistant mutants. A survey on mitochondrial enzyme levels in a wild type and three

Table 1. Enzyme activities in glucose-repressed and derepressed wild type and glucose-resistant CCR80^r strains. r = log-growth on YEP8%glucose, dr = log growth on YEP2%ethanol. SDH = succinate dehydrogenase, NADH:cyt.c = antimycin A-sensitive NADH:cytochrome c oxidoreductase, MDH = malate dehydrogenase; malate synthase is considered as a cytoplasmic enzyme. Values in nmoles/h·mg protein.

genotype	growth	SDH	NADH:cyt.c	MDH	Malate synthase
wild type	r	6	7	1210	1.5
	dr	73	161	7330	450
CCR80-1^r	r	54	46	3480	45
	dr	59	137	8280	543
CCR80-2^r	r	65	80	4240	24
	dr	84	127	7610	344
CCR80-3^r	r	65	75	4316	45
	dr	80	153	7000	501
CCR80-1^r/CCR80	r	51	-	4230	11
	dr	84	-	7450	490

representative mutant strains is given in table 1. Up to 90% of the derepressed SDH activity was found in the mutants, 34% - 63% for antimycin A-senstive NADH:cytochrome c oxidoreductase. The effect on MDH activity was less pronounced. Very probably this reflects the occurence of two or three differently regulated MDH isozymes in yeast (1). Synthesis of the cytoplasmic MDH seemed to be less stimulated in CCR80^r strains when grown on glucose. An analogous situation was found for other cytoplasmic glucose-repressible enzymes also. The values for malate synthase are given in table 1. According to the strongly reduced sensitivity of the synthesis of mitochondrial enzymes to glucose repression, no decrease in O_2-uptake could be observed in CCR80^r mutants after addition of glucose (fig.1).

The nature of the CCR80^r mutation was further elucidated by analysis of CCR80^r/CCR80 heterozygous cells. The significant

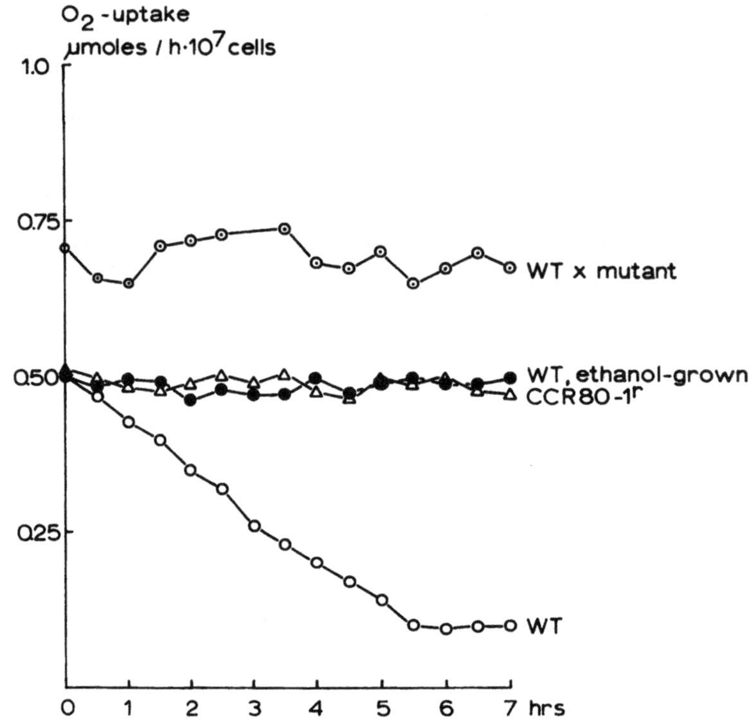

Fig.1. Oxygen-uptake of wild type (WT), $CCR80^r$ glucose-resistant mutant and of a WT x mutant heterozygote during growth on glucose. An ethanol-grown WT is given as a control. Cells were precultured on YEP2%ethanol, providing derepressed cells. Glucose was added at time 0h. O_2-uptake was determined manometrically.

increase of mitochondrial enzyme activities and the apparent lack of repression of respiration (Table 1 and Fig. 1) suggested that $CCR80^r$ was largely dominant over the wild type. Remarkably, dominance was less pronounced with respect to cytoplasmic enzymes (not shown). This fact was used for allelism studies. By this way, all mutants described here were found to be allelic to CCR80. Glucose resistance was obviously not caused by a defect in glucose uptake because of growth on glucose minimal medium. Furthermore, glucose stimulated growth of YEP-ethanol precultured cells, although not as good

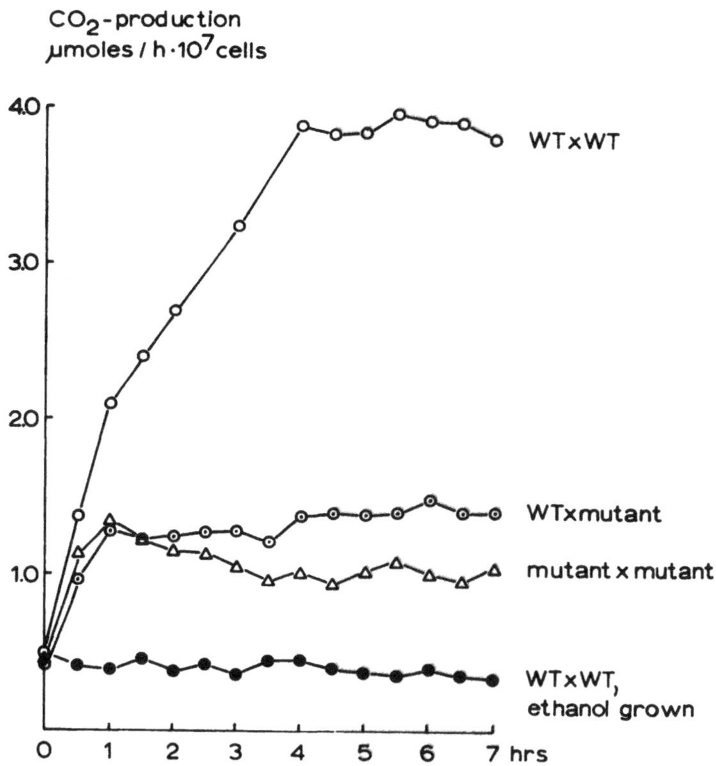

Fig.2. Effect of glucose on CO_2-production of derepressed WT x WT, WT x mutant and mutant x mutant diploids. Glucose was added at time 0h. Ethanol-grown WT x WT (•) as control. CO_2 was determined manometrically.

as in wild type cells. In order to test whether $CCR80^r$ influences aerobic glucose fermentation (5), CO_2-production was measured (Fig.2). A rapid increase was observed in all cases after glucose addition. Most remarkably, in both mutant x mutant and mutant x wild type cells, this increase was immediately stopped after one hour and the fermentation rate remained rather constant for the folling hours. At the first sight, this would suggest a defect in the glycolytic pathway. However, no differences could be observed for glucokinase, hexokinase and phosphofructokinase as compared to the wild type. Moreover, a mere defect of a glycolytic enzyme could be ruled out, because of the dominance of all CCR80^r mutants.

The pleiotropic nature of CCR80^r mutations on the expression of mitochondrial functions normally sensitive to carbon catabolite repression provides evidence for the existence of a regulatory gene. The concomitant effect on the synthesis of cytoplasmic gluconeogenic enzymes and of invertase even suggests a common regulatory element for the expression of all glucose-sensitive functions in yeast (yet unpublished). However, the precise role of CCR80 and of his product remains still unknown. A defect in a repressor gene can be ruled out because of almost complete dominance. Alternatively, the latter trait would fit into a model of positive control considering CCR80 as an activator gene. In the glucose-resistant mutants the function of CCR80 or of its gene product would be independent of the presence or absence of the unknown low molecular effector(s) (13) when the cells grow on glucose. However, the drastic inhibition of aerobic glucose fermentation in CCR80^r mutants does not support such a simple explanation. In fact, the mutants are altered in a dual way: (i) glucose repression of the synthesis of respiratory enzymes is alleviated. This can be considered as a defective gene regulation. (ii) Inhibition of aerobic fermentation shows that metabolic regulation is affected as well. Because of the above mentioned CCR80^r dominance it can be ruled out that the latter effect is caused by a mere enzyme defect. We are currently investigating the flow and intracellular levels of the glycolytic intermediates and of c-AMP in CCR80^r mutants. By this way, we hope to find out the causal regulatory defect leading to glucose resistance.

References:

1. Atzpodien, W., Gancedo, J.M., Duntze, W., Holzer, H.: Isoenzymes of malate dehydrogenase in Saccharomyces cerevisiae. Europ. J. Biochem. 7, 58-62 (1968).

2. Ciriacy, M.: Genetics of alcohol dehydrogenase in Saccharomyces cerevisiae. I. Isolation and genetic analysis of adh mutants. Mutation Res. 29, 315-326 (1975).
3. Ciriacy, M.: Cis-dominant regulatory mutations affecting the formation of glucose-repressible alcohol dehydrogenase (ADHII) in Saccharomyces cerevisiae.Molec. gen. Genet. 145, 327-333 (1976).
4. Ciriacy, M.: Isolation and characterization of yeast mutants defective in intermediary carbon metabolism and in carbon catabolite derepression. Molec. gen. Genet. 154, 213-220 (1977).
5. Deken, R.H. de: The Crabtree effect: a regulatory system in yeast. J. gen. Microbiol. 44, 149-156 (1966).
6. Holzer, H.: Catabolite inactivation in yeast. Trends Biochem. Sci. 1, 178-180 (1976).
7. Montenecourt, B.C., Kuo, S.-C., Lampen, J.O.: Saccharomyces mutants with invertase formation resistant to repression by hexoses. J. Bact. 114, 223-228 (1973).
8. Perlmann, P.S., Mahler, H.R.: Derepression of mitochondria and their enzymes in yeast: Regulatory aspects. Arch. Biochem. Biophys. 162, 248-271 (1974).
9. Polakis, E.S., Bartley, W.: Changes in enzyme activities in Saccharomyces cerevisiae during aerobic growth on different carbon sources. Biochem. J. 97, 284-297 (1965).
10. Polakis, E.S., Bartley, W., Meek, G.A.: Changes in the activities of respiratory enzymes during aerobic growth of yeast on different carbon sources. Biochem. J. 97, 298-302 (1965).
11. Rytka, J., Sledziewski, A., Litwinska, J., Bilinski, T.: Haemoprotein formation in yeast.II. Isolation of catalase regulatory mutants. Molec. gen. Genet. 145, 37-42 (1976).
12. Schamhart, D.J.H., Berge, A.M.A. ten, Poll, K.W. van de: A spontaneous catabolite repression insensitive mutant in Saccharomyces. J. Bact.121, 747-760 (1975).
13. Wijk, R.van, Slavenburg, J.H., Poll, K.W. van: Levels of glycolytic intermediates, adenosine phosphates and polyphosphates in Saccharomyces carlsbergensis under various degrees of catabolite repression. Proc. Kon. Ned. Akad. v. Wetenschap. C76, 512-527 (1973).
14. Zimmermann, F.K., Eaton, N.R.: Genetics of induction and catabolite repression of maltase synthesis in Saccharomyces cerevisiae. Moec gen. Genet. 134, 261-272 (1974).
15. Zimmermann, F.K., Scheel, I.: Mutants of Saccharomyces cerevisiae resistant to carbon catabolite repression. Molec. gen. Genet. 154, 75-82 (1977).

Acknowledgements: This work was supported by Deutsche Forschungsgemeinschaft. The author wishes to thank W. Bandlow, R.J. Schweyen, E. Böker (Munich), F.K. Zimmerman and K.-D. Entian for helpful discussions, and Mrs. Ingrid Scheel for her excellent technical assistance.

The Role of Mitochondria in the Mechanism of Initiation of Differentiation in *Saccharomyces cerevisiae*

H. J. Rhaese, R. Groscurth, and R. Scheckel
Arbeitsgruppe Molekulare Genetik, Institut für Mikrobiologie, Universität Frankfurt, D-6000 Frankfurt, Germany

Deprivation of nutrients of *Saccharomyces cerevisiae* induces ascospore formation (for review see ref.1). The frequency of sporulation is related to the previous growth conditions (2), indicating that sporulation requires a large internal pool of proteins and energy sources to form spores in the absence of extracellular nutrients. Induction of sporulation, however, takes place upon the transfer to sporulation medium, and not at the end of vegetative growth, when most of the nutrients have been exhausted (3). The molecular mechanism responsible for the apparent abrupt change in transcription after starvation, which leads to the formation of ascospores, is completely unknown. It is known, however, for a long time, that respiratory-deficient yeast mutants are unable to sporulate (4). Inhibitors of respiration, like 2,4-dinitrophenol also inhibit sporulation (5).

In *Bacillus subtilis*, where sporulation takes place under similar conditions, we have postulated (6) that a "sensory system" must exist, which translates the information "starvation" into a regulatory substance or substances. This or these substances in turn, should then be responsible for the initiation of a cascade of metabolic changes, known as sporulation. We were able to show in a series of experiments (7, 8) including genetic analysis of asporogenous mutants (9), that sporulation in *B.subtilis* requires the prior synthesis of an unusual highly phosphorylated nucleotide, adenosine 5',3'(2')-bis-triphosphate, p_3Ap_3. The enzyme system, synthesizing this substance is located in the cytoplasmic membrane (6), as shown by in vitro synthesis using membrane vesicles. Two other substances with apparent regulatory functions are synthesized by ribosome associated factors of sporulating cells of B.subtilis (10). These are adenosine 5',3'(2')-bis-diphosphate, p_2Ap_2 and adenosine 5'-tri-3'(2') -diphosphate, p_3Ap_2.

In this communication we would like to present evidence that the same highly phosphorylated nucleotides found in B.subtilis seem to be associated with sporulation in yeast and furthermore, that mitochondria are necessary for their synthesis and apparently therefore for sporulation.

MATERIALS AND METHODS

The diploid strain 1-2[d] (SMC-1B,a his4 MAL2-8[c] MAL3 SUC3 x SMC-5A,α leu1 MAL2-8[c] MAL3 SUC3) kindly provided by K.F. Zimmermann, Darmstadt, was used throughout this study. Growth was monitored by counting cells with a phase-contrast microscope. Yeast extract-peptone-dextrose-medium (YEPD) and sporulation medium III as described by Esposito et al. (11) were used in all experiments. Induction of ascospore formation was achieved by growing cells in YEPD from 1-5 x 10^6 cells/ml to 2-8 x 10^7/ml. After rapid centrifugation in an Eppendorf microcentrifuge, cells were resuspended in the same volume of sporulation medium III containing 0.2-0.6 mCi/ml of $[^{32}P]$ H_3PO_4. Low molecular weight substances were extracted by 1 M formic acid as described before (8). Separation of nucleoside triphosphates and of highly phosphorylated nucleotides by thin layer chromatography has also been described before (8).

RESULTS

The similarity in the conditions for the induction of sporulation in B.subtilis and in S.cerevisiae by starvation for nutrients prompted us to search for highly phosphorylated nucleotides in the eukaryote S.cerevisiae and, if present, to look for a possible correlation of synthesis of these substances with sporulation.

We, therefore, grew up cells to late log-phase and replaced the rich growth medium (YEPD) by 1 % potassium acetate (sporulation medium) containing $[^{32}P]$ H_3PO_4 as described in materials and methods. Sporulation was observed 24 hrs later. Extraction of low molecular weight substances followed by one-dimensional thin layer chromatography and subsequent autoradiography revealed that starved S.cerevisiae cells incorporate among others $[^{32}P]$ into all nucleoside triphosphates and also in at least three substances migrating slower than GTP. The chromatogram appears to be similar to those obtained from sporulating cells of B.subtilis (7). We anticipated, therefore, that adenosine-tetra-, penta-, and hexaphosphates (p_2Ap_2, p_3Ap_2, and p_3Ap_3) are synthesized in response to nutrient starvation (see Fig.1).

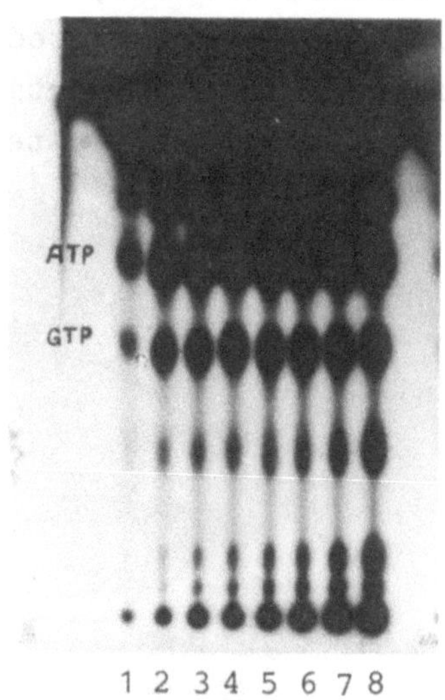

Fig.1. Autoradiogram of a 1 M formic acid extract of S.cerevisiae analyzed by one-dimensional chromatography. Samples were taken at 0-time (1) and at 30 min intervals up to three hrs. Column 8 represents a 6 hrs sample.

However, since it is known that S.cerevisiae synthesizes polyphosphates (12), even though there is ample evidence that polyphosphates disappear during sporulation (13, 14), we analyzed the extracts by two-dimensional thin layer chromatography.

As can be seen in Fig.2, there are mainly three radioactively labelled substances migrating almost as far as ATP in the first dimension, which separates mainly according to the base moiety. All three substances migrate in the second dimension (1.5 M potassium phosphate, pH 3.4) slower than GTP. There are, however, some other substances labelled with [^{32}P], especially those close to the origin. These may be polyphosphates, which apparently smear out of the origin in the first dimension. We have not attempted to characterize all these substances, especially since the concentration of most of them is low.

As will be published elsewhere (Rhaese, Groscurth, and Scheckel, in preparation), we have isolated these substances and analyzed their chromatographic properties in three different two-dimensional chromatography systems. In all

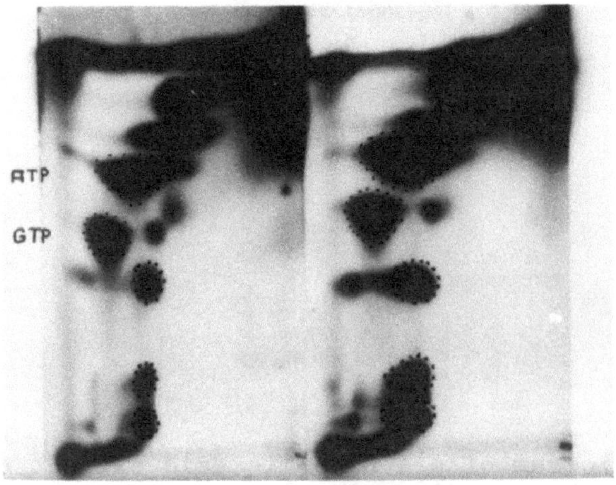

Fig.2. Autoradiogram of a two-dimensional chromatogram of an extract shown in figure 1, taken 180 min (left panel) and 240 min (right panel) after shift into sporulation medium.

systems studied, we found no difference between p_2Ap_2 and p_3Ap_2 synthesized by ribosomal factors from B.subtilis (10) and the two substances migrating directly underneath GTP in the one-dimensional chromatogram (see Fig.1).

Likewise, there was no difference in chromatographic behavior of p_3Ap_3 synthesized by membrane bound enzymes from B.subtilis and the third substance migrating close to the origin in Fig.1. Furthermore, isolation of larger quantities of the apparent p_3Ap_3 and subsequent UV spectroscopy followed by acid hydrolysis show that adenine is the base moiety in this substance (Rhaese, Groscurth, and Scheckel, unpublished data).

It is likely, therefore, that the three substances synthesized by sporulating S.cerevisiae cells are p_2Ap_2, p_3Ap_2, and p_3Ap_3.

In order to see, when and to which extent these highly phosphorylated nucleotides are being synthesized, we took samples of 20 µl at different times after shift to sporulation medium, added 20 µl of 2 M formic acid, and analyzed the extracts for unusual nucleotides by thin layer chromatography.

As can be seen in Fig.3, p_2Ap_2 (●—●), p_3Ap_2 (▲—▲), and p_3Ap_3 (■—■) are synthesized shortly after resuspension of 4 x 10^7 cells/ml, grown in YEPD (⊙—⊙) into sporulation medium (dotted line indicates that no increase in cell titer is observed). Accumulation of the nucleotides continues linearly for three hrs, but then increases logarithmically (notice the drastic increase between 3 and 4 hrs).

Experiments, which will be published elsewhere (Rhaese, Groscurth, and Scheckel, in preparation) show that the increase continues and reaches a peak around 18 hrs after shift to sporulation medium. The increase between 3 and 18 hrs is about 5fold above the concentration at 4 hrs.

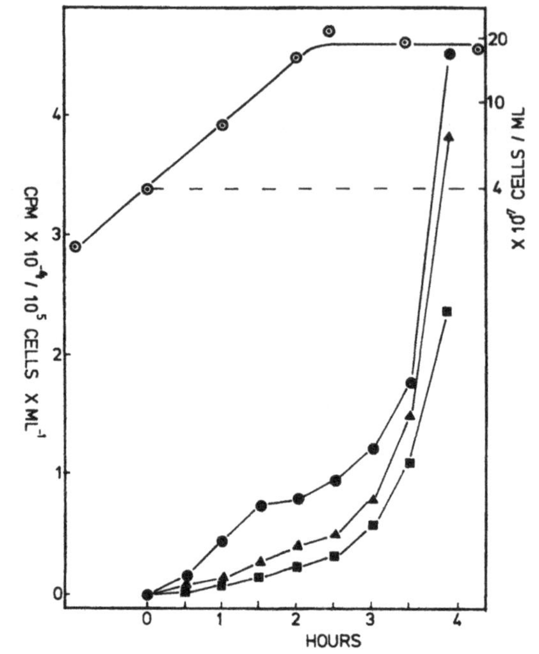

Fig.3. Kinetics of the synthesis of highly phosphorylated nucleotides in strain 1-2[d] after shift to sporulation medium.

These data are similar to those obtained in B.subtilis. But in yeast, synthesis of highly phosphorylated nucleotides continues for much longer. Since without synthesis of p_3Ap_3 sporulation in B.subtilis does not take place (9), we were looking for conditions, which do not allow sporulation in S.cerevisiae in order to correlate synthesis of p_3Ap_3 with the differentiation process in this organism. Mutants, with defective mitochondria, are known to be asporogenous (4). We, therefore, isolated a series of mutants with defective mitochondria and found, that most of them synthesize p_3Ap_3 between 5 to 10 times less than the corresponding wild type.

Fig.4 shows, that the same unusual nucleotides accumulating in the wild type upon starvation are also synthesized in a ρ mutant but at a concentration, which is 10 times lower than that in the wild type (see also Fig.3, notice the different ordinate).

As in the wild type, increase in cell titer (⊙—⊙) stops immediately when cells are resuspended in sporulation medium (dotted line). The highly phosphorylated nucleotides p_2Ap_2 (●—●), p_3Ap_2 (▲—▲), and p_3Ap_3 (■—■) increase only slightly with time. Only p_3Ap_2 seems to increase somewhat faster between 3.5 and 4 hrs. However, all unusual nucleotides remain low and even decrease at a time, when the same substances are at their maximum in the wild type (data not shown).

To show that the highly phosphorylated nucleotides are indeed synthesized to a much lesser extent in mutants with defective mitochondria than in the wild type, and to show that this difference is not due to a lower incorporation of [^{32}P]

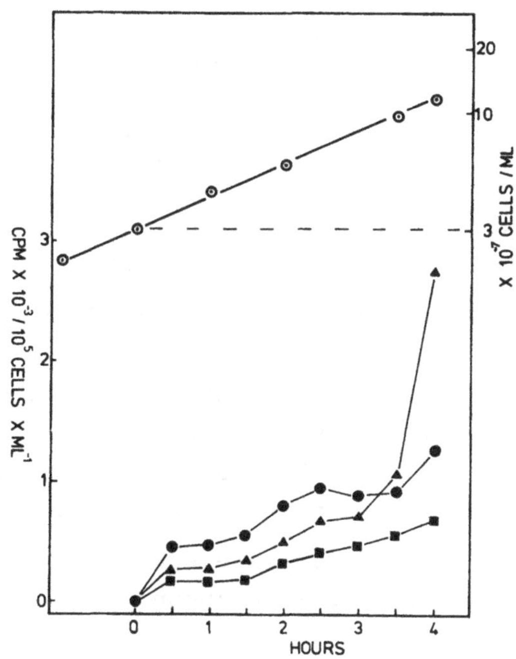

Fig.4. Kinetics of the synthesis of highly phosphorylated nucleotides in a ρ -mutant after shift to sporulation medium.

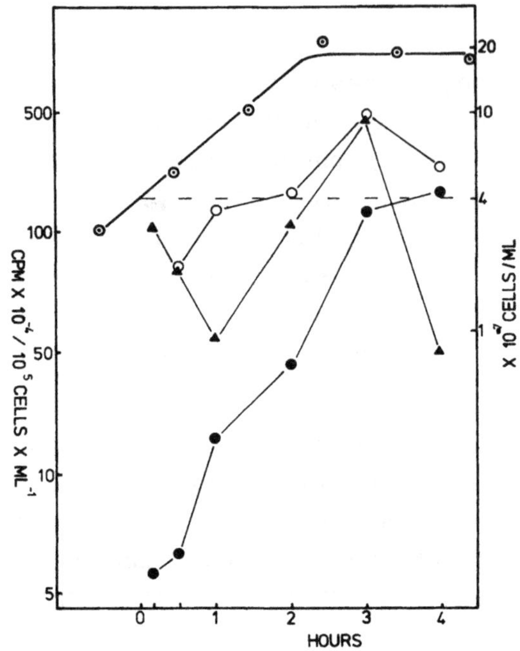

Fig.5. Incorporation of $[^{32}P]$ into whole cells of strain $1\text{-}2^{d}$ and ρ mutant in sporulation medium.

H_3PO_4 in the mutant, we measured the uptake of $[^{32}P]$ into whole cells both in wild-type and mutant.

Fig.5 shows the incorporation of $[^{32}P]$ from H_3PO_4 added to sporulation medium into the wild-type (strain $1\text{-}2^{d}$, ●——●), into strain $1\text{-}2^{d}$ in which sporulation is inhibited by the addition of 1 % ammonium sulfate (○——○), and into the ρ mutant (▲——▲). As can be seen, both strains incorporate $[^{32}P]$ quite efficiently and to the same extent.

Why the ρ mutant seems to excrete some $[^{32}P]$ after three hrs in sporulation medium, is unknown.

DISCUSSION

The experiments reported in this paper show that the same highly phosphorylated nucleotides associated with sporulation in B.subtilis are found in S.cerevisiae, when sporulation is induced by nutrient starvation. This finding indicates that a similar mechanism for the initiation of sporulation may exist in yeast, as has been shown to be responsible for initiation of sporulation in B.subtilis. Data, which will be published elsewhere (Rhaese, Groscurth, and Scheckel, in preparation) also show that inhibition of sporulation in yeast by nitrogen sources prevents synthesis of highly phosphorylated nucleotides as is true for B.subtilis, when sporulation is inhibited by excess glucose.

Since p_3Ap_3 is synthesized by membrane bound enzymes in B.subtilis (6), we also investigated the possibility to synthesize p_3Ap_3 in vitro using S.cerevisiae cytoplasmic membrane vesicles. Preliminary data show (Rhaese, Feller, and Groscurth, unpublished) that p_3Ap_3 can only be synthesized from cytoplasmic membrane vesicles of S.cerevisiae in the presence of mitochondria. Petites (ρ^- mutants) which cannot sporulate, also cannot synthesize p_3Ap_3 or only to a much lesser extent. This seems to indicate the importance of p_3Ap_3 for sporulation. However, mutations leading to strains with defective mitochondria could, among others, affect p_3Ap_3 synthesis (pleiotrophic effect). More work is needed to show the importance of p_3Ap_3 for sporulation and the role of mitochondria in this process.

ACKNOWLEDGEMENT

We would like to dedicate this paper to Prof.Dr.R.W.Kaplan at the occasion of his 65th birthday.
This work was supported by the Deutsche Forschungsgemeinschaft.

REFERENCES

1. Fowell, R.R.: Sporulation and hybridization of yeasts in "The Yeasts" (A.H.Rose and J.S.Harrison, eds.) Academic Press, New York, 1, 303-383 (1969)

2. Haber, J.E., Halvorson, H.O.: Cell cycle dependency of sporulation in Saccharomyces cerevisiae. J.Bacteriol.109 1027-1033 (1972)

3. Croes, A.F.: Induction of meiosis in yeast II. Metabolic factors leading to meiosis. Planta, Berlin, 76, 227-237 (1967)

4. Ephrussi, B., Hottinger, H.: Cytoplasmic constituents of heredity on an unstable state in yeast. Cold Spring Harbor Symp.Quant.Biol.16, 75-85 (1951)

5. Miller, J.J., Halpern, C.: The metabolism of yeast sporulation. Effect of certain metabolites and inhibitors. Can.J.Microbiol.2, 519-537 (1956)

6. Rhaese, H.-J., Groscurth, R.: Control of development, role of regulatory nucleotides synthesized by membranes of Bacillus subtilis in initiation of sporulation. Proc. Natl.Acad.Sci., USA, 73, 331-335 (1976)

7. Rhaese, H.-J., Dichtelmüller, H., Grade, R., Groscurth, R.: High phosphorylated nucleotides involved in regulation of sporulation in Bacillus subtilis. In "Spores VI" (P.Gerhardt, R.N.Costilow, and H.L.Sadof, eds.) American Society for Microbiology, 335-340 (1975)

8. Rhaese, H.-J., Grade, R., Dichtelmüller, H.: Studies on the control of development, correlation of initiation of differentiation with synthesis of highly phosphorylated nucleotides in Bacillus subtilis. Eur.J.Biochem.64, 205-213 (1976)

9. Rhaese, H.-J., Hoch, J.A., Groscurth, R.: Studies on the control of development, isolation of Bacillus subtilis mutants blocked early in sporulation and defective in synthesis of highly phosphorylated nucleotides. Proc. Natl.Acad.Sci., USA, 74, 1125-1129 (1977)

10. Rhaese, H.-J., Groscurth, R.: Studies on the control of development, in vitro synthesis of HPN and MS nucleotides by ribosomes from either sporulating or vegetative cells of Bacillus subtilis. FEBS-LETTERS 44, 87-93 (1974)

11. Esposito, M.S., Esposito, R.E., Arnaud, M., Halvorson, H.O.: Acetate utilization and macromolecular synthesis during sporulation of yeast. J.Bacteriol.100, 180-186 (1969)

12. Wiame, J.M.: The occurrence and physicological behavior of two metaphosphate fractions in yeast. J.Biol.Chem. 178, 919-929 (1949)

13. Mundkur, B.: The nature of basophile particles and vesicular nuclei in Saccharomyces. Expl.Cell Res.25, 1-23 (1961)

14. Mundkur, B.: Formation of the tetrad in Saccharomyces. Expl.Cell.Res.25, 24-40 (1961)

Mitochondria-Cell Relationships in Saccharomyces cerevisiae: Aspects of Sporulation.

E.S. Diala and D. Wilkie
Department of Botany & Microbiology, University College, London, England.

Sporulation in yeast involves a complex series of events apparently initiated by starvation conditions, leading to meiosis and the formation of a tetrad of ascospores. From studies of sporulation-defective mutants, it has been estimated that some 48 nuclear genes control the process (1) but the participation of functional mitochondria is also necessary. Hence the well known facts that respiratory-deficient petite (ρ^-) diploids cannot sporulate and that sporulation cannot proceed in normal diploids under anaerobic conditions (2). Not only is oxidative activity required but mitochondrial biogenesis apparently must also proceed since erythromycin, specifically blocking mitochondrial protein synthesis, prevents sporulation (3) as also does ethidium bromide and acriflavine (4).

Four stages in the meiotic cycle have been recognised (5) but it has not been established at which stage or stages mitochondrial activity is required. Petite mutants, which have neither respiratory activity nor mitochondrial protein-synthesizing capacity, would be expected to block meiosis at the first point of requirement. With the recent discovery of <u>mit</u>$^-$ mutants, which, although respiratory deficient, are ρ+ and are capable of mitochondrial protein synthesis (6), there is the possibility of studying effects on sporulation in diploid cells containing mitochondria with "minimum lesion".

There is also available a number of mitochondrial inhibitors which are known to act either at the level of function or of biogenesis. These can be used in an extension of the studies cited above to gain insight into the role of the mitochondrion in sporulation.

MATERIALS AND METHODS

Yeast strain D6 ($\underline{\alpha}$, $\underline{arg}$,$\underline{met}$) was from the collection of this laboratory. Strain 41/161 ($\underline{a}$, $\underline{ade_2}$) and the $\underline{mit}^-$ mutant C245 derived from it were kindly provided by Dr. H.R. Mahler. C245 lacks sub-unit I of cytochrome oxidase (7). For sporulation assays, strains were pregrown in YEPD medium (1% yeast extract 1% peptone, 2% glucose) to mid-log phase, washed and transferred to a liquid sporulation medium containing 1% potassium acetate, 0.25% yeast extract and 0.1% glucose. Shake culture was carried out at 30°.

Cells were sampled from the sporulating medium at intervals of 3.5h, 7h, 12h, 24h, 36h, 48h, 72h, 96h, and 120h and prepared for electron microscopy. Fixation was in 5% aqueous sodium permanganate for 90 min at room temperature and dehydration in a graded alcohol series followed by embedding in an epoxy araldite resin mixture (EM Scopes Laboratories, London). Sections were routinely post-stained with 2% aqueous uranyl acetate and lead citrate and examined in a Hitachi HS-9 electron microscope operated at 75KV.

ρ^- petites were induced by overnight culture in YEPD containing 10% ethidium bromide.

Mitochondrial inhibitors potassium cyanide, Oligomycin (Sigma), Antimycin A (Sigma), Erythromycin (Abbott) and Adriamycin (gift of Farmitalia, Milano) were added to the sporulation medium at time zero in concentrations which inhibited growth of diploid A (from the cross 41/161 X D_6) in YEG (yeast extract, glycerol) medium but not in YED (yeast extract, glucose).

Electron Microscopy Diploid cultures were obtained as follows

Cross 1 41/161 X D_6
↓
Diploid A —ethidium br.→ Diploid B (ρ^-)

Cross II C245 ($\underline{mit}^-$) X D_6
↙ ↘
ρ+ (80%) $\underline{mit}^-$ (20%) Diploid C

Sporulation of diploid A proceeded through the early stages of nuclear enlargement, fragmentation of the vacuole, proliferation of mitochondria, granular appearance of cytoplasm and the tendency for mitochondria to aggregate around the nucleus.

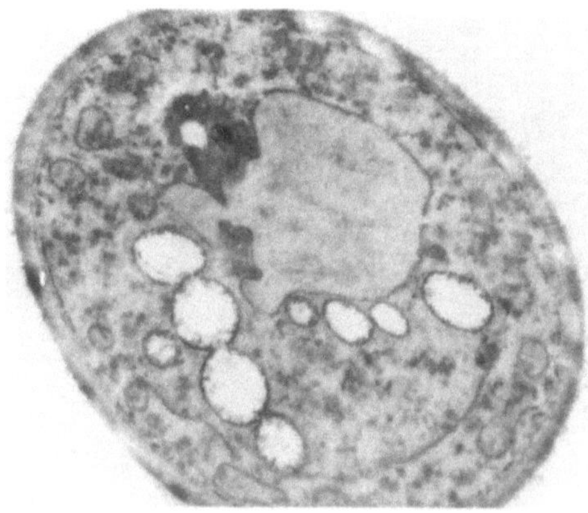

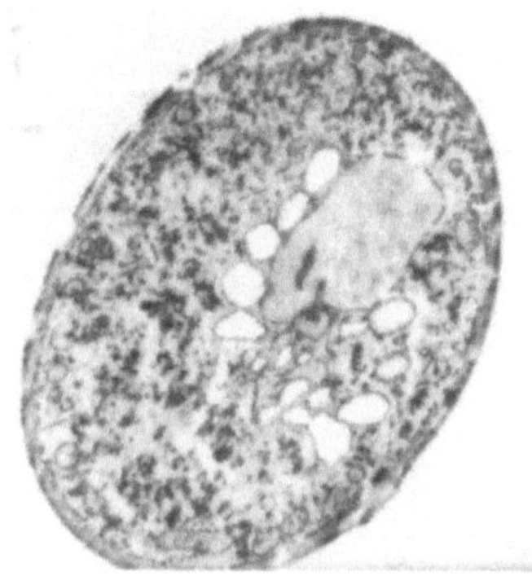

(b)

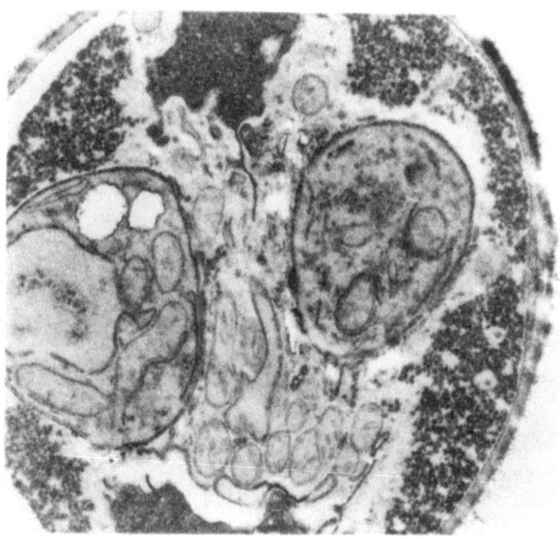

(c)

Figure 1 Stages in sporulation of Diploid A, ρ^+. (1) 24h sample - vacuole fragmentation, enlargement of nucleus. X 17,800. (2) 48h - granular cytoplasm, decrease in size of vacuoles. X 9,000. (3) 72h - ascospore formation, mitochondrial proliferation. X 27,000.

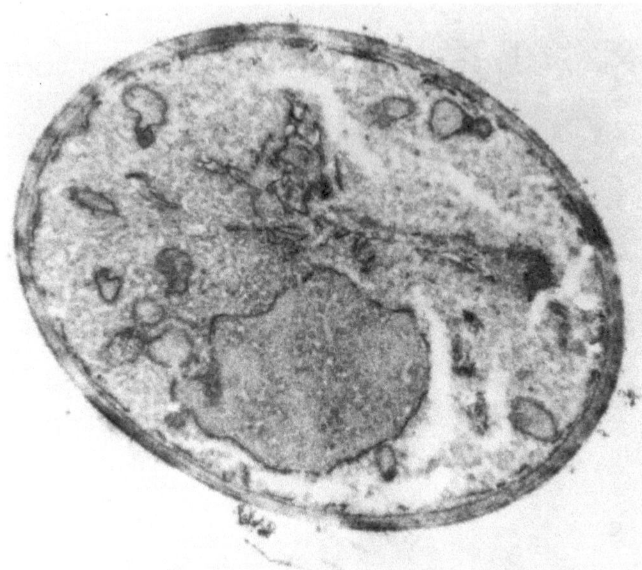

(a)

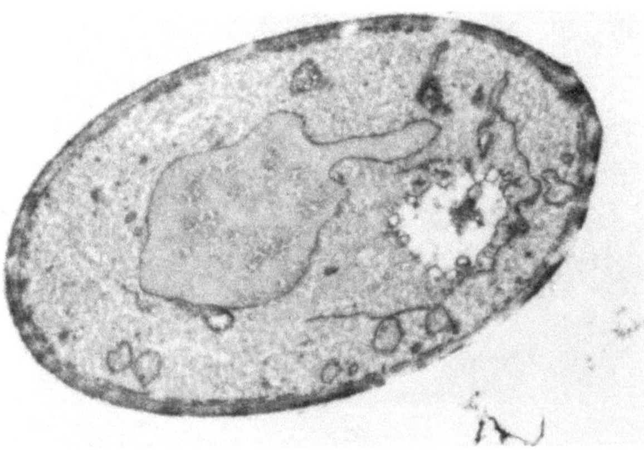

(b)

Figure 2 Diploid B, ρ^-. (1) 72h - dispersion of endoplasmic reticulum and mitochondria. X 31,000.
(2) 120h - apparent nuclear enlargement and elongation, possible indication of stage I of meiosis. X 18,000.

(Fig 1a and b). Later stages of nuclear division and spore formation can be seen in Fig 1c. These are essentially as described by other investigators (1, 5, 8, 9).

The ρ^- diploid B did not sporulate and, when viewed in the electron microscope, showed evidence of a blockage occurring during stage 1 or meiosis (Fig 2). In most respects, the general appearance of cells was similar to the early stages in diploid A, except that mitochondria of diploid B showed membrane aberrations characteristic of petite mutants (10), lacking inner membrane invaginations, and/or showing concentric multiple loops of inner membrane.

Diploid C failed to sporulate but the appearance of cells in the electron microscope was different from those of diploids A and B. Arrest of meiosis seemed to occur at the earliest initiation stage. Nuclei remained unaltered throughout the incubation period of 120h while the cytoplasm became generally devoid of vacuoles, lipid granules and of endoplasmic reticulum. Another peculiar aspect was a tendency to a peripheral distribution of mitochondria, apparently outside the plasma membrane (Fig 3).

Mitochondrial inhibitors. A range of concentrations of inhibitors was used in the sporulation tests but the lowest amounts were found to be effective in blocking ascus formation in diploid A. These were as follows for the respective drugs:- Potassium cyanide, ImM; Antimycin A, 0.5μg/ml; Oligomycin, 1μg/ml; Erythromycin, 1mg/ml and Adriamycin, 100μg/ml. Ascus formation was scored in the light microscope. Maximum sporulation of diploid A was about 30% under the standard conditions and this was reached after 96h incubation.

Teuboi et al (4) reported that addition of 0.1% glucose to their sporulation medium (a solution of potassium acetate) reversed the inhibition of sporulation by ethidium bromide but not inhibition by Antimycin A. Although 0.1% glucose was a constituent in our sporulation medium, none of the inhibitors used (including Antimycin A) was affected in their blockage of sporulation.

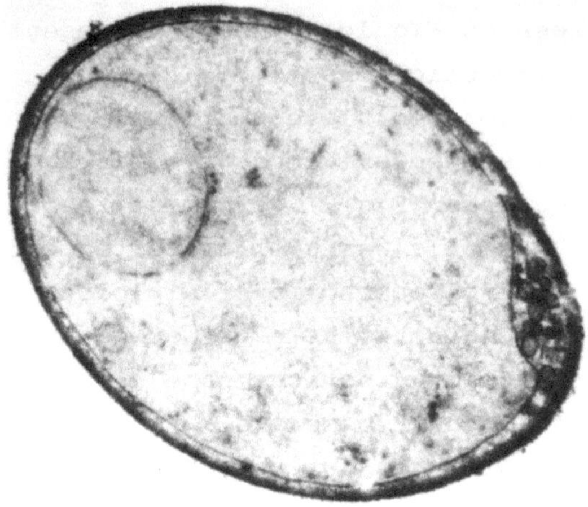

(a)

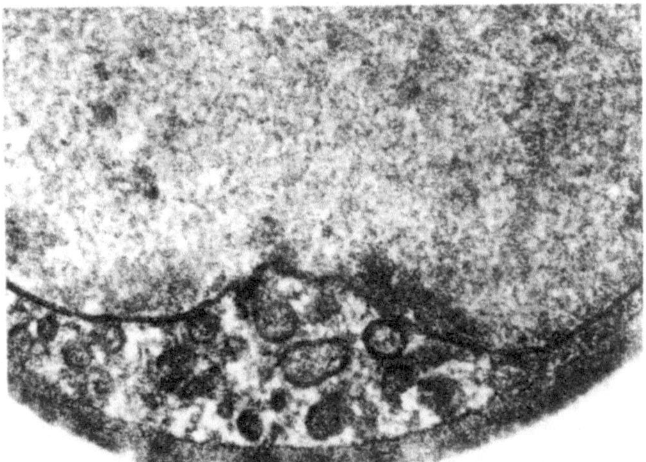

(b)

Figure 3 Diploid C, mit⁻. Both samples at 120h with apparent total blockage of meiosis. Note peculiar distribution of mitochondria. (1) X 27,000. (2) X 78,000.

Discussion. Our studies with the electron microscope show that, as expected, respiratory deficient diploids are blocked in the sporulation process. The results also indicate that the stage at which the blockage occurs, depends on the nature of the lesion in the mitochondria. Apparently, the mit⁻ mutation blocks meiosis at an earlier stage than the petite mutation as well as causing a more general disturbance in the organisation of cytoplasmic structures including the mitochondria. This was rather the reverse of expectation and difficult to explain. The work with specific inhibitors of mitochondria may be helpful when results of electron microscope studies with these inhibitors become available. It is possible for example, that inhibitors of mitochondrial function would be different from inhibitors of biogenesis in their mode of action in blocking sporulation. The effect of the mitochondrial state on sporulation is likely to be influenced by the genetic constitution of the nucleus and experiments are now proceeding to assess this possibility.

Acknowledgement. We would like to thank Mrs. Doreen Collier for expert technical assistance.

References

1 Tingle, M., Singh Klar, A.J., Henry, S.A. and Halvorson, H.O.: Ascospore formation in Yeast. Symp. Soc. Gen. Microbiol., 23, 209-243 (1973).

2 Ephrussi, B. Nucleocytoplasmic Relations in Microorganisms. Oxford, Clarendon Press (1953).

3 Puglisi, P.P. and Zennaro, E. Erythromycin inhibition of sporulation in Saccharomyces serevisiae. Experientia, 27 963-964 (1971).

4 Tsuboi, M., Kondo, K. and Yanagashima, N. Inhibition of sporulation by ethidium bromide and its reversal by fermentable sugars in Saccharomyces cerevisiae. Arch. Microbiol., 99, 295-305 (1974).

5 Moens, P.B. and Rapport, E. Synaptic structures in the nuclei of sporulating yeast, Saccharomyces cerevisiae. J. Cell Sci., 9, 665-677 (1971)

6 Tzagoloff, A., Akai, A. and Needleman, R.B. Properties of Cytoplasmic Mutants of Saccharomyces cerevisiae with specific lesions in Cytochrome Oxidase, Proc. of the Natl. Acad. of Sciences (U.S.A.) 72, 2054-2057, (1975).

7 Mahler, H.R., Bilinski, T., Miller, D., Hanson, D., Perlman, P.S. and Demko, C.A. Respiration deficient mutants with intact genomes: casting a wider net. In: Genetics and Biogenesis of chloroplasts and mitochondria (Th. Bucher, W. Neupert, W. Sebald and S. Werner, eds) pp 857-864 (1976) North Holland, New York.

8 Osumi, M., Sando, N., and Miyake, S. Morphological Changes in Yeast Cells During Sporogenesis, Japan Women's University Journal 13: 70-79 (1966).

9 Guth, E., Hashimoto, T., and Conti, S. Morphogenesis of Ascospores in Saccharomyces cerevisiae. J. of Bacteriology 109: 869-880 (1972)

10 Smith, D.G., Marchant, R., Maroudas, N. and Wilkie, D. A comparative study of the structure of mitochondria in petite strains of yeast. J. Gen. Microbiol., 56, 47-54 (1969).

Note added in proof: It has been found that strain C245 has a large deletion in mitDNA (see Mahler et al., this volume)

DIFFERENCES IN MITOCHONDRIAL ISOACCEPTING tRNAs FROM *S.cerevisiae* AS A FUNCTION OF GROWTH CONDITIONS.

G.Baldacci, C.Falcone, L.Frontali, G.Macino and C.Palleschi
Centro Acidi Nucleici, Istituto di Fisiologia Generale, University of Rome, Rome (Italy).

Chromatographic analysis of mitochondrial tRNA from *Saccharomyces cerevisiae* has shown the existence of multiple isoaccepting species for several aminoacids (1-5). However the problem of the number of tRNA species coded for by mt DNA is still open: differences in primary structure as demonstrated by additivity of plateaux and competition experiments have been reported for cystein (2) and methionine (3), in other cases isoaccepting species seem to have the same primary structure (1), while several cases have not yet been studied. Most of these results have been obtained from yeast cells grown on galactose and collected at late exponential phase. We have previously reported some data (5) indicating the existence on mt DNA of two genes for $tRNA_{ser}$ and of different chromatographic elution profiles of mitochondrial seryl-tRNA from cells collected in different physiological conditions. Owing to better resolution the presence of at least three mitochondrially coded species (m 1, m 2, m 3) is now ascertained, even if the majority of preparations shows the presence of only one or two mitochondrial species (see figures).
The presence of the different mitochondrial $tRNA_{ser}$ species was tested in mitochondrial preparations from cells grown in widely different cultural conditions. Strain 42 (DM) was grown on 1% peptone, 1% yeast extract medium containing 2% glucose, or galactose, or aethanol. Growth was followed both by turbidity measuring and hemocytometer counting: cells were collected at different growth phases and mitochondria were prepared from protoplasts according to the method reported by Faye *et al.* (6). RPC5 chromatography was performed by the method of Pearson *et al.* (7). All other methods have been reported previously (8). Studies of the addivity of plateaux are difficult because of the low hybridization levels which are obtained with mitochondrial seryl-tRNA but indicate lack of addivity between species m 1 and m 2 while additivity is obtained between these and m 3. The presence of the cytoplasmic species (c 1, c 2, c 3) is obviously dependent on the purification procedure but it is always much lower for cells grown on galactose and on aethanol (see Fig.2).

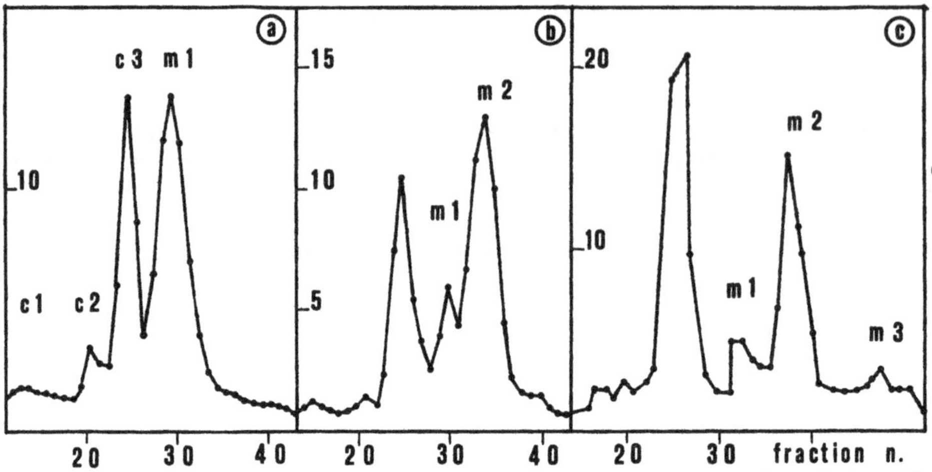

Fig.1 - Elution profiles of seryl-tRNA from mitochondria of cells grown on 2% glucose, a) to mid exponential; b) to late exponential; c) to early stationary phase. c = cytoplasmic, m = mitochondrially coded.

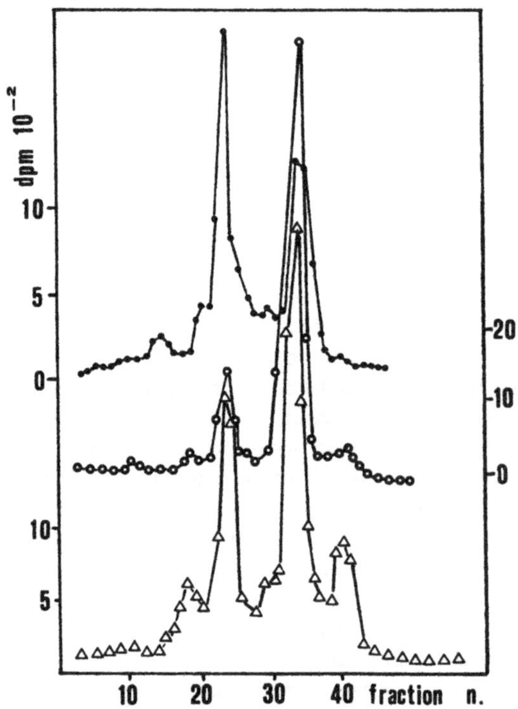

Fig.2 - Elution profiles of seryl-tRNA from mitochondria of cells collected at late exponential phase on different substrates: • glucose; o galactose; △ aethanol.

Chromatographic results show that when cells are collected at different points of the growth curve on 2% glucose (Fig.1), a shift from species m 1 to m 2 is observed, m 1 being the only detectable during the early exponential phase and m 2 becoming predominant in late exponential and in stationary phase (Fig. 1); in the latter a small proportion of a new species, m 3, appears.

Comparison of seryl tRNA prepared from mitochondria of cells grown to the end of exponential phase on different carbon sources shows that the amount of species m 3 is much higher on aethanol and on galactose than on glucose (Fig.2). The shift from peak m 1 to m 2 has been observed also passing from early to late exponential phase on aethanol, but on this carbon source peak m 3 is present also in mitochondria from cells collected at mid exponential phase.

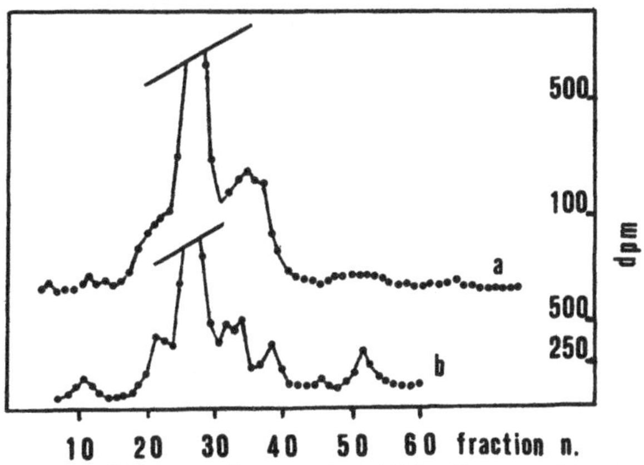

Fig.3 - Elution profiles of seryl tRNA from mitochondria of cells grown to mid exponential phase on 8% glucose (a) and of the same cells aerated for 9 h in the presence of 0.3% glucose (b).

These results indicate a shift from species m 1 to m 2 during the growth curve, probably related to post-transcriptional modifications, while species m 3 seems to be present only in derepressed cells. Further data are necessary to check this point;a first experiment is reported in Fig.3: cells grown to mid exponential phase on high glucose were aerated for 9 h in 0.1M phosphate buffer pH 7.0 containing 0.3% glucose. The elution profile of seryl-tRNA shows the appearance of peak m 3. Differences in the presence of isoacceptors have been observed in bacteria as a function of growth conditions and/or growth phase and between spores and vegetative cells (for a review see 9). The present results suggest that in yeast the metabolic state of cells has a very important influence on the population of aminoacyl-tRNAs available for mitochondrial protein synthesizing machinary which is dependent both on the growth phase and on the repressed or derepressed condition of cells.

REFERENCES

1) Schneller,J.M., Stahl,A.,Fukuhara,H.: Coding origin of isoaccepting tRNA in yeast mitochondria. Biochemie 57, 1051-1057 (1975).
2) Rabinowitz,M.,Jakovcic,S.,Martin,N.,Hendler,F.,Halbreich,A., Lewin,A.,Marimoto,R.: Transcription and organization of yeast mitochondrial DNA. In "The Genetic Function of Mitochondrial DNA", 219-230 (1976) (C.Saccone & A.M.Kroon,eds.), Elsevier/North Holland Biomedical Press, Amsterdam.
3) Martin,R., Schneller,J.M.,Stahl,A.J.C., Dirheimer,G.: Isoacceptor tRNA species in yeast mitochondria. Methionine and formylmethionine specific tRNAs coded by mitochondrial DNA. In "Genetics and Biogenesis of Chloroplasts and Mitochondria", 755-758 (1976) (Bücher et al., eds.), Elsevier/North Holland Biomedical Press, Amsterdam.
4) Martin,N.C. & Rabinowitz,M.: Transfer RNAs of yeast mitochondria. In: "Genetics and Biogenesis of Chloroplasts and Mitochondria", 749-754 (1976) (Bücher et al. eds.) Elsevier North Holland Biomedical Press, Amsterdam.
5) Baldacci,G.,Falcone,C.,Frontali,L., Macino,G.,Palleschi,C.: Isoaccepting $tRNA_{ser}$ in mitochondria from Saccharomyces cerevisiae: mitochondrially coded and cytoplasmic species. In: "Genetics and Biogenesis of Chloroplasts and Mitochondria", 759-762 (1976) (Bücher et al., eds.) Elsevier/North Holland Biomedical Press, Amsterdam.
6) Faye,G., Kujawa,C., Fukuhara,H.: In vivo transcription products of mitochondrial DNA and localization of 23S ribosomal RNA in Petite mutants of Saccharomyces cerevisiae. J. Mol.Biol. 88, 185-203 (1974).
7) Pearson,R.L., Weiss,J.F., Kelmers,A.D.: Improved separation of transfer RNA's on polychlorotrifluoroethylene-supported reversed-phase chromatography columns. Biochim.Biophys.Acta 228, 770-774 (1971).
8) Baldacci,G., Carnevali,F.,Frontali,L.,Leoni,L.,Macino,G., Palleschi,C.:Heterogeneity of mitochondrial DNA from Saccharomyces cerevisiae and genetic information for tRNA.Nucleic Acid Res.2, 1777-1786 (1975).
9) Littauer,U.Z., Inouye,H.: Regulation of tRNA. Ann.Rev.of Biochemistry 42, 439-470 (1973).

Analysis of Mitochondrial Ribosomal Proteins of Paromomycin and Chloramphenicol Resistant Mutants of *Saccharomyces cerevisiae*

G. Faye
Fondation Curie-Institut du Radium, Section de Biologie, Bât. 110,
91405 - ORSAY, France.

The mitochondrial ribosomes of Saccharomyces cerevisiae appear to be composed of two ribosomal RNAs (23S and 16S rRNA) and seventy one structural ribosomal proteins (1,2). The genes of the two rRNA have been localized on the mitochondrial DNA (3,4,5). Although it seems clear that the majority of the mitochondrial ribosomal proteins (r-proteins) are encoded by nuclear DNA, it has not been unambiguously demonstrated that at least one protein that is required for ribosomal functions, is encoded by the mitochondrial DNA (6). The mitochondrial erythromycin resistance mutation (E^R) seems to be in the 23S rRNA gene. Concerning the mitochondrial chloramphenicol and paromomycin resistance mutations (C^R and P^R), it is not clear what functions are affected by these mutations.

Chloramphenicol resistant mutants possess altered mitochondrial ribosomes as revealed by *in vitro* functional tests (7). The ribosomal function of paromomycin resistant mutants has not yet been examined *in vitro*. By analogy with bacteria, we may suppose that the mitochondrial ribosomal apparatus is modified in paromomycin resistant mutants. One of the possibilities is that C^R or P^R mutations might concern mitochondrial r-protein genes, assuming that a few, if any, r-proteins are specified by the mitochondrial genome. I have examined this hypothesis by analyzing the r-proteins of these resistant mutants. The r-proteins were fractionated by two dimensional polyacrylamide gel electrophoresis. The method has a high capacity of resolution for the r-proteins from various organelles including yeast mitochondria (2).

The strains used are listed in Table 1. In order to compare electrophoretic patterns of the r-proteins of wild type and mutant strains, it is important to use isochromosomic strains because some spots show slight variations in position in non-isochromosomic strains.

Two heteroalleles each of the RIB 1 locus, C^R_{321} and C^R_{323}, (frequency of C^S recombinants between them : 0.1 % (8)) and two paromomycin resistance mutations P^R_{454} and a new mutation P^R_{440} (isolated by B. Dujon & L. Weill, unpublished) were analysed. Allelism test performed between the two P^R mutations gave 2.6 % of wild type recombinants (P^S). These two mutations represent therefore two different sites separable by recombination but closely linked to each other (data kindly communicated to me by B. Dujon & L. Weill). Mitochondrial ribosomal subunits from each strain were purified by centrifugation of the mitochondrial lysate in isokinetic sucrose gradients containing 500 mM NH_4Cl, 20 mM Tris-HCl, pH 7.4, 25 mM $MgCl_2$ (2). The purified subunits were treated with ribonucleases. The ribosomal proteins were then analyzed by the two-dimensional polyacrylamide gel electrophoresis method described by Mets and Bogorad (10). The first dimension was in 8 M urea, pH 5.0, the second dimension was in sodium dodecylsulfate

(SDS), pH 6.75. Ribosomal proteins were stained with Coomassie blue.

No differences have been detected in the two-dimensional electrophoregrames of the 37S and 50S ribosomal subunits of the strains M12 and M12/54 (Figure 1) as well as for the strains IL871-3C and IL871-3C/440.

TABLE I

55 R5-3C	ω^-	C^S	E^S	O^S	P^S		
55 R5-3C / 321	ω^-	C^R_{321}	"	"	"	Dujon et al. 1976 (8)	
55 R5-3C / 323	ω^-	C^R_{323}	"	"	"		
M12	ω^-	C^S	E^S	O^S	P^S	Kutzleb et al. 1973 (9)	
M12 / 54	ω^-	"	"	"	P^R_{454}		
IL 871-3C	ω^-	C^R_{321}	E^R_{221}	O^R_I	P^S	Dujon & Weill	
IL 871-3C / 440	ω^-	"	"	"	P^R_{440}	(unpublished results)	

or a difference in methylation in the r-proteins of wild type and mutant strains. In any case, if any r-proteins should be modified by the C^R and P^R mutations, the modifications appear to be extremely small in terms of molecular charge, size and conformation in all the mutants examined. C^R and P^R mutations could concern (i) a protein factor involved in the mitochondrial protein synthesis, (ii) a small RNA acting as 5S rRNA (not yet identified), (or an extremity of the 23S rRNA working as a 5S rRNA), (iii) a protein linking the mitochondrial ribosomes to the mitochondrial membranes , or (iv) a protein involved in the maturation of ribosomes. The first possibility seems to me the most accessible to experimentation, at present.

For the chloramphenicol resistant mutants, no more differences were observed in the r-proteins from the 37S and 50S ribosomal subunits. The mutations C^R_{321} and C^R_{323} have been studied previously by Grivell et al. (7) who have detected no modification of r-proteins by a different approach.

I conclude from these data that the paromomycin and chloramphenicol resistance mutations studied here, are not localized in the genes of the structural proteins of mitochondrial ribosomes. However, the method employed may not detect a substitution of aminoacids without major change in charge

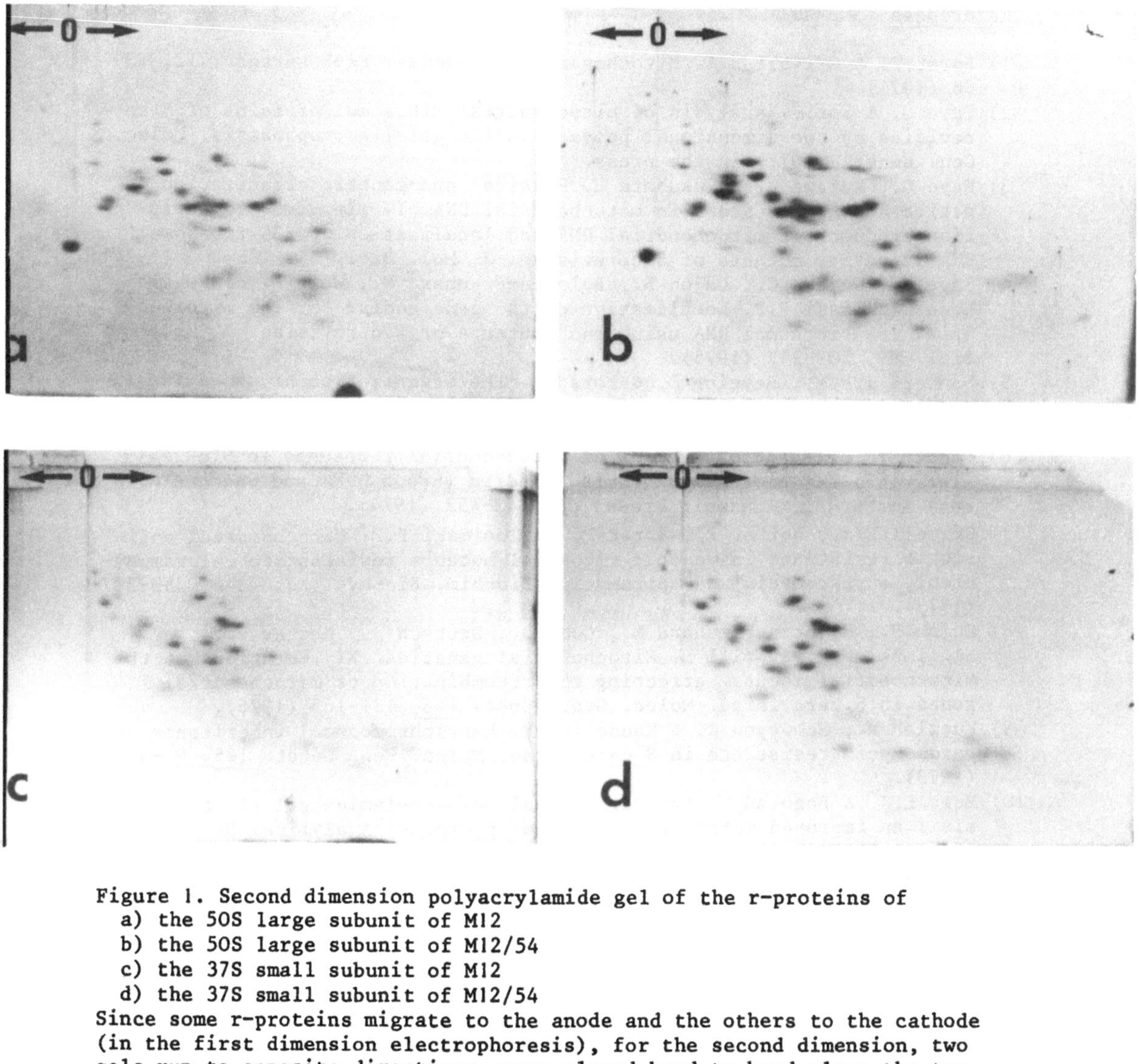

Figure 1. Second dimension polyacrylamide gel of the r-proteins of
a) the 50S large subunit of M12
b) the 50S large subunit of M12/54
c) the 37S small subunit of M12
d) the 37S small subunit of M12/54

Since some r-proteins migrate to the anode and the others to the cathode (in the first dimension electrophoresis), for the second dimension, two gels run to opposite directions, were placed head to head along the top of the second dimension gel slab (2). "0" shows the origin of the first dimension electrophoresis, arrows indicate the directions of migration in the first dimension. In the second dimension, migration was from top of bottom.

Acknowledgments

I thank Hiroshi Fukuhara for critical suggestions, Bernard Dujon and Laurent Weill, Centre de Génétique Moléculaire, for giving me yeast strains. Supported by DGRST 77.7.0306 and 0307.

References

1) Borst P. & Grivell L.A. Mitochondrial ribosomes. FEBS Letters, 13, 73-88 (1971).
2) Faye G. & Sor F. Analysis of mitochondrial ribosomal proteins of S.cerevisiae by two dimensional polyacrylamide gel electrophoresis. Molec. Gen. Genet. (1977) in the press.
3) Faye G., Kujawa C. & Fukuhara H. Physical and genetic organization of petite and grande yeast in mitochondrial DNA. IV. In vivo transcription products of mitochondrial DNA and localization of 23S ribosomal RNA in petite mutants of S.cerevisiae. J. Mol. Biol. 88, 185-203 (1974).
4) Faye G., Kujawa C., Dujon B., Bolotin-Fukuhara M., Wolf K., Fukuhara H. & Slonimski P.P. Localization of the gene coding for the mitochondrial 16S ribosomal RNA using rho$^-$ mutants of S.cerevisiae. J. Mol. Biol. 99, 203-217 (1975).
5) Sanders J.P.M., Heyting C. & Borst P. The organization of genes in yeast mitochondrial DNA. I. The genes for large and small ribosomal RNA are far apart. Biochem. Biophys. Res. Commun. 65, 699-707 (1975).
6) Groot G.S.P. The biosynthesis of mitochondrial ribosomes in S.cerevisiae. In : The biogenesis of mitochondria (Kroon A.M. and Saccone C. eds) Amsterdam, Academic Press, pp. 443-452 (1974).
7) Grivell L.A., Netter P., Borst P. & Slonimski P.P. Mitochondrial antibiotic resistance in yeast : ribosomal mutants resistant to chloramphenicol, erythromycin and spiramycin. Biochim. Biophys. Acta 312, 358-367 (1973).
8) Dujon B., Bolotin-Fukuhara M., Coen D., Deutsch J., Netter P., Slonimski P.P. & Weill L. Mitochondrial genetics. XI. Mutations at the mitochondrial locus affecting the recombination of mitochondrial genes in S.cerevisiae. Molec. Gen. Genet. 143, 131-165 (1976).
9) Kutzleb R., Schweyen R. & Kaudewitz F. Extrachromosomal inheritance of paromomycin resistance in S.cerevisiae. Molec. Gen. Genet. 125, 91-98 (1973).
10) Mets L.J. & Bogorad L. Two dimensional polyacrylamide gel electrophoresis : an improved method for ribosomal proteins. Analytical Biochem. 57, 200-210 (1974).

THE MITOCHONDRIAL GENETIC MAP OF SACCHAROMYCES CEREVISIAE : COMPILATION OF MUTATIONS, GENES, GENETIC AND PHYSICAL MAPS.

B. DUJON, A.M. COLSON (†) and P.P. SLONIMSKI
Centre de Génétique Moléculaire du CNRS, 91190 Gif sur Yvette, France, and (†) Laboratoire d'Enzymologie, Université de Louvain, 1348 Louvain-la-Neuve, Belgium.

Contents of the article :

- INTRODUCTION AND GENERAL COMMENTS .... page 580
- TABLE 1, LIST OF MUTATIONS AND GENES .... page 585
  . mutations selected as conferring drug resistance to
    chloramphenicol .... page 585
    erythromycin .... page 589
    spiramycin .... page 595
    paromomycin .... page 597
    oligomycin (rutamycin) .... page 598
    venturicidin .... page 603
    antimycin A .... page 604
    funiculosin .... page 605
    diuron (DCMU) .... page 607
    mucidin .... page 608
  . mutations conferring deficiencies in mitochondrial functions
    oxidase .... page 609
    reductase .... page 628
    ATPase .... page 637
    multiple deficiencies .... page 638
    conditional deficiencies for various or undefined functions .... page 639
    undefined deficiencies .... page 640
  . other genes
    ω locus .... page 641
    polymorphic proteins .... page 642
    ribosomal RNAs .... page 643
    transfer RNAs .... page 645
  . legend of table 1 .... page 648
  . list of laboratories quoted in table 1 ...... page 649
- FIG.1 "AVERAGE" GENETIC MAP OF MITOCHONDRIAL DNA. page 651
- FIG.2 DETAILED MAPS OF EACH SECTOR .... page 652
- FIG.3 OVERALL RESTRICTION MAPS OF MITDNA MOLECULES FROM 6 YEAST STRAINS .... pages 657 & 658
- FIG.4 DETAILED RESTRICTION MAPS OF THE LARGE rRNA and ω REGION OF THE MITDNA. .... pages 657 & 659
- REFERENCES .... page 660

INTRODUCTION AND GENERAL COMMENTS.

During the last few years our knowledge of the mitochondrial genome of *Saccharomyces cerevisiae* has developped considerably. Many new mutations located in the mitDNA and conferring various phenotypes have been found by different laboratories and several genes have been identified. New chemical and physical methods of analysis of DNA molecules have been successfully applied. This combined genetic and molecular approach has lead to a rapid progress of knowledge as witnessed by numerous recent publications and symposia (cf. 1, 2, 3, 4). Concomitantly with this progress some diversity, if not confusion,has arisen : e.g. allelism tests between similar mutations used by different laboratories (or even in the same laboratory) have not been performed; several genetic and/or physical maps have been published some of them stating the overall arrangement of genes, loci, or mutational sites, others showing fine structure analyses of particular regions, the maps however being oriented in almost all possible manners; different nomenclatures of genes and mutants have been used.

In view of such a proliferation of data, which is the price paid for progress, some uniformity seems necessary and possible. The purpose of this article is threefold. First to present an up to date compilation and classification of mutations and genes using the original nomenclatures and descriptions as published in the corresponding references. Second to present a comprehensive and up to date compilation of genetic and physical maps with the corresponding literature references. Third to propose an "average map" of the mitochondrial genome integrating the overall arrangements of the different maps, for which agreement is generally found, with the fine structure analysis of particular regions as they were described by different authors. In this work we have not attempted to uniformize the nomenclature. A proposal by BIRKY, COLSON, PERLMAN et al. (in preparation) should be consulted on this subject.

1. Mutations and genes localized on the mitDNA of *Saccharomyces cerevisiae*. Mutations and genes are listed in table 1. In order to establish this list we have adopted the following principles :

(1) Only mutations or genes which have been unambiguously localized on the mitDNA have been considered. All the non-mendelian mutations and/or genes of *S. cerevisiae* which have not been definitely allocated to the mitDNA or which have been allocated to other cytoplasmic factors, have been disregarded (e.g. psi, Killer, π etc ..). In addition, the ρ^- mutations, although extremely useful in mitochondrial genetics have not been listed since almost an unlimited number of them can be created, each one corresponding to large and various deletions of the mitDNA and reiteration of retained segments.

(2) Only mutations or genes which have already appeared in the literature at the time of this meeting (August 13th, 1977), including those presented at this meeting, have been considered. For the latter they are quoted as shown on the posters or preprints of manuscripts kindly communicated to us by the authors.

(3) In the list, mutations and genes are referred to by the symbols originally used by the authors. When different nomenclatures have been used for the same mutation (a single isolate) the equivalence is given only if explicitly stated in original articles.

(4) When explicitly stated, the parental strain of the original mitochondrial mutation as well as the strains used for the identification of the genes are mentioned for reasons discussed in section 3.

(5) Mutations have been subdivided in two classes : mutations selected as conferring drug resistance to the cell and mutations isolated as deficient in mitochondrial functions. Each class has been further subdivided into subclasses according to similarities of phenotype, function(s) affected or genetic location. Such a subdivision is used for convenience of the classification only and does not imply that all the mutations listed in one given subclass have been proven to concern the same function or the same gene. In each subclass, the mutations have been classified for convenience according to laboratories or groups of authors (a list of laboratories is given at the end of table 1). They have been ordered as far as possible by the year of appearance of the first publication of the group clearly stating the mitochondrial determinism

of at least one of the mutations belonging to that subclass.
Other genes have been subdivided in four classes according to their nature : genes for ribosomal RNA, transfer RNA, ω, and polymorphic proteins.

(6) Mutations and genes have been assigned to genetic loci or sectors of the genetic map according to the original articles (see section 3).

(7) When known, the probable gene product is indicated.

2. Methods used for mapping of mitochondrial mutations, genes, loci or clusters of mutational sites.

Several methods (genetic and physical) have been used by different authors (see table 1) in order to establish the map of the mitochondrial DNA of various strains. The methods used to construct the maps presented here are abbreviated as follows.

G1 measurement of the frequency of recombinants in crosses (ρ^+ by ρ^+ or ρ^+ by ρ^-) heterozygous for two or more mitochondrial markers.

G2 measurement of the frequency of separation of two or more mitochondrial markers by deletions (ρ^+ to ρ^-; ρ^- to ρ^-; ρ^- to ρ^o).

G3 marker separation by ρ^- deletions. The distinction between G2 and G3 is of quantity rather than of quality. When a large sample of deletions is analyzed in a quantitative manner the method is referred to as G2; an analysis of one or a few ρ^- clones is G3.

P1 RNA/DNA hybridization with total mitDNA and/or restriction fragments isolated from ρ^+ or ρ^- strains.

P2 DNA/DNA hybridization with total mitDNA and/or restriction fragments isolated from ρ^+ or ρ^- strains.

P3 mapping of restriction endonuclease cleavage sites in mitDNA.

P4 electron microscopy, partial denaturation mapping and/or heteroduplex mapping of mitDNA.

P5 analysis of in vitro transcription or translation products of mitDNA.

P6 analysis of buoyant density and/or high resolution melting of mitDNA.

It should be remembered that this classification is rather schematic and for the reader's guidance only. The methods are not necessarily independent one from another.

3. Definition of the "average map", sectors and boundaries.

There is increasing evidence (12A, 20A, 24, 73, 82, 86A, 103, 129A, 129B, 143A, 143B, 154, 154A, 154C, 190) in favor of the idea that various yeast strains, each regarded as wild type, differ significantly in their mitDNA sequence. Such strains are thus non-isomitochondrial or anisomitochondrial (36). Although the relative order of the major genetic markers (especially the boundaries, see later) studied so far, seems to be coherent from strain to strain, it is clear that distances between the same two genes in two different strains may not always be the same. In such a situation there may exist as many different fine structure maps as there exist different strains. However a complete fine structure mapping of the entire mitDNA has not been performed as yet for any strain. Furthermore, the situation becomes very rapidly complex if one considers that the same mutation is, in most cases, transferred through successive crosses and recombinations between nonisomitochondrial strains. In order to take into account this problem we have adopted the idea of presenting a unique map as the "average map", using sectors delimited by boundaries. Five boundaries have been defined. They have been termed B1 to B5. They represent the genetic location for the most classically studied drug-resistance mutations (chloramphenicol, erythromycin, oligomycin and paromomycin). Each boundary is defined by a series of homo- or heteroallelic mutations shown to belong to the same genetic locus or cluster of mutational sites by usual criteria (see BIRKY, COLSON, PERLMAN *et al.* in prep.). We have choosen as boundaries these loci for the following reasons :

i) if one considers that the following mutations (see table 1 for details) are included in each locus, general agreement is found among the authors as to their relative order on the map :

boundary B1 : locus CAP : mutations : C^R_{36}, C^R_{37}, C^R_{61}, C^R_{321}, C^R_{323}, C^R_{324}, C^R_{325}, C^R_{326}, C^R_{327}, C^R_{328}, C^R_{329}, C^R_{330}, C^R_{336}, C^R_{517}, cap1-r, cap2-r, spi3-r.

boundary B2 : locus ERY : mutations ery1-r, ery2-r, E^R_{221}, E^R_{236}, E^R_{353}, E^R_{514}, S^R_{352}, spi2-r.

boundary B3 : locus OLI1 : mutations O^R_1, O^R_2, O^R_4, O^R_5, O^R_6, O^R_7, O^R_8, O^R_{10}, O^R_{145}, O^R_{146}, O^R_{147}, O^R_{148}, O^R_{149}, O^R_{152}, O^R_{155}, O^R_{156}, oli17-r(B), OLG2(B), O^R_{625}, O^R_{626}, O^R_{627}, O^R_{628}, O^R_{629}, V^R_{60}.

boundary B4 : locus OLI2 : mutations O^R_{143}, O^R_{144}(C), O^R_{150}, O^R_{151}, O^R_{153}.

boundary B5 : locus PAR : mutations P^R_{454}, P^R_{455}, par1-r.

ii) The choice of the above mentioned loci is mnemomic because C (for CAP), E (for ERY), O_1 (for OLI1), O_2 (for OLI2) and P (for PAR) follow the alphabetical order. To facilitate memorizing this alphabetical order will be always presented from left to right when placed on linear maps and clockwise when placed on circular maps, point C, or boundary B1, being set at 12 o'clock.

Five sectors are delimited by the boundaries. They have been termed S1 to S5 (clockwise on the map of Fig.1). Each sector is defined as a continuous segment between its two boundaries. All genes or mutations which do not belong to the 5 most classically studied drug resistance loci, taken as boundaries, will necessarily be placed in the sectors.

This presentation permits to place in the same sector the different mutations or genes which have not always been ordered relatively to each other. However, it should be stressed that it has not been proven that all the mutations listed above as presumably belonging to the same locus are true homo- or heteroalleles. Furthermore, general agreement is found among the authors as to their relative order on the map, but not necessarily as to their relative distances, since the same sector may differ from one strain to another in terms of length, genetic composition or arrangement of genes or mutations. Further work is obviously necessary to solve this problem.

4. Conclusion.

This article represents the first attempt to present a complete list of all mutations or genes already known in the mitochondrial genome of *Saccharomyces cerevisiae* (and *S. carlsbergensis*). At the same time we have tried, using a uniform system, to compare the different maps presented by the various authors. We are aware of the fact that this attempt cannot be perfect in all points. Undoubtedly as learning in this field will progress in the future the number of mutations, genes and functions investigated will increase and simultaneously the comparison between maps will become easier and more precise. Nevertheless it is our hope that the present compilation can be of some help for the specialists working in the field as well as for anyone interested.We apologize for all errors and omissions which could have occured during this literature survey.

TABLE 1 - Page 1

CLASS		ORIGINAL NAME	PARENTAL STRAIN	MAIN PHENOTYPE	MAP & LOCUS ASSIGNMENT	MAPPING METHOD	PROBABLE GENE PRODUCT	REFERENCES
CHLORAMPHENICOL	GIF	C^R_{36}	LM650-2C	chloramphenicol resistance	B1 RIB1(=RI)	G1		16, 36, 47, 48, 201
		C^R_{37}	LM650-2C	chloramphenicol resistance	B1 RIB1(=RI)	G1		36
		C^R_{61}	LM650-2C	chloramphenicol resistance	B1 RIB1(=RI)	G1		36
		C^R_{321} renamed cml^R in (50) cap 321 (204) [CHL]R321 in (138)	55R5-3C	chloramphenicol resistance	B1 RIB1(=RI)	G1, G2, G3, P1, P2, P3 P4, P6	ribosome (72) ribosomal RNA (?) (129A, 154)	10, 16, 17, 30, 31, 34, 36, 42, 42A, 46, 47, 48, 49, 49A, 50, 51, 52, 53, 54, 55, 62, 63, 65, 72, 86A, 94, 96, 100, 102, 110, 110A, 111, 119, 121, 121A, 129, 134, 137, 138, 159, 164, 171, 173 185, 187, 194, 201, 203, 204
		C^R_{323}	55R5-3C	chloramphenicol resistance	B1 RIB1(=RI)	G1	ribosome (72)	36, 47, 49, 72
		C^R_{324}	55R5-3C	chloramphenicol resistance	B1 RIB1(=RI)	G1, P3		36, 47, 86A, 134
		C^R_{325}	55R5-3C	chloramphenicol resistance	B1 RIB1(=RI)	G1		36, 47
		C^R_{326}	55R5-3C	chloramphenicol resistance	B1 RIB1(=RI)	G1		36, 47, 49
		C^R_{327}	55R5-3C	chloramphenicol resistance	B1 RIB1(=RI)	G1		36, 47, 49
		C^R_{328}	55R5-3C	chloramphenicol resistance	B1 RIB1(=RI)	G1		36, 47, 49

TABLE 1 - Page 2

CLASS		ORIGINAL NAME	PARENTAL STRAIN	MAIN PHENOTYPE	MAP & LOCUS ASSIGNMENT	MAPPING METHOD	PROBABLE GENE PRODUCT	REFERENCES
CHLORAMPHENICOL	GIF	C^R_{329}	55R5-3C	chloramphenicol resistance	B1 RIB1(=RI)	G1		36, 47, 49
		C^R_{330}	55R5-3C	chloramphenicol resistance	B1 RIB1(=RI)	G1		36, 47
		C^R_{336}	55R5-3C	chloramphenicol resistance	B1 RIB1(=RI)	G1		36, 47
		C^R_{517} renamed [CHL]R517 in (138)	DP1-1B	chloramphenicol resistance	B1 RIB1(=RI)	G1		36, 47, 134, 138
	CLAYTON	cap1-r		chloramphenicol resistance	B1 RIB1(=RI) (109)	G1, G2, G3, P1, P2	ribosome (124)	21, 86, 107,109,124,126 132, 133, 166, 181, 188
		cap2-r		chloramphenicol resistance	B1	G1, G2 G3, P2	ribosome (124)	107, 124, 126
		CLr	N1300	chloramphenicol lincomycin, tetracycline resistance				21
	MINNEAPOLIS (A)	3725C	X97923	chloramphenicol resistance				90
		3729C	X97923	chloramphenicol resistance				90
		4107C	S1-3C	chloramphenicol resistance				59, 90
		4111C	S1-3C	chloramphenicol resistance				90

TABLE 1 - Page 3

CLASS		ORIGINAL NAME	PARENTAL STRAIN	MAIN PHENOTYPE	MAP & LOCUS ASSIGNMENT	MAPPING METHOD	PROBABLE GENE PRODUCT	REFERENCES
CHLORAMPHENICOL	MINNEAPOLIS (A)	4121C	S1-3C	chloramphenicol resistance				59
		4419C	55R5-3C	chloramphenicol resistance				59
		4422C	55R5-3C	chloramphenicol resistance				59
		4507C	DP1-1B	chloramphenicol resistance				59
		4532C	DP1-1B	chloramphenicol resistance				90
		4533C	DP1-1B	chloramphenicol resistance				90
		5115C	51	chloramphenicol resistance				59
		5403C	54	chloramphenicol resistance				59
		6808C	68	chloramphenicol resistance				59
		6820C	68	chloramphenicol resistance				59
		cap1	X97929	chloramphenicol resistance				89
	SASKATOON	chl^{R-1}	GR3	chloramphenicol resistance				148
		chl^{R2}(=GR62)	GR3	chloramphenicol resistance				12, 146, 147, 148

TABLE 1 - Page 4

CLASS		ORIGINAL NAME	PARENTAL STRAIN	MAIN PHENOTYPE	MAP & LOCUS ASSIGNMENT	MAPPING METHOD	PROBABLE GENE PRODUCT	REFERENCES
CHLORAMPHENICOL	NARA KOBE	CHR50 (=CHL^r)	C814-40D	chloramphenicol resistance				174, 175, 188, 189
	BROOKLYN	D587-4B-4	D587-4B	chloramphenicol resistance				200
		D6-3	D6	chloramphenicol resistance				200
		E290-3	E290	chloramphenicol resistance				198, 199
		D585-11C-3	D585-11C	chloramphenicol resistance				199
		62-25A		chloramphenicol resistance				199
	LONDON UNIV.COL.	C_3^R		chloramphenicol resistance				201, 202
		C_9^R		chloramphenicol resistance				201, 202
	NEW YORK EINSTEIN	chl-8	SaC	chloramphenicol resistance				67
		chl-9	SaC	chloramphenicol resistance				67
	YOKO-HAMA	G706C	G706	chloramphenicol resistance				77, 78
	MIN*	cap^r	37	chloramphenicol resistance				5, 6

* MIN : MINNEAPOLIS (B)

TABLE 1 - Page 5

CLASS		ORIGINAL NAME	PARENTAL STRAIN	MAIN PHENOTYPE	MAP & LOCUS ASSIGNMENT	MAPPING METHOD	PROBABLE GENE PRODUCT	REFERENCES
ERYTHROMYCIN, SPIRAMYCIN	CLAYTON	ER^r(=L3301)	L3300	erythromycin, lincomycin, carbomycin, spiramycin, resistance				21
		ES^r(=L3001)	L3000	erythromycin, spiramycin resistance				21
		ER^{r-1}	L410	erythromycin, spiramycin resistance				104, 145
		ER^{r-2}	L410	erythromycin, spiramycin resistance				104
		ER^{r-3}	L410	erythromycin, spiramycin resistance				104
		ery1-r (several homoalleles)	L410	erythromycin, spiramycin, carbomycin, lincomycin resistance	B2 RIB3 (109)	G1, G2, G3, P1, P2	21S ribosomal RNA (35,107, 132, 166)	80, 84, 85, 86, 106, 107, 108,109,125,126, 132, 133, 166, 167, 178, 181
		ery2-r	L410	erythromycin resistance	B2	G1, G2, G3 P1, P2		106, 107, 178
		ER^r (=L2001)	L2000	erythromycin resistance				153

TABLE 1 - Page 6

CLASS		ORIGINAL NAME	PARENTAL STRAIN	MAIN PHENOTYPE	MAP & LOCUS ASSIGNMENT	MAPPING METHOD	PROBABLE GENE PRODUCT	REFERENCES
ERYTHROMYCIN, SPIRAMYCIN	CLAYTON	ER^r(=L601)	L600	erythromycin resistance				153
		ER^r(=L2201)	L2200	erythromycin resistance				153
		ER^r(=L411) (ERY-R)	L410	erythromycin resistance				66, 105, 130, 131, 133, 153
	GIF	E^R_{221} renamed [ERY]R221 in (138)	55R5-3C	erythromycin resistance	B2 RIB3(=RIII)	G1, G2, G3, P1, P3		10, 16, 17, 34, 36, 42, 47, 48, 49, 49A, 54, 94, 129, 134, 137, 138, 171, 173, 201, 203.
		E^R_{236}	55R5-3C	erythromycin resistance	B2 RIB3(=RIII)	G1		47, 134
		E^R_{353}	55R5-3C	erythromycin resistance	B2 RIB3(=RIII)	G1		47, 72, 134
		E^R_{354}	55R5-3C	erythromycin spiramycin resistance	S1 or S5 RIB2(=RII)	G1 G3	ribosome (72)	10, 46, 47, 72, 134
		E^R_{514} renamed ery^R in (50)	DP1-1B	erythromycin resistance	B2 RIB3(=RIII)	G1, G2, G3, P1, P2, P3, P4, P6	ribosome (72) ribosomal RNA(53,82, 101,103,154)	10, 16, 17, 30, 31, 34, 36, 42, 42A, 46, 47, 48, 49, 50, 51, 53, 54, 55, 62, 63, 65, 72, 86A, 94, 96, 100, 102, 110, 110A, 111, 119, 121, 121A, 129, 134, 164, 185, 201
		E^R_{553}	DP1-1B	erythromycin resistance	S1 or S5 RIB2(=RII)	G1, P3	ribosome (72)	47, 48, 49, 72, 86A, 134

TABLE 1 - Page 7

CLASS		ORIGINAL NAME	PARENTAL STRAIN	MAIN PHENOTYPE	MAP & LOCUS ASSIGNMENT		MAPPING METHOD	PROBABLE GENE PRODUCT	REFERENCES
ERYTHROMYCIN, SPIRAMYCIN	GIF	E^R_{554}	DP1-1B	erythromycin resistance	S1 or S5	RIB2 ?	G1		47
ERYTHROMYCIN, SPIRAMYCIN	LONDON UNIV. COLLEGE	E^R		erythromycin resistance					176, 201
ERYTHROMYCIN, SPIRAMYCIN	LONDON UNIV. COLLEGE	E^R_9		erythromycin resistance					201, 202
ERYTHROMYCIN, SPIRAMYCIN	LONDON UNIV. COLLEGE	E^R_{76}		erythromycin resistance					201, 202
ERYTHROMYCIN, SPIRAMYCIN	LONDON UNIV. COLLEGE	741	74	erythromycin resistance					177
ERYTHROMYCIN, SPIRAMYCIN	LONDON UNIV. COLLEGE	742	74	erythromycin resistance					177
ERYTHROMYCIN, SPIRAMYCIN	LONDON UNIV. COLLEGE	761	76	erythromycin resistance					177
ERYTHROMYCIN, SPIRAMYCIN	LONDON UNIV. COLLEGE	261	26	erythromycin resistance					177
ERYTHROMYCIN, SPIRAMYCIN	LONDON UNIV. COLLEGE	6-81C		erythromycin resistance					73A
ERYTHROMYCIN, SPIRAMYCIN	MINNEAPOLIS (A)	4113E	S1-3C	erythromycin resistance					59
ERYTHROMYCIN, SPIRAMYCIN	MINNEAPOLIS (A)	4120E	S1-3C	erythromycin resistance					59
ERYTHROMYCIN, SPIRAMYCIN	MINNEAPOLIS (A)	4120CE	S1-3C	erythromycin resistance					59

TABLE 1 - Page 8

CLASS		ORIGINAL NAME	PARENTAL STRAIN	MAIN PHENOTYPE	MAP & LOCUS ASSIGNMENT	MAPPING METHOD	PROBABLE GENE PRODUCT	REFERENCES
ERYTHROMYCIN, SPIRAMYCIN	MINNEAPOLIS (A)	4401E	55R5-3C	erythromycin resistance				59
		4405E	55R5-3C	erythromycin resistance				59
		4502E	DP1-1B	erythromycin resistance				59
		5102E	X97923	erythromycin resistance				59
		5103E	X97923	erythromycin resistance				59
		5105E	X97923	erythromycin resistance				59
		5106E	X97923	erythromycin resistance				59
		6801E	68	erythromycin resistance				59
		6802E	68	erythromycin resistance				59
		6804E	68	erythromycin resistance				59
		6805E	68	erythromycin resistance				59
		ery1	S1-3C	erythromycin resistance				89

TABLE 1 - Page 9

CLASS		ORIGINAL NAME	PARENTAL STRAIN	MAIN PHENOTYPE	MAP & LOCUS ASSIGNMENT	MAPPING METHOD	PROBABLE GENE PRODUCT	REFERENCES
ERYTHROMYCIN, SPIRAMYCIN	SASKATOON	ERY^{R1}	GR3	erythromycin resistance				149
		ERY^{R2}	GR3	erythromycin resistance				147, 149
		ERY^{R3}($=ery^{R3}$)	GR3	erythromycin resistance				12, 122, 146, 147, 149, 151.
		ERY^{R4}($=ery^{R4}$)	GR3	erythromycin resistance				12, 149
		ERY^{R5}	GR3	erythromycin resistance				149
		ERY^{R6}	GR3	erythromycin resistance				149
		ERY^{R7}	GR3	erythromycin resistance				149
		ERY^{R8}	GR3	erythromycin resistance				149
		ERY^{R9}	GR3	erythromycin resistance				149
		ERY^{R10}	GR3	erythromycin resistance				147, 149
	NARA KOBE	ERY^{r}(=R102)	C814-36B	erythromycin resistance				174, 175, 188, 189

TABLE 1 - Page 10

CLASS		ORIGINAL NAME	PARENTAL STRAIN	MAIN PHENOTYPE	MAP & LOCUS ASSIGNMENT	MAPPING METHOD	PROBABLE GENE PRODUCT	REFERENCES
ERYTHROMYCIN, SPIRAMYCIN	BROOKLYN	D587-4B-2	D587-4B	erythromycin resistance				198, 199, 200
		D22-1	D22	erythromycin resistance				200
		N11C-1	N11C	erythromycin resistance				199
	TOKYO	G706E	G706	erythromycin resistance				77, 78
		G102E(=ery102)	G102D	erythromycin resistance				77, 78, 193, 194, 196
		M1-7C/E	M1-7C	erythromycin resistance				77, 78
		ery	R8	erythromycin resistance				197
	NEW YORK EINSTEIN	ery-8	SaC	erythromycin resistance				67
		ery-9	SaC	erythromycin resistance				67
		ery-28	R28	erythromycin resistance				67
	MIN*	Ery^r	41	erythromycin resistance				5, 6
	COLUMBUS	E^R_2	D22	erythromycin resistance				14, 137, 138
		E^R_{16}	D6	erythromycin resistance				13

* MIN : MINNEAPOLIS (B)

TABLE 1 - Page 11

CLASS		ORIGINAL NAME	PARENTAL STRAIN	MAIN PHENOTYPE	MAP & LOCUS ASSIGNMENT	MAPPING METHOD	PROBABLE GENE PRODUCT	REFERENCES
ERYTHROMYCIN, SPIRAMYCIN	TALLIN	E_1^R		erythromycin resistance				87, 88
		E_2^R		erythromycin resistance				87, 88
		E_4^R		erythromycin resistance				87, 88
		E_5^R		erythromycin resistance				87, 88
		E_7^R		erythromycin resistance				87, 88
		E_8^R		erythromycin resistance				87, 88
		E_{14}^R		erythromycin resistance				87, 88
		E_{15}^R		erythromycin resistance				87, 88
		E_{18}^R		erythromycin resistance				87, 88
	NEW YORK PUBLIC HEALTH	E_{621}^R	D273-10B	erythromycin resistance				185, 187
		E_{624}^R	D273-10B	erythromycin resistance				185, 187
	LONDON UNIV.CO.	S^R		spiramycin resistance				176, 201

TABLE 1 - Page 12

CLASS		ORIGINAL NAME	PARENTAL STRAIN	MAIN PHENOTYPE	MAP & LOCUS ASSIGNMENT	MAPPING METHOD	PROBABLE GENE PRODUCT	REFERENCES
ERYTHROMYCIN, SPIRAMYCIN	CLAYTON	spi2-r	L410	spiramycin resistance	B2	G1, G2, G3, P1, P2		84, 106, 107, 178
		spi3-r	L410	spiramycin resistance	B1	G1, G2, G3, P1, P2		84, 106, 107, 178
		spi4-r	L410	spiramycin resistance	B1 or S5			84, 106, 107, 178
	ADELAIDE	SpirF3-R	A11-5	spiramycin resistance				28
		SpirF6-R	A11-5	spiramycin resistance				28
		SpirF25-R(=SPR2)	A11-5	spiramycin resistance				26, 27, 28
		SpirF27-R	A11-5	spiramycin resistance				28
		SpirG3-R	A11-4	spiramycin resistance				28
		SpirG6-R	A11-4	spiramycin resistance				28
		SpirG7-R(=SPR1)	A11-4	spiramycin resistance				26, 27, 28
		SpirG8-R	A11-4	spiramycin resistance				28
	GIF	S^R_{352}	55R5-3C	spiramycin resistance	B2 RIB3(=RIII)	G1, P3		47, 72, 86A, 134
		S^R_{551}	DP1-1B	spiramycin resistance (weak)	S1 or S5 RIB2(=RII)	G1		47, 134

TABLE 1 - Page 13

CLASS		ORIGINAL NAME	PARENTAL STRAIN	MAIN PHENOTYPE	MAP & LOCUS ASSIGNMENT	MAPPING METHOD	PROBABLE GENE PRODUCT	REFERENCES
	GIF	S^R_{552}	DP1-1B	spiramycin resistance (weak)	S1 RIB2(=RII) or S5	G1		47
PAROMOMYCIN(=MONOMYCIN),NEOMYCIN	LONDON UNIV.COL.	P^R		paromomycin resistance				176, 201
		P^R_{701}		paromomycin resistance				201, 202
	MIN*	par 1 renamed PAR R1 in (138) (204)	48	paromomycin resistance				89, 137, 138, 204
	MUNICH	P^R_{454}	M12	paromomycin resistance	B5 PAR1(=PI)	G1, G2, G3, P1, P2, P3, P4		17, 34, 42A, 47, 48, 49A, 54, 55, 63, 65, 95, 96, 102, 129, 159, 164, 185, 187, 203
		P^R_{455}	M12	paromomycin resistance	B5 PAR1(=PI)	G1		47, 95, 203
	TALLIN	M^R_2		monomycin resistance				87
		M^R_6		monomycin resistance				87
		M^R_{15}		monomycin resistance				87
		N^R_1		neomycin resistance				87, 88
		N^R_2		neomycin resistance				87, 88

* MIN : MINNEAPOLIS (A)

TABLE 1 - Page 14

CLASS		ORIGINAL NAME	PARENTAL STRAIN	MAIN PHENOTYPE	MAP & LOCUS ASSIGNMENT		MAPPING METHOD	PROBABLE GENE PRODUCT	REFERENCES
PAROMOMYCIN, NEOMYCIN	TALLIN	N_4^R		neomycin resistance					87, 88
		N_5^R		neomycin resistance					87, 88
	CLAYTON	par-r		paromomycin resistance	B5	par	G1, G2, G3, P1, P2		5, 107, 181
		par1-r		paromomycin resistance	B5	par	G2, P1, P2	15S rRNA (32, 166)	32, 126, 166, 167
	NEW YORK PUBLIC HEALTH	P_{628}^R	D273-10B	paromomycin resistance					185, 187
		P_{626}^R	D273-10B	paromomycin resistance					185, 187
OLIGOMYCIN, RUTAMYCIN	LONDON MILL HILL	6α01	6	oligomycin, rutamycin resistance					168
	TOKYO	oli	R8	oligomycin resistance					197
		102P7	102D	oligomycin resistance					195
		N11	1586-2B	oligomycin resistance					193
		N15	1586-2B	oligomycin resistance					193
		G706R(=oli706) OLG^r(=706R1)	G706 706	oligomycin resistance					77, 78, 174, 175, 188, 189, 191, 193, 194, 196
		G102R	G102D	oligomycin resistance					77, 78

TABLE 1 - Page 15

CLASS		ORIGINAL NAME	PARENTAL STRAIN	MAIN PHENOTYPE	MAP & LOCUS ASSIGNMENT		MAPPING METHOD	PROBABLE GENE PRODUCT	REFERENCES
OLIGOMYCIN	CLAYTON	oli1-r (several homoalleles)	L410	oligomycin resistance	S2	Class A (=OIII)	G1, G2, G3, P1, P2	ATPase subunit 6 (75)	12, 75, 84, 85, 86, 106, 107, 108, 126, 132, 133, 166, 167, 180, 181
		tso-r(=L748)	L410	oligomycin resistance and cold sensitivity for growth on glycerol	S2	Class A (OIII)	G1, G2, G3, P1, P2	ATPase subunit 6 (75)	75, 107, 179, 180, 181
		oli17-r		oligomycin resistance	B3	Class B (=OI)	G1, G2, G3, P1, P2	ATPase subunit 9 (75)	75, 107, 181
		L4100	1300	oligomycin resistance					123
	GIF and WARWICK	0^R_1	55R5-3C	oligomycin resistance	B3	OLI1(=OI)	G1, G2, G3, P1, P2, P3, P4	ATPase subunit 9 (187)	10, 17. 34, 42, 42A. 47, 49, 49A, 54, 55, 63, 69, 102, 129, 134, 159, 164, 171, 173, 185, 187.
		0^R_2	55R5-3C	oligomycin resistance	B3	OLI1(=OI)	G1		10, 47
		0^R_4	DP1-1B	oligomycin resistance	B3	OLI1(=OI)	G1		10, 47
		0^R_5	55R5-3C	oligomycin resistance	B3	OLI1(=OI)	G1		10
		0^R_6	55R5-3C	oligomycin resistance	B3	OLI1(=OI)	G1		10
		0^R_7	DP1-1B	oligomycin resistance	B3	OLI1(=OI)	G1		10, 47, 48

TABLE 1 - Page 16

CLASS		ORIGINAL NAME	PARENTAL STRAIN	MAIN PHENOTYPE	MAP & LOCUS ASSIGNMENT	MAPPING METHOD	PROBABLE GENE PRODUCT	REFERENCES
OLIGOMYCIN	GIF and WARWICK	O^R_8	IL16-11D	oligomycin resistance	B3 OLI1(=O1)	G1		10
		O^R_9	FL496-2B	oligomycin resistance	S3 OLI4(=OIV)	G1		34
		O^R_{10}	IL16-11D	oligomycin resistance	B3 OLI1(=OI)	G1		10
		O^R_{143}(=D22A13)	D22	oligomycin resistance	B4 OLI2(=OII)	G1		8, 9, 10, 34, 47, 68, 69, 70, 83
		O^R_{144}(=D22A15)	D22	oligomycin resistance	B4 OLI2(=OII) Class C (181)	G1, G2, G3, P2, P3, P4		8, 9, 10, 34, 47, 48, 54,63, 68, 69, 70, 83, 96, 102, 129, 138, 165, 181, 185, 187
		O^R_{145}(=D22A16)	D22	oligomycin resistance	B3 OLI1(=OI)	G1, G2, G3, P3, P4	ATPase subunit 9 (187)	8, 9, 10, 17, 34, 47, 48, 65, 68, 69, 70, 83, 102, 129, 164, 185, 187
		O^R_{146}(=D22A21)	D22	oligomycin resistance	B3 OLI1(=OI) Class B (181)	G1, G2, G3		8, 9, 10, 34, 47, 68, 69, 70, 83, 97, 138, 165, 181
		O^R_{147}(=D6A1)	D6	oligomycin resistance	B3 OLI1(=OI)	G1		7, 8, 9, 10
		O^R_{148}(=D6B2)	D6	oligomycin resistance	B3 OLI1(=OI)	G1		7, 8, 9, 10
		O^R_{149}(=D6B3)	D6	oligomycin resistance	B3 OLI1(=OI)	G1		8, 9, 10
		O^R_{150}(=D22A14)	D22	oligomycin resistance	B4 OLI2(=OII)	G1		10, 47, 69, 70

TABLE 1 - Page 17

CLASS		ORIGINAL NAME	PARENTAL STRAIN	MAIN PHENOTYPE	MAP & LOCUS ASSIGNMENT	MAPPING METHOD	PROBABLE GENE PRODUCT	REFERENCES
OLIGOMYCIN	GIF and WARWICK	0^R_{151}(=D22C4)	D22	oligomycin resistance	B4 OLI2(=OII)	G1		8, 9, 10, 69
		0^R_{152}(=D6A5)	D6	oligomycin resistance	B3 OLI1(=OI)	G1		10
		0^R_{153}(=D22A18)	D22	oligomycin resistance	B4 OLI2(=OII)	G1		7, 8, 9, 10, 69, 70
		0^R_{154}(=D22A19)	D22	oligomycin resistance	S2 OLI3(=OIII) =Class $A_{(181)}$	G1		10, 69, 98, 181
		0^R_{155}(=D22B1)	D22	oligomycin resistance	B3 OLI1(=OI)	G1		10, 70
		0^R_{156}(=D22B21)	D22	oligomycin resistance	B3 OLI1(=OI)	G1		7, 8, 9, 10, 68, 69
	SASKATOON	oli^{R1}(=OLI^{R1})	GR3	oligomycin resistance				146, 149
		oli^{R2}(=OLI^{R2})	GR3	oligomycin resistance				146, 149
		oli^{R3}(=OLI^{R3})	GR3	oligomycin resistance				146, 149
		oli^{R4}(=OLI^{R4})	GR3	oligomycin resistance				12, 146, 149
	LONDON UNIV.COL	0^R_1		oligomycin resistance				201, 202
		0^R_6		oligomycin resistance				201, 202

TABLE 1 - Page 18

CLASS		ORIGINAL NAME	PARENTAL STRAIN	MAIN PHENOTYPE	MAP & LOCUS ASSIGNMENT		MAPPING METHOD	PROBABLE GENE PRODUCT	REFERENCES
OLIGOMYCIN	ADELAIDE	OligD8-R(= OLG1)	A11-1	oligomycin resistance	S2	Class A (=OIII)	G1, G2, G3 P1, P2		26, 27, 28, 75, 107, 181
		OligE15-R(= OLG2)	A11-2	oligomycin resistance	B3	Class B (=OI)	G1, G2, G3 P1, P2		26, 27, 28, 75, 107, 133, 181
	BAL*	OLI^R1(G3)	D311-3A	oligomycin resistance					39
	NEW YORK EINSTEIN	oliA-8	SaC	oligomycin resistance		oliA	G1		67
		oliA-9	SaC	oligomycin resistance		oliA	G1		67
		oliB-28	R28	oligomycin resistance		oliB	G1		67
	NEW YORK PUBLIC HEALTH	O^R_{622}	D273-10B	oligomycin resistance		OLI4 ?	G1		185, 187
		O^R_{625}	D273-10B	oligomycin resistance	B3	OLI1	G1	ATPase subunit 9 (187)	185, 187
		O^R_{626}	805-4A/A7	oligomycin resistance	B3	OLI1	G1		185
		O^R_{627}	AA3-1C/A10	oligomycin resistance	B3	OLI1	G1		185

* BAL : BALTIMORE

TABLE 1 - Page 19

CLASS		ORIGINAL NAME	PARENTAL STRAIN	MAIN PHENOTYPE	MAP & LOCUS ASSIGNMENT		MAPPING METHOD	PROBABLE GENE PRODUCT	REFERENCES
OLIGOMYCIN	NEW YORK PUBLIC HEALTH	O^R_{628}	AA4-2C/A5	oligomycin resistance	B3	OLI1	G1		185
OLIGOMYCIN	NEW YORK PUBLIC HEALTH	O^R_{629}	AA3-2B/A6	oligomycin resistance	B3	OLI1	G1		185
OLIGOMYCIN	DAVIS	O^R_4 renamed OLI_1 in (204, 138)	D243-4A	oligomycin resistance					13, 14, 137, 138, 163, 204
OLIGOMYCIN	N.Y.*	oli^r_1	A664/18A	oligomycin resistance					122
VENTURICIDIN, OLIGOMYCIN	WARWICK	V^R_{60}	D22	venturicidin resistance	B3	OLI1(=OI)	G1, G3		69, 71, 96, 97, 98
VENTURICIDIN, OLIGOMYCIN	WARWICK	V^R_{61}	D22	venturicidin + oligomycin resistance	S2	OLI3(=OIII)	G1, G3		69, 71, 97, 98
VENTURICIDIN, OLIGOMYCIN	WARWICK	V^R_{62}	D22	venturicidin + oligomycin resistance	S2	OLI3(=OIII)	G1, G3		71, 98
VENTURICIDIN, OLIGOMYCIN	WARWICK	V^R_{64}	D22	venturicidin + oligomycin resistance	S2	OLI3(=OIII)	G1		71

* NEW YORK (EINSTEIN)

TABLE 1 - Page 20

CLASS		ORIGINAL NAME	PARENTAL STRAIN	MAIN PHENOTYPE	MAP & LOCUS ASSIGNMENT		MAPPING METHOD	PROBABLE GENE PRODUCT	REFERENCES
ANTIMYCIN	WÜRZBURG	G13	D8-22A	antimycin resistance	S3	AI	G1		118, 140
		G1a	D8-22A	antimycin resistance	S3	AI	G1, G3		118, 140, 142
		G25	D8-22A	antimycin resistance	S3	AII	G1		118, 140
		A32	D243-4A	antimycin resistance	S3	AI	G1		140, 141
		A41	D243-4A	antimycin resistance	S3	AI	G1		141
		A83	D243-4A	antimycin resistance	S3	AI	G1		141
		A84	D243-4A	antimycin resistance	S3	AI	G1		140, 141
		A86	D243-4A	antimycin resistance	S3	AI	G1		141
		A103	D243-4A	antimycin resistance	S3	AI	G1		141
		A106	D243-4A	antimycin resistance	S3	AI	G1		141
		A108	D243-4A	antimycin resistance	S3	AI	G1		141
		A111	D243-4A	antimycin resistance	S3	AI	G1		141
		A112	D243-4A	antimycin resistance	S3	AI	G1		141

TABLE 1 - Page 21

CLASS		ORIGINAL NAME	PARENTAL STRAIN	MAIN PHENOTYPE	MAP & LOCUS ASSIGNMENT		MAPPING METHOD	PROBABLE GENE PRODUCT	REFERENCES
ANTIMYCIN, MIKAMYCIN	WÜRZBURG	A115	D243-4A	antimycin resistance	S3	AI	G1		141
		A321	D243-4A	antimycin resistance	S3	AI	G1		141
	MUNICH	ana^r-101	SM11-6C	antimycin resistance	S3		G2, G3		22, 23, 99, 162
		ana^r-104		antimycin resistance	S3		G2, G3		22
	CLAYTON	N1301	N1300	(mikamycin) antimycin resistance	S3		G3, P1		21, 76
FUNICULOSIN	WÜRZBURG	FUN^R_5 (renamed /fun5 in 138)	D243-4A	funiculosin resistance	S3	AII	G1		140, 141
		FUN^R_9 renamed /fun9 in 138)	D243-4A	funiculosin resistance	S3	AII	G1		140, 141
		/Fun2	D243-4A	funiculosin resistance	S3	AII	G1		141
		/Fun3	D243-4A	funiculosin resistance	S3	AII	G1		141
		/Fun4	D243-4A	funiculosin resistance	S3	AII	G1		141
		/Fun6	D243-4A	funiculosin resistance	S3	AII	G1		141

TABLE 1 - Page 22

CLASS		ORIGINAL NAME	PARENTAL STRAIN	MAIN PHENOTYPE	MAP & LOCUS ASSIGNMENT		MAPPING METHOD	PROBABLE GENE PRODUCT	REFERENCES
FUNICULOSIN	WÜRZBURG	/Fun7	D243-4A	funiculosin resistance	S3	AII	G1		141
		/Fun8	D243-4A	funiculosin resistance	S3	AII	G1		141
		/Fun10	D243-4A	funiculosin resistance	S3	AII	G1		141
		/Fun11	D243-4A	funiculosin resistance	S3	AII	G1		141
		/Fun12	D243-4A	funiculosin resistance	S3	AII	G1		141
		/Fun13	D243-4A	funiculosin resistance	S3	AII	G1		141
		/Fun14	D243-4A	funiculosin resistance	S3	AII	G1		141
		/Fun15	D243-4A	funiculosin resistance	S3	AII	G1		141
		/Fun16	D243-4A	funiculosin resistance	S3	AII	G1		141
		/Fun17	D243-4A	funiculosin resistance	S3	AII	G1		141
		/Fun18	D243-4A	funiculosin resistance	S3	AII	G1		141
		/Fun19	D243-4A	funiculosin resistance	S3	AII	G1		141
		/Fun20	D243-4A	funiculosin resistance	S3	AII	G1		141

TABLE 1 - Page 23

CLASS		ORIGINAL NAME	PARENTAL STRAIN	MAIN PHENOTYPE	MAP & LOCUS ASSIGNMENT		MAPPING METHOD	PROBABLE GENE PRODUCT	REFERENCES
FUNICULOSIN	MUNICH	fun^r-202		funiculosin resistance					99
FUNICULOSIN	MUNICH	fun^r-209		funiculosin resistance	S3		G3		22
FUNICULOSIN	MUNICH	fun^r-210		funiculosin resistance	S3		G3		22
FUNICULOSIN	MUNICH	fun^r-211		funiculosin resistance	S3		G3		22
FUNICULOSIN	MUNICH	fun^r-213		funiculosin resistance	S3		G3		22
FUNICULOSIN	MUNICH	fun^r-214		funiculosin resistance	S3		G3		22
DIURON	GIF and LOUVAIN-LA-NEUVE	diu^r-1 (renamed diu 1-721 (= D^R_{721} in 41)	IL125-2B	diuron resistance	S3	DIU1	G1		40, 41
DIURON	GIF and LOUVAIN-LA-NEUVE	diu^r-34	IL125- 2B	diuron resistance	S3				40
DIURON	GIF and LOUVAIN-LA-NEUVE	diu^r-64 (renamed diu 2-764 (= D^R_{764} in 41)	IL125-2B	diuron resistance	S3	DIU2	G1		40, 41
DIURON	GIF and LOUVAIN-LA-NEUVE	diu1-724(= D^R_{724})	D273-10B/A21	diuron resistance	S3	DIU1	G1		41
DIURON	GIF and LOUVAIN-LA-NEUVE	diu1-725(= D^R_{725})	KL14-4A	diuron resistance	S3	DIU2	G1		41

TABLE 1 - Page 24

CLASS		ORIGINAL NAME	PARENTAL STRAIN	MAIN PHENOTYPE	MAP & LOCUS ASSIGNMENT		MAPPING METHOD	PROBABLE GENE PRODUCT	REFERENCES
DIURON	GIF and LOUVAIN-LA NEUVE	diu2-726(= D^R_{726})	KL14-4A	diuron resistance	S3	DTU2	G1		41
		diu1-727(= D^R_{727})	KL14-4A	diuron resistance	S3	DIU1	G1		41
		diu2-728(= D^R_{728})	KL14-4A	diuron resistance	S3	DIU2	G1		41
	WURZBURG	diu1-731(= D^R_{731})	D243-4A	diuron resistance	S3	DIU1	G1		41
		diu2-732(= D^R_{732})	D243-4A	diuron resistance	S3	DIU2	G1		41
MUCIDIN	BRATISLAVA	M^R_{3D} (renamed muc1-771 = M^R_{771} in 41)	D225-5A	mucidin resistance	S3	MUC1	G1		41, 171, 172, 173
		M^R_{101} (renamed muc2-772 = M^R_{772} in 41)	D225-5A	mucidin resistance	S3	MUC2	G1		41, 171, 172, 173

TABLE 1 - Page 25

CLASS		ORIGINAL NAME	PARENTAL STRAIN	MAIN PHENOTYPE	MAP & LOCUS ASSIGNMENT		MAPPING METHOD	PROBABLE GENE PRODUCT	REFERENCES
OXIDASE	NEW YORK, PUBLIC HEALTH	E3-19	D273-10B	cyt a (-) oxi (-)	S5	OXI2	G1,2,3		164, 182, 183
		M3-9	D273-10B	cyt a (-) oxi (-)	S4	OXI3	G1,2,3	cyt.oxidase subunit 1	164, 182 , 25
		M5-16	D273-10B	cyt a (-) oxi (-)	S4	OXI3	G1,2,3	cyt. oxidase subunit 1	128A, 164,
		M5-70	D273-10B	cyt a (-) § oxi (-)	S5	OXI1	G1,2,3		164
		M5-85	D273-10B	cyt a (-) oxi (-)	S4	OXI3	G1,2,3	cyt.oxidase subunit 1	164, 25
		M5-121	D273-10B	cyt a (-) oxi (-)	S4	OXI3	G1,2,3		164,
		M6-183	D273-10B	cyt a (-) oxi (-)	S4	OXI3	G1,2,3	cyt.oxidase subunit 1	164, 25
		M8-139	D273-10B	cyt a (-) oxi (-)	S4	OXI3	G1,2,3		164
		M8-162	D273-10B	cyt a (-) oxi (-)	S4	OXI3	G1,2,3		164, 184,
		M8-171	D273-10B	cyt a (-) oxi (-)	S5	OXI1	G1,2,3	cyt.oxidase subunit 2	25, 164
		M8-227	D273-10B	cyt a (-) oxi (-)	S4	OXI3	G1,2,3		164,
		M9-3	D273-10B	cyt a (-) oxi (-)	S5	OXI2	G1,2,3		164
		M9-9	D273-10B	cyt a (-) oxi (-)	S4	OXI3	G1,2,3		164
		M9-27	D273-10B	cyt a (-) oxi (-)	S4	OXI3	G1,2,3		164

§ see page 64 of this table

TABLE 1 - Page 28

CLASS		ORIGINAL NAME	PARENTAL STRAIN	MAIN PHENOTYPE	MAP & LOCUS ASSIGNMENT		MAPPING METHOD	PROBABLE GENE PRODUCT	REFERENCES
OXIDASE	NEW YORK PUBLIC HEALTH	M16-201	D273-10B	cyt a (-) oxi (-)	S4	OXI3	G1,2,3		164
		M18-65			S4	OXI3			113
		M36-35			S4	OXI3			113
		M37-62			S4	OXI3			113
	GIF + WROCLAW	G20	777-3A	cyt a (-)	S4	OXI3	G3		93
		G74	777-3A	cyt a (-)	S4	OXI3	G3		93
		G107	777-3A	cyt a (-) cyt b (-)			G3		93
		G108	777-3A	cyt a (-)	S4	OXI3	G3		93
		G110	777-3A	cyt a (-)	S4	OXI3	G3		93
		G112	777-3A	§	S4	OXI3	G3		93
		G192	777-3A	cyt a (-)	S4	OXI3	G3		93
		G199	777-3A	cyt a (-)	S5	OXI2	G3		93
		G205	777-3A	cyt a (-) §	S4	OXI3	G3		93
		G225	777-3A	cyt a (-)	S4	OXI3	G3		93

§ see page 64 of this table

TABLE 1 - Page 29

CLASS		ORIGINAL NAME	PARENTAL STRAIN	MAIN PHENOTYPE	MAP & LOCUS ASSIGNMENT		MAPPING METHOD	PROBABLE GENE PRODUCT	REFERENCES
OXIDASE	GIF + WROCLAW	G247	777-3A	cyt a (-) §	S4	OXI3	G3		93
		G291	777-3A	cyt a (-)	S4	OXI3	G3		93
		G341	777-3A	cyt a (-) cyt b (-)	S4	OXI3	G3		93
		G389	777-3A	§	S5	OXI2	G3		93
		G390	777-3A		S4	OXI3	G3		93
		G421	777-3A	cyt a (-)	S4	OXI3	G3		93
		G433	777-3A	cyt a (-)	S4	OXI3	G3		93
		G450	777-3A	cyt a (-)	S4	OXI3	G3		93
		G481	777-3A	cyt a (-)	S4	OXI3	G3		93
		G565	777-3A	cyt a (-)	S5	OXI1	G3		93
		G588	777-3A	cyt a (-)	S4	OXI3	G3		93
		G656	777-3A	cyt a (-)	S4	OXI3	G3		93
		G689	777-3A	cyt a (-)	S4	OXI3	G3		93

§ see page 64 of this table

TABLE 1 - Page 30

CLASS		ORIGINAL NAME	PARENTAL STRAIN	MAIN PHENOTYPE	MAP & LOCUS ASSIGNMENT	MAPPING METHOD	PROBABLE GENE PRODUCT	REFERENCES
OXIDASE	GIF + WROCLAW	G696	777-3A	cyt a (-)	S4 OXI3	G3		93
		G728	777-3A	cyt a (-)	S4 OXI3	G3		93
		G789	777-3A	cyt a (-)	S4 OXI3	G3		93
		G922	777-3A	cyt a (-)	S4 OXI3	G3		93
		G950	777-3A	§	S5 OXI2	G3		93
		G1003	777-3A	cyt a (-)	S4 OXI3	G3		93
		G1055	777-3A	cyt a (-)	S4 OX13	G3		93
		G1089	777-3A	cyt a (-)	S4 OXI3	G3		93
		G1099	777-3A	cyt a (-)	S4 OXI3	G3		93
		G1160	777-3A	cyt a (-)	S4 OXI3	G3		93
		G1181	777-3A	cyt a (-)	S4 OXI3	G3		93
		G1186	777-3A	cyt a (-)	S4 OXI3	G3		93
		G1194	777-3A	cyt a (-)	S4 OXI3	G3		93

§ see page 64 of this table

TABLE 1 - Page 31

CLASS		ORIGINAL NAME	PARENTAL STRAIN	MAIN PHENOTYPE	MAP & LOCUS ASSIGNMENT		MAPPING METHOD	PROBABLE GENE PRODUCT	REFERENCES
OXIDASE	GIF + WROCLAW	G1264	777-3A	cyt a (-)	S4	OXI3	G3		93
		G1273	777-3A	cyt a (-) cyt b (-)	S4	OXI3	G3		93
		G1280	777-3A	cyt a (-) §	S4	OXI3	G3		93
		G1287	777-3A	cyt a (-)	S4	OXI3	G3		93
		G1305	777-3A	cyt a (-) §	S4	OXI3	G3		93
		G1342	777-3A	cyt a (-)	S5	OXI1	G3		93
		G1374	777-3A	cyt a (-)	S5	OXI1	G3		93
		G1375	777-3A	cyt a (-)	S4	OXI3	G3		93
		G1547	777-3A	cyt a (-)	S4	OXI3	G3		93
		G1683	777-3A	cyt a (-)	S4	OXI3	G3		93
		G1734	777-3A	cyt a (-)	S4	OXI3	G3		93
		G1765	777-3A	cyt a (-)	S4	OXI3	G3		93
		G1810	777-3A	cyt a (+)	S5	OXI2	G3		93

§ see page 64 of this table

TABLE 1 - Page 32

CLASS		ORIGINAL NAME	PARENTAL STRAIN	MAIN PHENOTYPE	MAP & LOCUS ASSIGNMENT		MAPPING METHOD	PROBABLE GENE PRODUCT	REFERENCES
OXIDASE	GIF + WROCLAW	G1859	777-3A	cyt a (-)	S4	OXI3	G3		93
		G1893	777-3A	cyt a (-)	S5	OXI1	G3		93
		G1921	777-3A	cyt a (-)	S4	OXI3	G3		93
		G1979	777-3A		S4	OXI3	G3		93
		G2027	777-3A	cyt a (-)	S4	OXI3	G3		93
		G2029	777-3A	cyt a (-)	S4	OXI3	G3		93
		G2084	777-3A	cyt a (-)	S4	OXI3	G3		93
		G2124	777-3A	cyt a (-)	S4	OXI3	G3		93
		G2133	777-3A	cyt a (-)	S4	OXI3	G3		93
		G2166	777-3A	cyt a (-)	S4	OXI3	G3		93
		G2190	777-3A	cyt a (-) §	S4	OXI3	G3		93
		G2220	777-3A	cyt a (-)	S4	OXI3	G3		93
		G2276	777-3A	cyt a (-)	S4	OXI3	G3		93

§ see page 64 of this table

TABLE 1 - Page 33

CLASS		ORIGINAL NAME	PARENTAL STRAIN	MAIN PHENOTYPE	MAP & LOCUS ASSIGNMENT		MAPPING METHOD	PROBABLE GENE PRODUCT	REFERENCES
OXIDASE	GIF + WROCLAW	G2283	777-3A	cyt a (-) cyt b (-)			G3		93
		G2291	777-3A	cyt a (-)	S4	OXI3	G3		93
		G2312	777-3A	cyt a (-)	S4	OXI3	G3		93
		G2342	777-3A	cyt a (-)	S4	OXI3	G3		93
		G2343	777-3A	cyt a (-)	S4	OXI3	G3		93
		G2354	777-3A	cyt a (-)	S4	OXI3	G3		93
		G2394	777-3A	cyt a (-)	S4	OXI3	G3		93
		G2399	777-3A	cyt a (-)	S4	OXI3	G3		93
		G2443	777-3A	cyt a (-)	S4	OXI3	G3		93
		G2461	777-3A	cyt a (-)	S4	OXI3	G3		93
		G2469	777-3A	cyt a (-)	S4	OXI3	G3		93
		G2496	777-3A	cyt a (-)	S4	OXI3	G3		93
		G2508	777-3A	cyt a (-)	S4	OXI3	G3		93

TABLE 1 - Page 34

CLASS		ORIGINAL NAME	PARENTAL STRAIN	MAIN PHENOTYPE	MAP & LOCUS ASSIGNMENT	MAPPING METHOD	PROBABLE GENE PRODUCT	REFERENCES
OXIDASE	GIF + WROCLAW	G2567	777-3A	cyt a (-) §	S4 OXI3	G3		93
		G2592	777-3A	cyt a (-) cyt b (-)	S4 OXI3	G3		93
		W22	777-3A	cyt a (-) cyt b (-)		G3		
		W48	777-3A	cyt a (-)	S4 OXI3	G3		93
		W69	777-3A	cyt a (-)	S4 OXI3	G3		93
		W109	777-3A		S4 OXI3	G3		93
		W113	777-3A		S4 OXI3	G3		93
		W164	777-3A		S4 OXI3	G3		93
		W165	777-3A	cyt a (-)	S4 OXI3	G3		93
		W201	777-3A	cyt a (-)	S4 OXI3	G3		93
		W235	777-3A	cyt a (-) §	S5 OXI2	G3		93
		W246	777-3A	cyt a (-)	S4 OXI3	G3		93
		W248	777-3A	cyt a (-)	S4 OXI3	G3		93

§ see page 64 of this table

TABLE 1 - Page 35

CLASS		ORIGINAL NAME	PARENTAL STRAIN	MAIN PHENOTYPE	MAP & LOCUS ASSIGNMENT		MAPPING METHOD	PROBABLE GENE PRODUCT	REFERENCES
OXIDASE	GIF + WROCLAW	W264	777-3A	cyt a (-) cyt b (-)	S5	OXI3	G3		93
OXIDASE	GIF + WROCLAW	W364	777-3A	cyt a (-)	S4	OXI3	G3		93
OXIDASE	GIF + WROCLAW	W373	777-3A	cyt a (-)	S4	OXI3	G3		93
OXIDASE	CLAYTON	45.1		cyt a (-) oxi (-)	S4	cya2	G3		152
OXIDASE	CLAYTON	45.2		cyt a (-) oxi (-)	S4	cya2	G3		152
OXIDASE	CLAYTON	45.3		cyt a (-) oxi (-)	S5	cya3	G3		152
OXIDASE	CLAYTON	46		cyt a (-) oxi (-)	S6	cya1	G3		152
OXIDASE	CLAYTON	46-3-1	J69	cyt a (-) oxi (-) §	S5	cya1	G3		35
OXIDASE	CLAYTON	45-1-2	J69	cyt a (-) oxi (-)	S4	cya2	G3		35
OXIDASE	CLAYTON	45-2-1	J69	cyt a (-) oxi (-)	S4	cya2	G3		35
OXIDASE	CLAYTON	10-15-1B	J69	cyt a (-) oxi (-)	S4	cya2	G3		35
OXIDASE	CLAYTON	45-3-2	J69	cyt a (-) oxi (-)	S5	cya3	G3		35
OXIDASE	CLAYTON	Mn46		cyt a (-) oxi (-) §					75

§ see page 64 of this table

TABLE 1 - Page 36

CLASS		ORIGINAL NAME	PARENTAL STRAIN	MAIN PHENOTYPE	MAP & LOCUS ASSIGNMENT	MAPPING METHOD	PROBABLE GENE PRODUCT	REFERENCES
OXIDASE	CLAYTON	Mn10-15		cyt a (-) oxi (-)				75
OXIDASE	CLAYTON	Mn18-28-3		cyt a (-) oxi (-)				75
OXIDASE	ORSAY	mit112	FF1210-6C		S5 OXI-1	G2, 3		19
OXIDASE	ORSAY	mit117	FF1210-6C		S5 OXI-1	G2, 3		19
OXIDASE	ORSAY	mit136	FF1210-6C		S5 OXI-1	G2, 3		19
OXIDASE	ORSAY	mit142	FF1210-6C		S5 OXI-1	G2, 3		19
OXIDASE	ORSAY	mit179	FF1210-6C		S5 OXI-1	G2, 3		19
OXIDASE	ORSAY	mit229	FF1210-6C		S5 OXI-I	G2, 3		19
OXIDASE	ORSAY	mit231	FF1210-6C		S5 OXI-II	G2, 3		19, 65
OXIDASE	ORSAY	mit116	FF1210-6C		S4 OXI-III DI	G2, 3		19
OXIDASE	ORSAY	mit122	FF1210-6C		S4 OXI-III DI	G2, 3		19
OXIDASE	ORSAY	mit123	FF1210-6C		S4 OXI-III DI	G2, 3		19
OXIDASE	ORSAY	mit125	FF1210-6C		S4 OXI-III DI	G2, 3		19

TABLE 1 - Page 37

CLASS		ORIGINAL NAME	PARENTAL STRAIN	MAIN PHENOTYPE	MAP & LOCUS ASSIGNMENT		MAPPING METHOD	PROBABLE GENE PRODUCT	REFERENCES
OXIDASE	ORSAY	mit127	FF1210-6C		S4	OXI-III DI	G2, 3		19
		mit133	FF1210-6C		S4	OXI-III DI	G2, 3		19
		mit139	FF1210-6C		S4	OXI-III DI	G2, 3		19
		mit140	FF1210-6C		S4	OXI-III DI	G2, 3		19
		mit143	FF1210-6C		S4	OXI-III DI	G2, 3		19
		mit c 2-2	FF1210-6C		S4	OXI-III DI	G2, 3		19
		mit d 2-1	FF1210-6C		S4	OXI-III DI	G2, 3		19
		mit178	FF1210-6C		S4	OXI-III DI	G2, 3		19
		mit124	FF1210-6C		S4	OXI-III D2	G2, 3		19
		mit129	FF1210-6C		S4	OXI-III D2	G2, 3		19
		mit134	FF1210-6C		S4	OXI-III D2	G2, 3		19
		mit135	FF1210-6C		S4	OXI-III D2	G2, 3		19
		mit137	FF1210-6C		S4	OXI-III D2	G2, 3		19

TABLE 1 - Page 38

CLASS		ORIGINAL NAME	PARENTAL STRAIN	MAIN PHENOTYPE	MAP & LOCUS ASSIGNMENT	MAPPING METHOD	PROBABLE GENE PRODUCT	REFERENCES
OXIDASE	ORSAY	mit200	FF1210-6C		S4 OXI-III D2	G2, 3		19
		mit207	FF1210-6C		S4 OXI-III D2	G2, 3		19
		mit218	FF1210-6C		S4 OXI-III D2	G2, 3		19
		mit228	FF1210-6C		S4 OXI-III D2	G2, 3		19
		mit111	FF1210-6C		S4 OXI-III D3	G2, 3		19
		mit118	FF1210-6C		S4 OXI-III D3	G2, 3		19
		mit121	FF1210-6C		S4 OXI-III D3	G2, 3		19
		mit153	FF1210-6C		S4 OXI-III D3	G2, 3		19
		mit215	FF1210-6C		S4 OXI-III D3	G2, 3		19
		mit221	FF1210-6C		S4 OXI-III D3	G2, 3		19
		mit206			S4 OXI3			65
		mit201			S5 OXI1			65
	MUNICH	1273	777-3A		S5 OXI1	G2, 3		160, 162

TABLE 1 - Page 39

CLASS		ORIGINAL NAME	PARENTAL STRAIN	MAIN PHENOTYPE	MAP & LOCUS ASSIGNMENT		MAPPING METHOD	PROBABLE GENE PRODUCT	REFERENCES
OXIDASE	MUNICH	1524	777-3A		S5	OXI1	G2, 3		162
		1702	777-3A		S4	OXI3	G2, 3		162
		2011	777-3A		S5	OXI1	G2, 3		162
		2031	777-3A		S4	OXI3	G2, 3		162
		2041	777-3A		S4	OXI3	G2, 3		162
		2111	777-3A		S4	OXI3	G2, 3		160, 162
		2131	777-3A		S4	OXI3	G2, 3		162
		2271	777-3A		S4	OXI3	G2, 3		162
		2481	777-3A		S5	OXI1	G2, 3		162
		2501	777-3A		S4	OXI3	G2, 3		162
		2532	777-3A		S5	OXI2	G2, 3		162
		2591	777-3A		S4	OXI3	G2, 3		162
		2621	777-3A		S4	OXI3	G2, 3		162
		3771	777-3A		S4	OXI3	G2, 3		162

TABLE 1 - Page 40

CLASS		ORIGINAL NAME	PARENTAL STRAIN	MAIN PHENOTYPE	MAP & LOCUS ASSIGNMENT		MAPPING METHOD	PROBABLE GENE PRODUCT	REFERENCES
OXIDASE	MUNICH	3781	777-3A		S4	OXI3	G2, 3		162
		3791	777-3A		S5	OXI2	G2, 3		162
		3792	777-3A		S4	OXI3	G2, 3		162
		3821	777-3A		S5	OXI1	G2, 3		162
		3831	777-3A		S4	OXI3	G2, 3		162
		3851	777-3A		S4	OXI3	G2, 3		162
		3861	777-3A		S4	OXI3	G2, 3		162
		4021	777-3A		S4	OXI3	G2, 3		162
		4051	777-3A		S4	OXI3	G2, 3		162
		4061	777-3A		S4	OXI3	G2, 3		162
		4071	777-3A		S4	OXI3	G2, 3		162
		4121	777-3A		S4	OXI3	G2, 3		162
		4131	777-3A		S4	OXI3	G2, 3		162

TABLE 1 - Page 41

CLASS		ORIGINAL NAME	PARENTAL STRAIN	MAIN PHENOTYPE	MAP & LOCUS ASSIGNMENT		MAPPING METHOD	PROBABLE GENE PRODUCT	REFERENCES
OXIDASE	MUNICH	4151	777-3A		S4	OXI3	G2, 3		162
		4171	777-3A		S4	OXI3	G2, 3		162
		4191	777-3A		S4	OXI3	G2, 3		162
		5431	777-3A		S4	OXI3	G2, 3		162
		6111	RM395-4C		S4	OXI3	G2, 3		162
		6201	RM395-4C		S4	OXI3	G2, 3		162
		6291	RM395-4C		S4	OXI3	G2, 3		162
		6321	RM395-4C		S4	OXI3	G2, 3		162
		6331	RM395-4C		S4	OXI3	G2, 3		162
		6351	RM395-4C		S4	OXI3	G2, 3		162
		6431	RM395-4C		S4	OXI3	G2, 3		162
		6511	RM395-4C		S5	OXI2	G2, 3		162
		6512	RM395-4C		S5	OXI2	G2, 3		162
		6761	RM395-4C		S5	OXI1	G2, 3		162

TABLE 1 - Page 42

CLASS		ORIGINAL NAME	PARENTAL STRAIN	MAIN PHENOTYPE	MAP & LOCUS ASSIGNMENT		MAPPING METHOD	PROBABLE GENE PRODUCT	REFERENCES
OXIDASE	MUNICH	6762	RM395-4C		S5	OXI1	G2, 3		162
		6771	RM395-4C		S5	OXI1	G2, 3		162
		6773	RM395-4C		S5	OXI1	G2, 3		162
		7181	RM395-4C		S4	OXI3	G2, 3		162
		7351	RM395-4C		S4	OXI3	G2, 3		162
	BLOOMINGTON + COLUMBUS	E3	41-6/161 ?	cyt a (-) oxi (-)	S5	OXI2	G3	cyt.oxidase subunit 3	113
		E20	1D41/161 ?	cyt a (-) oxi (-)	S5	OXI2	G3	cyt.oxidase subunit 3	113
		E37	41-6/161 ?	cyt a (-) oxi (-)	S5	OXI1	G3	cyt.oxidase subunit 1	113
		E65	41-6/161 ?	cyt a (-) oxi (-)	S5	OXI1	G3	cyt.oxidase subunit 1	113
		E69	41-6/161 ?	cyt a (-) oxi (-)	S5	OXI2	G3	cyt.oxidase subunit 3	113
		CAD245	41-6/161 ?	cyt a (-) oxi (-)	S4	OXI3	G3	cyt.oxidase subunit 1	113
		C245	1D41/161	cyt a (-)					112

TABLE 1 - Page 43

CLASS		ORIGINAL NAME	PARENTAL STRAIN	MAIN PHENOTYPE	MAP & LOCUS ASSIGNMENT		MAPPING METHOD	PROBABLE GENE PRODUCT	REFERENCES
OXIDASE	BLOOMINGTON + COLUMBUS	M36	1D41/161	cyt a (-)					112
OXIDASE	BLOOMINGTON + COLUMBUS	M37	1D41/161	cyt a (-)					112
OXIDASE	BLOOMINGTON + COLUMBUS	M38	1D41/161	cyt a (-)					112
OXIDASE	BLOOMINGTON + COLUMBUS	M44	1D41/161	cyt a (-)					112
OXIDASE	BLOOMINGTON + COLUMBUS	M70	1D41/161	cyt a (-)					112
OXIDASE	BLOOMINGTON + COLUMBUS	M80	1D41/161	cyt a (-)					112
OXIDASE	BLOOMINGTON + COLUMBUS	M117	1D41/161	cyt a (-)					112
OXIDASE	BLOOMINGTON + COLUMBUS	M357	1D41/161	cyt a (-)					112
OXIDASE (CONDITIONAL)	AMHERST	hs69	S288C	heat sensitivity for growth on glycerol (oligomycin resistance)	S4	OXI3	G1		117
OXIDASE (CONDITIONAL)	ORSAY	ts628	PF1210-6C	§	S5	OXI1	G2, 3		19
OXIDASE (CONDITIONAL)	ORSAY	ts936	FF1210	§	S4	OXI3D1	G2, 3		19
OXIDASE (CONDITIONAL)	ORSAY	cs899	FF1210-6C	§	S4	OXI3D2	G2, 3		19
OXIDASE (CONDITIONAL)	ORSAY	tscs975	FF1210-6C	§	S4	OXI3D2	G2, 3		19
OXIDASE (CONDITIONAL)	ORSAY	ts360	FF1210-6C	§	S4	OXI3D3	G2, 3		19
OXIDASE (CONDITIONAL)	BLOOMINGTON + COLUMBUS	E33	41-6/161 ?	cyt a (-) oxi (-) §	S4	OXI3	G3		113
OXIDASE (CONDITIONAL)	BLOOMINGTON + COLUMBUS	E56	41-6/161 ?	cyt a (-) oxi (-) §	S4	OXI3	G3		113

§ see page 64 of this table

TABLE 1 - Page 44

CLASS		ORIGINAL NAME	PARENTAL STRAIN	MAIN PHENOTYPE	MAP & LOCUS ASSIGNMENT		MAPPING METHOD	PROBABLE GENE PRODUCT	REFERENCES
REDUCTASE	NEW YORK, PUBLIC HEALTH	M6-185	D273-10B	cyt a (-) cyt b (-) oxi (-) red (-)	S3 and S4	COB1 and OXI3	G1		61
		M6-200	D273-10B	cyt b (-) red (-)	S3	COB1	G1, 2, 3		92, 142, 164, 184, 185, 187A
		M7-40	D273-10B	cyt b (-) red (-)	S3	COB1	G1, 2, 3		92, 164, 185, 187A
		M8-13	D273-10B	cyt a (-) cyt b (-)	S3	OXI3	G1		61
		M8-53	D273-10B	cyt b (-) red (-)	S3	COB1	G1, 2, 3		92, 164, 184, 185, 187A
		M8-130	D273-10B	cyt a (-) cyt b (-)	S3 and S4	COB2 and OXI3	G1		61, 183
		M8-147	D273-10B	cyt a (-) cyt b (-) oxi (-)	S3 and S4	COB1 and OXI3	G1, 2, 3		61
		M8-181	D273-10B	cyt b (-)	S3	COB1	G1, 2, 3		185, 187A
		M8-219	D273-10B	cyt a (-) cyt b (-) oxi (-) red (-)	S3	COB2	G1, 2, 3		164
		M9-226	D273-10B	cyt a (-) cyt b (-)	S3	COB2	G1		61, 183
		M9-228	D273-10B	cyt a (-) cyt b (-) oxi (-) red (-)	S3	COB2(185) relocated in COB1 in (92)	G1, 2, 3	cyt. b	92, 164, 185, 187A

TABLE 1 - Page 45

CLASS		ORIGINAL NAME	PARENTAL STRAIN	MAIN PHENOTYPE	MAP & LOCUS ASSIGNMENT		MAPPING METHOD	PROBABLE GENE PRODUCT	REFERENCES
REDUCTASE	NEW YORK, PUBLIC HEALTH	M10-152	D273-10B	cyt b (-) red (-) §	S3	COB2	G1, 2, 3	cyt. b	164, 185, 187A
		M13-101	D273-10B	cyt b (-) red (-)	S3	COB1	G1, 2, 3		92, 128A, 164, 184 185, 187A
		M15-207	D273-10B	cyt b (-)	S3	COB1	G1, 2, 3		185, 187A
		M16-99 ? or M6-199 ?	D273-10B	cyt a (-) cyt b (-)	S3 and S4	COB2 and OXI3	G1		61
		M17-162	D273-10B/A1	cyt b (-)	S3	COB2	G1, 2, 3	cyt b	92, 162, 185, 187A
		M17-231	D273-10B/A1	cyt b (-)	S3	COB1	G1, 2, 3		185, 187A
		M18-68	D273-10B/A	cyt a (-) cyt b (-)	S3 and S4	COB2 and OXI3	G1	cyt. b	61, 187A
		M21-12	D273-10B/A21	cyt a (-) cyt b (-)	S3 and S4	COB2 and OXI3	G1		61
		M21-71	D273-10B/A1	cyt b (-)	S3	COB2	G1, 2, 3	cyt b	92, 185, 187A
		M24-213	D273-10B/A16	cyt a (-) cyt b (-)	S3	COB1	G1, 2, 3		61
		M24-241	D273-10B/A1	cyt b (-)	S3	COB1	G1, 2, 3		92, 162, 185, 187A

§ see page 64 of this table

TABLE 1 - Page 46

CLASS		ORIGINAL NAME	PARENTAL STRAIN	MAIN PHENOTYPE	MAP & LOCUS ASSIGNMENT		MAPPING METHOD	PROBABLE GENE PRODUCT	REFERENCES
REDUCTASE	NEW YORK, PUBLIC HEALTH	M26-216	D273-10B/A16	cyt a (-) cyt b (-)	S3 and S4	COB2 and OXI3	G1, 2, 3		61
		M33-119	D273-10B/A1	cyt b (-)	S3	COB2	G1, 2, 3	cyt b	185, 187A
		M33-236	D273-10B/A	cyt a (-) cyt b (-)	S3 and S4	COB2 and OXI3	G1		61
		M34-171	D273-10B/A	cyt a (-) cyt b (-)	S3 and S4	COB2 and OXI3	G1, 2, 3		61
		M35-175	D273-10B/A1	cyt b (-)					185
		M35-201	D273-10B/A	cyt a (-) cyt b (-)	S3 and S4	COB2 and OXI3	G1		61
		M37-7	D273-10B/A	cyt a (-) cyt b (-)	S3 and S4	COB2 and OXI3	G1		61
	GIF + WROCLAW	G55	777-3A	cyt a (-) cyt b (-) oxi (-) red (-)	S3	BOX3 COB2	G1, 2, 3	cyt b	33, 92, 93, 136
		G171	777-3A	cyt a (-) cyt b (-) oxi (-) red (-)	S3	BOX4 COB2	G1, 2, 3	cyt b	33, 92, 93, 136

TABLE 1 - Page 47

CLASS		ORIGINAL NAME	PARENTAL STRAIN	MAIN PHENOTYPE	MAP & LOCUS ASSIGNMENT		MAPPING METHOD	PROBABLE GENE PRODUCT	REFERENCES
REDUCTASE	GIF + WROCLAW	G392	777-3A	cyt a (-) cyt b (-) oxi (-) red (-)	S3	BOX5 COB2	G1, 2, 3	cyt b	33, 92, 93, 136
		G410	777-3A	cyt b (-) §	S3	BOX	G3		93
		G442	777-3A	cyt a (-) §	S3	BOX	G3		93
		G476	777-3A	cyt a (-) cyt b (-)	S3	BOX	G3		93
		G625	777-3A	cyt a (-) cyt b (-) oxi (-) red (-)	S3	deletion BOX 1 to 6	G1, 2, 3	cyt b	33, 92, 93, 136
		G706	777-3A	cyt a (-) cyt b (-) oxi (-) red (-)	S3	BOX5 COB2	G1, 2, 3	cyt b	33, 92, 93, 136
		G839	777-3A	cyt a (-) cyt b (-) oxi (-) red (-)	S3	BOX COB2	G1, 2, 3	cyt b	33, 92, 93, 136
		G925	777-3A	cyt b (+) red (-)	S3	BOX6 COB1	G1, 2, 3	cyt b	33, 92, 93, 136
		G1334	777-3A	cyt b (-) § red (-)	S3	BOX4 COB2	G1, 2, 3	cyt b	33, 41, 91, 92, 93, 135, 136, 162

§ see page 64 of this table

TABLE 1 - Page 48

CLASS		ORIGINAL NAME	PARENTAL STRAIN	MAIN PHENOTYPE	MAP & LOCUS ASSIGNMENT		MAPPING METHOD	PROBABLE GENE PRODUCT	REFERENCES
REDUCTASE	GIF + WROCLAW	G1370	777-3A	cyt a (-) cyt b (-) oxi (-) red (-)	S3	BOX3 COB2	G1, 2, 3	cyt b	33, 41, 91, 92, 93, 135, 136, 162
		G1659	777-3A	cyt a (-) cyt b (-) oxi (-) red (-)	S3	BOX COB1	G1, 2, 3	cyt b	33, 41, 92, 93, 136
		G1919	777-3A	cyt a (-) cyt b (-) oxi (-) red (-)	S3	BOX3 COB2	G1, 2, 3	cyt b	33, 91, 92, 93, 135, 136
		G1988	777-3A	cyt b (-) § red (-)	S3	BOX1 COB1	G1, 2, 3	cyt b	33, 41, 91, 92, 93 135, 136
		G2026	777-3A	cyt a (-) § cyt b (-)	S3		G3		93
		G2286	777-3A	cyt b (-) § red (-)	S3	BOX1 COB1	G1, 2, 3	cyt b	33, 41, 91, 92, 93, 135, 136, 164
		G2457	777-3A	cyt b (-) § red (-)	S3	BOX1 COB1	G1, 2, 3	cyt b	33, 91, 92, 93, 135; 136
		G2494	777-3A	cyt b (-) § red (-)	S3	BOX2 COB1	G1, 2, 3	cyt b	33, 41, 91, 92, 93, 135, 136, 162
		G2590	777-3A	cyt b (-) § red (-)	S3	BOX2 COB1	G1, 2, 3	cyt b	33, 91, 92, 93, 135, 136
		W7	777-3A	cyt b (+) red (-)	S3	BOX4 COB2	G1, 2, 3	cyt b	33, 92, 93, 136

§ see page 64 of this table

TABLE 1 - Page 49

CLASS		ORIGINAL NAME	PARENTAL STRAIN	MAIN PHENOTYPE	MAP & LOCUS ASSIGNMENT		MAPPING METHOD	PROBABLE GENE PRODUCT	REFERENCES
REDUCTASE	GIF + WROCLAW	W19	777-3A	cyt b (-) red (-)	S3	BOX2 COB1	G1, 2, 3	cyt b	33, 92, 93, 136
		W91	777-3A	cyt a (-) cyt b (-)	S3	BOX3 COB2	G2, 3	cyt b	92, 93
		W231	777-3A	cyt a (-) cyt b (-)	S3 and S4	BOX and OXI3	G3		93
		W267	777-3A	cyt b (-) red (-)	S3	BOX6 COB1	G1, 2, 3	cyt b	33, 92, 93, 136
		W300	777-3A	cyt b (-) red (-)	S3	BOX2 COB1	G1, 2, 3	cyt b	33, 92, 93, 136
		W401	777-3A	cyt a (-) cyt b (-) red (-)	S3	BOX3 COB2	G1, 2, 3	cyt b	33, 92, 93, 136
	CLAYTON	41.2		cyt b (-) red (-)	S3	cyb1	G1, 3, P2		152
		37.16		cyt b (-) red (-)	S3	cyb1	G1, 3, P2		152
		17.62.2			S3	cyb1	G1, 3, P2		152
		41-2-4	J69	cyt b (-) red(-)	S3	cyb1	G1, 3, P2		35
		17-62-1	J69	cyt b (-) red (-)	S3	cyb1	G1, 3, P2		35

TABLE 1 - Page 50

CLASS		ORIGINAL NAME	PARENTAL STRAIN	MAIN PHENOTYPE	MAP & LOCUS ASSIGNMENT		MAPPING METHOD	PROBABLE GENE PRODUCT	REFERENCES
REDUCTASE	CLAYTON	I7-62-11	J69	cyt b (-) red (-)	S3	cybl	G1, 3, P2		35
REDUCTASE	CLAYTON	37-16-6	J69	cyt b (+) red (-)	S3	cybl	G1, 3, P2		35
REDUCTASE	ORSAY	mit226	FF1210-6C		S3	cob (sub1)	G2, 3		19
REDUCTASE	ORSAY	mit148	FF1210-6C		S3	cob (sub1)	G2, 3		19
REDUCTASE	ORSAY	mit141	FF1210-6C		S3	cob (sub1)	G2, 3		19
REDUCTASE	ORSAY	mit119	FF1210-6C		S3	cob (sub2)	G2, 3		19
REDUCTASE	ORSAY	mit185	FF1210-6C		S3	cob (sub2)	G2, 3		19
REDUCTASE	ORSAY	mit199	FF1210-6C		S3	cob (sub2)	G2, 3		19
REDUCTASE	MUNICH	1232	777-3 A		S3	COB2	G2, 3		160, 162
REDUCTASE	MUNICH	1233	777-3 A		S3	COB2	G2, 3		162
REDUCTASE	MUNICH	1282	777-3 A		S3	COB2	G2, 3		162
REDUCTASE	MUNICH	1301	777-3 A		S3	COB2	G2, 3		162

TABLE 1 - Page 51

CLASS		ORIGINAL NAME	PARENTAL STRAIN	MAIN PHENOTYPE	MAP & LOCUS ASSIGNMENT		MAPPING METHOD	PROBABLE GENE PRODUCT	REFERENCES
REDUCTASE	MUNICH	1302	777-3A		S3	COB2	G2, 3		162
		1321	777-3A		S3	COB1	G2, 3		162
		1354	777-3A		S3	COB1	G2, 3		162
		1391	777-3A		S3	COB2	G2, 3		162
		1392	777-3A		S3	COB2	G2, 3		162
		1431	777-3A		S3	COB1	G2, 3		162
		1432	777-3A		S3	COB1	G2, 3		162
		1622	777-3A		S3	COB2	G2, 3		162
		1623	777-3A		S3	COB2	G2, 3		162
		1625	777-3A		S3	COB2	G2, 3		162
		1651	777-3A		S3	COB2	G2, 3		162
		2011	777-3A		S3	COB2	G2, 3		162
		2034	777-3A		S3	COB2	G2, 3		162

TABLE 1 - Page 52

CLASS		ORIGINAL NAME	PARENTAL STRAIN	MAIN PHENOTYPE	MAP & LOCUS ASSIGNMENT		MAPPING METHOD	PROBABLE GENE PRODUCT	REFERENCES
REDUCTASE	MUNICH	2041	777-3A		S3	COB2	G2, 3		162
		2044	777-3A		S3	COB2	G2, 3		162
		2045	777-3A		S3	COB2	G2, 3		162
		2046	777-3A		S3	COB2	G2, 3		162
		2048	777-3A		S3	COB2	G2, 3		162
		2151	777-3A		S3	COB2	G2, 3		162
		2161	777-3A		S3	COB2	G2, 3		162
		2282	777-3A		S3	COB1	G2, 3		162
		2351	777-3A		S3	COB2	G2, 3		162
		2352	777-3A		S3	COB2	G2, 3		162
		2411	777-3A		S3	COB2	G2, 3		162
		2521	777-3A		S3	COB2	G2, 3		162
		2531	777-3A		S3	COB2	G2, 3		162

TABLE 1 - Page 53

CLASS		ORIGINAL NAME	PARENTAL STRAIN	MAIN PHENOTYPE	MAP & LOCUS ASSIGNMENT		MAPPING METHOD	PROBABLE GENE PRODUCT	REFERENCES
REDUCTASE	MUNICH	2551	777-3A		S3	COB2	G2, 3		162
		3881	777-3A		S3	COB1	G2, 3		162
		4111	777-3A		S3	COB2	G2, 3		162
		4151	777-3A		S3	COB2	G2, 3		162
	BLOOMIN-GTON + COLUMBUS	EM17	41-6/161 ?	cyt b (-)	S3	COB2	G3, P3		113
		EM25	41-6/161 ?	cyt b (-)	S3	COB2	G3, P3		113
REDUCTASE (CONDITIONAL)	ORSAY	cs937	FF1210-6C	§	S3	cob (sub1)	G2, 3		19
		tscs632	FF1210-6C	§	S3	cob (sub1)	G2, 3		19
		ts234	FF1210-6C	§	S3	cob (sub2)	G2, 3		19
		cs361	FF1210-6C	§	S3	cob (sub2)	G2, 3		19
ATPase	NEW YORK PUBLIC HEALTH	M6-239	D273-10B	ATPase (-)	S4	pho1	G1, 3		60, 184
		M19-229	D273-10B/A	ATPase (-)	S2 or S3	pho2 ?	G1, 3		60

§ see page 64 of this table

TABLE 1 - Page 54

CLASS		ORIGINAL NAME	PARENTAL STRAIN	MAIN PHENOTYPE	MAP & LOCUS ASSIGNMENT	MAPPING METHOD	PROBABLE GENE PRODUCT	REFERENCES
ATPase	NEW-YORK PUBLIC HEALTH	M28-81	D273-10B/A	ATPase (-)	S4 pho1	G1, 3		60
		M28-82	D273-10B/A	ATPase (-)	S4 pho1	G1, 3		60
MULTIPLES DEFICIENCIES	BLOOMINGTON + COLUMBUS	UF73/1	D243-4A	cyt a (-) red (-) ATPase oli^R				58
		UF20/1	D243-4A	cyt a (-) red (-) ATPase oli^R				58
		UF32/1	D243-4A	cyt a (-) red (-) ATPase oli^R				58
		UF42/1	D243-4A	cyt a (-) red (-) ATPase oli^R				58
		M45	ID41/161	cyt a (-) cyt b (-) red (-)				112
	PUBLIC HEALTH	E1-9	D273-10B	oxi (-) red (-) ATPase (-)				183
		M1-234	D273-10B	oxi (-) red (-) ATPase (-)				183
		M7-37	D273-10B	cyt a (-) cyt b (-) oxi (-) red (-) ATPase (-)				184
		M15-153	D273-10B	cyt a (-) cyt b (-) oxi (-) red (-) ATPase (-)				184

TABLE 1 - Page 55

CLASS		ORIGINAL NAME	PARENTAL STRAIN	MAIN PHENOTYPE	MAP & LOCUS ASSIGNMENT	MAPPING METHOD	PROBABLE GENE PRODUCT	REFERENCES
CONDITIONAL DEFICIENCIES FOR VARIOUS FUNCTIONS OR UNDEFINED FUNCTIONS	MUNICH	tsm8	M12	temperature sensitive for growth on glycerol and ρ^- production	S5 ile tRNA vicinity	G1, 2, 3	ile tRNA ?	11, 15, 19, 81, 127, 158, 161.
	N.Y *	aa53	2-38(5-6)C	inability to grow on glycerol, ethanol at 36°C				169
	WARWICK	D6/5	D6	cold sensitivity for growth on glycerol	S5	G1, 3		96
		D6/26L	D6	heat sensitivity for growth on glycerol	S4	G1, 3		96
		D6/N302	D6	cold sensitivity for growth on glycerol	S3	G1, 3		96
		D22/96	D22	cold sensitivity for growth on glycerol	S1	G1, 3		96
		D22/107	D22	cold sensitivity for growth on glycerol	S2	G1, 3		96
	AMHERST	cs23	S288C	cold sensitivity for growth on glycerol (erythromycin resistance)		G1		117
	ORSAY	tscs902	FF1210-6C		S1	G2, 3		19, 57
		tscs983	FF1210-6C		S2	G2, 3		19
		cs901	FF1210-6C		B1, S1 or S5	G2, 3		19, 57

* N.Y : NEW YORK (EINSTEIN)

TABLE 1 - Page 56

CLASS		ORIGINAL NAME	PARENTAL STRAIN	MAIN PHENOTYPE	MAP & LOCUS ASSIGNMENT	MAPPING METHOD	PROBABLE GENE PRODUCT	REFERENCES
CONDITIONAL DEFICIENCIES FOR VARIOUS FUNCTIONS OR UNDEFINED FUNCTIONS	ORSAY	cs909	FF1210-6C		B1, S1 or S2	G2, 3		19, 57
		cs990	FF1210-6C		S2	G2, 3		19, 57
		ts832	FF1210-6C		S5 asp tRNA	G2, 3	asp tRNA ?	19, 57
		ts984	FF1210-6C		S5 metItRNA vicinity	G2, 3	metI tRNA ?	19, 57
		ts982	FF1210-6C		S5 met_ItRNA	G2, 3	met_I tRNA ?	19
	CLAYTON	tsr1	L410	cold sensitive respiratory deficiency	S1	G1, 3	21S RNA	109 , 35
UNDEFINED DEFICIENCIES	BLOOMINGTON + COLUMBUS	M113	1D41/161					112
		M375	1D41/161					112
	ORSAY	C2-3	FF1210-6C		tightly linked to O_{II}	G2, 3		19
		mit_{b2-1}	FF1210-6C		tightly linked to O_{II}	G2, 3		19
		mit175	FF1210-6C		S3	G2, 3		19
	GIF + WROCLAW	G179	777-3A					93
		G183	777-3A					93
		G2118	777-3A	cyt a (+) cyt b (+)				93

TABLE 1 - Page 57

CLASS : ω LOCUS	MAP ASSIGNMENT : S1
Original name of the allele	Parental strain and references
ω^+ ($)	DP1-1B (10, 13, 14, 15, 16, 17, 20A, 34, 36, 40, 42, 46, 47, 48, 49, 49A, 60, 61, 77, 78, 82, 86A, 94, 98, 117, 134, 137, 138, 154, 158, 159, 164, 171, 173, 185, 187, 193, 194, 197, 201, 203); D 22 (10, 13, 14, 34, 48, 60, 70, 71, 96, 98, 137, 138, 181, 185, 187, 200, 204); 650-2C (15, 16, 34, 36, 42, 46, 47, 48, 49, 134, 158, 201, 203); D 6 (10, 13, 14, 58, 86A, 96, 137, 200, 204); D243-4A (13, 58, 137, 138, 141, 142, 204); D273-10B (40, 60, 61, 185, 187); N1300 (86, 167, 180, 181); 770-7B (125, 126, 167, 181); FF1301-17B (40, 60, 141, 164); 4810 (137, 138, 204); FF1123-19C (141, 164); D 41 (71, 96); L410 (109, 180); L748(180, 181); 2-36-1 (13,204); 761-7A (125, 126); 863-2C (126, 181); S2/6B (15); SM1211-1B (15); FL496-2B (34); MB5-2B (34); K10 (86A); 198 (94); DL160 (96); DL191 (96); E23 (117); OL29 (117); 432-31 (125); 829-5B (125); 750-17-2C (126); AMF2-9 (138); D8-22A (141); D225-5A (173); A11 (181); D311-3A (181); T108 (181); T131-3D (181); T245-3A (181); T247 (181); T248-1D (181); 56-28D (180); 690/13 (181); 9003-2D (181); 805-4A (185); AA3 (185); AA4-2C (185); N11 (197); N15 (197); A364-A (204);
ω^- ($)	55R5-3C (10, 13, 14, 15, 16, 17, 34, 36, 40, 45, 46, 47, 48, 49, 52, 77, 78, 86A, 98, 117, 134, 137, 138, 158, 159, 164, 170, 171, 173, 185, 193, 197, 200, 201, 203, 204); G706 (77, 78, 86A, 193, 194, 197); M12 (48, 52, 86A, 158, 203); G102D (77, 78, 193, 194); N123 (13, 14, 86A, 137); D253-3C (86, 167, 180); 41-6 (44, 117, 170); SM50 (127, 159); 711 (197); 413-16 (125); 652-1-2 (126); 5L2-4 (58); M1-7C (78); M2-8C (78); E10-10/10 (204); D515-1B-47-3 (126); S. carlsbergensis (154);
ω^n (₤)	55R5-3C (36, 46, 47, 49); IL324-3A (46, 47, 49, 154); IL324-4B (49); IL324-6B (49); IL174/48 (49)

$,₤ see page 64 of this table.

TABLE 1 - Page 58

CLASS		ORIGINAL NAME	PARENTAL STRAIN	MAIN PHENOTYPE	MAP & LOCUS ASSIGNMENT	MAPPING METHOD	PROBABLE GENE PRODUCT	REFERENCES
POLYMORPHIC PROTEINS	DALLAS	var1 (fast)	41-6/19		S2	G1, G3 P3	protein 40,000 m.w. (43)	24, 43, 44, 45
		var1 (slow)	55R5-3C/221		S2	G1, G3 P3	protein 44,900 m.w. (24)	24, 43, 44, 45
		var1	5DSS		S2	P3	protein 46,500 m.w. (24)	24
		var1	D6		S2	P3	protein 45,800 m.w. (24)	24
		var1	NP1-3D		S2	P3	protein 44,400 m.w. (24)	24
		var1	41-6/161		S2	P3	protein 42,400 m.w. (24)	24
		var1	1-1/1		S2	P3	protein 42,100 m.w. (24)	24
		var2	55R5-3C/221		S2	G1, G3	protein 24,000-25,000 m.w. (43)	43, 44, 45
		var3	41-6/19		vicinity of 0_I	G1, G3	protein 46,000 m.w. (43)	43, 44, 45

TABLE 1 - Page 59

CLASS	ORIGINAL NAME	STRAIN TESTED	MAP & LOCUS ASSIGNMENT	MAPPING METHOD	REFERENCES
RIBOSOMAL RNA GENES	21S rRNA (†)	S. carlsbergensis	S1, B1, B2	P1, P3	154B, 190
	21S rRNA (†)	JS1-3D and KL14-4A	S1, B2	P1, P3	20, 82, 154, 154B
	23S rRNA (†)	IL8-8C	B1, S1, B2	P3, P4, P6	51, 53, 55, 62, 64, 101
	21S rRNA (†)	770-7B and 863-2C	S1, B2	P1	35, 107, 132, 166
	23S rRNA (†)	IL8-8C	S1	P1, P4	110A
	21S rRNA (†)	MH41-7B	S1, B2	P1, P3, P4	103, 129A
	15S rRNA (†)	770-7B and 863-2C	B5 par	P1	32, 35, 107, 132, 166
	14S rRNA (†)	S. carlsbergensis	S4	P1, P3	154A, 154B
	14S rRNA (†)	JS1-3D and KL14-4A	S4	P1, P3	154, 154A, 154B

† see page 64 of this article

TABLE 1 - Page 60

CLASS	ORIGINAL NAME	STRAIN TESTED	MAP AND LOCUS ASSIGNMENT	MAPPING METHOD	REFERENCES
RIBOSOMAL RNA GENES	16S rRNA (†)	KL14-4A IL8-8C TR3-15A MH32-12B M41-78/R12	S4 or S5 in the vicinity of PAR	P3, P4	54, 56, 64
	14S rRNA (†)	MH41-7B	S4	P1, P3, P4	103, 129A
	15S rRNA (†)	S. carlsbergensis	S4	P1, P3	190

† see page 64 of this article

TABLE 1 - Page 61

CLASS	ORIGINAL NAME OF THE GENE	NUMBER OF ISO - ACCEPTING SPECIES (Ψ)	STRAIN TESTED	MAP & LOCUS ASSIGNMENT (Ψ)	MAPPING METHOD	REFERENCES
TRANSFER RNA GENES	alanyl tRNA	1 (115)	IL8-8C MH32-12D MH41-7B D243-2B/R1	S5	G3, P1	19,30, 38, 63, 103, 115, 144A
	arginyl tRNA	2? (115)	MH32-12D MH41-7B	S5	G3, P1	19, 103, 115
	aspartyl tRNA	1 (115)	IL8-8C MH32-12D MH41-7B	S5	G3, P1	19, 30, 31, 63, 103,115, 144A
	cysteyl tRNA	2 (115)	MH32-12D MH41-7B	S5	G3, P1	19, 103, 115
	glutamyl I tRNA (= glutamic acid)	1 (115)	IL8-8C MH32-12D MH41-7B	S3	G3, P1	19, 30, 31, 63, 103, 110A, 114, 115, 144A
	glutamyl II tRNA (= glutamin)	1 (115)	IL8-8C MH32-12D MH41-7B	S5	G3, P1	19, 30, 31, 63, 103, 110A, 114, 115, 144A
	glycyl tRNA	2? (115)	D243-2B/R1	S5	G3, P1	38, 103, 115, 144A
	histidyl tRNA	1 (115)	IL8-8C MH32-12D MH41-7B	S5	G3, P1	19, 30, 31, 63, 103, 110A, 115, 144A

Ψ see page 64 of this article

TABLE 1 - Page 62

CLASS	ORIGINAL NAME OF THE GENE	NUMBER OF ISO - ACCEPTING SPECIES (Ψ)	STRAIN TESTED	MAP & LOCUS ASSIGNMENT (Ψ)	MAPPING METHOD	REFERENCE
TRANSFER RNA GENES	isoleucyl tRNA	1 (115)	IL8-8C MH32-12D MH41-7B D243-2B/R1	S5 vicinity of tsm8	G3, P1	19, 30, 38, 63, 103, 115, 144A
	leucyl tRNA	2? (115) or 1 (156)	IL8-8C MH32-12D MH41-7B D243-2B/R1	S5	G3, P1, P2	19, 29, 30, 31A, 37, 63, 67A, 103, 110A, 115, 144A, 156
	lysyl tRNA	1 (115)	IL8-8C MH32-12D MH41-7B	S5	G3, P1	19, 30, 31, 63, 103, 115, 144A
	methionyl I tRNA (†)	1 (115)	IL8-8C	S5	G3, P1	19, 103, 115
	methionyl II tRNA (†)	1 (115)	D243-2B/R1 IL8-8C	S5	G3, P1	19, 103, 115
	methionyl M tRNA (†)	1 (116)	IL8-8C		P1	116
	methionyl F tRNA (†) (or formyl methionyl tRNA)	1 (116)	IL8-8C D243-2B/R1		P1	79, 115, 116, 144A, 190
	phenylalanyl tRNA	2 (115)	IL8-8C MH32-12D MH41-7B D243-2B/R1	S5	G3, P1	19, 28A, 30, 38, 63,103, 115, 144A

† Ψ see page 64 of this article

TABLE 1 - Page 63

CLASS	ORIGINAL NAME OF THE GENE	NUMBER OF ISO-ACCEPTING SPECIES (Ψ)	STRAIN TESTED	MAP & LOCUS ASSIGNMENT (Ψ)	MAPPING METHOD	REFERENCES
TRANSFER RNA GENES	prolyl tRNA	1 (115)	IL8-8C MH32-12D MH41-7B	S5	G3, P1	19, 30, 31, 56, 63, 103, 115, 144A
	seryl tRNA	1 (115)	IL8-8C	S3	G3, P1	10A, 28A, 31, 103, 115, 144A
	threonyl tRNA	2? (115)	MH32-12D MH41-7B	S5	G3, P1	19, 103, 115
	tryptophan tRNA	3? (115)	MH32-12D MH41-7B	S5	G3, P1	19, 103, 115
	tyrosyl tRNA	1 (156) or 4 (115)	IL8-8C MH32-12D MH41-7B	S5	G3, P1	19, 30, 31, 63, 103, 115, 144A, 156
	valyl tRNA	2 (115) or 3 (156)	IL8-8C MH32-12D MH41-7B D243-2B/R1	S5	G3, P1	19, 29, 30, 37, 56, 63, 103, 115, 144A, 156

Ψ see page 64 of this article

TABLE 1 - Page 64

In all this table the main phenotype of each mutation is either explicitely stated according to original publications or abbreviated as follows : cyt a and cyt b:respectively cytochrome a + a_3 and cytochrome b spectra ; oxi , red and ATPase respectively cytochrome c oxidase, coenzyme QH_2 cytochrome c reductase and ATPsynthetase activities ;(+) wild type; (-) deficiencies. For map and locus assignment refer to Fig. 1 and 2. For mapping method refer to section 2 of the Introduction.

§ Means that additional information concerning partial deficiencies, leakiness, temperature dependance or catabolite repression dependant expression of the phenotype is given in the original publication.

$ Hundreds if not thousands of different ω^+ or ω^- strains are altogether cited in the literature. However, most of these strains are not independant and are derived by crosses from a relatively small number of strains which can be considered as parental strains of the ω^+ or ω^- alleles transmitted to the progeny. In cases where it was possible to trace the genealogy, only the parental strain is listed, (e.g. the strain IL8-4C ω^+ is not listed because its ω^+ allele is derived from the ω^+ parental strain DP1-1B which is listed (49). It is certain that most of the strains remaining on the list are still not of independant origin. It was however impossible to trace their genealogy from the publication cited. Readers are encouraged to provide information on the pedigree of the listed strains. An interesting possibility remains that the list of parental strains for the two ω alleles will eventually be reduced to two, one for ω^+ and one for ω^-.

£ The parental strain number corresponds to the name of the strain in which the ω^n mutation has been found. All parental strains carry the ω^- allele.

† Ribosomal RNA. In all publications cited, it is stated or implied that there is one gene for the large rRNA and one gene for the small rRNA per mitochondrial DNA molecule. The large rRNA is referred to as 21S or 23S and the small one as 14S, 15S or 16S depending on publications.

† tRNA met. In all publications cited, it is stated or implied that there are two genes for tRNA met or fmet. In different publications, these genes are referred to as I, II or F, M. The correspondance between these nomenclatures is unclear.

Ψ For all tRNA's the number of isoaccepting species has been indicated along with the corresponding literature citation. When more than one species is indicated and only one map and locus assignment is given, it is unclear whether all the genes for different isoaccepting species have been mapped or only one of them.

TABLE 1 - Page 65

ADDRESSES OF LABORATORIES QUOTED IN TABLE 1

Ref. in table 1	Full name and address
ADELAIDE	Department of Genetics, University of Adelaide, South Australia, Australia.
AMHERST	Department of Biochemistry, University of Massachusetts, Amherst, Massachusetts 01002, USA.
AMSTERDAM	Section for Medical Enzymology and Molecular Biology, Laboratory of Biochemistry, University of Amsterdam, Eerste Constantijn Huygenstraat, 20, Amsterdam, The Netherlands.
BALTIMORE	Department of Physiological Chemistry, The Johns Hopkins University, School of Medicine, Baltimore, Maryland 21205, USA.
BLOOMINGTON	Chemistry Department, Indiana University, Bloomington, Indiana 47401, USA.
BRATISLAVA	Food Research Institute, 89813 Bratislava, Czechoslovakia.
BROOKLYN	Department of Biology, Brooklyn College, City University of New-York, New-York, USA.
CHICAGO	Department of Medicine, Biochemistry and Biology, The University of Chicago and the Franklin McLean Research Institute, Chicago, Illinois 60637, USA.
CLAYTON	Department of Biochemistry, Monash University, Clayton, Victoria 3168, Australia.
COLUMBUS	Department of Genetics and Developmental Biology Program, The Ohio State University, Columbus, Ohio 43210, USA.
DALLAS	Department of Biochemistry, University of Texas, Health Science Center at Dallas, Dallas, Texas, USA.
DAVIS	Department of Biochemistry and Biophysics, University of California, Davis, California, USA.
GIF	Centre de Génétique Moléculaire du C.N.R.S., F-91190, Gif sur Yvette, France.
KOBE	Biology Division, College of General Education, Kobe University, Kobe 657, Japan.
LONDON (Univ.Col.)	Department of Botany and Microbiology, University College London, England.
LONDON (Mill Hill)	National Institute for Medical Research, Mill Hill, London, N.W.7, England.
LOUVAIN-LA-NEUVE	Laboratoire d'Enzymologie, Université de Louvain, 1348, Louvain-La-Neuve, Belgique.
MINNEAPOLIS (A)	Department of Agronomy and Plant Genetics, University of Minnesota, St.Paul, Minnesota, USA.

TABLE 1 - Page 66

MUNICH	Genetisches Institut der Universität Munchen, Maria-Ward Strasse, 1a, D-8000 Munchen 19, Germany.
NARA	Biological Laboratory, Nara University of Education, Takabatake, Nara 630, Japan.
NEW-YORK (EINSTEIN)	Departments of Genetics and Biochemistry, Albert Einstein College of Medicine, Bronx, New-York, USA.
NEW-YORK (PUBLIC HEALTH)	The Public Health Research Institute of the City of New-York, Inc., New-York, N.Y 10016, USA.
ORSAY	Fondation Curie, Institut du Radium, Section de Biologie, and Laboratoire de Biologie Générale, Université de Paris-Sud., Orsay 91405, France.
ROME	Centro Acidi Nucleici. Instituto di Fisiologia Generale, University of Rome, Italie.
SASKATOON	Department of Biology, University of Saskatchewan, Saskatoon, Saskatchewan, Canada.
STRASBOURG	Laboratoire de Biochimie, Faculté de Pharmacie, 2, rue Saint-Georges, 67 Strasbourg, France.
TALLIN	B.P. Konstantinov Leningrad Institute of Nuclear Physics, Academy of Sciences of the USSR and Institute of Experimental Biology, Academy of Sciences of the Estonian SSR, Tallin, USSR.
TOKYO	Faculty of Medicine, Department of Biochemistry, University of Tokyo, Tokyo, Japan.
WARWICK	Department of Molecular Sciences, University of Warwick, Coventry CV4 7AL, England.
WROCLAW	Institute of Microbiology, University of Wroclaw, Wroclaw, Poland.
WURZBURG	Physiologisch-Chemisches Institut der Universität, Koellikerstr. 2, D-87, Wurzburg, West Germany.
YOKOHAMA	Central Research Laboratory, Mitsubishi Chemical Industries Limited, Yokohama, Japan.

FIG. 1. "AVERAGE" GENETIC MAP OF MITOCHONDRIAL DNA

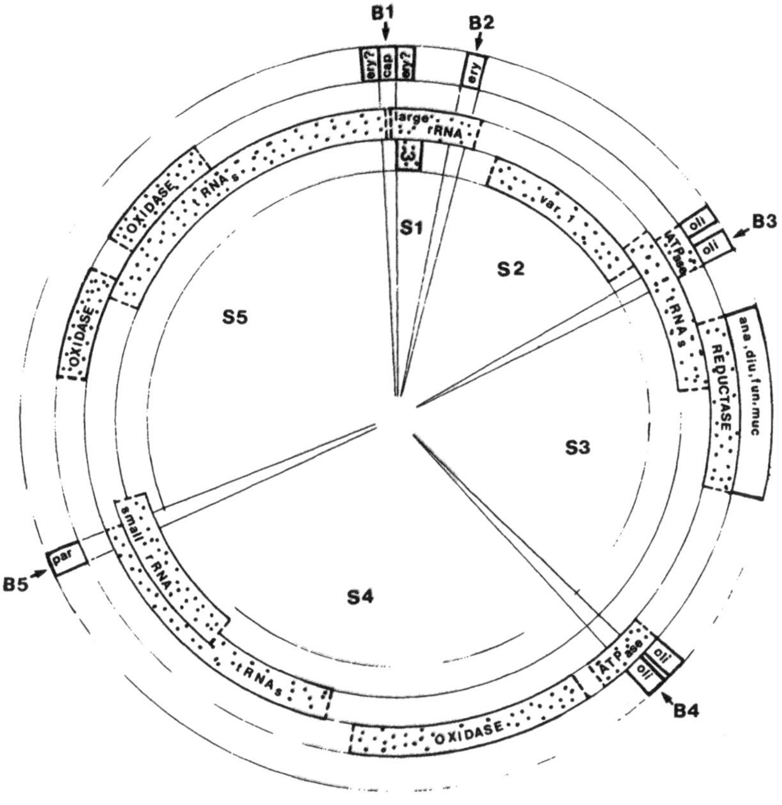

- B1 to B5 correspond to boundaries and S1 to S5 to sectors of the average map (see introduction, section 3).
- The main types of mutations and genes have been placed on concentric circles.
  First circle (external) : mutations and genes conferring drug resistance;
  Second circle : mutations and genes affecting respiratory or phosphorylating functions.
  Third circle : genes of rRNAs and tRNAs.
  Fourth circle : other mutations and genes.
- This average map is schematic and gives mainly the general topology of the molecule. In its construction we have tried to represent the relative size of sectors S1-S5 as close as possible to the published genetic and physical data by making an average estimation of the positions of boundaries and lengths of sectors. Because there are no major discrepancies between different methods, authors and yeast strains, it is probable that the positions of markers used as boundaries and lengths of sectors may be rather close to reality. The positions and lengths of segments where other mutations or genes are localized do not represent the actual relative sizes of the genes and are mainly given for the reader's guidance (e.g. detailed information on oxidase mutations and the small rRNA gene will be found in the maps of sectors S4 and S5 shown in next pages).

FIG. 2 DETAILED MAPS OF EACH SECTOR

For each sector, genetic (upper panels) and physical (lower panels) maps are shown with the corresponding literature citations.

Physical maps are represented by blocks. They are drawn to the same scale according to the original publications,the size of the sector being normalized for the common distance between two boundaries. The length of the block represents either the actual size of a gene or mutation or symbolizes the limits of a segment where a gene or a mutational site is located. The original publications should be consulted. The genetic maps always observe the names and the topology given in the original publication but there is no common scale. Thus, the distances drawn between genes or mutational sites may reflect neither the genetic nor the physical distances. When the same mutational site is mentioned in different maps its symbol is underlined. Δ indicates a deletion whose limits are shown by a bar. Other horizontal bars symbolize the limits of the segment where a gene or a mutational site is located. The width of the boundaries is not on the same scale as the corresponding sector. The names of mutations indicated in boundaries correspond to the mutations used in the original publication to establish each map.

Comments on individual sectors.

Sector S1 - A detailed restriction map of this region is presented p. 81

- * indicates that altogether 12 different C^R mutations have been localized at this locus (36, 47, 49, 134).
- ▲ indicates that the 21S RNA gene covers the cap1 mutational site in some publications quoted but not in the others.

Sector S3 - because of a limited space, the names of mutations 6-200, 7-40, 8-53, 13-101, 9-228, 10-152, 15-207, 17-162, 17-231, 18-68, 18-181, 21-71, 24-241, 33-119 should read M6-200, M7-40 etc ..

- when several vertical bars are indicated below a cluster of mutational sites it means that these sites are heteroallelic as shown by recombination analysis. When only one bar is indicated it means that either the sites are homoallelic or the analysis has not been performed. Original publications should be consulted.

Sector S5 - Because of limited space the tRNA genes are indicated by the name of the amino acid only.

- ▲ indicates that the 16S RNA gene covers this position in some of the publications quoted but not in others.

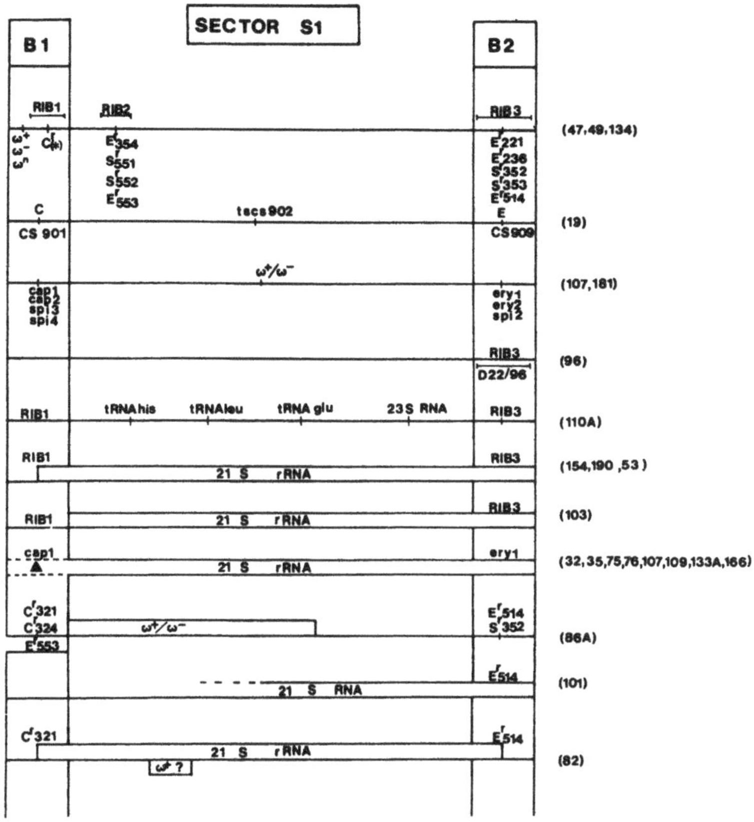
SECTOR S1
B1
B2
RIB1
RIB2
RIB3
(47,49,134)
tscs902
CS 901
CS909
(19)
(107,181)
cap1
cap2
spi3
spi4
ery1
ery2
spi2
D22/96
(96)
tRNAhis
tRNAleu
tRNA glu
23 S RNA
(110A)
21 S rRNA
(154,190 ,53)
(103)
cap1
ery1
(32,35,75,76,107,109,133A,166)
(86A)
21 S RNA
(101)
(82)

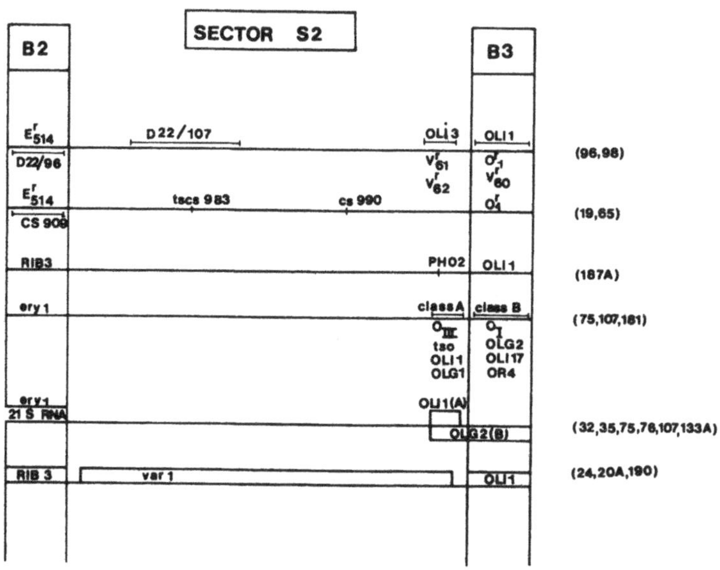
SECTOR S2
B2
B3
D22/107
OLI3
OLI1
(96,98)
D22/96
CS 909
tscs 983
cs 990
(19,65)
RIB3
PHO2
OLI1
(187A)
ery 1
class A
class B
(75,107,181)
tso
OLI1
OLG1
OLG2
OLI17
OR4
ery1
21 S RNA
OLI1(A)
OLG2(B)
(32,35,75,76,107,133A)
RIB 3
var 1
OLI1
(24,20A,190)

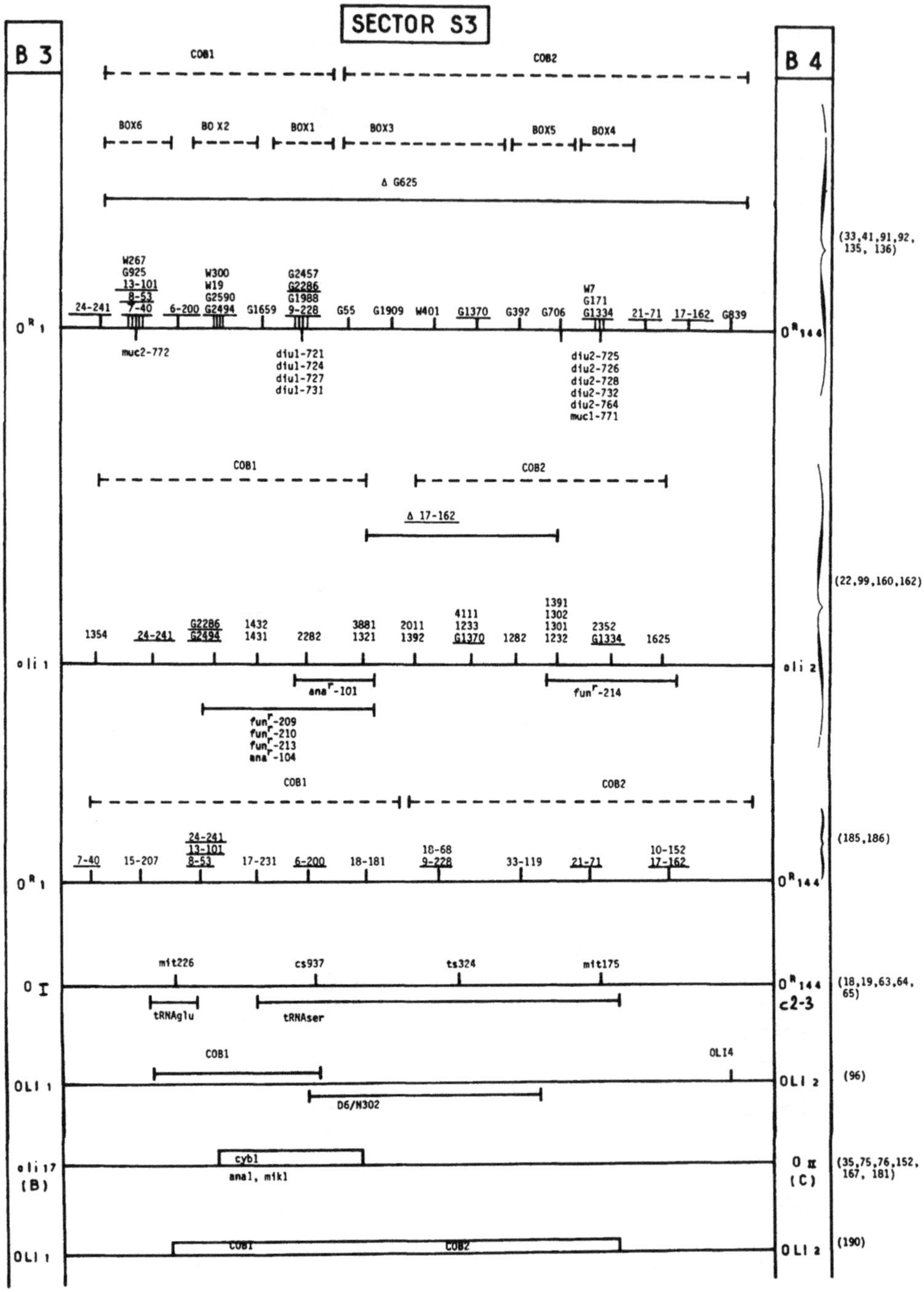

SECTOR S3
B 3
B 4
COB1
COB2
BOX6
BO X2
BOX1
BOX3
BOX5
BOX4
Δ G625
(33,41,91,92, 135, 136)
W267
G925
13-101
8-53
7-40
W300
W19
G2590
G2494
G2457
G2286
G1988
9-228
W7
G171
G1334
24-241
6-200
G1659
G55
G1909
W401
G1370
G392
G706
21-71
17-162
G839
O^R 1
O^R 144
muc2-772
diu1-721
diu1-724
diu1-727
diu1-731
diu2-725
diu2-726
diu2-728
diu2-732
diu2-764
muc1-771
Δ 17-162
(22,99,160,162)
1354
1432
1431
2282
3881
1321
2011
1392
4111
1233
1282
1391
1302
1301
1232
2352
1625
oli 1
oli 2
ana^r-101
fun^r-214
fun^r-209
fun^r-210
fun^r-213
ana^r-104
(185,186)
10-152
18-68
15-207
17-231
18-181
33-119
mit226
cs937
ts324
mit175
O I
c2-3
(18,19,63,64, 65)
tRNAglu
tRNAser
OLI4
OLI 1
OLI 2
(96)
D6/N302
cybl
anal, mikl
oli 17
(B)
O II
(C)
(35,75,76,152, 167, 181)
(190)

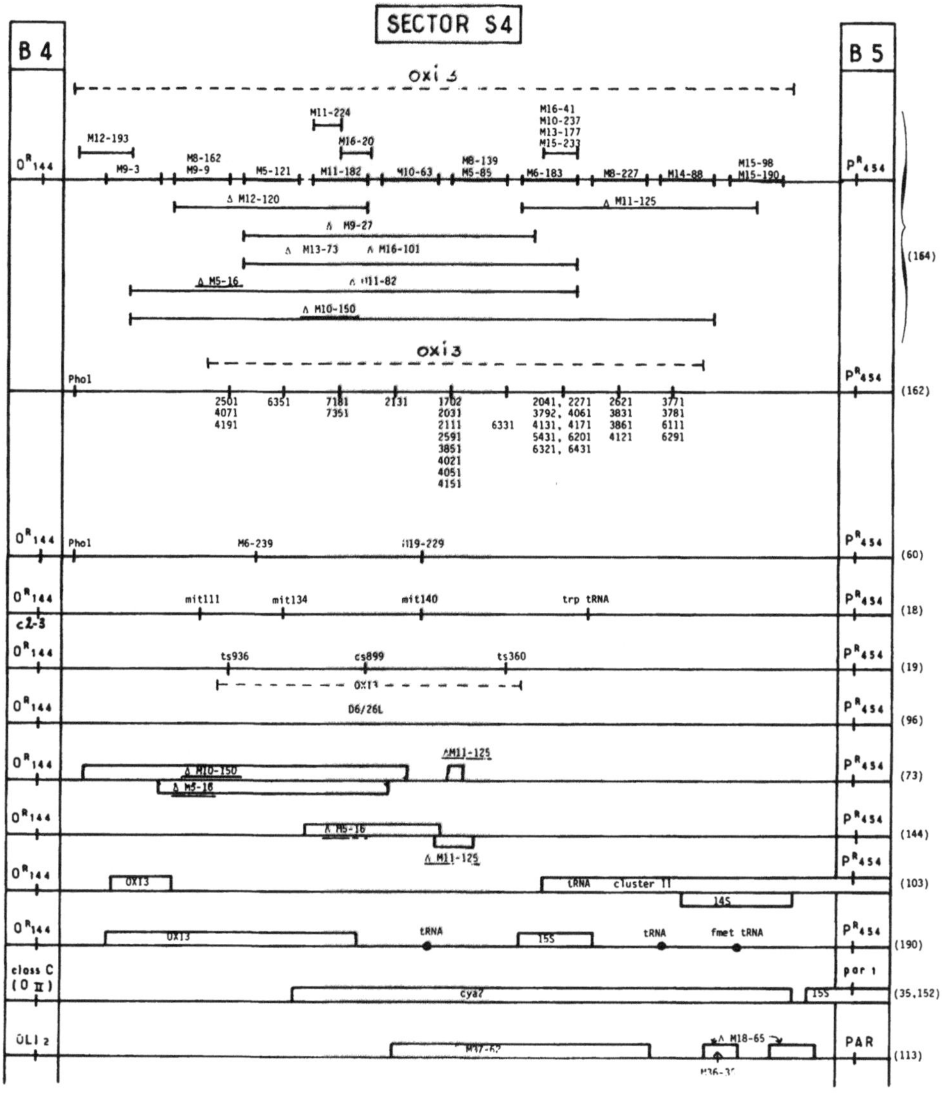
SECTOR S4
B 4
B 5
oxi 3
M11-224
M16-20
M12-193
M16-41
M10-237
M13-177
M15-233
M8-162
M9-9
M9-3
M5-121
M11-182
M10-63
M8-139
M5-85
M6-183
M8-227
M14-88
M15-98
M15-190
Δ M12-120
Δ M11-125
Δ M9-27
Δ M13-73
Δ M16-101
Δ M5-16
Δ M11-82
Δ M10-150
(164)
oxi3
Phol
2501 4071 4191
6351
7181 7351
2131
1702 2031 2111 2591 3851 4021 4051 4151
6331
2041, 2271 3792, 4061 4131, 4171 5431, 6201 6321, 6431
2621 3831 3861 4121
3771 3781 6111 6291
(162)
M6-239
M19-229
(60)
mit111
mit134
mit140
trp tRNA
(18)
c2-3
ts936
cs899
ts360
OXI3
(19)
D6/26L
(96)
ΔM11-125
Δ M10-150
Δ M5-16
(73)
Δ M5-16
Δ M11-125
(144)
OXI3
tRNA cluster II
14S
(103)
OXI3
tRNA
15S
tRNA
fmet tRNA
(190)
class C (O II)
par 1
cya7
15S
(35,152)
OLI 2
M17-62
M18-65
M36-35
PAR
(113)

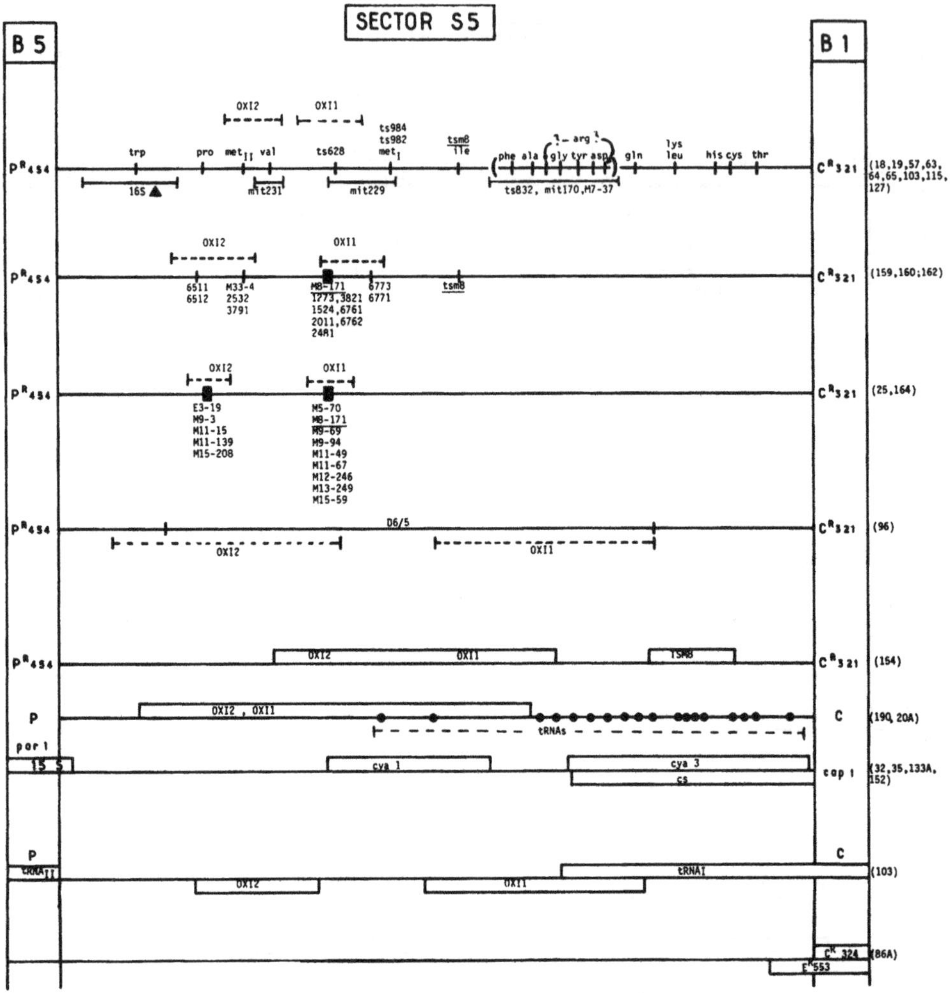
SECTOR S5
B 5
B 1
OXI2
OXI1
ts984
ts982
trp pro met II val ts628 met I
tsm8
ile
phe ala gly tyr asp arg gln lys leu his cys thr
16S
mit231
mit229
ts832, mit170, M7-37
P R 454
C R 321
(18,19,57,63, 64,65,103,115, 127)
6511
6512
M33-4
2532
3791
M8-171
1273,3821
1524,6761
2011,6762
2481
6773
6771
tsm8
(159,160;162)
E3-19
M9-3
M11-15
M11-139
M15-208
M5-70
M8-171
M9-69
M9-94
M11-49
M11-67
M12-246
M13-249
M15-59
(25,164)
D6/5
(96)
TSM8
(154)
P
C
OXI2 , OXI1
tRNAs
(190, 20A)
par 1
15 S
cya 1
cya 3
cs
cap 1
(32,35,133A, 152)
tRNA II
tRNA I
(103)
C R 324
E R 553
(86A)

FIG.3 OVERALL RESTRICTION MAPS OF mitDNA MOLECULES FROM SIX YEAST STRAINS

All maps are redrawn from the original publications and normalized to the same scale according to the total length of the molecule which is indicated on the right with the name of the strain. They were aligned with respect to each other using the first from the left EcoRI site present in all molecules examined (indicated on the figure by the arrow ➡).Maps have been linearized by cutting them at the SalI site or the corresponding HindII site in order to help in the comparison of the different strains and to obtain congruency between genetic and physical maps following, from left to right, the mnemomic order C - E - O - P. The limits of positions of genes or mutational sites are indicated by dotted blocks. Positions of insertions relative to S. carlsbergensis are indicated by hatched blocks and roman numbers. Restriction endonucleases are indicated as follows :

a = EcoRI, b = HindIII, c = HindII, c' = HincII, d = HhaI, e = BamH1, f = HpaI, g = XbaI, j = AvaI, k = XhoI, p = PstI, s = SalI.

FIG.4 DETAILED RESTRICTION MAPS OF THE LARGE rRNA and ω REGION OF THE mitDNA

All maps are redrawn from the original publications and normalized to the same scale as indicated at the bottom of the figure. The limits of positions of genes or mutated sites are indicated by dotted blocks. The position of the ω^+ insertion is indicated by a hatched block and related to the corresponding ω^- sequence by broken lines. For full details see original publications. Restriction endonucleases are indicated as follows :

b = HindIII, c = HindII, d = HhaI, h = HaeIII, i = HapII, l = AluI, s = SalI.

FIG.3 OVERALL RESTRICTION MAPS OF mitDNA MOLECULES FROM SIX YEAST STRAINS

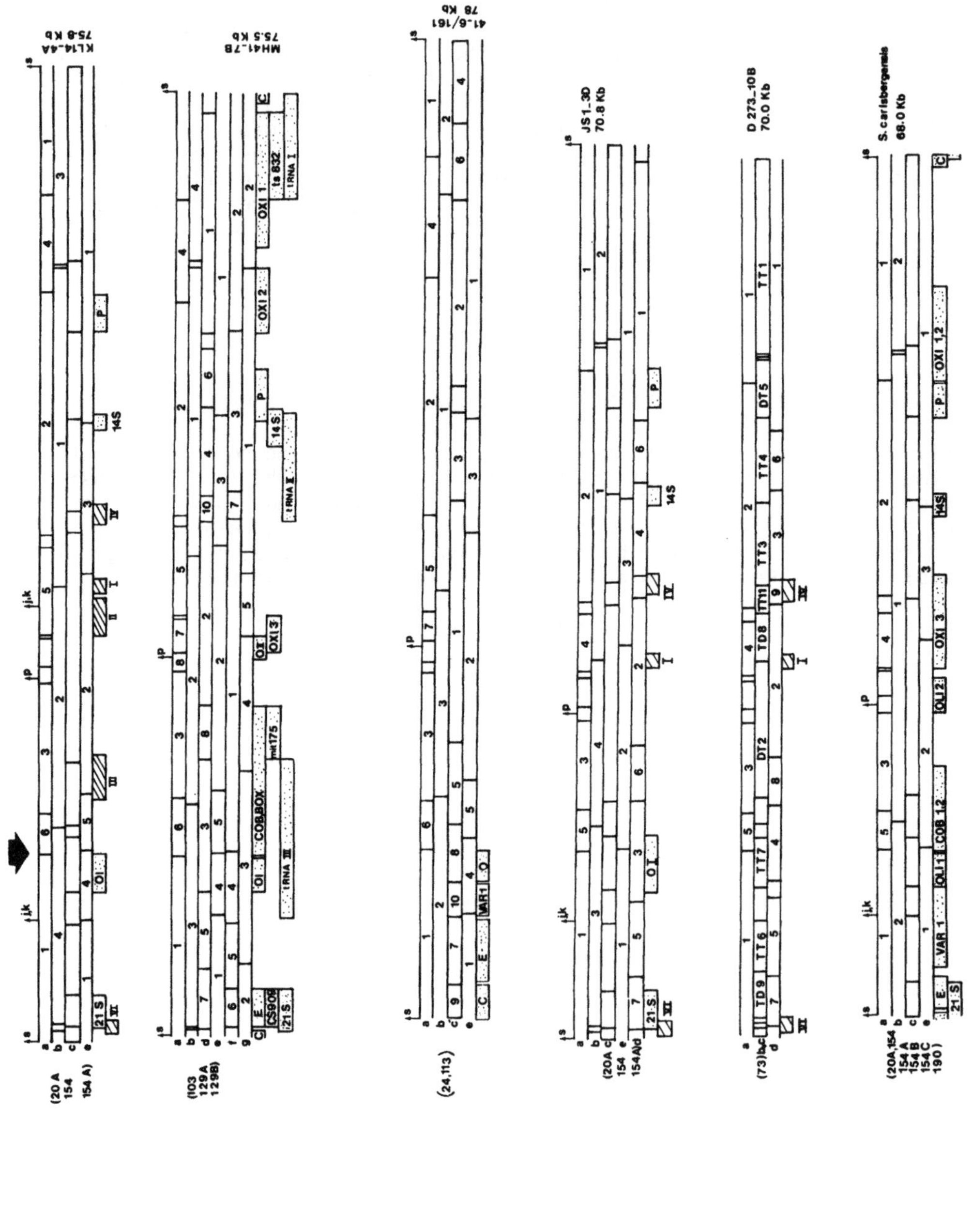

FIG.4 DETAILED RESTRICTION MAPS OF THE LARGE rRNA and ω REGION OF THE mitDNA

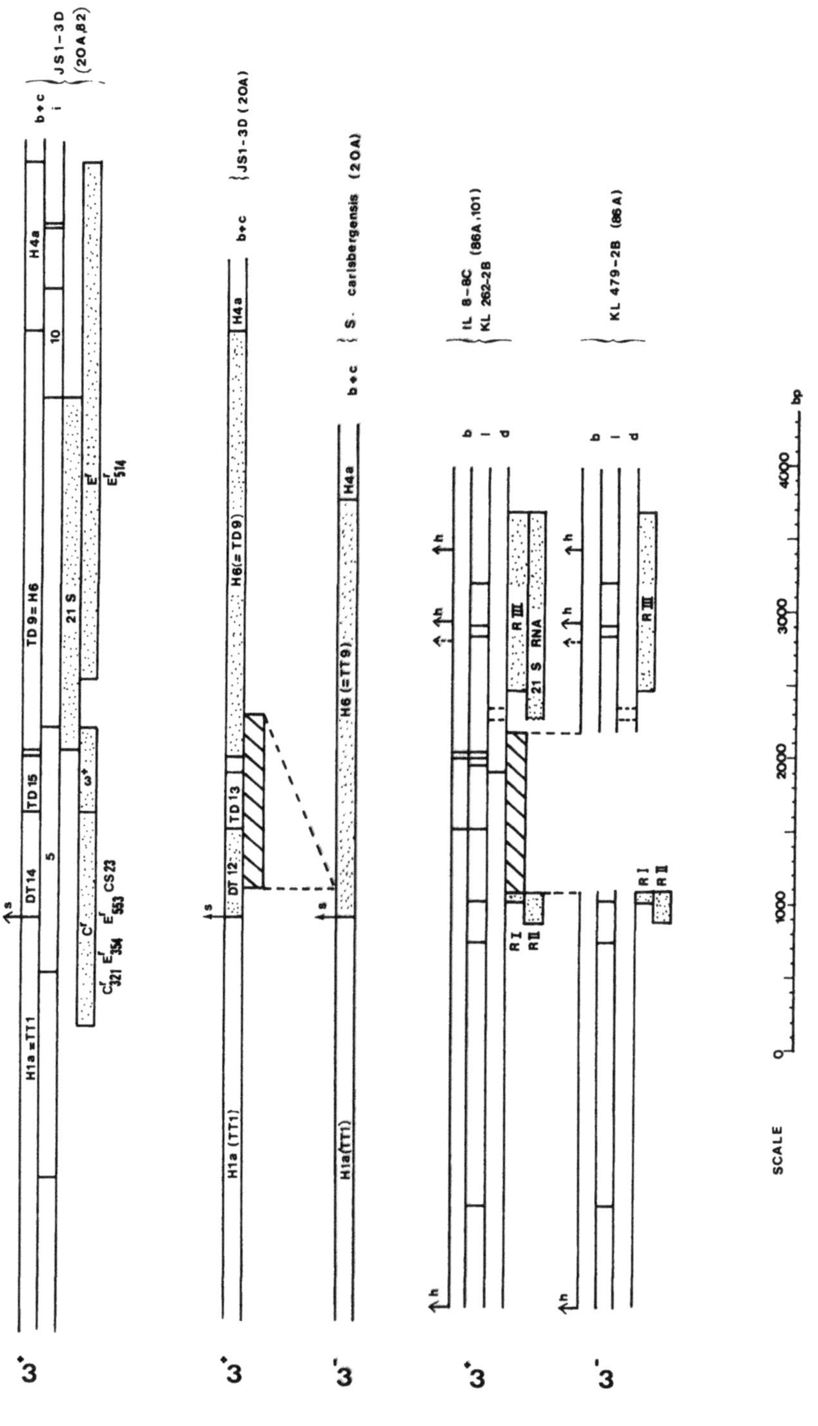

References

1. Genetics, Biogenesis and Bioenergetics of Mitochondria. (W. BANDLOW, R.J. SCHWEYEN, D.Y. THOMAS, K. WOLF, F. KAUDEWITZ, eds.), Walter de Gruyter, Berlin, New York (1976).
2. The Genetic Function of Mitochondrial DNA (C. SACCONE and A.M. KROON, eds.), Elsevier/North-Holland Biomedical Press, Amsterdam (1976).
3. Genetics and Biogenesis of Chloroplasts and Mitochondria (Th. BÜCHER, W. NEUPERT, W. SEBALD and S. WERNER, eds.), Elsevier/North-Holland (1976).
4. This Volume (1977).
5. AUFDERHEIDE, K.J. (1975) Molec. gen. Genet. 140, 231-241.
6. AUFDERHEIDE, K.J. and JOHNSON, R.G. (1976) Molec. gen. Genet. 144, 289-299.
7. AVNER, P.R. and GRIFFITHS, D.E. (1970) FEBS Letters, 10, 202-207.
8. AVNER, P.R. and GRIFFITHS, D.E. (1973) Eur. J. Biochem. 32, 301-311.
9. AVNER, P.R. and GRIFFITHS, D.E. (1973) Eur. J. Biochem. 32, 312-321.
10. AVNER, P.R., COEN, D., DUJON, B. and SLONIMSKI, P.P. (1973) Molec. gen. Genet. 125, 9-52.
10A. BALDACCI, G., FALCONE, C., FRONTALI, L., MACINO, G., PALLESCHI, C. in (3), pp. 759-762.
11. BANDLOW, W. (1976) in (1), pp. 179-202.
12. BECH-HANSEN, N.T. and RANK, G.H. (1973) Can. J. Genet. Cytol. 15, 381-387.
12A. BERNARDI, G., PRUNELL, A., KOPECKA, H. in "Molecular Biology of Nucleo-cytoplasmic relationships" (S. PUISEUX DAO edt.) Elsevier Amsterdam New-York (1975) p. 85-90.
13. BIRKY, C.W. (1975) Genetics, 80, 695-709.
14. BIRKY, C.W. (1975) Molec. gen. Genet. 141, 41-58.
15. BÖKER, E., KAUDEWITZ, F., RICHMOND, V., SCHWEYEN, R.J., THOMAS, D.Y. (1976) in (1), pp. 99-115.
16. BOLOTIN, M., COEN, D., DEUTSCH, J., DUJON, B., NETTER, P., PETROCHILO, E. and SLONIMSKI, P.P. (1971) Bull. Inst. Pasteur, 69, 215-239.
17. BOLOTIN-FUKUHARA, M. and FUKUHARA, H. (1976) Proc. Nat. Acad. Sci. U.S.A. 73, 4608-4612.
18. BOLOTIN-FUKUHARA, M., FAYE, G. and FUKUHARA, H. (1976) in (2), pp. 243-250.
19. BOLOTIN-FUKUHARA, M., FAYE, G. and FUKUHARA, H. (1977) Molec. gen. Genet. 152, 295-305.
20. BORST, P. and GRIVELL, L.A. (1971) FEBS Letters, 13, 73-88.
20A. BORST P., BOS, J.L., GRIVELL, L.A., GROOT, G.S.P., HEYTING, C., MOORMAN, A.F.M., SANDERS, J.P.M., TALEN, J.L. VAN KREIJL, C.F., VAN OMMEN G.J.B. in 4.
21. BUNN, C.L., MITCHELL, C.H., LUKINS, H.B. and LINNANE, A.W. (1970) Proc. Nat. Acad. Sci. U.S.A. 67, 1233-1240.
22. BURGER, G., LANG, B., BACKHAUS, B., BANDLOW, W. and KAUDEWITZ, F. (1977) in (4).

23. BURGER, G., LANG, B., BANDLOW, W., SCHWEYEN, R.J., BACKHAUS, B. and KAUDEWITZ, F. (1976) Biochem. Biophys. Res. Commun. 72, 1201-1208.

24. BUTOW, R.A., VINCENT, R.D., STRAUSBERG, R.L., ZANDERS, E. and PERLMAN, P.S. (1977) in (4).

25. CABRAL, F., SOLIOZ, M., DETERS, D., RUDIN, Y., SCHATZ, G., CLAVILIER, L. GROUDINSKY, O. and SLONIMSKI, P.P. (1977) in (4).

26. CALLEN, D.F. (1974) Molec. gen. Genet. 134, 65-76.

27. CALLEN, D.F. (1974) Molec. gen. Genet. 134, 49-63.

28. CALLEN, D.F. (1974) Molec. gen. Genet. 128, 321-329.

28A. CARNEVALI, F., FALCONE, C., FRONTALI, L., LEONI, L., MACINO, G., and PALLESCHI, C. (1973) Biochem. Biophys. Res. Commun. 51, 651-658.

29. CASEY, J., COHEN, M., RABINOWITZ, M., FUKUHARA, H. and GETZ, G.S. (1972) J. Mol. Biol. 63, 431-440.

30. CASEY, J., HSU, H.J., RABINOWITZ, M., GETZ, G.S., FUKUHARA, H. (1974) J. Mol. Biol. 88, 717-733.

31. CASEY, J., HSU, H.J., GETZ, G.S., RABINOWITZ, M. and FUKUHARA, H. (1974) J. Mol. Biol. 88, 735-747.

31A. CASEY, J.,GORDON, P. and RABINOWITZ,M. (1974) Biochemistry, 13, 1059-1067.

32. CHOO, K.B., NAGLEY, P., LUKINS, H.B., LINNANE, A.W. (1977) Molec. gen. Genet. 153, 279-288.

33. CLAISSE, M.L., SPYRIDAKIS, A. and SLONIMSKI, P.P. (1977) in (4).

34. CLAVILIER, L. (1976) Genetics, 83, 227-243.

35. COBON, G.S., GROOT-OBBINK, D.J., HALL, R.M., MAXWELL, R., MURPHY, M., RYTKA, J. and LINNANE, A.W. (1976) in (3), pp. 453-460.

36. COEN, D., DEUTSCH, J., NETTER, P., PETROCHILO, E. and SLONIMSKI, P.P. (1970) in Symp. of the Society for Experiment Biology, Cambridge, Univ. Press, 24, 449-496.

37. COHEN, M., CASEY, J., RABINOWITZ, M. and GETZ, G.S. (1972) J. Mol. Biol. 63, 441-451.

38. COHEN, M. and RABINOWITZ, M. (1972) Biochim. Biophys. Acta, 281,192-201.

39. COLSON, A.M., GOFFEAU, A., BRIQUET, M., WEIGEL, P. and MATTOON, J.R. (1974) Molec. gen. Genet. 135, 309-326.

40. COLSON, A.M., LUU THE VAN, CONVENT, B., BRIQUET, M. and GOFFEAU, A. (1977) Eur. J. Biochem. 74, 521-526.

41. COLSON, A.M. and SLONIMSKI, P.P. (1977) in (4).

42. DEUTSCH, J., DUJON, B., NETTER, P., PETROCHILO, E., SLONIMSKI, P.P., BOLOTIN-FUKUHARA, M. and COEN, D. (1974) Genetics, 76, 195-219.

42A. DI FRANCO, A., SANDERS, J.P.M., HEYTING, C., BORST, P., SLONIMSKI, P.P. in (2), pp. 291-304.

43. DOUGLAS, M.G. and BUTOW, R.A. (1976) Proc. Nat. Acad. Sci. U.S.A. 73, 1083-1086.

44. DOUGLAS, M.G., KENDRICK, E., BOULIKAS, P., PERLMAN, P. and BUTOW, R.A. (1976) in (2), pp. 199-208.

45. DOUGLAS, M.G., STRAUSBERG, R.L., PERLMAN, P.S. and BUTOW, R.A. (1976) in (3), pp. 435-442.

46. DUJON, B. and MICHEL, F. (1976) in (2), pp. 175-184.

47. DUJON, B., SLONIMSKI, P.P. and WEILL, L. (1974) Genetics, 78, 415-437.

48. DUJON, B., KRUSZEWSKA, A., SLONIMSKI, P.P., BOLOTIN-FUKUHARA, M., COEN, D., DEUTSCH, J., NETTER, P. and WEILL, L. (1975) Molec. gen. Genet. 137, 29-72.

49. DUJON, B., BOLOTIN-FUKUHARA, M., COEN, D., DEUTSCH, J., NETTER, P., SLONIMSKI, P.P. and WEILL, L. (1976) Molec. gen. Genet. 143, 131-165.

49A. DUJON, B., BARANOWSKA, H., DUJARDIN, G. in (4).

50. FAUMAN, M.A. and RABINOWITZ, M. (1974) Eur. J. Biochem. 42, 67-71.

51. FAYE, G. (1976) FEBS Letters, 69, 167-170.

52. FAYE, G. (1977) in (4).

53. FAYE, G., KUJAWA, C. and FUKUHARA, H. (1974) J. Mol. Biol. 88, 185-203.

54. FAYE, G., KUJAWA, C., DUJON, B., BOLOTIN-FUKUHARA, M., WOLF, K., FUKUHARA, H. and SLONIMSKI, P.P. (1975) J. Mol. Biol. 99, 203-217.

55. FAYE, G., FUKUHARA, H., GRANDCHAMP, C., LAZOWSKA, J., MICHEL, F., CASEY, J., GETZ, G.S., LOCKER, J., RABINOWITZ, M., BOLOTIN-FUKUHARA, M. COEN, D., DEUTSCH, J., DUJON, B., NETTER, P. and SLONIMSKI, P.P. (1973) Biochimie, 55, 779-792.

56. FAYE, G., KUJAWA, C., FUKUHARA, H. and RABINOWITZ, M. (1976) Biochem. Biophys. Res. Commun. 68, 476-482.

57. FAYE, G., BOLOTIN-FUKUHARA, M. and FUKUHARA, H. (1976) in (3), pp. 547-555.

58. FLURY, U., MAHLER, H.R. and FELDMAN, F. (1974) J. Biol. Chem. 249, 6130-6137.

59. FORSTER, J. and KLEESE, R.A. (1975) Molec. gen. Genet. 139, 341-355.

60. FOURY, F. and TZAGOLOFF, A. (1976) Eur. J. Biochem. 68, 113-119.

61. FOURY, F. and TZAGOLOFF, A. (1976) Molec. gen. Genet. 149, 43-50.

62. FUKUHARA, H., FAYE, G., MICHEL, F., LAZOWSKA, J., DEUTSCH, J., BOLOTIN-FUKUHARA, M., and SLONIMSKI, P.P. (1974) Molec. gen. Genet. 130, 215-238.

63. FUKUHARA, H., BOLOTIN-FUKUHARA, M., HSU, H.J. and RABINOWITZ, M. (1976) Molec. gen. Genet. 145, 7-17.

64. FUKUHARA, H., FAYE, G., BOLOTIN-FUKUHARA, M., HSU, H.J., MARTIN, N. and RABINOWITZ, M. (1976) in (1), pp. 57-68.

65. FUKUHARA, H. and WESOLOWSKI, M. (1977) in (4).

66. GINGOLD, E.B., SAUNDERS, G.W., LUKINS, H.B. and LINNANE, A.W. (1969) Genetics, 62, 735-744.

67. GOLDTHWAITE, C.D., CRYER, D.R. and MARMUR, J. (1974) Molec. gen. Genet. 133, 87-104.

67A. GORDON, P., CASEY, J., RABINOWITZ, M. (1974) Biochemistry, 13, 1067-1075.

68. GRIFFITHS, D.E., AVNER, P.R., LANCASHIRE, W.E. and TURNER, J.R. (1972) in Biochemistry and Biophysics of Mitochondrial Membranes. Acad.Press, Inc., New York and London, pp. 505-521.

69. GRIFFITHS, D.E., HOUGHTON, R.L. and LANCASHIRE, W.E. (1974) in The Biogenesis of Mitochondria (A.M. KROON and C. SACCONE, eds.) Acad. Press. Inc. London, pp. 215-223.

70. GRIFFITHS, D.E. and HOUGHTON, R.L. (1974) Eur. J. Biochem. 46, 157-167.

71. GRIFFITHS, D.E., HOUGHTON, R.L., LANCASHIRE, W.E. and MEADOWS, P.A. (1975) Eur. J. Biochem. 51, 393-402.

72. GRIVELL, L.A., NETTER, P., BORST, P. and SLONIMSKI, P.P. (1973) Biochim. Biophys. Acta, 312, 358-367.

73. GRIVELL, L.A. and MOORMAN, A.F.M. (1977) in (4).

73A. GRIVELL, L.A., REIJNDERS, L., DE VRIES, H. (1971) FEBS Letters 16, 159-163.

74. GROOT, G.S.P. (1974) in the Biogenesis of Mitochondria (A. KROON and C. SACCONE, eds.) Acad. Press, New York, pp. 443-452.

75. GROOT-OBBINK, D.J., HALL, R.M., LINNANE, A.W., LUKINS, H.B., MONK, B.C., SPITHILL, T.W. and TREMBATH, M.K. (1976) in (2), pp. 163-173.

76. GROOT-OBBINK, D.J., SPITHILL, T.W., MAXWELL, R.J. and LINNANE, A.W. (1977) Molec. gen. Genet. 151, 127-132.

77. GUNGE, N. (1975) Molec. gen. Genet. 139, 189-202.

78. GUNGE, N. (1976) Molec. gen. Genet. 146, 5-16.

79. HALBREICH, A. and RABINOWITZ, M. (1971) Proc. Nat. Acad. Sci. U.S.A. 68, 294-298.

80. HALL, R.M., NAGLEY, P. and LINNANE, A.W. (1976) Molec. gen. Genet. 145, 169-175.

81. HANDWERKER, A., SCHWEYEN, R.J., WOLF, K. and KAUDEWITZ, F. (1973) J. Bacteriol. 113, 1307-1310.

82. HEYTING, C. and SANDERS, J.P.M. (1976) in (2), pp. 273-279.

83. HOUGHTON, R.L., LANCASHIRE, W.E. and GRIFFITHS, D.E. (1974) Biochem. Soc. Trans. 2, 210-213.

84. HOWELL, N., TREMBATH, M.K., LINNANE, A.W. and LUKINS, H.B. (1973) Molec. gen. Genet. 122, 37-51.

85. HOWELL, N., MOLLOY, P.L., LINNANE, A.W. and LUKINS, H.B. (1974) Molec. gen. Genet. 128, 43-54.

86. HOWELL, N., HALL, R.M., LINNANE, A.W. and LUKINS, H.B. (1974) J. Bacteriol. 119, 1063-1065.

86A. JACQ, C., KUJAWA, C., GRANDCHAMP, C., NETTER, P. (1977) in (4).

87. KALDMA, J.A. (1975) Genetika 11 N°3, 111-121.

88. KALDMA, J.A. (1975) Genetika 11 N°8,

89. KLEESE, R.A., GROTBECK, R.C. and SNYDER, J.R. (1972) J. Bacteriol. 112, 1023-1025.

90. KLEESE, R.A., GROTBECK, R.C. and SNYDER, J.R. (1972) Can. J. Genet. Cytol. 14, 713-715.

91. KOTYLAK, Z. and SLONIMSKI, P.P. (1976) in (2), pp. 143-153.

92. KOTYLAK, Z. and SLONIMSKI, P.P. (1977) in (4).

93. KOTYLAK, Z. and SLONIMSKI, P.P. (1977) in (4).

94. KRUSZEWSKA, A., SZCZESNIAK, B. and GAJEWSKI, W. (1976) Molec. gen. Genet. 148, 65-77.

95. KUTZLEB, R., SCHWEYEN, R.J. and KAUDEWITZ, F. (1973) Molec. gen. Genet. 125, 91-98.

96. LANCASHIRE, W.E. (1976) in (3), pp. 481-490.

97. LANCASHIRE, W.E., HOUGHTON, R.L. and GRIFFITHS, D.E. (1974) Biochem. Soc. Trans. 2, 213-214.

98. LANCASHIRE, W.E. and GRIFFITHS, D.E. (1975) Eur. J. Biochem. 51, 403-413.

99. LANG, B., BURGER, G., BANDLOW, W., KAUDEWITZ, F. and SCHWEYEN, R.J. (1976) in (3), pp. 461-465.

100. LAZOWSKA, J., MICHEL, F., FAYE, G., FUKUHARA, H. and SLONIMSKI, P.P. (1974) J. Mol. Biol. 85, 393-410.

101. LAZOWSKA, J., JACQ, C., CEBRAT, S. and SLONIMSKI, P.P. (1976) in (2), pp. 325-336.

102. LAZOWSKA, J. and SLONIMSKI, P.P. (1976) Molec. gen. Genet. 146, 61-78.

103. LEWIN, A., MORIMOTO, R., MERTEN, S., MARTIN, N., BERG, P., CHRISTIANSON, T., LEVENS, D. and RABINOWITZ, M. (1977) in (4).

104. LINNANE, A.W., SAUNDERS, G.W., GINGOLD, E.B. and LUKINS, H.B. (1968) Proc. Nat. Acad. Sci. U.S.A. 59, 903-910.

105. LINNANE, A.W., LAMB, A.J., CHRISTODOULOU, D. and LUKINS, H.B. (1968) Proc. Nat. Acad. Sci. U.S.A. 59, 1288-1293.

106. LINNANE, A.W., HOWELL, N. and LUKINS, H.B. (1974) in The Biogenesis of Mitochondria (A.M. KROON and C. SACCONE, eds.) Acad. Press, Inc. London, pp. 193-213.

107. LINNANE, A.W., LUKINS, H.B., MOLLOY, P.L., NAGLEY, P., RYTKA, J., SRIPRAKASH, K.S., TREMBATH, M.K. (1976) Proc. Nat. Acad. Sci. U.S.A. 73, 2082-2085.

108. LUKINS, H.B., TATE, J.R., SAUNDERS, G.W. and LINNANE, A.W. (1973) Molec. gen. Genet. 120, 17-25.

109. LUKINS, H.B., ENGLISH, J.T., SPITHILL, T.W., DEVENISH, R.J., HALL, R.M. NAGLEY, P. and LINNANE, A.W. (1977) in (4).

110. LOCKER, J., RABINOWITZ, M. and GETZ, G.S. (1974) J. Mol. Biol. 88, 489-507.

110A. LOCKER, J., RABINOWITZ, M. (1976) in (2) pp. 313-324.

111. LOCKER, J., RABINOWITZ, M. and GETZ, G.S. (1974) Proc. Nat. Acad. Sci. U.S.A. 71, 1366-1370.

112. MAHLER, H.R., BILINSKI, T., MILLER, D., HANSON, D., PERLMAN, P.S. and DEMKO, C.A. (1976) in (3), pp. 857-863.

113. MAHLER, H.R., HANSON, D., MILLER, D., BILINSKI, T., ELLIS, D.M., ALEXANDER, N.O. and PERLMAN, P.S. (1977) in (4).

114. MARTIN, N.C., RABINOWITZ, M. and FUKUHARA, H. (1976) J. Mol. Biol. 101, 285-296.

115. MARTIN, N.C. and RABINOWITZ, M. (1976) in (3), pp. 749-754.

116. MARTIN, R., SCHNELLER, J.M., STAHL, A.J.C., DIRHEIMER, G. (1976) in (3), pp. 755-758.

117. MASON, T., BOERNER, P., BIRON, C. (1976) in (3), pp. 239-246.

118. MICHAELIS, G. (1976) Molec. gen. Genet. 146, 133-138.

119. MICHAELIS, G., PETROCHILO, E. and SLONIMSKI, P.P. (1973) Molec. gen. Genet. 123, 51-65.

120. MICHAELIS, G., PRATJE, E., DUJON, B. and WEILL, L. (1976) in (1), pp. 49-56.

121. MICHAELIS, G., MICHEL, F., LAZOWSKA, J. and SLONIMSKI P.P. (1976) Molec. gen. Genet. 149, 125-130.

121A. MICHEL, F., LAZOWSKA, J., FAYE, G., FUKUHARA, H. and SLONIMSKI, P.P. (1974) J. Mol. Biol. 85, 411-431.

122. MICHELS, C.A., BLAMIRE, J., GOLDFINGER, B. and MARMUR, J. (1974) J. Mol. Biol. 90, 431-449.

123. MITCHELL, C.H., BUNN, C.L., LUKINS, H.B. and LINNANE, A.W. (1972) Bioenergetics, 4, 363-379.

124. MOLLOY, P.L., HOWELL, N., PLUMMER, D.T., LINNANE, A.W. and LUKINS, H.B. (1973) Biochem. Biophys. Res. Commun. 52, 9-14.

125. MOLLOY, P.L., LINNANE, A.W. and LUKINS, H.B. (1975) J. Bacteriol. 122, 7-18.

126. MOLLOY, P.L., LINNANE, A.W. and LUKINS, H.B. (1976) Genet. Res. 26, 319-325.

127. MONNEROT, M., SCHWEYEN, R.J. and FUKUHARA, H. (1977) Molec. gen. Genet 152, 307-309.

128. MOORMAN, A.F.M., VERKLEY, F.N., ASSELBERGS, F.A.M. and GRIVELL, L.A. (1977) in (4).

128A. MOORMAN, A.F.M., GRIVELL, L.A. (1976) in (3), pp. 539-546.

129. MORIMOTO, R., LEWIN, A., HSU, H.J., RABINOWITZ, M. and FUKUHARA, H. (1975) Proc. Nat. Acad. Sci. U.S.A. 72, 3868-3872.

129A. MORIMOTO, R., LEWIN, A., MERTEN, S., RABINOWITZ, M. (1976) in (3), pp. 519-524.

129B. MORIMOTO, R., LEWIN, A., RABINOWITZ, M. (1977) Nucl. Acid. Res. 4, 2331-2351.

130. NAGLEY, P. and LINNANE, A.W. (1970) Biochem. Biophys. Res. Commun. 39, 986-996.

131. NAGLEY, P., GINGOLD, E.B., LUKINS, H.B. and LINNANE, A.W. (1973) J. Mol. Biol. 78, 335- 350.

132. NAGLEY, P., MOLLOY, P.L., LUKINS, H.B. and LINNANE, A.W. (1974) Biochem. Biophys. Res. Commun. 57, 232-239.

133. NAGLEY, P., GINGOLD, E.B. and LINNANE, A.W. (1974) in the Biogenesis of Mitochondria (A.M. KROON and C. SACCONE, eds.) Acad. Press. Inc. London, pp. 157-168.

133A.NAGLEY, P., SRIPRAKASH, K.S., RYTKA, J., CHOO, K.B., TREMBATH, M.K., LUKINS, H.B., LINNANE, A.W. in (2) pp. 231-242.

134. NETTER, P., PETROCHILO, E., SLONIMSKI, P.P., BOLOTIN-FUKUHARA, M., COEN, D., DEUTSCH, J. and DUJON, B. (1974) Genetics, 78, 1063-1100.

135. PAJOT, P., WAMBIER-KLUPPEL, M.L., KOTYLAK, Z. and SLONIMSKI, P.P. (1976) in (3),pp. 443-451.

136. PAJOT, P., WAMBIER-KLUPPEL, M.L. and SLONIMSKI, P.P. (1977) in (4).

137. PERLMAN, P.S. and BIRKY, C.W. (1974) Proc. Nat. Acad. Sci. U.S.A. 71, 4612-4616.

138. PERLMAN, P.S. (1976) Genetics, 82, 645-663.

139. PERLMAN, P.S., BIRKY, C.W., DEMKO, C.A. and STRAUSBERG, R.L. (1976) in (3), pp. 405-414.

140. PRATJE, E. and MICHAELIS, G. (1976) in (3), pp. 467-471.

141. PRATJE, E. and MICHAELIS, G. (1977) Molec. gen. Genet. 152, 167-174.

142. PRATJE, E. and MICHAELIS, G. (1977) in (4).

143. PUTRAMENT, A., POLANOWSKA, R., BARANOWSKA, H. and EJCHART, A. (1976) in (3), pp. 415-418.

143A.PRUNELL, A., KOPECKA, M., STRAUSS, F., BERNARDI, G. (1977)J. Mol. Biol. 110, 17-52.

143B.PRUNELL, A., BERNARDI, G. (1977) J. Mol. Biol.110, 53-74.

144. RABINOWITZ, M., JAKOVCIC, S., MARTIN, N., HENDLER, F., HALBREICH, A., LEWIN, A. and MORIMOTO, R. (1976) in (2), pp. 219-230.

144A. RABINOWITZ, M., CASEY, J., GORDON, P., LOCKER, J., HSU, H.J., GETZ, G.S. (1974) in The Biogenesis of Mitochondria (A. KROON and C. SACCONE, eds.) Acad. Press London, New York, pp. 89-105.

145. RANK, G.H. (1970) Can. J. Genet. Cytol. 12, 129-136.

146. RANK, G.H. (1973) Heredity, 30, 265-271.

147. RANK, G.H. and BECH-HANSEN, N.T. (1972) Genetics, 72, 1-15.

148. RANK, G.H. and BECH-HANSEN, N.T. (1972) Can. J. Microbiol. 18, 1-7.

149. RANK, G.H. and MARTIN, R. (1972) Can. J. Genet. Cytol. 14, 197-199.

150. REIJNDERS, L. and BORST, P. (1972) Biochem. Biophys. Res. Commun. 47, 126-133.

151. ROBERTSON, A.J., BECH-HANSEN, N.T. and RANK, G.H. (1975) Can. J. Genet. Cytol. 17, 381-389.

152. RYTKA, J., ENGLISH, K.J., HALL, R.M., LINNANE, A.W. and LUKINS, H.B. (1976) in (3), pp. 427-434.

153. SAUNDERS, G.W., GINGOLD, E.B., TREMBATH, M.K., LUKINS, H.B. and LINNANE, A.W. (1971) in Autonomy and Biogenesis of Mitochondria and Chloroplasts. (N.K. BOARDMAN, A.W. LINNANE, R.M. SMILLIE, eds.) North-Holland, Amsterdam, pp. 185-193.

154. SANDERS, J.P.M., HEYTING, C., DI FRANCO, A., BORST, P. and SLONIMSKI, P.P. (1976) in (2), pp. 259-272.

154A. SANDERS, J.P.M., HEYTING, C., BORST, P. (1976) in (3),pp.511-517.

154B. SANDERS, J.P.M., HEYTING, C. and BORST, P. (1975) Biochem. Biophys. Res. Commun. 65, 699-707.

154C.SANDERS, J.P.M., BORST, P., WEIJERS, P.J. (1975) Mol. Gen. Genet. 143, 53-64.

155. SCHNELLER, J.M., SCHNELLER, C. and STAHL, A.J.C. (1976) in (3), pp. 775-778.

156. SCHNELLER, J.M., STAHL, A. and FUKUHARA, H. (1975) Biochimie, 57, 1051-1057.

157. SCHNELLER, J.M., FAYE, G., KUJAWA, C. and STAHL, A.J.C. (1975) Nucleic acid Research, 3, 831-838.

158. SCHWEYEN, R.J. and KAUDEWITZ, F. (1976) Molec. gen. Genet. 149, 311-322.

159. SCHWEYEN, R.J., STEYRER, U., KAUDEWITZ, F., DUJON, B. and SLONIMSKI, P.P. (1976) Molec. gen. Genet. 146, 117-132.

160. SCHWEYEN, R.J., WEISS-BRUMMER, B., BACKHAUS , B. and KAUDEWITZ, F. (1976) in (2), pp. 251-258.

161. SCHWEYEN, R.J., BACKHAUS , B., MATTEWS, S. and KAUDEWITZ, F. (1976) in (1), pp. 7-22.

162. SCHWEYEN, R.J., WEISS-BRUMMER, B., BACKHAUS , B. and KAUDEWITZ, F. (1977) in (4).

163. SHANNON, C., ENNS, R., WHEELIS, L., BURCHIEL, K., CRIDDLE, R.S. (1973) J. Biol. Chem. 248, 3004-3011.

164. SLONIMSKI, P.P. and TZAGOLOFF, A. (1976) Eur. J. Biochem. 61, 27-41.

165. SOMLO, M., AVNER, P.R., COSSON, J., DUJON, B. and KRUPA, M. (1974) Eur. J. Biochem. 42, 439-445.

166. SRIPRAKASH, K.S., CHOO, K.B., NAGLEY, P. and LINNANE, A.W. (1976) Biochem. Biophys. Res. Commun. 69, 85-91.

167. SRIPRAKASH, K.S., MOLLOY, P.L., NAGLEY, P., LUKINS, H.B. and LINNANE, A.W. (1976) J. Mol. Biol. 104, 485-503.

168. STUART, K.D. (1970) Biochem. Biophys. Res. Commun. 39, 1045-1051.

169. STORM, E.M. and MARMUR, J. (1975) Biochem. Biophys. Res. Commun. 64, 752-759.

170. STRAUSBERG, R.L.,and BUTOW, R.A. (1977) Proc. Nat. Acad. Sci. U.S.A. 74, 2715-2719.

171. SUBIK, J. (1976) in (3), pp. 473-477.

172. SUBIK, J. (1975) FEBS Letters, 59, 273-276.

173. SUBIK, J., KOVACOVA, V. and TAKACSOVA, G. (1977) Eur. J. Biochem. 73, 275-286.

174. SUDA, K. and UCHIDA, A. (1972) Jap. J. Genet. 47, 441-444.

175. SUDA, K. and UCHIDA, A. (1974) Molec. gen. Genet. 128, 331-339.

176. THOMAS, D.Y. and WILKIE, D. (1968) Biochem. Biophys. Res. Commun. 30, 368-372.

177. THOMAS, D.Y. and WILKIE, D. (1968) Genet. Res. 11, 33-41.

178. TREMBATH, M.K., BUNN, C.L., LUKINS, H.B. and LINNANE, A.W. (1973) Molec. gen. Genet. 121, 35-48.

179. TREMBATH, M.K., MONK, B.C., KELLERMAN, G.M. and LINNANE, A.W. (1975) Molec. gen. Genet. 140, 333-337.

180. TREMBATH, M.K., MONK, B.C., KELLERMAN, G.M. and LINNANE, A.W. (1975) Molec. gen. Genet. 141, 9-22.

181. TREMBATH, M.K., MOLLOY, P.L., SRIPRAKASH, K.S., CUTTING, G.J., LINNANE, A.W. and LUKINS, H.B. (1976) Molec. gen. Genet. 145, 43-52.

182. TZAGOLOFF, A., AKAI, A. and NEEDLEMAN, R.B. (1975) Proc. Nat. Acad. Sci. U.S.A. 72, 2054-2057.

183. TZAGOLOFF, A., AKAI, A. and NEEDLEMAN, R.B. (1975) J. Bacteriol. 122, 826-831.

184. TZAGOLOFF, A., AKAI, A., NEEDLEMAN, R.B. and ZULCH, G. (1975) J. Biol. Chem. 250, 8236-8242.

185. TZAGOLOFF, A., FOURY, F. and AKAI, A. (1976) Molec. gen. Genet. 149, 33-42.

186. TZAGOLOFF, A., FOURY, F. and AKAI, A. (1976) in (2), pp. 155-161.

187. TZAGOLOFF, A., AKAI, A.,and FOURY, F. (1976) FEBS Letters 65, 391-395.

187A. TZAGOLOFF, A., FOURY, F., AKAI, A. (1976) in (3), pp. 419-426.

188. UCHIDA, A. and SUDA, K. (1973) Mutat. Res. 19, 57-63.

189. UCHIDA, A. and SUDA, K. (1976) Molec. gen. Genet. 145, 159-163.

190. VAN OMMEN, G.J.B. and GROOT, G.S.P. (1976) in (4).

191. WAKABAYASHI, K. (1972) J. Antibiotics, 25, 475-476.

192. WAKABAYASHI, K. (1976) in (3), pp. 535-538.

193. WAKABAYASHI, K. (1974) J. Antibiotics, 27, 373-378.

194. WAKABAYASHI, K. (1974) Proc. Jap. Acad. Sci. 50, 396-399.

195. WAKABAYASHI, K. and GUNGE, N. (1970) FEBS Letters, 6, 302-304.

196. WAKABAYASHI, K. and KAMEI, S. (1973) FEBS Letters, 33, 263-265.

197. WAKABAYASHI, K. and KAMEI, S. (1974) J. Antibiotics, 27, 729-732.

198. WAXMAN, M.F., EATON, N. and WILKIE, D. (1973) Molec. gen. Genet. 127, 277-284.

199. WAXMAN, M.F. and EATON, N.R. (1974) Molec. gen. Genet. 133, 37-45.

200. WAXMAN, M.F. (1975) Molec. gen. Genet. 141, 285-290.

201. WILKIE, D. (1972) Genetic Aspects of Mitochondria. pp. 85-94.

202. WILKIE, D. and THOMAS, D.Y. (1973) Genetics, 73, 367-377.

203. WOLF, K., DUJON, B. and SLONIMSKI, P.P. (1973) Molec. gen. Genet. 115, 53-90.

204. YOUNG, R.A. and PERLMAN, P.S. (1975) J. Bacteriol. 124, 290-295.

205. ZAKHAROV, I. Genetika.

Acknowledgments.

We wish to express our gratitude to Dr. A. Goffeau for his encouraging belief in the usefulness of this work, to Mrs D. Lapointe and M. Even for the composition of the manuscript and to Dr. R. Lang for the scrutiny of its English version.

This work could not have been completed without the kindness of the authors of original articles in this volume who have communicated to us their manuscripts prior to publication.

Finally we wish to thank the organisers of the meeting and the publishers of this book for offering us facilities for publication.

Walter de Gruyter
Berlin · New York

W. Bandlow
R. J. Schweyen
D. Y. Thomas
K. Wolf
F. Kaudewitz
(Editors)

Genetics, Biogenesis and and Bioenergetics of Mitochondria

Proceedings of a Symposium held at the Genetisches Institut der Universität München, Munich, September 11-13, 1975

1976. Large-octavo. XV, 418 pages. With many illustrations. Bound DM 140,–; $56.00
ISBN 3110068656

A comprehensive view of the present status of mitochondrial research; this book is a one-volume source of basic information on mitochondrial genetics, biogenesis and the function of respiratory chain and oxidative phosphorylation as well as an introduction into recent developments and current problems in this exiting field. It demonstrates how many approaches are now being used together to tackle problems in the development and function of this organelle.
The contributions consist of 40 original papers by leading workers in this field.

$ price is subject to change without further notice

Bei Fragen zur Produktsicherheit wenden Sie sich bitte an:
If you have any questions regarding product safety,
please contact:

Walter de Gruyter GmbH
Genthiner Straße 13
10785 Berlin
productsafety@degruyterbrill.com